Mechanical Behaviors of Carbon Nanotubes

Mechanical Behaviors of Carbon Nanotubes

Theoretical and Numerical Approaches

K.M. Liew
City University of Hong Kong, Kowloon Tong, Hong Kong

Jian-Wei Yan
Jinan University, Guangdong, China

Lu-Wen Zhang
Shanghai Jiao Tong University, Shanghai, China

ELSEVIER
William Andrew
Applied Science Publishers
elsevier.com

William Andrew is an imprint of Elsevier
The Boulevard, Langford Lane, Kidlington, Oxford, OX5 1GB, United Kingdom
50 Hampshire Street, 5th Floor, Cambridge, MA 02139, United States

British Library Cataloguing-in-Publication Data
A catalogue record for this book is available from the British Library

Library of Congress Cataloging-in-Publication Data
A catalog record for this book is available from the Library of Congress

ISBN: 978-0-323-43137-8

For Information on all William Andrew publications
visit our website at https://www.elsevier.com

Publisher: Matthew Deans
Acquisition Editor: Simon Holt
Editorial Project Manager: Sabrina Webber
Production Project Manager: Susan Li
Designer: Greg Harris

Typeset by MPS Limited, Chennai, India

Contents

Preface

The great discovery of carbon nanotubes has sparked an innovation in material science because of their superior physical, chemical, electronic, and mechanical properties. A good understanding of mechanical behaviors is a key factor for the usage of carbon nanotubes in practical engineering. Traditional experimental research can provide intuitive but rough awareness of mechanical behaviors of carbon nanotubes. Furthermore, experimental research unavoidably encounters challenges when entering into the nanoscale regime. Therefore, theoretical study plays an important role in this kind of nanomaterial.

This book originally states various theoretical and numerical studies for mechanical behaviors of a current hot research topic—carbon nanotubes—by our research team. On the one hand, the main theoretical aspects in this book include: Classical Molecular Dynamics Simulation, Atomistic-Continuum Theory, Atomic Finite Element Method, and Coupling it with Atomistic-Continuum Method, Continuum Plate and Shell Models, Nonlocal Continuum Plate and Shell Models. On the other hand, the main research contents include: Structural and Elastic Properties, Trace of Large Deformation, Buckling and Postbuckling Behaviors, Fracture, Vibration Characteristics, Wave Propagation, and the most promising applications in engineering. The target of this book is to provide to our peers the available theoretical and numerical methods, and useful computational results of carbon nanotubes for reference.

This book is chiefly prepared for academics. In order to achieve this target, the book introduces the microstructure information of carbon nanotubes and the derivation process of our theories at full length. The advantages and disadvantages of these theories are also discussed in detail. Then numerical simulation results and existent experiments are also provided for comparison. These valuable methods lay the theoretical foundations for the analysis of mechanical behaviors and computational results are helpful for the design and process of carbon nanotubes. Finally, based on the aforementioned study, we investigate the most promising applications such as Fluid Conveying, Driving Water Molecules, Hydrogen Storage, and Mass Detection. Also, a special effort has been made to have the chapters as independent from one another as possible so that the interested chapters can be individually selected for self-study.

This book covers 10 chapters and is organized in the following manner. Chapter 1, Introduction, introduces the background knowledge of carbon nanotubes, including the history of discovery and commonly synthesized techniques. The general development and commercialization of carbon nanotubes in nanoscience and nanotechnology are also presented. Chapter 2, Experimental Aspect, comprehensively introduces experimental tools in preparing and testing carbon nanotubes as well as the wide range of promising applications, including electronics, optical communications, and biological systems. Chapter 3, Classical Molecular Dynamics Simulations, gives many examples of classical molecular dynamics simulation in the study of mechanical properties of carbon nanotubes, such as Young's modulus, buckling, fracture, and thermal stability behaviors. Chapter 4, Atomistic-Continuum Theory, presents an atomistic-continuum theory based on the higher-order gradient theory which links the deformation of the lattice vectors of crystal to that of a continuum deformation field. Chapter 5, Atomic Finite Element Method and Coupling With Atomistic-Continuum Method, presents a multiscale method by coupling the developed atomistic-continuum theory with an atomic finite element method and shows a highly efficient performance for localized problems. Chapter 6,

Continuum Models, develops a new van der Waals interaction model for not only the postbuckling pressure but also the prebuckling pressure between any two layers of a multiwalled carbon nanotube. Then a continuum cylindrical shell model is presented to study the buckling of a multiwalled carbon nanotube, while Donnell's shell model is applied to study the vibration. Chapter 7 introduces the nonlocal elasticity theory, including nonlocal beam, nonlocal rod, and nonlocal shell models, to consider the scale effect in constitutive equations for vibrations and wave propagation by simply introducing a material parameter. Chapter 8, Technologically Relevant Applications, introduces technologically relevant applications of carbon nanotubes. Chapter 9, 2-D Graphene and White Graphene, introduces the background knowledge of two-dimensional graphene, including the history of discovery, its remarkable material properties, and its common synthesis and characterizing techniques. Graphenes can serve as fillers in polymer nanocomposites and are allowed to be used as field effect transistors, sensors, and clean energy devices. Finally, Chapter 10, Arrangements of Carbon-Based Structures, shows other kinds of carbon-based nanostructures, such as carbon nanorings, nanosprings, and nanocones.

A great effort has been made to emphasize the physical meanings of the above-mentioned theories. The accuracy of numerical results obtained with the usage of these theories and possible areas of their application have been discussed. The principal purpose is to help the reader to well understand the current widely used theories in the nanoscale area and to express the physical meanings rather than purely mathematical aspects. Since the failure of carbon nanotubes is more often caused by buckling. The issue of the buckling and postbuckling analysis of carbon nanotubes is given much attention in this book. Particular emphasis is placed on the critical strain and buckling patterns for both single-walled and multiwalled carbon nanotubes. The book also emphasizes the understanding of basic phenomena in vibration and wave propagation of carbon nanotubes. We believe that the capture of delicate mechanical behaviors and the size effect of carbon nanotubes by these theories presented in this book are helpful for designing nanodevices.

CHAPTER

INTRODUCTION

1

1.1 GENERAL

In 1991, Iijima [1] reported his discovery of needle-like carbon nanotubes (CNTs) using an arc-discharge evaporation method in NEC corporation. Shortly thereafter, Iijima and Ichihashi [2] and Bethune et al. [3] separately reported the growth of single-walled CNTs (SWCNTs) which opened a fresh chapter on nanoscience and nanotechnology. CNTs were marked as one of the top ten advances in "Materials Today" [4]. From the viewpoint of chemical composition, CNTs are the simplest materials, having pure C–C atomic bonding configuration. Nevertheless, the simple covalently bonded carbon network leads to an extremely unique nanostructure and makes CNTs perhaps the most diverse and richest among nanomaterials in regard to structures and structure–property relations [5]. In the past two decades, CNTs have attracted intense interest due to their exceptional mechanical, chemical, electrical, and thermal properties along with specific low density, being regarded as promising for a broad range of potential applications. For example, SWCNTs have exhibited great electrical conductivity, closely related with their chirality, which allows the miniaturization of electronics. From the viewpoint of chemical composition, CNTs are the simplest materials, having a pure C–C atomic bonding configuration. Nevertheless, the simple covalently bonded carbon network leads to an extremely unique nanostructure and makes CNTs perhaps the most diverse and richest among nanomaterials in regard to structures and structure–property relations [5].

The molecular structure of an SWCNT is regarded as a seamless hollow cylindrical structure formed by rolling up a graphene along a (m, n) index which determines its diameter and chirality. An SWCNT is an ideal slender structure with ultrahigh aspect ratio, typically approximately 1.5 nm in diameters and up to dozens of micrometers in lengths. An individual SWCNT with an atomically monolayered surface and curved π-boding configuration can be either metallic or semiconducting, depending on its chirality (the chiral angle between hexagons and the tube axis) and diameter [6]. As far as SWCNTs are concerned, their band gaps can be relatively large ($\sim$0.5 eV) or small ($\sim$10 meV) even if they have nearly identical diameters (typical diameter of 1.5 nm) [5]. More interestingly, the band gap of the same-chirality semiconducting SWCNT is inversely proportional to its diameter. Such an intriguing character of an SWCNT strongly motivates worldwide efforts to study its fundamental nature and explore the potential application. For instance, various kinds of SWCNTs have been promising for advanced electrical interconnects on the basis of their low resistivity [7–10], high thermal conductivities ($\sim$3500 W/m/K), and current-carrying capacities ($\sim 10^9$ A/cm^2) [11,12].

The covalently carbon hexagonal network creates an extremely stable structure. Arguably, CNTs are the strongest nanomaterial in nature. Elastic properties of CNTs have been measured by

Mechanical Behaviors of Carbon Nanotubes. DOI: http://dx.doi.org/10.1016/B978-0-323-43137-8.00001-9

various means. Both experimental measurements and theoretical calculations reveal that the axial Young's modulus of CNTs is typically in the range from 1 to 1.8 TPa [13–20], while the breaking strength of SWCNT bundles is up to 50 GPa [20,21], corresponding to ~50 times that of density-normalized strength of steel wires [17]. Simultaneously, various derived geometrical patterns of CNTs, such as rings [22,23], X-, T-, and Y-junctions [24,25], quantum dots [26], and crossing objects [27] also have been experimental observed. Such abovementioned geometrical shapes give rise to novel properties which are desirable for the successful application of SWCNTs to the design and manufacture of devices. What's more, material properties of CNTs are greatly sensitive to chemical modification or atomic doping which suggests a promising remarkable application for brand-new quantum materials. Therefore, there are infinite possibilities in these kinds of CNTs.

In the past two decades, intensive theoretical and experimental efforts have been made in order to illuminate the effect of atomic structures on the electronic properties of CNTs [28–30]. CNTs can be synthesized by several means, such as arc-discharge, laser ablation, chemical vapor deposition (CVD) methods, and metallic and semiconducting CNTs have been found. As an ideal quasi-1D structure, metallic SWCNTs have been selected as model systems to explore the rich quantum phenomena, involving single-electron charging, Luttinger liquid, quantum interference, weak localization, and ballistic transport [28,29,31,32]. Besides, semiconducting CNTs have been utilized as novel building blocks in the field of nanoelectronics, such as transistors and logic, memory, and sensory devices [28–30].

The recently rapid development and great progress in the synthesis of CNTs have greatly facilitated CNTs characterizations and nanodevice exploitations. Conversely, the strong requirements of various aspects of fundamental research and practical application have been positively motivating synthetic methods in order to acquire more perfect materials. They are mutually complementary and promote each other. Nowadays, the synthesis of CNTs becomes more focused in consciously controlling the growth rather than, as previously, enabling growth. As mentioned, the electronic properties are extremely sensitive to the atomic structure of CNTs which has led to rich scientific research. However, it encounters a significant challenge to precisely control the CNT diameter and chirality for chemical synthesis methods. It is necessary to exquisitely control atomic arrangements along the CNTs in order to synthesize CNTs with predictable properties. In reality, it is the ultimate task for chemists in the CNT synthesis field.

Various approaches, such as arc-discharge, laser ablation, and CVD, have been mainly adopted to synthesize CNTs [33]. The first two approaches use solid-state carbon precursors to provide carbon sources for CNT growth and vaporize carbon at thousands of degrees Celsius. These methods are well established to produce high-quality and ideal CNT structures, in spite of a mass of by-products associated with them. Hydrocarbon gases are used to provide sources for carbon atoms, while metal catalyst particles are utilized as "seeds" for CNTs growth by means of CVD which takes place at relatively lower temperatures (500–1000°C) [33]. However, as far as SWCNTs synthesis is concerned, none of the abovementioned three synthesis methods are capable of yielding bulk materials with homogeneous diameter and chirality thus far. It is noteworthy that SWCNTs with impressively narrow diameter distributions averaging ~1.4 nm have been produced by arc-discharge and laser ablation techniques. By utilizing electrical transport and microscopy and spectroscopy measurement, CVD methods have experienced a long process from producing carbon fibers, filaments, and MWCNTs to synthesizing SWCNTs [33–39] with high crystallinity and perfection comparable to those of arc [40] and laser [41] materials.

1.2 ATOMIC STRUCTURE OF CNTs

A graphene is a carbon-based repeated hexagonal cellular structure, in which each carbon atom is connected to three neighboring atoms. An SWCNT can be conceptualized as a three-dimensional seamless hollow cylindrical graphitic membrane, with a thickness of only one atom, by rolling up a graphene, while MWCNTs can be regarded as a form by rolling up multiple layers of graphene. The ends of CNTs can be either capped or not.

The chirality of an SWCNT is determined by the rolling direction $\tilde{\Gamma}$ on the graphene (shown in Fig. 1.1). The chirality and diameter of an SWCNT can be characterized by a (n, m) index and thus chiral vector $\tilde{\Gamma}$ is expressed by:

$$\tilde{\Gamma} = n\tilde{\mathbf{a}}_1 + m\tilde{\mathbf{a}}_2, \quad n \geq m \tag{1.1}$$

where $\tilde{\mathbf{a}}_1$ and $\tilde{\mathbf{a}}_2$ are unit vectors in the graphene. The length of the chiral vector is $l = a\sqrt{n^2 + m^2 + nm}$ and the diameter is l/π, where $a = \sqrt{3}l_b$ with l_b being the C–C bond length. The chiral angle θ is calculated by:

$$\theta = \arctan\left(\frac{\sqrt{3}m}{2n + m}\right) \tag{1.2}$$

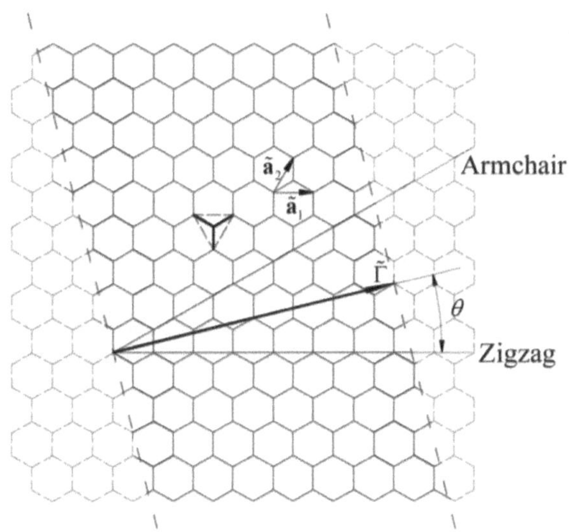

FIGURE 1.1

Illustration of an SWCNT formed by rolling up a planar graphene. θ is the chiral angle which is in the range of 0–30 degrees. Each atom is connected to three neighboring atoms by covalent bonds (shown in a dashed triangle).

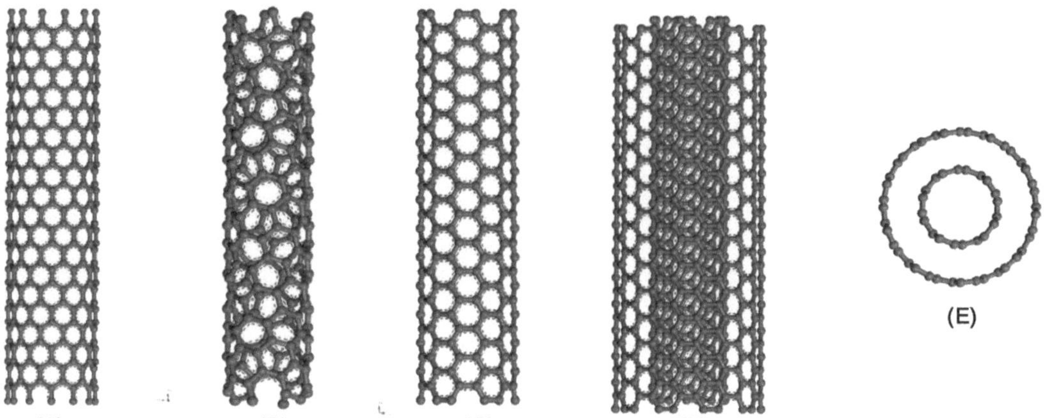

FIGURE 1.2

Schematic of atomic structures of zigzag (A), chiral (B), armchair SWCNTs (C), and a double-walled CNTs side view (D) and top view (E).

CNTs can be classified into three categories according to the chirality. The chiral angle θ is 0 degrees when $m = 0$ and 30 degrees when $n = m$, and these tubes are corresponding to zigzag and armchair, respectively. The others are chiral tube with a chiral angle 0 degree $< \theta <$ 30 degrees. The atomic structures of different kinds of SWCNTs are shown in Fig. 1.2A–C. Fig. 1.2D and E are the side view and top view of a DWCNT. It contains two layers of coaxial SWCNTs which correlate with each other by the weak vdW forces. The spacing between the outer and inner layers is determined by vdW force and approximately 0.34 nm slightly influenced by the total number of layers. Although graphene is always a semiconductor, CNTs can exhibit either metallic or semiconducting properties. The electrical properties are largely dependent on the chirality and diameter of CNTs [42–45]. In general, CNTs are semiconductive. Nevertheless, CNTs exhibit a metallic property when the (n, m) index satisfies $n - m = 3I$, where I is an integer [45]. Note that different from other kinds of CNTs, armchair CNTs are always metallic. As illustrated in Fig. 1.3, circled dots are metallic CNTs, while small dots are semiconducting CNTs.

On the other hand, capped CNTs are regarded as elongated fullerenes, in which the walls of the tubes consist of cylindrical graphenes with their ends capped by fullerene hemispheres [1]. We can specify an SWCNT by bisecting a C_{60} molecule at the equator and joining the two resulting hemispheres with a cylindrical tube one layer thick and of the same diameter as the C_{60} molecule. If the C_{60} molecule is bisected normal to a fivefold axis, an armchair nanotube is formed (see Fig. 1.4B), while if it is bisected normal to a threefold axis, then a zigzag nanotube is formed (see Fig. 1.4C) [46]. Armchair and zigzag CNTs with larger diameters and correspondingly larger caps can likewise be defined, as shown in Fig. 1.4B and C. In addition, a large number of capped chiral CNTs can be formed with a variety of hemispherical caps. A representative chiral CNT is shown in Fig. 1.4D.

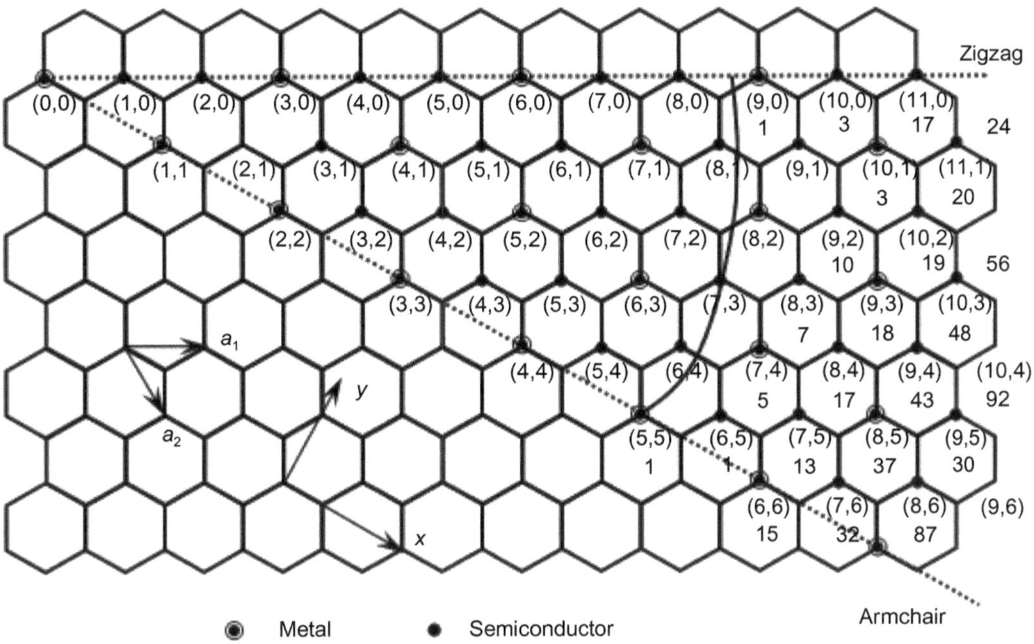

FIGURE 1.3

The pairs of integers (n, m) in the 2D grapheme sheet specify chiral vectors C_h (see Fig. 2.1) for CNTs, including zigzag, armchair, and chiral nanotubes [45].

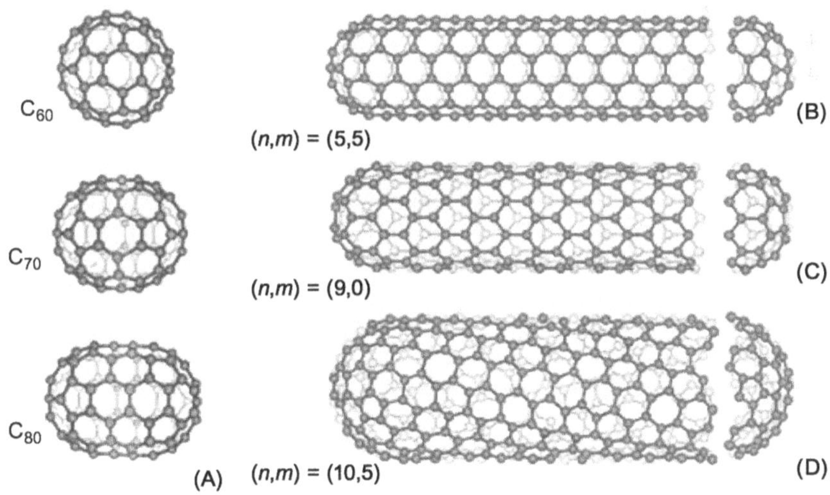

FIGURE 1.4

Sketches of (A) fullerenes: C_{60}, C_{70}, and C_{80} molecules; and capped SWCNTs: (B) (5, 5) armchair CNT; (C) (9, 0) zigzag CNT; (D) (10, 5) chiral CNT [46].

1.3 GENERAL DEVELOPMENT AND CURRENT SITUATION OF CNTs IN NANOSCIENCE AND NANOTECHNOLOGY

In the mechanical properties testing of individual MWCNTs, an elastic modulus of ~1 TPa and a tensile strength of ~100 GPa, which is over 10-fold higher than any industrial fiber in nature, has been measured [47]. As far as the electrical properties are concerned, SWCNTs can be either metallic or semiconducting, relying on the their chirality, while MWCNTs are always metallic and can carry currents of up to 10^9 A/cm^2 [48]. The thermal conductivity of an individual SWCNT can reach up to 3500 W/m/K at room temperature [11] which largely exceeds that of diamond (2000 W/m/K). Based on a more and more comprehensive understanding of CNTs, numerous potential applications have been proposed to full make use of their prominent superiority, including heat conductive and ultrahigh-strength composites, energy storage and conversion devices, sensors and high frequency oscillator, field emission displays and radiation sources, hydrogen storage media and drug encapsulation, nanometer-sized semiconductor devices and probes, interconnects, etc. So far, some of these applications have already been realized in products and others are still in the initial stage of exploring advanced devices.

So far, the low production output of CNTs has inhibited mass consumer products in spite of diverse potential applications reported to be utilizing CNTs. According to the statistics, CNT-related commercial activity has substantially and constantly increasingly grown during the past decade. The data demonstrates that the worldwide annual production capacity of CNTs has increased at least 10-fold in 2011, compared to that produced in 2006 [49]. Although there are more and more commercial suppliers of CNTs, such as BlueNano (USA), Chengdu Organic Chemical Co. Ltd. (China), Hyperion Catalysis (USA), NanoLab (USA), Nanocyl S.A. (Belgium), Southwest Nanotechnology Inc. (USA), Nanointegris (USA), Thomas & Swan Co. Ltd. (UK) and so on, it is still a serious challenge to meet the demand for CNTs. The main reason is that CNTs are too expensive, ranging from US\$2.50 g^{-1} for MWCNTs (NanoLab) to US\$2500 g^{-1} for SWCNTs (CNI) [50,51]. It is noted that since 2005 the low-cost production of CNTs were supplied by Cheap Tubes, Inc. [52] (US\$275–300 g^{-1} for 90% pure SWCNT), prices of high-quality CNTs have little changed in the past 10 years. The high cost is argued to be related to the low production capacity. Similar to the other high-value materials in history such as silicon wafers and aluminum, the price of CNTs is expected to drop to an acceptable level or even much lower with the improvement of manufacture technology and increment [53]. Tables 1.1–1.3 list the worldwide known industrial activity and commercial products.

1.4 FUNDAMENTAL PROPERTIES AND GENERAL BEHAVIORS OF CNTs

Many experiments have been conducted to excavate the intrinsic property of CNTs. These kinds of methods can provide intuitive knowledge of fundamental properties and general behaviors of CNTs. Pan et al. [54] produced a very long (approximately 2 mm) MWCNTs using pyrolysis of acetylene over iron/silica substrates method. In accordance with the theoretical prediction, Qin et al. [55] reported the possible smallest CNT with a diameter of 0.4 nm which is imprisoned inside an MWCNT. They found that these smallest CNTs are always metallic no matter what their

Table 1.1 Producers of CNT Powders and Dispersions [51]

Company	Country	URL
Arkema	France/USA	http://www.arkema-inc.com/ http://www.graphistrength.com
Bayer Material Science AG	Germany	www.bayer.com http://www.baytubes.com/
Blue Nano	USA	www.bluenanoinc.com
Catalytic Materials	USA	http://www.catalyticmaterials.com
Chengdu Organic Chemical Co. Ltd.	China	www.timesnano.com
Cnano	China/USA	http://www.cnanotechnology.com
Eden Energy	Australia/India	http://www.edenenergy.com.au/
Eikos	USA	www.eikos.com
Hanwha Nanotech Corporation	South Korea	www.hanwhananotech.com
Hodogaya	Japan	http://www.hodogaya.co.jp
Hyperion Catalysis	USA	www.hyperioncatalysis.com
Hythane Co.	USA	http://hythane.net/
Idaho Space Materials	USA	www.idahospace.com
Klean Carbon	Canada	http://www.kleancarbon.com/
Meijo-nano Carbon	Japan	www.meijo-nano.com
Mitsubishi Rayon Co.	Japan	http://www.mrc.co.jp
Mitsui	Japan	www.mitsui.com
Nanocyl S.A.	Belgium	www.nanocyl.com
Nanointegris	USA	www.nanointegris.com
Nanolab	USA	http://www.nano-lab.com/
Nanothinx	Greece	http://www.nanothinx.com
Nano-C	USA	http://www.nano-c.com
Raymor Industries Inc.	Canada	www.raymor.com
Rosseter Holdings Ltd.	Cyprus/USA	www.e-nanoscience.com
Shenzhen Nanotech Port Co. Ltd.	China	www.nanotubes.com.cn
Showa Denko K.K.	Japan	www.sdk.co.jp
SouthWest NanoTechnologies Inc.	USA	www.swentnano.com
Sun Nanotech Co. Ltd.	China	www.sunnano.com
Thomas Swan & Co. Ltd.	England	www.thomas-swan.co.uk
Toray	Japan	www.toray.com
Ube Industries	Japan	www.ube-ind.co.jp
Unidym Inc.	USA	www.unidym.com
Zyvex	USA	www.zyvex.com

helicities or chiralities are. Inspired by the good toughness of spider silk, Dalton et al. [56] spun 100 μm-long CNT fibers to make supercapacitors and wove them into textiles. They predicted these supercapacitors CNT textiles show promising to be applied in sensors, electronic interconnects, electromagnetic shields, antennas, and batteries. Dillon et al. [57] found that hydrogen was

Table 1.2 Manufacturers of CNT Synthesis Systems [51]

Company	URL
Aixtron	www.aixtron.com
First Nano	www.firstnano.com
Oxford Instruments	www.oxford-instruments.com
Tokyo Electron Limited (TEL)	www.tel.com

condensed to high density inside SWCNTs from temperature-programmed desorption spectroscopy. This suggests that CNTs have potential to be used in hydrogen-storage. Iijima and Ichihashi [2] synthesized single-shell tubes with diameters of approximately 1 nm in abundance. Ruoff et al. [58] exhibited flat morphology deformed by vdW forces between adjacent nanotubes and predicted this effect is observable. The experimental study of Dai et al. [59] showed the potential of CNTs to be used for SPM because of their flexibility and slenderness. As materials would exhibit unique properties owing to their morphologies, Ajayan and Tour [60] coaxed carbon atoms into several topologies to study their application in composite. Cumings et al. [61] successfully peeled and removed the outer layers at the end of MWCNTs to produce a sharpened structure. Under the help of an improved electric arc technique, Sun et al. [62] created a CNT with a diameter of 0.5 nm. Journet et al. [40] produced abundant SWCNTs using the electric-arc technique and suggested that the growth mechanism of SWCNTs was independent on the technique. Treacy et al. [18] measured the Young's modulus of individual CNTs by transmission electron microscopy and found it could reach an order of terapascals. Odom et al. [63] reported the atomic structure and electronic properties of SWCNTs using scanning tunneling microscopy (STM), and suggested that SWCNTs exhibit different structures and the electronic properties of CNTs are dependent on diameter and helicity. Wildöer et al. [64] examined electronic properties of individual SWCNTs by STM and spectroscopy. Both metallic and semiconducting CNTs are observed and electronic properties are in fact dependent on wrapping angle. Falvo et al. [65] reported that MWCNTs could be repeatedly bent over large angles by means of AFM which confirms CNTs are remarkably flexible and resilient. Tsang et al. [66] opened CNTs at the end and filled them with a variety of metal oxides using wet chemical techniques. Lu [67] obtained an average Young's modulus of approximately 1 TPa by testing a series of SWCNTs. The experimental results indicated that the modulus is insensitive to the tube's radius and chirality. Wagner et al. [68] observed fragmentation in SWCNTs and obtained a tensile strength of 55 GPa. Walters et al. [21] conducted large elastic strain tests and applied a tensile load on the lateral wall in ropes of SWCNTs using an AFM. The elastic strains of the ropes were investigated when freely supported with tension proportional to lengthening. The results produced by elastic deformation for more than 10 cycles indicate that the maximum strain of SWCNTs ropes is 5.8% ($\pm$0.9%), which corresponds to a lower circumscription of 45 ($\pm$7) GPa for the tensile strength. Yu et al. [20] investigated the mechanical properties of 15 separate SWCNT bundles by applying a tensile load on the perimeter of each bundle. The results demonstrated that the average tensile strength of SWCNTs ranges from 13 to 52 GPa with the maximum tensile strain of 5.3%. Yu et al. [13] measured the mechanical properties of 19 individual MWCNTs by tensile loading. They found that the tensile strain of MWCNTs breaking is

Table 1.3 Companies Developing and/or Selling CNT Products [51]

Company	URL	Field of Application	Notes
Adidas	www.adidas.com	Composites	Running shoe sole http://www.sweatshop.co.uk/Details.cfm?ProdID = 9007&category = 0
Aldila	http://www.aldila.com	Composites	Golf shafts http://www.aldila.com/products/vs-proto/
Amendment II	http://www.amendment2.com/	Composites	Armor vests
Amroy	http://www.amroy.fi/	Composites	Partnerships with Yachts, sports goods, and wind turbine blades manufacturers
Aneeve	http://aneeve.com	Microelectronics Biotechnology	Printed FET; RFID Sensing and diagnostics
ANS Synthetic fibers; EMI shielding; lightening protection	http://www.appliednanostructuredsolutions.com	Composites Energy	Synthetic fibers; EMI shielding; lightning protection http://www.appliednanostructuredsolutions.com/archives/4 CNT based powder for battery electrodes
Axson	www.axson-group.com	Composites	EMI shielding; spark protection Structural composites (Nanoledge)
Baltic	http://www.balticyachts.com/	Composites	Sailng yachts
BASF	www.basf.com	Composites	Conductive POM for fuel lines and filter housing (with Audi) http://www.basf.com/group/corporate/de/literature-document:/Brand + Ultraform-Case + Studys–Fuel + filter + housing-English.pdf
BlueNano	www.bluenanoinc.com	Energy	CNT-based powder for battery electrodes http://www.bluenanoinc.com/nanomaterials/carbon-nanomaterials.html
BMC	www.bmc-racing.com	Composites	Bicycles (with Easton-Zyvex)
Canatu	www.canatu.com	Coatings	Transparent conductor (nanobuds); touch screens; touch sensors
Canon	www.canon.com	Microelectronics	Field emission display; SED TV
Eagle Windpower	-	Energy	Wind turbine blades
Easton	www.easton.com	Composites	Archery arrows (with Amroy) http://www.eastonarchery.com/ Baseball bat (with Zyvex)
Eikos	www.eikos.com	Coatings Energy	Transparent conductors Photovoltaics; copper indium gallium selenide (CIGS) thin film solar cells

(Continued)

Table 1.3 Companies Developing and/or Selling CNT Products [51] *Continued*

Company	URL	Field of Application	Notes
Evergreen	-	Energy	Wind turbine blades
Fujitsu	www.fujitsu.com	Microelectronics	Interconnect vias; thermal interfaces
General Electric	www.ge.com	Coatings	Thermal sensing and imaging
General Nano	http://www.generalnanollc.com/	Composites	CNT forests; dry-spun yarns and sheets
Hexcel	www.hexcel.com	Composites	Conductive aerospace composites
		Energy	Wind turbine blades
Hyperion Catalysis	www.hyperioncatalysis.com	Composites	Automotive fuel line parts; electrostatic painting
Iljin Nanotech	www.iljin.co.kr	Coatings Microelectronics	Transparent conductors Field emission display
Imec	www.imec.be	Microelectronics	Interconnect via
Intel	www.intel.com	Microelectronics	Electronics devices and switches; FET
Meijo-nano carbon	www.meijo-nano.com	Composites	Yarns, sheets, and tapes
NanOasis	http://www.nanoasisinc.fogcitydesign.com/	Energy	Filtration membranes
Nanocomp	www.nanocomptech.com	Composites	CNT yarns and sheets made directly from floating CNT by CVD
			EMI shielding; spark protection flame retardant; ballistic shields
Nanocyl S.A.	www.nanocyl.com	Composites	EMI shielding for electronic packages; prepreg antifouling paint; flame retardant coating
		Coatings	
NanoIntegris	www.nanointegris.com	Coatings	Transparent conductors
		Microelectronics	FET; LED; IR sensing
		Biotechnology	Chemical sensing and diagnostics
Nanomix	www.nano.com	Biotechnology	Sensing and diagnostics
Nantero	www.nantero.com	Microelectronics	Electromechanical non-volatile memory
		Coatings	Chemical sensing and diagnostics; IR sensing (with Brewer Science)
NEC Corp.	www.nec.co.jp	Microelectronics	Printed electronics; FET
Nokia	www.nokia.com	Coatings	Transparent conductor (KINETIC with Toray)
Panasonic Boston Labs	www.panasonic.com	Coatings	Transparent conductor (with SWeNT); touch screen http://swentnano.com/news/index.php?subaction = showfull&id = 1309490173&archive =
Paru Corporation	-	Microelectronics	FET; RFID

Plasan Ltd.	www.plasansasa.com	Composites	Yarns (Cambridge method) http://www.plasansasa.com/node/151
Porifera	http://poriferanano.com/	Energy	Filtration membranes
Q-flo	www.q-flo.com	Composites	Yarns; conductive polymer composites (Cambridge start-up)
Renegade	http://www.renegadematerials.com/	Composites	Fuzzy fibers; Field emission display
Samsung	www.samsung.com	Coatings	Transparent conductor (with Unidym)
Seldon	http://seldontechnologies.com/	Energy	Water purification systems http://seldontechnologies.com/products/
Showa Denko K.K	www.sdk.co.jp	Energy	CNT based powder for battery electrode
Takiron Co.	http://www.takiron.co.jp/	Coatings	Electrostatic dissipative windows
Teco Nanotech Co. Ltd	http://wwwe.teconano.com.tw/	Coatings	Field emission display; touch sensor
Tesla nanocoating Ltd.	www.teslanano.com	Coatings	Anti-corrosion coatings (lower Zinc and higher duability)
Top Nanosys	www.topnanosys.com	Coatings	Transparent conductor; transparent displays
Toray	www.toray.com	Coatings	Transparent conductor; anticorrosion; thermal sensing
Ube Industries	www.ube-ind.co.jp	Energy	CNT based powder for battery electrode
Unidym Inc.	www.unidym.com	Coatings	Transparent conductor (for resistive touch screen); Organic photovoltaics
Yonex	www.yonex.com	Composites	Badminton rackets; tennis rackets http://www.yonex.com/tennis/technology/racquets.html
Zoz GmbH	http://www.zoz-group.de	Composites	Al-CNT alloys for sport equipment, machine parts, and aerospace http://www.zoz-group.de/zoz.engl/zoz.main/content/view/147/165/lang,en/
Zyvex	www.zyvex.com	Composites	Light weight composites for speedboats; Sporting goods (Epovex with Easton and BMC); prepreg (Arovex)

approximately 12%, corresponding to a tensile strength in the outermost layer ranging from 11 to 63 GPa. Li et al. [69] reported an average tensile strength of 22 GPa by measuring a resin-based SWCNT composite using the hypothesis of treating the interfacial load applied on the composite. Demczyk et al. [70] reported the results of pulling and bending tests on individual CNTs, demonstrating that the tensile strength of breaking a CNT is approximately 150 GPa. Considering the temperature effect on the mechanical properties of CNTs, Huang et al. [71] conducted large deformation tests of individual SWCNTs at high temperatures, demonstrating that they can undergo superplastic deformation. The CNTs can be lengthened by nearly 280% and narrowed 15 times before fracturing. These results are useful for the fabrication of composite materials and the toughening of ceramics at high temperatures. In addition, experimental studies of vibration characteristics of carbon nanostructures [72–75] show CNTs are good candidates as electromechanical oscillator and resonators.

1.5 THEORIES FOR MECHANICAL BEHAVIORS OF CNTs

The aforementioned experimental results show that CNTs have higher levels of stiffness and strength than other known structures. However, there is a broad discrepancy in these results on the stiffness and strength. This is mainly due to the very small size of CNTs. The variations reveal that, despite their visualization, experimental tools can provide only rough results and are limited to the study of certain mechanisms and lack universality. In addition, it is difficult to accurately conduct and control experiments when the specimen size reaches to the nanometer level. Therefore, strenuous efforts have been made by theoretical researchers in an effort to explore the inherent mechanisms underlying the mechanical behaviors of nanomaterials. Theoretical analysis is an accurate tool to investigate the material characteristics of these carbon nanostructures as it is easy to conduct and control the boundary conditions. In general, theoretical approaches mainly include three categories: atomistic simulations, continuum models, and hybrid approaches.

1.5.1 ATOMISTIC SIMULATIONS

Atomistic simulations, the most widely used methods in the nanomechanics field, are important numerical methods for the investigation of magnetic, electronic, chemical, and mechanical properties of carbon nanostructures since these modeling approaches can accurately trace atomic position and precisely capture the microscale physical mechanism, such as buckling. There has been already much research of carbon nanostructures using atomistic simulation.

Yakobson et al. [76] studied the large deformation of CNTs using MD simulation. The change of each morphological pattern occurs accompanying an abrupt release of energy. Liew et al. [77–81] investigated the thermal stability, the elastic and plastic properties of CNTs and twisting effect, as well as buckling on CNTs bundles using MD simulation which employs the second-generation of reactive empirical bond-order potential and Lennard–Jones potential to describe the atomic interaction. Liu et al. [82,83] proposed an atomic-scale finite element method, which has the same framework as the traditional finite element method, for numerical simulation of CNTs. Feng and Liew [84,85] employed MD simulation to analyze the stability and buckling of carbon nanorings by the definition of stability as no buckling occurring when mapping nanorings from CNTs. The temperature effect on the elastic properties has been examined by Hsieh et al. [86] and Zhang et al. [87].

For the purpose of showing the promising application as nanoresonators and mass detectors, Li and Chou [88–90] presented a molecular–structural–mechanics method to simulate vibration behaviors of CNT. Their results demonstrated that SWCNTs have ultrahigh fundamental frequency, a level of 10 GHz–1.5 THz for cantilevered or bridged boundary conditions. It was also observed that the fundamental frequencies of double-walled CNTs (DWCNTs) are 10% lower than those of SWCNTs of the same outer diameter and the noncoaxial vibration started at the third vibration mode. Chowdhury et al. [91] adopted a molecular mechanics approach to investigate the vibration properties of two kinds of SWCNTs and found natural frequencies of zigzag CNTs are higher than those of the counterpart armchair cases. Hashemnia et al. [92] determined the fundamental frequency of CNTs and graphene sheets by the means of molecular–structural–mechanics approach. It is observed that fundamental frequencies of CNTs are greater than those of graphene sheets. Reddy et al. [93] employed atomistic simulations and continuum shell modeling to analyze free vibration of SWCNTs. They pointed out that a minimum potential configuration of CNT was important for the computation of the stiffness matrix.

In spite of the precise trace of the atomic displacement of CNTs, these atomistic simulations are highly expensive in terms of computational resources. Therefore, their applications are extremely limited to a simple system with a small number of atoms, and do not meet the demand of engineering application. For examples, Liew et al. [77,81] spent 36 h studying the buckling behavior of a (10, 10) SWCNT containing 2000 atoms with MD simulation in a single SGI origin 2000 CPU, whereas the investigation of a four-walled CNT containing 15,097 atoms required 4 months.

1.5.2 CONTINUUM MODELS

In view of these aforementioned problems of MD simulation, there is an urgent need to establish an effective and efficient computational method. Researchers have sought to adopt alternative approaches, e.g., continuum simulation methods [94–101], instead of working from an atomic perspective. Continuum simulation has been found to be computationally very efficient. These continuum simulation methods, which can largely reduce the degrees of freedom, are much faster solution techniques than atomistic simulations for analyzing nanostructures. Thus, they are much more attractive in the practical application.

Wang and Hu [100] examined the flexural wave propagation in SWCNTs using continuum beam models such as nonlocal elastic Timoshenko beam, traditional Timoshenko beam and Euler beam, and MD simulation. Yoon et al. [102] suggested a multielastic beam model to determine noncoaxial vibration frequencies and mode shapes of an isolated MWCNT. Wu et al. [103] employed a beam-bending model to investigate the vibration frequency of a cantilevered CNT-based mass sensor. He et al. [104] derived an explicit formula to describe the vdW interaction between any two layers and analyze vibration behaviors of multilayered graphene sheets using a stacked plate model. Govindjee and Sackman [94] applied the Euler beam theory to simulating mechanical properties of CNTs. Ru [95,96] treated an SWCNT as a single-layer shell with an effective bending stiffness and modeled mechanical responses of CNTs under axial compression. Li and Chou [97] employed a molecular–structural–mechanics method to predict elastic properties of SWCNTs. He et al. [98] and

Liew et al. [99] employed a continuum cylindrical shell model to account for the vdW interaction between different walls of CNTs. Wang and Hu [100] and Wang et al. [101] investigated the longitudinal and flexural wave propagation in CNTs by modeling a CNT as a nonlocal elastic cylindrical shell and employing continuum mechanics. Wei and Srivastava [105] used a continuum elastic theory and derived a general analytical expression that there is a relationship of $\cos^4\theta$ between Young's modulus of an SWCNC and that of an equivalent SWCNT. He et al. [106] employed an elastic multiple shell model and considered vdW interaction between any two layers as the radius-dependent function for vibration analysis of MWCNTs. For a MWCNT with a small radius, their numerical results demonstrated that the interlayer vdW interaction exhibited a significant influence on vibration behaviors. Natsuki et al. [107] presented a wave propagation approach for vibration analysis of CNTs as well as embedded CNTs and Yoon et al. [108] studied a short CNT using Timoshenko beam model. Their results showed that the order of vibration frequencies of CNTs reached a level of terahertz but in a wide range. Wang and Varadan [109] developed a nonlocal continuum mechanics model based on elastic beam theories to analyze the vibration characteristic of both SWCNTs and DWCNTs. Lu et al. [110] provided consistent equations of motion for the nonlocal Timoshenko beam model to study wave and vibration properties of CNTs. Hsu et al. [111] derived an analytical solution of fundamental frequency for chiral SWCNTs subjected to thermal vibration, using the Timoshenko beam model. On comparing results obtained by the Timoshenko beam model with the Euler beam model, it is found that the former's frequency is lower than the latter. Considering two nanotubes shells coupled by vdW interaction, Natsuki et al. [112] employed Euler–Bernoulli beam theory to analyze vibration characteristics of DWCNTs and studied the vibration of fluid-filled CNTs by employing wave propagation approach [113]. Yang et al. [114] applied nonlocal Timoshenko beam theory to study nonlinear free vibration of SWCNTs and found the nonlocal parameter to have a slight effect on the nonlinear mode shape but considerable effect on frequencies. Wang et al. [115] developed an elastic Bernoulli–Euler beam model based on the thermal elasticity mechanics theory to analyze the vibration and instability of fluid-conveying SWCNTs. It was found that the temperature change plays a significant role in buckling stability and natural frequency of a fluid-conveying SWCNT. Later, Yan et al. [116] treated triple-walled CNTs (TWCNTs) as Euler–Bernoulli beam, considering the interlayer vdW interactions, to study natural frequencies and stability problems of the fluid-conveyed TWCNTs. It was found that interlayer vdW interactions have a large effect on natural frequencies and the internal moving fluid on instability of TWCNTs. Using the Timoshenko beam model, Chang and Lee [117] analyzed flexural vibration behaviors of SWCNTs which contain a fluid flow. A significant effect of rotary inertia and transverse shear deformation on vibration frequencies is observed. Georgantzinos and Anifantis [118] modeled MWCNTs using springs and lumped masses to study the vibration characteristic. It is observed that fundamental frequencies of CNTs are far larger than those of graphene sheets. Firouz-Abadi et al. [119] adopted a nonlocal continuum model to examine the free vibration of CNCs. Kitipornchai et al. [120] developed a continuum-plate model by a explicitly derived formal to reflect the vdW interaction between any two layers of multilayered graphene sheets. Ansari et al. [121] introduced a nonlocal elastic plate model considering the small scale to study vibration behaviors of multilayered graphite sheets under various boundary conditions.

Nevertheless, while continuum mechanics is capable of studying large atomic systems, severe challenges can be encountered in considering the information of atomic structure. Therefore, the practicability of continuum mechanics depends on additional introduced material properties so that it is restricted to certain mechanical behaviors of nanomaterials. Moreover, it is noteworthy that the size effect becomes significant in the mechanical behaviors of graphene as the size of specimens approach to the nanometer level.

1.5.3 HYBRID APPROACHES

Most continuum models are restricted to small deformations. In addition, as the general continuum models directly employ the theory of continuum mechanics, the material parameters need to be fitted based on the available data. In recent years, a series of hybrid techniques have emerged as feasible and efficient tools, possessing the advantages of both atomistic simulation and continuum mechanics, for analyzing large-size system problems within the fields of nanoscience and nanotechnology.

Tadmor et al. [122,123] analyzed complex Bravais crystals using an FEM framework incorporating atomic interaction. Shenoy et al. [124] proposed a quasicontinuum method by mixed atomistic simulations and continuum methods which gives rise to a reduction of atomistic degrees of freedom. Liew et al. [125] examined the elastic modulus and fracturing of SWCNTs using an atomistic-based continuum method and found a good agreement with the results of previous experiments and atomistic simulations.

One newly hybrid approach in the study of CNTs can be traced back to the atomistic-continuum method [126], in which the constitutive model is constructed based on the Cauchy–Born rule. Due to the lack of centrosymmetry of the hexagonal lattice structures in CNTs, internal relaxation needs to be adopted to ensure the minimum energy of the atomic system. Zhang et al. [126–128] introduced this internal relaxation method when studying CNTs. They recommended a combination of interatomic potentials with a continuum model and studied linear elastic modulus fracture nucleation. However, Arroyo and Belytschko [129–131] argued that this application of the standard Cauchy–Born rule may be inadequate. Essentially, SWCNTs are curved crystalline sheets with a one-atom thickness. Therefore, the curved effect should not be neglected. Arroyo and Belytschko's numerical simulation demonstrated that the constitutive relationship constructed based on the Cauchy–Born rule fails to truly describe buckling behaviors of CNTs because of the neglect of bending effect. Sunyk and Steinmann [132] noted that deformations of the underlying crystal must be sufficiently homogeneous when the standard Cauchy–Born rule is to be applied. Later, Guo et al. [133] and Wang et al. [134] proposed a higher-order Cauchy–Born rule for analysis of elastic properties of SWCNTs. In the higher-order gradient theory, the second deformation gradient can accurately describe the bending effect which makes the results much closer to the actual case. Liew and Sun [135] and Sun and Liew [135–137] developed a mesh-free method based on the higher-order deformation gradient continuum theory to simulate the buckling behaviors of SWCNTs under various loadings.

One remarkable advantage here is that this atomistic-continuum approach avoids the difficulties of the thickness issue. A significant degree of debate remains regarding the definition of the thickness of CNTs. There is currently no agreed thickness concept in the open literature [138–140] which gives rise to a great variety of elastic properties. In the study of CNTs based on the continuum mechanism, the stiffness is related to the moment of inertia; as such, remarkably different results will be obtained when a scattering of thicknesses are used. In the present atomistic-continuum approach, the constitutive model is constructed on the basis of the atomic interaction so that its stiffness matrix is entirely unrelated to thickness. Therefore, the present atomistic-continuum method can effectively avoid the thickness issue. Moreover, in the atomistic-continuum approach, the selected representative cell can reflect the atomic structure so that the chiral angle can be incorporated. Thus, its effect on structural parameters and elastic properties can be studied directly. As the atomic interaction of CNTs is included within the constitutive model, the constitutive relationship of CNTs updates at each iterative step; thus the material nonlinearity can be exactly captured in this approach. Furthermore, based on the higher-order gradient continuum theory, the contribution of the second-order gradient to the stiffness matrix is considered, reflecting the geometrical nonlinearity, and this is where a further advantage of this approach lies. Therefore, by combining the constitutive model with an appropriate numerical method, this hybrid approach is capable of accurately capturing the physical behaviors, including buckling and postbuckling behaviors.

REFERENCES

[1] Iijima S. Helical microtubules of graphitic carbon. Nature 1991;354(6348):56–8.

[2] Iijima S, Ichihashi T. Single-shell carbon nanotubes of 1-nm diameter. Nature 1993;363(6430):603–5.

[3] Bethune DS, Klang CH, Vries MSD, Gorman G, Savoy R, Vazquez J, et al. Cobalt-catalysed growth of carbon nanotubes with single-atomic-layer walls. Nature 1993;363(6430):605–7.

[4] Wood J. The top ten advances in materials science. Mater Today 2008;11(07):40–5.

[5] Dresselhaus MS, Dresselhaus G, Eklund PC. Science of fullerenes and carbon nanotubes. San Diego, CA: Academic Press; 1996.

[6] Hamada N, Si S, Oshiyama A. New one-dimensional conductors: graphitic microtubules. Phys Rev Lett 1992;68(10):1579–81.

[7] White CT, Todorov TN. Carbon nanotubes as long ballistic conductors. Nature 1998;393(6682):240–2.

[8] Quinn BM, Lemay SG. Single-walled carbon nanotubes as templates and interconnects for nanoelectrodes. Adv Mater 2006;18(18):855–9.

[9] Naeemi A, Meindl JD. Design and performance modeling for single-walled carbon nanotubes as local, semiglobal, and global interconnects in gigascale integrated systems. IEEE Trans Electron Devices 2007;54(1):26–37.

[10] Cho H, Koo KH, Kapur P, Saraswat KC. Performance comparisons between cu/low-κ, carbon-nanotube, and optics for future on-chip interconnects. IEEE Elect Device Lett 2008;29(1):122–4.

[11] Pop E, Mann D, Wang Q, Goodson KE, Dai HJ. Thermal conductance of an individual single-wall carbon nanotube above room temperature. Nano Lett 2006;6(1):96–100.

[12] Yao Z, Kane CL, Dekker C. High-field electrical transport in single-wall carbon nanotubes. Phys Rev Lett 2000;84(13):2941–4.

[13] Yu MF, Lourie O, Dyer MJ, Moloni K, Kelly TF, Ruoff RS. Strength and breaking mechanism of multiwalled carbon nanotubes under tensile load. Science 2000;287(5453):637–40.

[14] Pan ZW, Xie SS, Lu L, Chang BH, Sun LF, Zhou WY, et al. Tensile tests of ropes of very long aligned multiwall carbon nanotubes. Appl Phys Lett 1999;74(21):3152–4.

[15] Poncharal P, Wang ZL, Ugarte D, Heer WAD. Electrostatic deflections and electromechanical resonances of carbon nanotubes. Science 1999;283(5407):1513–16.

[16] Salvetat JP, Briggs GAD, Bonard JM, Bacsa RR, Kulik AJ, Stöckli T, et al. Elastic and shear moduli of single-walled carbon nanotube ropes. Phys Rev Lett 1999;82(5):944–7.

[17] Wong EW, Sheehan PE, Lieber CM. Nanobeam mechanics: elasticity, strength, and toughness of nanorods and nanotubes. Science 1997;277(5334):1971–5.

[18] Treacy MMJ, Ebbesen TW, Gibson JM. Exceptionally high Young's modulus observed for individual carbon nanotubes. Nature 1996;381(6584):678–80.

[19] Khang DY, Xiao J, Kocabas C, Maclaren S, Banks T, Jiang H, et al. Molecular scale buckling mechanics in individual aligned single-wall carbon nanotubes on elastomeric substrates. Nano Lett 2008;8(1):124–30.
[20] Yu MF, Files BS, Arepalli S, Ruoff RS. Tensile loading of ropes of single wall carbon nanotubes and their mechanical properties. Phys Rev Lett 2000;84(24):5552–5.
[21] Walters DA, Ericson LM, Casavant MJ, Liu J, Colbert DT, Smith KA, et al. Elastic strain of freely suspended single-wall carbon nanotube ropes. Appl Phys Lett 1999;74(25):3803–5.
[22] Watanabe H, Manabe C, Shigematsu T, Shimizu M. Dual-probe scanning tunneling microscope: Measuring a carbon nanotube ring transistor. Appl Phys Lett 2001;78(78):2928–30.
[23] Cohen G, Hod O, Rabani E. Constructing spin interference devices from nanometric rings. Phys Rev B Condens Matter 2007;76(23):4692.
[24] Terrones M, Banhart F, Grobert N, Charlier JC, Terrones H, Ajayan PM. Molecular junctions by joining single-walled carbon nanotubes. Phys Rev Lett 2002;89(7):075505.
[25] Wei D, Liu Y. ChemInform abstract: the intramolecular junctions of carbon nanotubes. Adv Mater 2008;20(15):2815–41.
[26] Kim H, Lee J, Kahng SJ, Son YW, Lee SB, Lee CK, et al. Direct observation of localized defect states in semiconductor nanotube junctions. Phys Rev Lett 2003;90(21):216107.
[27] Gao B, Komnik A, Egger R, Glattli DC, Bachtold A. Evidence for Luttinger-liquid behavior in crossed metallic single-wall nanotubes. Phys Rev Lett 2004;92(21):216804.
[28] Dekker C. Carbon nanotubes as molecular quantum wires. Phys Today 1999;52(5):22–8.
[29] Mceuen PL. Single-wall carbon nanotubes. Physics World 2000;13(6):31–6.
[30] Dai H. Controlling nanotube growth. Phys World 2000;13:43–7.
[31] Liang W, Bockrath M, Bozovic D, Hafner JH, Tinkham M, Park H. Fabry–Perot interference in a nanotube electron waveguide. Nature 2001;411(6838):665–9.
[32] Kong J, Yenilmez E, Tombler TW, Kim W, Dai H, Laughlin RB, et al. Quantum interference and ballistic transmission in nanotube electron waveguides. Phys Rev Lett 2001;87(10):106801.
[33] Dai H. Nanotube growth and characterization: carbon nanotubes. Berlin: Springer; 2001.
[34] Kong J, Soh HT, Cassell AM, Quate CF, Dai H. Synthesis of individual single-walled carbon nanotubes on patterned silicon wafers. Nature 1998;395(6705):878–81.
[35] Hafner JH, Bronikowski MJ, Azamian BR, Nikolaev P, Rinzler AG, Colbert DT, et al. Catalytic growth of single-wall carbon nanotubes from metal particles. Chem Phys Lett 1998;296(1):195–202.
[36] Satishkumar BC, Govindaraj A, Sen R, Rao CNR. Single-walled nanotubes by the pyrolysis of acetylene-organometallic mixtures. Chem Phys Lett 1998;293(1-2):47–52.
[37] Nikolaev P, Bronikowski MJ, Bradley RK, Rohmund F, Colbert DT, Smith KA, et al. Gas-phase catalytic growth of single-walled carbon nanotubes from carbon monoxide. Chem Phys Lett 1999;313(1–2):91–7.
[38] Su M, Zheng B, Liu J. A scalable CVD method for the synthesis of single-walled carbon nanotubes with high catalyst productivity. Chem Phys Lett 2000;322(5):321–6.
[39] Colomer JF, Stephan C, Lefrant S, Tendeloo GV, Willems I, Kónya Z, et al. Large-scale synthesis of single-wall carbon nanotubes by catalytic chemical vapor deposition (CCVD) method. Chem Phys Lett 2000;317(1–2):83–9.
[40] Journet C, Maser WK, Bernier P, Loiseau A, delaChapelle ML, Lefrant S, et al. Large-scale production of single-walled carbon nanotubes by the electric-arc technique. Nature 1997;388(6644):756–8.
[41] Thess A, Lee R, Nikolaev P, Dai H, Petit P, Robert J, et al. Crystalline ropes of metallic carbon nanotubes. Science 1996;273(5274):483–7.
[42] Dai H, Wong EW, Lieber CM. Probing electrical transport in nanomaterials: conductivity of individual carbon nanotubes. Science 1996;272(5261):523–6.

[43] Ebbesen T, Lezec H, Hiura H, Bennett J, Ghaemi H, Thio T. Electrical-conductivity of individual carbon nanotubes. Nature 1996;382(6586):54–6.
[44] Calvert P. Nanotube composites: a recipe for strength. Nature 1999;399(6733):210–11.
[45] Dresselhaus MS, Dresselhaus G, Saito R. Physics of carbon nanotubes. Carbon N Y 1995;33(7):883–91.
[46] Dresselhaus M, Dresselhaus G, Saito R. Carbon fibers based on C 60 and their symmetry. Phys Rev B 1992;45(11):6234.
[47] Peng B, Locascio M, Zapol P, Li S, Mielke SL, Schatz GC, et al. Measurements of near-ultimate strength for multiwalled carbon nanotubes and irradiation-induced crosslinking improvements. Nat Nanotechnol 2008;3(10):626–31.
[48] Wei BQ, Vajtai R, Ajayan PM. Reliability and current carrying capacity of carbon nanotubes. Appl Phys Lett 2001;79(8):1172–4.
[49] De Volder MF, Tawfick SH, Baughman RH, Hart AJ. Carbon nanotubes: present and future commercial applications. Science 2013;339(6119):535–9.
[50] NanoLab. Nanolab website. http://www.nano-lab.com/.
[51] CNI. Carbon nanotechnologies website. http://www.cnanotech.com/pages/store/6-0_online_store.html.
[52] Mike. Cheap Tubes Inc. website. http://www.cheaptubesinc.com/.
[53] Baughman RH, Zakhidov AA, de Heer WA. Carbon nanotubes—the route toward applications. Science 2002;297(5582):787–92.
[54] Pan ZW, Xie SS, Chang BH, Wang CY, Lu L, Liu W, et al. Very long carbon nanotubes. Nature 1998;394(6694):631–2.
[55] Qin LC, Zhao XL, Hirahara K, Miyamoto Y, Ando Y, Iijima S. Materials science—the smallest carbon nanotube. Nature 2000;408(6808):50.
[56] Dalton AB, Collins S, Munoz E, Razal JM, Ebron VH, Ferraris JP, et al. Super-tough carbon-nanotube fibres—these extraordinary composite fibres can be woven into electronic textiles. Nature 2003;423 (6941):703.
[57] Dillon AC, Jones KM, Bekkedahl TA, Kiang CH, Bethune DS, Heben MJ. Storage of hydrogen in single-walled carbon nanotubes. Nature 1997;386(6623):377–9.
[58] Ruoff RS, Tersoff J, Lorents DC, Subramoney S, Chan B. Radial deformation of carbon nanotubes by Van-Der-Waals Forces. Nature 1993;364(6437):514–16.
[59] Dai HJ, Hafner JH, Rinzler AG, Colbert DT, Smalley RE. Nanotubes as nanoprobes in scanning probe microscopy. Nature 1996;384(6605):147–50.
[60] Ajayan PM, Tour JM. Materials science—nanotube composites. Nature 2007;447(7148):1066–8.
[61] Cumings J, Collins PG, Zettl A. Materials—peeling and sharpening multiwall nanotubes. Nature 2000;406(6796):586.
[62] Sun LF, Xie SS, Liu W, Zhou WY, Liu ZQ, Tang DS, et al. Materials—creating the narrowest carbon nanotubes. Nature 2000;403(6768):384.
[63] Odom TW, Huang JL, Kim P, Lieber CM. Atomic structure and electronic properties of single-walled carbon nanotubes. Nature 1998;391(6662):62–4.
[64] Wildoer JWG, Venema LC, Rinzler AG, Smalley RE, Dekker C. Electronic structure of atomically resolved carbon nanotubes. Nature 1998;391(6662):59–62.
[65] Falvo MR, Clary GJ, Taylor RM, Chi V, Brooks FP, Washburn S, et al. Bending and buckling of carbon nanotubes under large strain. Nature 1997;389(6651):582–4.
[66] Tsang SC, Chen YK, Harris PJF, Green MLH. A simple chemical method of opening and filling carbon nanotubes. Nature 1994;372(6502):159–62.
[67] Lu JP. Elastic properties of carbon nanotubes and nanoropes. Phys Rev Lett 1997;79(7):1297–300.
[68] Wagner HD, Lourie O, Feldman Y, Tenne R. Stress-induced fragmentation of multiwall carbon nanotubes in a polymer matrix. Appl Phys Lett 1998;72(2):188–90.

[69] Li F, Cheng HM, Bai S, Su G, Dresselhaus MS. Tensile strength of single-walled carbon nanotubes directly measured from their macroscopic ropes. Appl Phys Lett 2000;77(20):3161–3.
[70] Demczyk BG, Wang YM, Cumings J, Hetman M, Han W, Zettl A, et al. Direct mechanical measurement of the tensile strength and elastic modulus of multiwalled carbon nanotubes. Mat Sci Eng a-Struct 2002;334(1-2):173–8.
[71] Huang JY, Chen S, Wang ZQ, Kempa K, Wang YM, Jo SH, et al. Superplastic carbon nanotubes—conditions have been discovered that allow extensive deformation of rigid single-walled nanotubes. Nature 2006;439(7074):281.
[72] Garcia-Sanchez D, Paulo AS, Esplandiu MJ, Perez-Murano F, Forro L, Aguasca A, et al. Mechanical detection of carbon nanotube resonator vibrations. Phys Rev Lett 2007;99(8):085501.
[73] Chiu HY, Hung P, Postma HWC, Bockrath M. Atomic-scale mass sensing using carbon nanotube resonators. Nano Lett 2008;8(12):4342–6.
[74] Sazonova V, Yaish Y, Ustunel H, Roundy D, Arias TA, McEuen PL. A tunable carbon nanotube electromechanical oscillator. Nature 2004;431(7006):284–7.
[75] Wang ZL, Poncharal P, de Heer WA. Measuring physical and mechanical properties of individual carbon nanotubes by in situ TEM. J Phys Chem Solids 2000;61(7):1025–30.
[76] Yakobson BI, Brabec CJ, Bernholc J. Nanomechanics of carbon tubes: Instabilities beyond linear response. Phys Rev Lett 1996;76(14):2511–14.
[77] Liew KM, He XQ, Wong CH. On the study of elastic and plastic properties of multi-walled carbon nanotubes under axial tension using molecular dynamics simulation. Acta Mater 2004;52(9):2521–7.
[78] Liew KM, Wong CH, Tan MJ. Twisting effects of carbon nanotube bundles subjected to axial compression and tension. J Appl Phys 2006;99(11):114312.
[79] Liew KM, Wong CH, Tan MJ. Buckling properties of carbon nanotube bundles. Appl Phys Lett 2005;87 (4):041901.
[80] Liew KM, Wong CH, He XQ, Tan MJ. Thermal stability of single and multi-walled carbon nanotubes. Phys Rev B 2005;71(7):075424.
[81] Liew KM, Wong CH, He XQ, Tan MJ, Meguid SA. Nanomechanics of single and multiwalled carbon nanotubes. Phys Rev B 2004;69(11):115429.
[82] Liu B, Huang Y, Jiang H, Qu S, Hwang KC. The atomic-scale finite element method. Comput Method Appl M. 2004;193(17–20):1849–64.
[83] Liu B, Jiang H, Huang Y, Qu S, Yu MF, Hwang KC. Atomic-scale finite element method in multiscale computation with applications to carbon nanotubes. Phys Rev B 2005;72(3):035435.
[84] Feng C, Liew KM. Energetics and structures of carbon nanorings. Carbon N Y 2009;47(7):1664–9.
[85] Feng C, Liew KM. A molecular mechanics analysis of the buckling behavior of carbon nanorings under tension. Carbon N Y 2009;47(15):3508–14.
[86] Hsieh JY, Lu JM, Huang MY, Hwang CC. Theoretical variations in the Young's modulus of single-walled carbon nanotubes with tube radius and temperature: a molecular dynamics study. Nanotechnology 2006;17 (15):3920–4.
[87] Zhang YC, Chen X, Wang X. Effects of temperature on mechanical properties of multi-walled carbon nanotubes. Compos Sci Technol 2008;68(2):572–81.
[88] Li CY, Chou TW. Single-walled carbon nanotubes as ultrahigh frequency nanomechanical resonators. Phys Rev B 2003;68(7):073405.
[89] Li CY, Chou TW. Vibrational behaviors of multiwalled-carbon-nanotube-based nanomechanical resonators. Appl Phys Lett 2004;84(1):121–3.
[90] Li CY, Chou TW. Mass detection using carbon nanotube-based nanomechanical resonators. Appl Phys Lett 2004;84(25):5246–8.

[91] Chowdhury R, Adhikari S, Wang CY, Scarpa F. A molecular mechanics approach for the vibration of single-walled carbon nanotubes. Comp Mater Sci 2010;48(4):730–5.
[92] Hashemnia K, Farid M, Vatankhah R. Vibrational analysis of carbon nanotubes and graphene sheets using molecular structural mechanics approach. Comp Mater Sci 2009;47(1):79–85.
[93] Reddy CD, Rajendran S, Liew KM. Vibration properties of single walled carbon nanotubes—a comparison between the atomistic simulations and continuum shell modelling. J Comp Theor Nanosci 2010;7(8):1400–6.
[94] Govindjee S, Sackman JL. On the use of continuum mechanics to estimate the properties of nanotubes. Solid State Commun 1999;110(4):227–30.
[95] Ru CQ. Effective bending stiffness of carbon nanotubes. Phys Rev B 2000;62(15):9973–6.
[96] Ru CQ. Axially compressed buckling of a doublewalled carbon nanotube embedded in an elastic medium. J Mech Phys Solids 2001;49(6):1265–79.
[97] Li CY, Chou TW. Elastic properties of single-walled carbon nanotubes in transverse directions. Phys Rev B 2004;69(7):073401.
[98] He XQ, Kitipornchai S, Liew KM. Buckling analysis of multi-walled carbon nanotubes: a continuum model accounting for van der Waals interaction. J Mech Phys Solids 2005;53(2):303–26.
[99] Liew KM, He XQ, Kitipornchai S. Buckling characteristics of embedded multi-walled carbon nanotubes. P Roy Soc a-Math Phy 2005;461(2064):3785–805.
[100] Wang LF, Hu HY. Flexural wave propagation in single-walled carbon nanotubes. Phys Rev B 2005;71(19):195412.
[101] Wang LF, Hu HY, Guo WL. Validation of the non-local elastic shell model for studying longitudinal waves in single-walled carbon nanotubes. Nanotechnology 2006;17(5):1408–15.
[102] Yoon J, Ru CQ, Mioduchowski A. Noncoaxial resonance of an isolated multiwall carbon nanotube. Phys Rev B 2002;66(23):233402.
[103] Wu DH, Chien WT, Chen CS, Chen HH. Resonant frequency analysis of fixed-free single-walled carbon nanotube-based mass sensor. Sensor Actuat a-Phys 2006;126(1):117–21.
[104] He XQ, Kitipornchai S, Liew KM. Resonance analysis of multi-layered graphene sheets used as nanoscale resonators. Nanotechnology 2005;16(10):2086–91.
[105] Wei CY, Srivastava D. Nanomechanics of carbon nanofibers: structural and elastic properties. Appl Phys Lett 2004;85(12):2208–10.
[106] He XQ, Eisenberger M, Liew KM. The effect of van der Waals interaction modeling on the vibration characteristics of multiwalled carbon nanotubes. J Appl Phys 2006;100(12): 124317-1-12.
[107] Natsuki T, Endo M, Tsuda H. Vibration analysis of embedded carbon nanotubes using wave propagation approach. J Appl Phys 2006;99(3).
[108] Yoon J, Ru CQ, Mioduchowski A. Terahertz vibration of short carbon nanotubes modeled as Timoshenko beams. J Appl Mech-T Asme 2005;72(1):10–17.
[109] Wang Q, Varadan VK. Vibration of carbon nanotubes studied using nonlocal continuum mechanics. Smart Mater Struct 2006;15(2):659–66.
[110] Lu P, Lee HP, Lu C, Zhang PQ. Application of nonlocal beam models for carbon nanotubes. Int J Solids Struct 2007;44(16):5289–300.
[111] Hsu JC, Chang RP, Chang WJ. Resonance frequency of chiral single-walled carbon nanotubes using Timoshenko beam theory. Phys Lett A 2008;372(16):2757–9.
[112] Natsuki T, Ni QQ, Endo M. Analysis of the vibration characteristics of double-walled carbon nanotubes. Carbon N Y 2008;46(12):1570–3.
[113] Natsuki T, Ni QQ, Endo M. Vibrational analysis of fluid-filled carbon nanotubes using the wave propagation approach. Appl Phys a-Mater 2008;90(3):441–5.

[114] Yang J, Ke LL, Kitipornchai S. Nonlinear free vibration of single-walled carbon nanotubes using nonlocal Timoshenko beam theory. Physica E 2010;42(5):1727–35.
[115] Wang L, Ni Q, Li M, Qian Q. The thermal effect on vibration and instability of carbon nanotubes conveying fluid. Physica E 2008;40(10):3179–82.
[116] Yan Y, He XQ, Zhang LX, Wang CM. Dynamic behavior of triple-walled carbon nanotubes conveying fluid. J Sound Vib 2009;319(3-5):1003–18.
[117] Chang WJ, Lee HL. Free vibration of a single-walled carbon nanotube containing a fluid flow using the Timoshenko beam model. Phys Lett A 2009;373(10):982–5.
[118] Georgantzinos SK, Anifantis NK. Vibration analysis of multi-walled carbon nanotubes using a spring-mass based finite element model. Comp Mater Sci 2009;47(1):168–77.
[119] Firouz-Abadi RD, Fotouhi MM, Haddadpour H. Free vibration analysis of nanocones using a nonlocal continuum model. Phys Lett A 2011;375(41):3593–8.
[120] Kitipornchai S, He XQ, Liew KM. Continuum model for the vibration of multilayered graphene sheets. Phys Rev B 2005;72(7):075443.
[121] Ansari R, Arash B, Rouhi H. Vibration characteristics of embedded multi-layered graphene sheets with different boundary conditions via nonlocal elasticity. Compos Struct 2011;93(9):2419–29.
[122] Tadmor EB, Smith GS, Bernstein N, Kaxiras E. Mixed finite element and atomistic formulation for complex crystals. Phys Rev B 1999;59(1):235–45.
[123] Tadmor EB, Ortiz M, Phillips R. Quasicontinuum analysis of defects in solids. Philos Mag A. 1996;73(6):1529–63.
[124] Shenoy VB, Miller R, Tadmor EB, Rodney D, Phillips R, Ortiz M. An adaptive finite element approach to atomic-scale mechanics—the quasicontinuum method. J Mech Phys Solids 1999;47(3):611–42.
[125] Liew KM, Chen BJ, Xiao ZM. Analysis of fracture nucleation in carbon nanotubes through atomistic-based continuum theory. Phys Rev B 2005;71(23):235424.
[126] Zhang P, Huang Y, Geubelle PH, Klein PA, Hwang KC. The elastic modulus of single-wall carbon nanotubes: a continuum analysis incorporating interatomic potentials. Int J Solids Struct 2002;39(13-14):3893–906.
[127] Zhang P, Jiang H, Huang Y, Geubelle PH, Hwang KC. An atomistic-based continuum theory for carbon nanotubes: analysis of fracture nucleation. J Mech Phys Solids 2004;52(5):977–98.
[128] Zhang P, Huang Y, Gao H, Hwang KC. Fracture nucleation in single-wall carbon nanotubes under tension: a continuum analysis incorporating interatomic potentials. J Appl Mech-T Asme 2002;69(4):454–8.
[129] Arroyo M, Belytschko T. An atomistic-based finite deformation membrane for single layer crystalline films. J Mech Phys Solids 2002;50(9):1941–77.
[130] Arroyo M, Belytschko T. Finite element methods for the non-linear mechanics of crystalline sheets and nanotubes. Int J Numer Meth Eng 2004;59(3):419–56.
[131] Arroyo M, Belytschko T. Finite crystal elasticity of carbon nanotubes based on the exponential Cauchy-Born rule. Phys Rev B 2004;69(11):115415.
[132] Sunyk R, Steinmann P. On higher gradients in continuum-atomistic modelling. Int J Solids Struct 2003;40(24):6877–96.
[133] Guo X, Wang JB, Zhang HW. Mechanical properties of single-walled carbon nanotubes based on higher order Cauchy-Born rule. Int J Solids Struct 2006;43(5):1276–90.
[134] Wang JB, Guo X, Zhang HW, Wang L, Liao JB. Energy and mechanical properties of single-walled carbon nanotubes predicted using the higher order Cauchy-Born rule. Phys Rev B 2006;73(11):115428.
[135] Liew KM, Sun YZ. Elastic properties and pressure-induced structural transitions of single-walled carbon nanotubes. Phys Rev B 2008;77(20):205437.

[136] Sun YZ, Liew KM. Mesh-free simulation of single-walled carbon nanotubes using higher order Cauchy-Born rule. Comp Mater Sci 2008;42(3):444–52.
[137] Sun YZ, Liew KM. The buckling of single-walled carbon nanotubes upon bending: the higher order gradient continuum and mesh-free method. Comput Method Appl M 2008;197(33–40):3001–13.
[138] Chowdhury R, Adhikari S, Scarpa F, Friswell MI. Transverse vibration of single-layer graphene sheets. J Phys D Appl Phys 2011;44(20).
[139] DiBiasio CM, Cullinan MA, Culpepper ML. Difference between bending and stretching moduli of single-walled carbon nanotubes that are modeled as an elastic tube. Appl Phys Lett 2007;90(20): 203116.
[140] Huang Y, Wu J, Hwang KC. Thickness of graphene and single-wall carbon nanotubes. Phys Rev B 2006;74(24):245413.

CHAPTER

EXPERIMENTAL ASPECT

2

2.1 INTRODUCTION

Until 1991, no previous materials had been found to exhibit such a perfect combination of remarkable mechanical, thermal, chemical, and electronic properties as carbon nanotubes (CNTs). Their unprecedented physical and chemical properties have generated a series of huge activities, covering most areas of science and engineering. The rapid developments of CNT-related researches in various fields in the past two decades have confirmed their broad application prospects. This great discovery of CNTs provides not only ideal candidates for engineering applications [1] but also novel test samples for fundamental scientific research [2].

In particular, the much more superlative mechanical properties of CNTs than traditional materials have already attracted intensive attention. This perfect combination of material properties makes them ideal candidates as advanced filler materials in composites. Researchers have envisaged conductive plastics with exceedingly low percolation thresholds produced by taking full advantage of their conductivity and high aspect ratio [3]. In addition, the excellent thermal conductivity of CNTs suggests a new strategy to exploit thermally conductive composites [4]. However, among various composites research, the mechanical enhancement of plastics with CNTs reinforcing fillers probably is the current most promising application.

The first one-dimensional filler used as a reinforcing agent probably is straw which has been used to strengthen mud bricks since about 4000 BC. From the 20th century, accompanying the rapid development of materials, fibers formed from alumina, glass, boron, carbon, silicon carbide, and mineral have been used as fillers. However, the dimensions of these traditional fibers are mesoscale (diameters and lengths of several microns and millimeters, respectively). In particular, carbon fibers exhibit the most impressive mechanical properties among all the fillers, typically with a stiffness and strength of 230–725 and 1.5–4.8 GPa, respectively [5]. More recently, the importance of carbon nanofibers in composites has been progressively manifested. Carbon nanofibers, with diameters and lengths of tens of nanometers and hundreds of micrometers, respectively, have been successfully generated from the vapor phase. Such a small dimension endows them with a much higher surface area than conventional carbon fibers and thus improves the contact with composite matrices. Many literatures demonstrate that their Young's moduli are approximately 100–1000 GPa and strengths 2.5–3.5 GPa [6].

Note CNTs may be the ultimate mechanical filler material whose diameters are 1–100 nm and lengths up to millimeters [7]. CNTs possess low densities of $\sim$1.3 g/cm^3 and Young's moduli of greater than 1 TPa which are much superior to all carbon fibers [8]. However, the strengths of the strongest CNTs were measured up to 63 GPa [9], an order of magnitude larger than high strength

Mechanical Behaviors of Carbon Nanotubes. DOI: http://dx.doi.org/10.1016/B978-0-323-43137-8.00002-0

carbon fibers, which really sets them apart from the latter. Other measured strengths for CNTs are also reported, such as tensile strengths of 55 GPa [10], 45 (±7) GPa [11], 13–52 GPa [12], 11–63 GPa [9], and 22 GPa. It reveals that even the weakest CNT have a strength of more than 10 GPa [13].

2.2 PREPARATION METHODS

Various approaches have been proposed to synthesize CNTs [14]. Commonly, three main techniques are employed: arc discharge [15], laser ablation [16], and chemical vapor deposition (CVD) [17]. There are also some other synthesized methods for CNTs such as plastic pyrolysis [18], flame synthesis [19], and liquid hydrocarbon synthesis [20], but these will not be further presented since they are not capable of synthesizing the current demand of large-scale CNTs. A comparison of established techniques is given in Table 2.1.

It is found that the product purification of CNT production growth by the abovementioned synthesis techniques is a key factor on the total cost and quality [21]. Product purification usually undergoes a multistep process, mainly including the controlled oxidation of amorphous carbon and subsequent elimination of catalyst particles by using mineral acids [22]. Up to now, many purification methods for CNT products have been proposed [23–27].

2.2.1 ARC DISCHARGE AND LASER ABLATION

In this section, arc discharge and laser ablation are discussed in detail. The arc discharge technique is originated from an electric arc between two closely spaced graphite electrodes, less than 1 mm apart, in an inert gas such as helium and argon [28]. Also usually a vacuum (50 and 700 mbar) is additionally required. A higher than 3000°C temperature plasma, between the two closely spaced graphite

Table 2.1 Comparison of the Established Techniques for CNT Synthesis

Method	Arc Discharge	Laser Ablation	Chemical Vapor Deposition
Description	Arc evaporation of graphite in the presence of inert gas; CNT formed on electrodes during quenching	Vaporization of graphite target by laser; CNT formed on receiver during quenching	Decomposition of hydrocarbons over transition metal catalyst to form CNT
Operating temperature	> 3000°C	> 3000°C	> 1200°C
Operating pressure	50–7600 Torr generally under vacuum	200–750 Torr generally under vacuum	760–7600 Torr
Advantages	Good quality CNTs	Good quality CNTs; single conformation SWNT formed (10, 10)	Easy scale up; synthesis on templates possible
Disadvantages	Difficult to scale up	Difficult to scale up; expensive	Quality of CNT not as good

electrodes, is generated by a direct current (between 50 and 120 A) with a driving potential of ~30 V [28]. Under such high temperature, the graphite electrodes sublime and condense rapidly to generate CNTs in the interelectrode plasma region. However, it is always accompanied with other carbonaceous by-products. Similar to the principle of arc discharge technique, a laser source is employed to create a high temperature environment on a carbon target in order to form vaporized carbon. A rapid cooling process of the vaporized carbon by introducing a carrier gas stream, e.g., helium, finally forms CNTs along with other carbonaceous by-products [16]. Although laser ablation is capable of producing CNTs with a certain chirality (10, 10), it has not been used in any specific application.

As far as the qualities of synthesized CNTs are concerned, arc discharge or laser ablation can ensure a better quality compared to CVD. To seek the balance between cost and quality, most current commercially available CNTs are synthesized using the arc discharge technique as a compromise. However, these techniques are unable to synthesize the large-scale production of CNTs due to their inherent demand. Usually, vacuum conditions are concomitantly required in arc discharge in order to avoid the interference of unwanted ions at the high temperature. However, it is quite difficult to scale up vacuum conditions, even at a high price. Moreover, graphite targets and electrodes require continuous supplementation as the synthesis proceeds, which raises challenges in continuous operation for these techniques.

2.2.2 CHEMICAL VAPOR DEPOSITION

Essentially, CVD is a thermal dehydrogenation reaction via a transition metal catalyst, i.e., iron, nickel, or cobalt, in order to lower the reaction temperature required in order to decompose a gaseous hydrocarbon feed into carbon and hydrogen. In most aspects, it is analogous to the large-scale synthesis of hydrocarbon fuels by such a way as catalytic cracking octane. Compared to the aforementioned arc discharge and laser ablation, CVD has already been successfully scaled up to high throughputs. Many other derivative CVDs have been reported in the literature, involving fixed beds [29–33], fluidized beds [34–38], aerosols [39,40], floating catalysts [41–44], and combination methods [45–50]. All of these techniques are capable of producing freestanding CNTs. Similar to laser ablation and arc discharge, these synthesized freestanding CNTs are grown without templates. For the sake of enhancing the quality and capacity of CNTs, hybrid systems combining CVD, arc discharge, and laser ablation have been proposed compared to traditional CVD techniques [45–50]. As expected, these hybrid techniques have achieved the desired results but have led to a complexity to illustrate the growth mechanism of CNTs. Several new variables, involving laser power, type of plasma, and ion strength, will be introduced when using these hybrid techniques and it cannot essentially solve the scale-up problems. The universality of CVD technique enables CNTs to be realized in a wide range of end-use applications [51], such as field emission devices, while arc discharge and laser ablation techniques are not easy to achieve.

The process of CVD synthesized technique contains two steps: catalyst preparation and actual reaction. Catalysts are usually prepared on a substrate and there are commonly four approaches, sol–gel [52], impregnation [53], metallo-organic CVD [54], and coprecipitation methods [55]. Without the assistance of a substrate, CNTs can also be formed in situ by such a way as the decomposition of volatile metallo-organic compounds within the reactor, namely, floating catalyst method [56].

2.2.3 CNTs GROWTH MECHANISM

In spite of particular concern paid to the growth mechanism of CNTs, the underlying growth mechanism is still not well understood. Observed and discussed from various synthesis techniques, it can be summarized that three critical elements, including a carbon source, a heat source to achieve the desired operating temperature, and a transition metal catalyst, are necessary. Analyzed from the observations reported in previous literatures, probably a single growth mechanism is most rational.

Single-walled (SW) CNTs are synthesized at a temperature of ~1500K using a quenching technique [57,58] in the arc discharge and laser ablation techniques. The generated morphologies of SWCNTs by the established techniques are also identical. Using CVD method, bundle SWCNTs were observed with averaging diameters of 20 nm, and individual SWCNTs with diameters of ~1.4 nm [59]. Note it is necessary to introduce the transition metal in order to induce SWCNT growths. These abovementioned similarities confirm the reasonability of the hypothesis for the similar CNTs growth mechanism in all the synthesized approaches.

A most agreed hypothesis is similar to the diffusion–precipitation mechanism in regard to the growth of carbon fibers which was proposed by Baker et al. [60,61]. They suggest that a hydrocarbon feedstock is initially decomposed into carbon atoms and hydrogen molecules on the active transition metal surface, and then followed by a diffusion of carbon into the metal particles until the solution becomes saturated. The supersaturated solution finally gives rise to precipitation and growth of solid carbon on the metal surface [61].

According to the placement of the catalyst particle at the tip of CNTs or the base, there are two growth patterns, namely, tip growth [60] and base growth [57], which are dependent on whether bulk diffusion or surface diffusion occupies the predominant position. This also relies on the adhesive strength of the catalyst adhesive to substrate. It is found that CNTs tend to grow by the means of a root-growth mechanism owing to strongly interacting metal catalyst and substrate pairs such as Fe and Al2O3 possessing a high bonding energy between them. However, the detachment of Fe from Al_2O_3 has also been observed [36], implying an effect of other factors, such as the reaction atmosphere, on the growth mechanism of CNTs.

There are still several remarkable observations that cannot be addressed by this growth mechanism, including the endothermic decomposition of methane and "Y-" carbon fibers [61]. For further details refer to Alvarez et al. [62], Gavillet and coworkers [58,63], Jost et al. [64], and Teo et al. [51].

2.2.4 CNT QUALITY

There is no specific definition for CNT quality in the literature. Although CNT characterization techniques such as transmission electron microscopy (TEM), thermogravimetric analysis (TGA), and Raman spectroscopy have been suggested for standardization, an international standard to assess the CNT quality is still absent. One of the most popular and widely used qualifications for CNT quality is crystallinity, describing the degree of graphitization, or defect density of CNTs. For example, Fig. 2.1 depicts high-resolution transmission electron micrographs of multiwalled (MW) CNTs synthesized using arc discharge and fluidized-bed CVD, respectively.

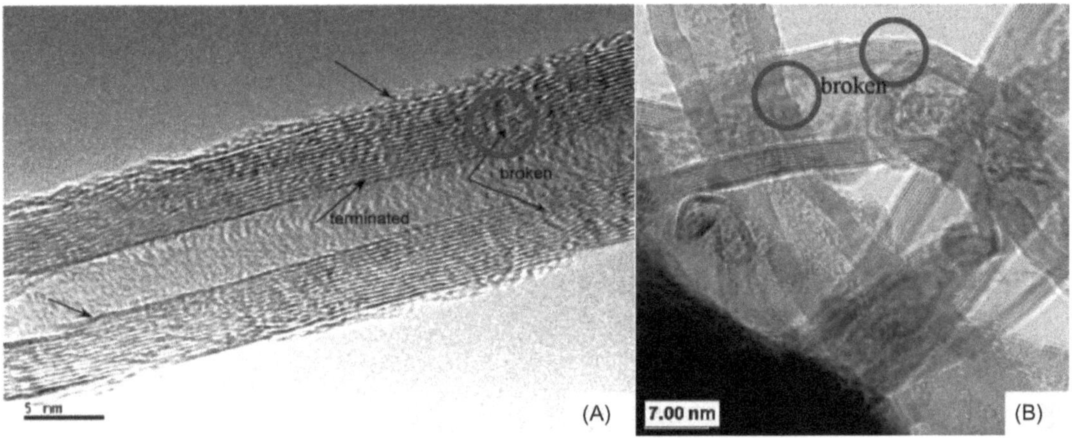

FIGURE 2.1

High-resolution transmission electron micrograph of typical CNT synthesized using (A) arc discharge [65] and (B) fluidized-bed CVD [66]. Defects are observed in the form of terminated growth areas and breakages.

As observed from Fig. 2.1A and B, the higher density of terminated growth and broken walls suggests a better graphitization and higher quality of CNTs using arc discharge synthesis technique rather than CVD. Although the abovementioned example cannot completely mirror the synthesized quality of CNTs by CVD, it can provide an illustration of this usual method to distinguish CNT quality to some extent. Recently, Lacerda et al. [67] reported that high-quality CNTs have been synthesized using CVD.

Purity, thermal stability, homogeneity, and dispersibility are adopted as the four critical elements to measure the quality of SWCNTs [68]. This protocol is not only restricted to the detection of SWCNTs but also can be extended to the measurement of MWCNTs. TGA is an important analytical tool which can quantitatively measure purity, thermal stability, and homogeneity. Several factors, including the defect density and the amount of catalyst particles, affect thermal stability of CNTs and their contribution may be further understood by the CNT oxidation mechanism.

2.3 TESTING TECHNOLOGIES

Many techniques are adopted to characterize the structure and properties of as-prepared SWCNTs [69] in spite of the challenge of quantification [70]. Recently, a related set of techniques has been applied to the study of covalent functionalization [71]. Some of these techniques are impressible to the selectivity of a reaction/separation process. Spectroscopic approaches are capable of studies of bulk samples to give population statistics. Although the characterization of individual SWCNTs is costly and tedious, it is often most relevant to device application. These tube-by-tube techniques can provide the gold standard for identifying (n, m) species but are highly challenging regarding expertise. Any of the techniques discussed below are insufficient to convincingly prove SWCNT selectivity. In reality, the relationships among all the techniques are competitive and mutually supportive as well offering full characterizations for samples. Optical techniques are commonly introduced to determine the (*n*, *m*) specificity of a covalent functionalization reaction, including Raman, UV-vis-nIR, and photoluminescence (PL) spectroscopies. These optical techniques play significant roles in characterizing SWCNT metrologics. The measurements by these techniques are simple, quick, nondestructive, and can be directly conducted at ambient conditions.

2.3.1 RAMAN SPECTROSCOPY

Raman spectroscopy can provide large amounts of information concerning the structures and properties of CNTs (refer to Dresselhaus et al. [72]). Both isolated and bundled CNTs, either in solution or solid phase, can be characterized by Raman spectroscopy. Typical Raman spectra display a series of different peaks which allow semiquantitative evaluation of the structures and properties of CNTs, including purity, diameter, metallicity, crystallinity, and chirality. Strong resonance Raman scattering can be clearly observed when the laser excitation energy (E_{laser}) satisfies the electronic state transitions between the vHS of specific (*n*, *m*) tubes. Radial breathing mode (RBM) peaks, corresponding to the concerted atomic vibration of the carbon atoms in the radial direction, generally appear at low wavenumber $<\sim 500\ cm^{-1}$. Based on RBMs, a simple equation, namely, $\omega_{RBM} = (c_1/d_t) + c_2$, can be applied to the determination of diameters of CNTs, in which c_1 and c_2 are experimentally determined constants. Typically, values of c_1 and c_2 are given as 223.5 nm/cm and $12.5\ cm^{-1}$ for HiPco SWCNTs [73], and 218.7 nm/cm and $15.3\ cm^{-1}$ for CoMoCAT SWCNTs [74], respectively. The combination of Kataura plot [75] of the allowed vHS transitions (E_{ii}) versus CNT diameter with experimental Raman spectra can be used to identify the types of SWCNT samples, using multiple laser excitations (Fig. 2.2A) [76]. By assigning the calculated transitions to experimentally determined S_{11}, S_{22}, and M_{11} values which were obtained by resonance Raman spectra, the Kataura plot can be obtained. Therefore, types of SWCNTs and even higher energy transitions can be distinguished by the variations of the plot. Since it is ambiguous over the dividing line between m-/sc-SWNT transitions, confidence in (*n*, *m*) indexing can be effectively modified by using a combination of these plots.

Several laser excitations are requested to match the nonlinearity of the resonance Raman scattering effect. For example, in order to characterize helicities and diameters for a mixture of SWCNT, three laser excitations are necessary, while many more lasers are required in the case of full characterization. Different from the PL excitation versus emission (PLE) map, the resonance Raman map [76] (Fig. 2.2A) includes m-SWCNTs although quantification of SWCNT species is less possible. Recently, Heeg et al. [77] discussed the differences of Raman and PL measurements in quantifying (*n*, *m*) tube type for three specific SWCNTs. It is observed that each (*n*, *m*) tube type inherently possesses a particular Raman resonance profile which has a strong influence on apparent abundances (Fig. 2.3). It is clearly evident that SWCNT resonance profiles react automatically to their local environment, such as chemical functionalization specifically, owing to electron transfer doping effects. In order to clearly illustrate this issue, take the bonded oxygen atoms on the side-wall as an example, resonance peaks are found to shift by as much as $7-8\ cm^{-1}$ [78]. What's more, the interband transitions of SWCNTs can be altered by the changes in sample morphology according to debundling and reaggregation [79].

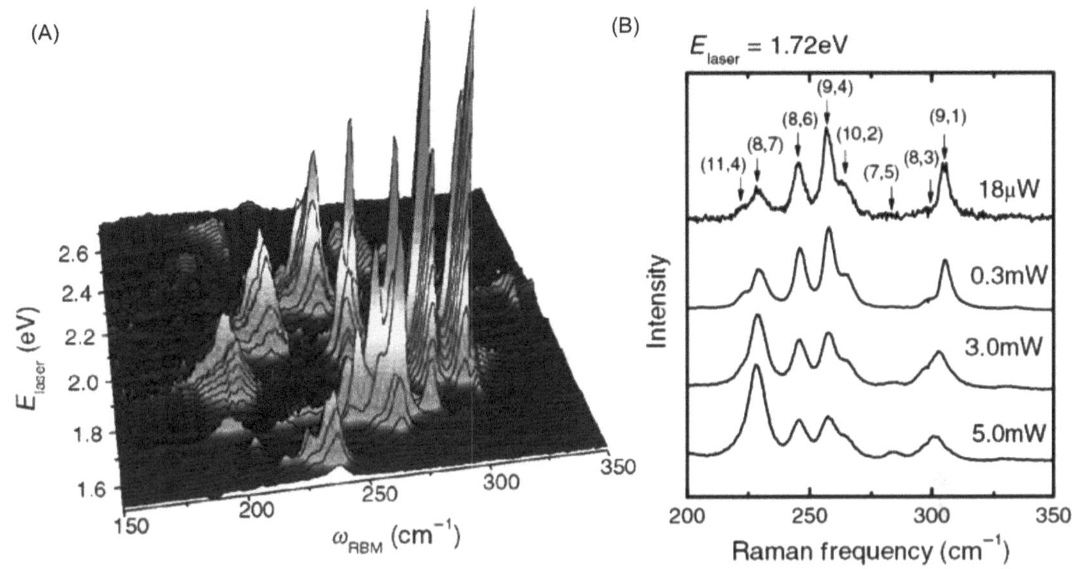

FIGURE 2.2

(A) RBM Raman measurements of HiPco SWCNTs dispersed in SDS aqueous solution, measured with 76 different laser lines. The nonresonance Raman spectrum from a CCl_4 solution is acquired after each RBM measurement for calibration of the spectral intensities and to check the frequency calibration [76]. (B) The dependence of the RBM spectra for HiPco SWNTs in bundles on the laser excitation power for a fixed Elaser (1.72 eV) [76].

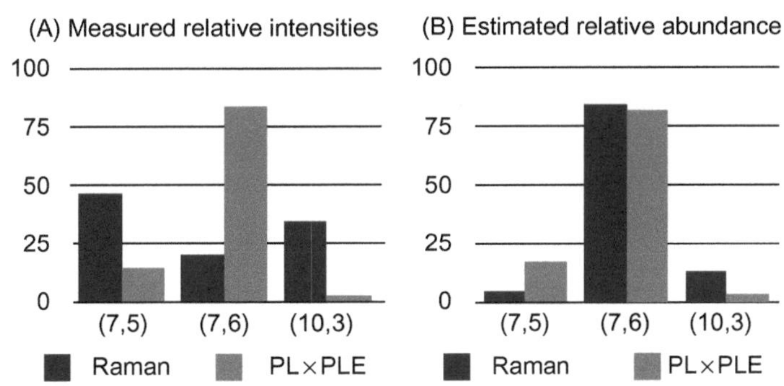

FIGURE 2.3

(A) Relative signal intensities obtained by Raman (*blue* (black in print versions), scaling to full resonance included) and PL (*red* (gray in print versions) measurements). (B) Estimated abundance including corrections [77].

The G band which is related to structural order is a tangential mode and for all graphitic materials it is observed in the frequency at a range of 1520–1630 cm^{-1}. This feature implies that there are well-established dependencies on both diameter and electronic type of SWCNTs. It reveals that there are two intense peaks which are the results of the symmetry breaking of tangential vibration by rolling the graphene sheet, G^+ and G^-. Changes in electronic character can be identified by the G^- peak. On the basis of a Breit–Wigner–Fano (BWF) form, sc-SWCNTs and m-SWCNTs possess a Lorentzian G^- lineshape and a broadened G^- lineshape, respectively [72]. One double resonance feature, the D band (~1350 cm^{-1}), is dispersive and is closely related to the reduction in the symmetry of SWCNTs in the outcome of functionalization or the existence of defects. The intensity of the D peak involving G peak (I_D/I_G) is widely used to evaluate the defect concentration and functionalization in SWCNTs. It is noteworthy that the I_D/I_G ratio remains unchanged at high degrees of functionalization which is in accordance with the loss of resonance enhancement because of functionalization [80]. The G′ peak (~2700 cm^{-1}), known as another double resonance feature, is also sensitive to the diameter and helicity of a SWCNT. In general, a single peak is observed for most graphitic materials. Nevertheless, a two-peak feature regarding the specific cutting lines (1D wave-vectors) of the Brillouin zone appears for isolated SWCNTs and this feature can be used to discriminate differences in SWCNT helicity [72].

Usually, a strong scatterer such as silicon is adopted to calibrate Raman spectrometers. However, well-established SWCNTs of one specific (n, m) tube type may be hopefully used as a good reference object in the future. In order to ensure credible statistical data, spectra should be gathered at plenty of independent locations on account of the heterogeneity of SWCNT samples. It's worth noting that the effect of laser power must be checked to ensure there is no heating or burning of the sample (Fig. 2.2B). To allow comparison, it is essential to prepare samples uniformly since excitation window, excitation energy, and degree of heating are influenced by the environment and the degree of nanotube aggregation. Sample preparation is typically simple [70]. For instance, measurements can be taken of raw, dry SWCNT powders, as a pellet, or dissolved into solution, and it is also possible to measure the Raman spectra of isolated individual SWCNTs.

2.3.2 UV-VIS-NIR ABSORPTION SPECTROSCOPY

The absorbance of light of individualized SWNTs can be measured by UV-vis-nIR absorption spectroscopy. Unlike other characterization techniques, UV-vis-nIR absorption spectroscopy can characterize the optical absorption of SWCNT suspensions with identifying information in regard to every CNT emerged in a sample. However, measurement issues emerge again, arising from the choice of sample preparation. Furthermore, the forms of CNTs bundles or ropes have a diminished effect on the absorption signals corresponding to the specific vHS. This principle is also suitable for CNTs with excessive defects probably caused by covalent modification. Additionally, absorption signals are also strongly disturbed by the charge-transfer, leading to a Fermi level shift in the SWCNTs [81]. As the diameters of CNTs increase, it is found that (n, m) resolution becomes more and more difficult owing to the fundamental peak width. Nevertheless, for typical diameter distributions, regions in which sc- (S_{ii}) and m- (M_{ii}) transitions take place can be detected. Take HiPco SWNTs as an example, M_{11} transitions can be found at ~400–600 nm; S_{11} and S_{22} transitions are viewed at ~900–1600 and ~600–900 nm, respectively (Fig. 2.4).

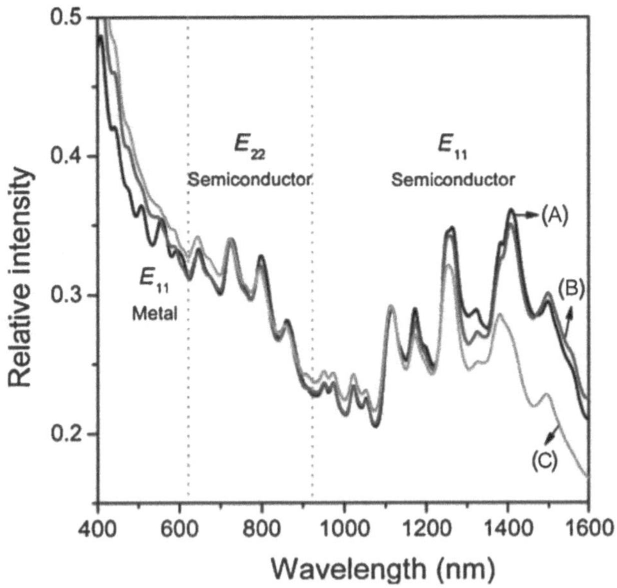

FIGURE 2.4

UV-vis-nIR absorption spectra of samples with different selectivities and reaction extents for metallic SWNT used in free solution electrophoresis: (A) control, (B) reaction 1, and (C) reaction 2 [82].

For UV-vis-nIR spectra, either thin films or solutions of SWCNT samples can be adopted. Thin film spectroscopy [70] is quite useful since SWCNTs are extremely difficult to disperse. In particular the dispersibility of long SWCNTs and purified SWCNTs which have been annealed in vacuum at high temperature becomes lower. There are many approaches, involving drop-casting, spin-coating, spray-coating, and using a draw-down bar, that can be adopted to synthesize thin films. However, these methods generally have poor reproducibility of thickness and encounter severe challenges in generating homogeneous nonscattering samples; this is more easily achieved in solution-phase by using ultracentrifugation. SWCNT solutions are commonly prepared at an initial concentration of around 1 mg/mL in D_2O solutions with 0.5–2 wt% surfactant. D_2O is employed to measure the absorption spectra of SWCNTs, attributing to the red-shifted peaks, instead of H_2O since H_2O peaks appear at 760 and 1000 nm [83]. Surfactant systems may be selective in themselves which results in the misinterpretation of spectra. Commonly, the best choice of surfactants [84] for SWCNT dissolution are ionic. Currently, sodium dodecylbenzenesulfonate (SDBS) is popularly selected because of its well-resolved spectral features. According to the enhancement of steric stabilization, nonionic surfactants, such as Pluronic, Triton-X, polyvinylpyrrolidone (PVP), can be modified with increased molecular weight. Organic solvent systems such as dimethylformamide (DMF), *N*-methylpyrrolidone (NMP), or cyclohexylpyrrolidone (CHP) [85] are alternatives. For the sake of high-quality dispersions, an ultrasonic processor with cup-horn or tapered microtip is employed to initially disperse the SWCNTs and then followed by ultracentrifugation process with rcf > 100,000 g, 1 h. Usually, a concentration of 0.001–0.2 mg/mL SWCNT can be achieved following centrifugation, and scanned in a standard 10 mm path length quartz cuvette. Absorbance values of these concentrations should be in the range of 0.033–6.53, on the basis of the average extinction coefficient of 3264 mL/mg/m at 660 nm for HiPco SWCNTs in NMP [85]. Extinction coefficients should principally be a function of the type of SWCNT as well as wavelength, though rather analogical average values have been presented for various polydispersed systems [86]. Note that the degrees of dispersability of SWCNTs are largely dependent on their original type and subsequent treatments; sample preparation methods may need to be adapted and are difficult to compare.

2.3.3 PL SPECTROSCOPY

In the study of SWCNTs, PL spectroscopy is widely used. However, it is difficult for PL spectroscopy to characterize m-SWCNTs which do not fluoresce. The overlapped spectral features in simple emission or absorption spectra are clearly separated in 2D excitation-emission spectra so that it is quite suitable to identify the sc-SWCNTs [87].

Excitation happens between the conduction and valence bands of sc-SWCNTs (S_{ii}) and the emission of S_{11} fluorescence can be detected. In general, SWCNT samples are also prefabricated as thin films or in D_2O solutions, which contains 0.5–2 wt% surfactant, as absorption spectroscopy, however, low SWCNT concentrations are measured (<0.020–0.025 mg/mL), so as to minimize fluorescence quenching from SWCNT bundles and m-SWCNTs. Similar to UV-vis-nIR, ultracentrifugation makes an essential effect on these high-quality dispersions and the selection of surfactant system must be prudent. A PL contour map is drawn according to the measurement of emission intensity on the dependence of both excitation wavelength and emission wavelength (Fig. 2.5) [87]. A nonemissive region observed lying between the regions of the S_{22} and S_{33} branches is attributed to m-SWCNTs. PL spectroscopy can also be used to detect defects and sites of covalent modification at CNT side-walls, leading to fluorescence quenching attributing to the relatively long exciton diffusion length, with high sensitivity [89].

2.3.4 OTHER CHARACTERIZATION TECHNIQUES

In addition, there are some other relevant bulk characterization techniques, including X-ray photoelectron spectroscopy (XPS), solid-state NMR [90], and electron paramagnetic resonance spectroscopy (EPR). Nevertheless, these techniques are less widely-used for SWCNTs since they often encounter technical challenges, and difficult interpretation. Moreover, larger quantities of material are usually required. XPS is adopted to identify the selectivity by changes in the valence band features. Solid-state NMR can discriminate distinct signals between m- and sc-SWCNTs, while EPR is capable of distinguishing electronic types.

All sorts of microscopies, directly imaging and identifying lengths, diameters, helicities, defects, purities of SWCNTs, are applied to direct tube-by-tube characterization [69,91–93]. Aberration-corrected TEM measures (n, m) distributions and identifies defects. Voltage-contrast SEM (VC-SEM) images electronic conduction pathways through isolated SWCNTs and networks, and combined AFM-Raman characterizes individual SWCNTs. STM and spectroscopy atomically determines the electronic structure of SWCNTs and identifies chiral angle, diameter, electronic DOS, and band gap energies, while selected-area electron diffraction (SAED) is capable of categorically identifying SWCNT helicity. Other commonly reported characterization techniques involve

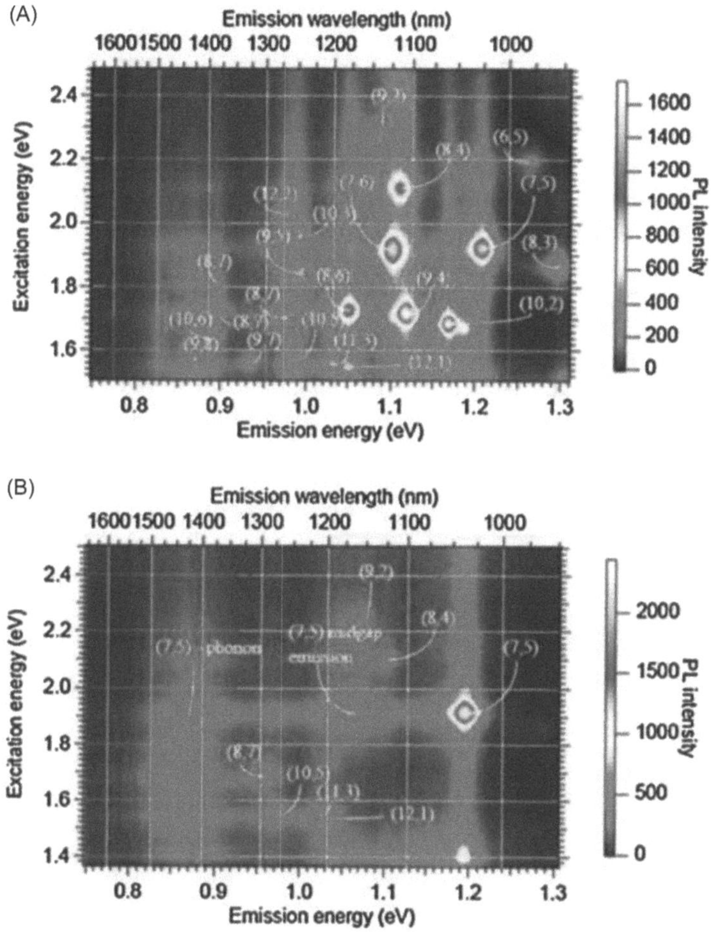

FIGURE 2.5

Excitation versus emission PL contour maps of (A) purified HiPco SWNTs suspended by sodium cholate and (B) the same suspension added to an equal part of H_2O_2 after 7 days. Many new phonon-assisted absorption and emission features are present in panel B, as discussed in the text and the Supporting Information [88].

performance-based monitoring of I–V characteristics of SWCNT network, film, or single-tube devices [94].

2.4 MECHANICAL PROPERTIES OF CNTs

As a derived structure of graphite, CNTs are expected to have the same remarkable mechanical properties as graphite. Graphite has been well-understood with an in-plane modulus of 1.06 TPa

[95]. Since the tensile strength of graphite is not easy to measure, Perepelkin estimated it from a viewpoint of the properties of C–C bonds and found it to be as high as 130 GPa [96]. Bacon reported a yield strength of 20 GPa by measuring the fabricated graphite whiskers [97].

For a long time, computer simulation was employed as an important tool to investigate the mechanical properties of CNTs due to the acute shortage of CNT products supplied for experimental measurements. Early in 1993, Overney et al. [98] employed ab initio local density calculations, determining the parameters in a Keating potential, to calculate the rigidity of short SWCNT and found a Young's modulus of 1500 GPa which is close to that of graphite. Subsequently, a range of papers have focused on the prediction of Young's modulus of CNTs and Lu et al. [99] obtained a value of close to 1 TPa independent of chirality and diameter.

As a matter of fact, MWCNTs synthesized by arc discharge were first selected as the samples and the early mechanical measurements were conducted by TEM since only small amounts were available. Treacy et al. [100] observed intrinsic thermal vibrations of CNTs by TEM. According to the measured amplitude, the Young's moduli can be determined and it is calculated to be a value of 0.41–4.15 TPa for a number of CNTs. Afterwards, electromechanical resonant vibrations were induced by Poncharal et al. [101] giving rise to moduli values in the range from 0.7 to 1.3 TPa. Moreover, what's exciting was the reversible bending of MWCNT, of which radii as low as ~25 nm were observed in Falvo et al.'s [102] experiments, with an indication of unprecedented flexibility.

The first direct measurement of the elastic properties of CNTs was conducted by Wong et al. in 1997 [8]. The stiffness constant of arc-MWCNTs pinned at one end was directly measured by utilizing an AFM and an average Young's modulus of 1.28 TPa was obtained. Moreover, the strength measurements were also first successfully carried out, giving an average bending strength of 14 GPa. The bending test of an arc-MWCNT pinned at each end over a hole was realized by utilizing AFM and an average modulus of 810 GPa was obtained [103]. In 2000, Yu et al. [9] managed to measure stress–strain relationships of individual arc-MWCNTs inside an electron microscope (Fig. 2.6). The Young's modulus was determined in a range of 0.27–0.95 TPa for a variety of MWCNTs and their experiments demonstrated that fracture of MWCNT occurred at a tensile strain of up to 12% with a corresponding tensile strength and toughness being in the range of 11–63 GPa and ~1240 J/g, respectively. It was found that the MWCNTs broke in the outermost layer by a "sword-in-sheath" breaking mechanism.

Since they are quite difficult to undertake, measurements on SWCNT were running for a long time. Salvetat et al. [104] first conducted mechanical measurements by AFM, observing a tensile modulus of ~1 TPa for SWCNT bundles with small diameters by bending tests. However, the elastic properties of SWCNT bundles with larger diameters were largely affected by shear slippage of individual CNTs. The same method which was utilized for the investigation of MWCNTs was also capable of measuring the tensile properties of SWCNT bundles [12]. Young's moduli were found to be ranging from 0.32 to 1.47 TPa and strengths were between 10 and 52 GPa. Failure occurred at a maximum strain of 5.3% with a corresponding toughness of ~770 J/g. It was also observed that failure only occurred on the perimeter of the SWCNT bundle with the rest of the tubes slipping apart. In actuality, intertube slippage within bundles has a large reduced effect on their mechanical properties. The low shear modulus implies a far less effective moduli and strengths for bundles than those of individual SWCNTs. As mentioned above, it is extremely difficult to de-bundle SWCNT. Forro and coworkers [105,106] reported a fusion of SWCNT in bundles by electron irradiation. The SWCNT bundle bending modulus can be increased to 750 GPa by fine-tuning the dose and irradiation energy, which is quite close to that of an individual SWCNT.

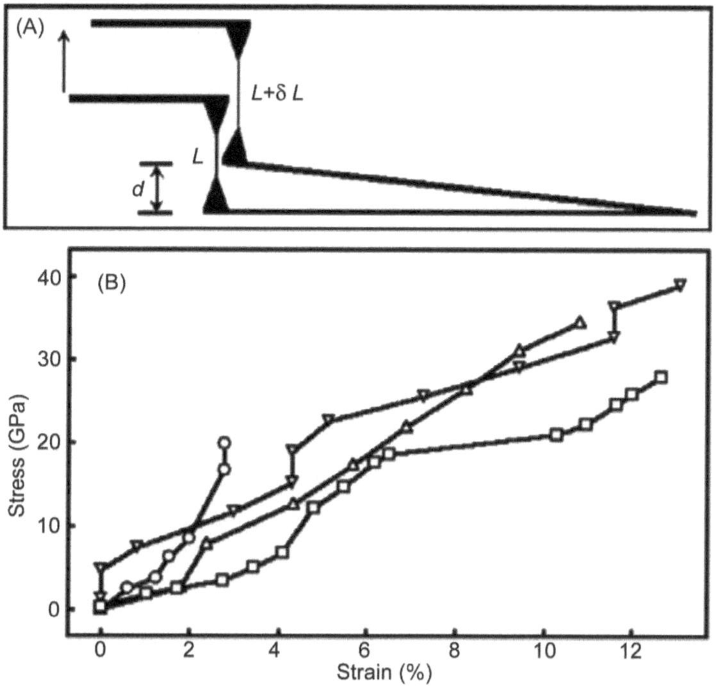

FIGURE 2.6

Stress–strain curves for individual MWCNTs [9].

Comparatively high values of modulus and strength of ~1 TPa and tens of GPa, respectively, have been achieved by measuring both the high-quality SWCNT and arc discharge MWCNT. However as discussed before, MWCNTs synthesized by CVD are forecasted to exhibit significantly reduced properties. Salvetat et al. [103] first measured CVD-MWCNTs using the AFM technique (Fig. 2.7) and found a Young's modulus of between 12 and 50 GPa. Sooner after, Xie et al. [13] acquired a stress–strain curve from measurements of CVD-MWCNT bundles. A modulus of 0.45 TPa and a tensile strength of ~4 GPa were extracted from the stress–strain curve. The comparison between the modulus of MWCNT and that of arc-MWCNT strongly confirms that the mechanical property is extremely sensitive to defect concentration and tube type.

2.5 APPLICATION PROSPECT AND RESEARCHING SIGNIFICANCE

CNTs, as one of the most important nanomaterials in the field of nanotechnology, offer an extremely wide range of promising applications including electronics, optical communications, and biological systems. Due to the rapid developments of nanotechnology, many possible applications have been explored and more and more potential applications are being proposed. Some representative examples are selected to illustrate the possibilities of CNTs in device fabrication and

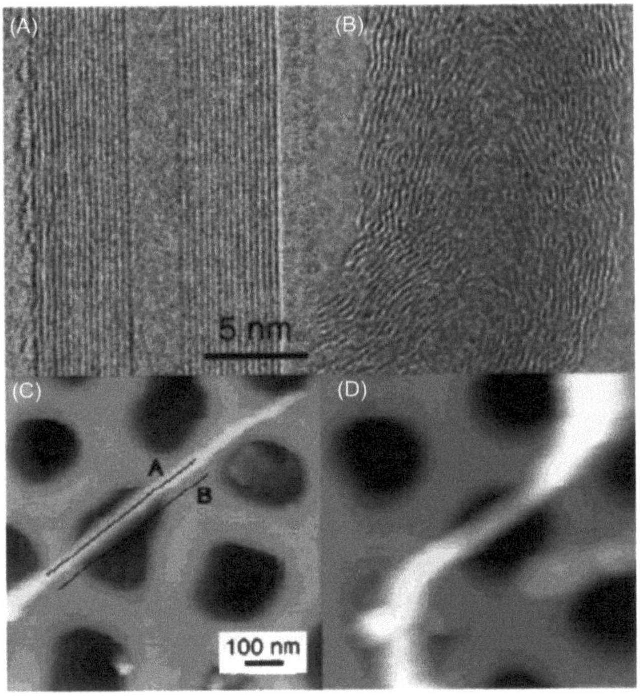

FIGURE 2.7

TEM images of (A) an arc-MWCNT, (B) a CVD-MWCNT. AFM images of (C) arc-MWCNT, and (D) CVD-MWCNT lying across a pore [103].

application since it is apparently impossible to cover all the as-synthesized and new predicted devices and applications.

Compared with the traditional materials, several prominent features listed as follows are the basis of applications of CNTs: (1) the peculiar physical properties since CNTs approach nanometer scale; (2) the specific surface area; and (3) the small size that supplies extra possibilities for manipulation and space for accommodation of multiple functionalities. Some typical examples are given to illustrate the explored applications of CNTs.

2.5.1 COMPOSITE MATERIALS

On the basis of the high aspect ratio, CNTs were first adopted as electrically conductive fillers in plastics to form a percolation network at concentrations as low as 0.01 wt%. Bauhofer and Kavocs [107] reported that the conductivities of randomly distributed 10 wt% CNT-polymer composites could reach up to 10,000 Sm^{-1}. In their application in the automotive industry, conductive CNT plastics have suggested a method of electrostatic-assisted painting for mirror housings, fuel lines, and filters. Besides, CNTs have also been applied in the microelectronics industry, i.e., as electromagnetic interference shielding packages and wafer carriers.

The stiffness, strength, and toughness of polymers can be markedly enhanced by mixing CNT powders [108]. For example, the stiffness and fracture toughness of ~1 wt% CNT-epoxy resin can be enhanced by 6% and 23%, respectively, with no influence on other mechanical properties [109]. Several elements, including CNT diameter, aspect ratio, alignment, dispersion, and interfacial interaction with the matrix, are critical to these enhancements. In actuality, many CNT manufacturers sell premixed resins with CNT loadings from 0.1 to 20 wt%. It is noteworthy that material damping increases accompanied with engineering nanoscale stick-slip among MWCNTs and CNT-polymer contacts [110]. Such a result can largely enhance sporting goods, e.g., tennis racquets, baseball bats, and bicycle frames.

CNT resins are also helpful for enhancing fiber composites [108,111], e.g., strong and lightweight wind turbine blades and hulls for maritime security boats made by using carbon fiber composite with CNT-enhanced resin. In addition, CNTs can also be used as additives in the organic precursors to develop carbon fibers. Since CNTs have an influence on the arrangement of carbon in the pyrolyzed fiber, it is possible to fabricate 1-μm diameter carbon fibers with over 35% increasing strength (4.5 GPa) and stiffness (463 GPa) compared with non-CNTs samples [112].

It is still a challenge to orderly organize CNTs at larger scales. Therefore, hierarchical fiber composites have been synthesized by growing aligned CNTs clusters onto glass, SiC, alumina, and carbon fibers [108,113,114], namely, fuzzy fibers. Compared with control specimens, it reveals that crack-opening and in-plane shear interlaminar toughness of fuzzy CNT-SiC fabric disseminated with epoxy are improved by 348% and 54%, respectively [113], while in-plane shear interlaminar toughness of CNT-alumina fabric is improved by 69% [114]. Some multifunctional applications such as lightning-strike protection, deicing, and structural health monitoring for aircraft are being explored [108,113].

In addition to the abovementioned polymer composites, adding small amounts of CNTs to metals can also increase tensile strength and modulus [115] which is suitable to be applied in aerospace and automotive structures. Commercial Al-MWCNT composites have an equivalent strength to stainless steel (0.7−1 GPa) and Al-Li alloys with only a fraction of density of steel (2.6 g/cm^3) but at a lower cost.

Lastly, CNTs are also being explored to be applied in the field of flame-retardant composites [116]. These CNT flame-retardant additives are commercially attractive as a candidate for conventional halogenated flame retardants which are environmentally hazardous and thus strictly limited.

2.5.2 COATINGS AND FILMS

CNTs are proposed as novel multifunctional coating materials with the assistance of the dispersion, functionalization, and large-area deposition techniques. For example, MWCNT-containing paints can prevent biofouling including algae and barnacles from depositing on ship hulls [117]. They are probable candidates to replace environmentally hazardous biocide-containing paints. CNTs possess superior mechanical properties and thus their incorporation in anticorrosion coatings for ship hulls can distinctly enhance coating stiffness and strength. Meanwhile, they can provide an electric pathway for cathodic protection.

Indium tin oxide (ITO) is commonly used in transparent conducting films. Nevertheless, due to the scarcity of indium, ITO is becoming more and more expensive which has led to a continuous effort on the preparation of CNT-based transparent conducting films in order to satisfy the growing demand for displays, touch-screen devices, and photovoltaics [118]. Besides, ITO is brittle and thus encounters a challenge for manufacturing flexible displays, while CNT transparent conductors have a prominent flexibility. Furthermore, CNT transparent conductors can be simply deposited from solution and patterned by cost-effective nonlithographic methods. Recently, SWCNT films with 90% transparency and a sheet resistivity of 100 ohm per square have been commercialized. Nevertheless, this surface resistivity is still substantially higher than that of equally transparent, optimally doped ITO coatings [119].

2.5.3 MICROELECTRONICS

According to the low electron scattering and bandgap, SWCNTs are expected to be used as transistors. Moreover, high-quality SWCNTs have aroused intensive attention since they are compatible with field effect transistor (FET) architectures and high-k dielectrics [120,121]. Since the first invention of the CNT transistor in 1998 [122], many researches have been conducted to exploit its potential application. SWCNT tunneling FET with a subthreshold swing of <60 mV/decade in 2004 [121,123] and later in 2007 CNT-based radios [124] were reported. What's more, a noteworthy point is the report of sub-10-nm channel lengths long SWCNT FETs with a normalized current density of 2.41 mA/μm at 0.5 V which is much higher than that of silicon-based devices [125].

In commercial manufacturing, it is still a challenge to accurately control the diameter, chirality, density, and placement, especially of large areas which are closely related to the electrical properties of individual SWCNTs, when the size approaches the nanometer scale. Correspondingly, devices such as transistors comprising patterned films with SWCNT arrays are more currently practical. The introduction of CNT arrays into the transistor can improve output current and neutralize the influence of defects and chirality so that the device can be fabricated uniformly and reproducibly [120]. Besides, the recent developed methods for precise high-density CNT film, enabling semiconductor fabrication of over 10,000 CNT devices in a single chip, provided the support for CNT arrays devices [126]. Cao et al. [127] reported that the transistor manufactured by using horizontally aligned CNT arrays achieved a mobility of 80 $cm^2/V/s$, a subthreshold slope of 140 mV/decade, and on/off ratio up to 10^5.

It demonstrates that CNT thin-film possesses a higher mobility than amorphous silicon [56] and can even be deposited at low temperature and nonvacuum conditions. Therefore, CNT thin-film transistors (TFTs) are particularly promising for driving organic light-emitting diode (OLED) displays. Sun et al. [128] recently reported that the mobility and on/off ratio of flexible CNT TFTs are 35 $cm^2/V/s$ and 6×10^6, respectively. A low voltage can provide sufficient current output for vertical CNT FET driving OLEDs [129], enabling red-green-blue emission by the OLED on the basis of a transparent CNT network. There also have been some other promising commercial developments of CNT electronics, such as low-cost printing of TFTs [130] and RFID tags [131].

2.5.4 ENERGY STORAGE AND ENVIRONMENT

MWCNTs have also been started to be studied for their potential application in lithium ion batteries for notebook computers and mobile phones [132,133]. These batteries fixed a small amount of MWCNT powder, such as 1 wt%, with active materials and a polymer binder can increase electrical connectivity and mechanical integrity so that the rate capability and cycle life can be effectively improved [132,134,135].

Gravimetric energy storage and power densities for unpackaged batteries and supercapacitors have been reported in the literature. It is extremely difficult to assess how gravimetric performance metrics depend on those for packaged cells [136,137] due to the frequent use of low areal densities for active materials. However, in order to realize high performance based on total cell weight or volume, high areal energy storage and power densities are required. In recent years, a remarkable performance of packaged cells has been realized for supercapacitors deploying forest-grown SWCNTs with no binder or additive [138]. It is found that a 40-F supercapacitor, with a maximum voltage of 3.5 V, possesses an energy density of 16 Wh/kg and a power density of 10 kW/kg. By conducting accelerated tests with a temperature up to 105°C, a 16-year lifetime was predicted.

CNTs may be a promising catalyst support candidate of carbon black in fuel cells. Compared with carbon black fuel cells, the usage of Pt can be reduced by 60% in CNT fuel cells [139]. In particular, if doped CNTs are adopted, maybe no Pt is required in the fuel cells [140,141]. Besides, continuous efforts are made to take full advantage of the properties of CNTs in order to reduce undesired carrier recombination and enhance resistance to photooxidation in the organic solar cells [142]. In the long run, transparent SWCNT electrodes and CNT-Si heterojunctions may achieve a great contribution to the future of photovoltaic technologies [143].

Lastly, CNT sheets can be tangled to provide mechanically and electrochemically robust networks with controlled nanoscale porosity and thus can be applied in water purification. A major commercial success is the portable filter which employs CNT meshes to purify contaminated drinking water. Moreover, permeable membranes consisting of aligned encapsulated CNTs with open ends allow flow through the interior of the hollow tubes. This characteristic gives rise to unprecedented low flow resistance for both gases and liquids and thus enhances the permeability [144]. Therefore, this allows lower energy costs for water desalination by this new kind of CNT-meshed permeable membrane in comparison to commercial polycarbonate membranes. However, it is necessary to select SWCNTs with quite small diameter in order to reject salt at seawater concentrations [145].

2.5.5 BIOTECHNOLOGY

In the field of biotechnology, there has been intensive interest in CNTs due to their dimensional and chemical compatibility with biomolecules such as DNA and proteins. Besides the CNT biosensors and medical devices, CNTs allow fluorescent [146] and photoacoustic imaging [147], and are capable of realizing localized heating using near-infrared radiation [148].

Electrical impedance [149] and optical properties [150] of SWCNT biosensors, typically affected by the adsorbate on the CNT surface, can be largely altered to adapt to the surrounding environment. Low detection limits and high selectivity are closely related to the design of the CNT surface, such as functional groups and coatings [149], and appropriate sensor design, such as field effects, capacitance, Raman spectral shifts, and PL [151,152]. Other application of CNTs on medical devices involve inkjet-printed test strips for estrogen and progesterone detection, microarrays for DNA and protein detection, and sensors for NO_2 and cardiac troponin [153]. Similarly, CNT sensors have also been used to detect gas and toxins in related fields, such as food industry, military, and environmental applications [151,154]. In the in vivo applications, CNTs can bind their tips to the receptors and thus be internalized by cells [155]. This behavior enables CNT transport of molecular cargo [156].

For instance, in the field of cancer, CNTs can load a cancer drug doxorubicin by up to 60 wt% [157] and molecular cargo can be released by using near-infrared radiation. However, the retention of CNTs within the body and possible undesirable accumulation are still disturbing [158]. For example, Poland et al. [159] reported that asbestos-like pathogenicity may be caused by the injection of large quantities of MWCNTs into the lungs of mice. Therefore, it is urgent to thoroughly understand the immune response, including the definition of exposure standards for various usages, such as inhalation, injection, ingestion, and skin contact, in order to accept CNTs in the future medicine.

REFERENCES

[1] Baughman RH, Zakhidov AA, de Heer WA. Carbon nanotubes—the route toward applications. Science 2002;297(5582):787–92.

[2] Cao J, Wang Q, Rolandi M, Dai HJ. Aharonov-bohm interference and beating in single-walled carbon-nanotube interferometers. Phys Rev Lett 2004;93(21):216803.

[3] Kilbride BE, Coleman JN, Fraysse J, Fournet P, Cadek M, Drury A, et al. Experimental observation of scaling laws for alternating current and direct current conductivity in polymer-carbon nanotube composite thin films. J Appl Phys 2002;92(7):4024–30.

[4] Biercuk MJ, Llaguno MC, Radosavljevic M, Hyun JK, Johnson AT, Fischer JE. Carbon nanotube composites for thermal management. Appl Phys Lett 2002;80(15):2767–9.

[5] Callister WD. Materials science and engineering, an introduction. New York, NY: Wiley; 2003.

[6] Tibbetts GG, Beetz Jr CP. Mechanical properties of vapour-grown carbon fibres. J Phys D Appl Phys 1987;20:292–7.

[7] Hata K, Futaba DN, Mizuno K, Namai T, Yumura M, Iijima S. Water-assisted highly efficient synthesis of impurity-free single-walled carbon nanotubes. Science 2004;306(5700):1362–4.

[8] Wong EW, Sheehan PE, Lieber CM. Nanobeam mechanics: elasticity, strength, and toughness of nanorods and nanotubes. Science 1997;277(5334):1971–5.

[9] Yu MF, Lourie O, Dyer MJ, Moloni K, Kelly TF, Ruoff RS. Strength and breaking mechanism of multiwalled carbon nanotubes under tensile load. Science 2000;287(5453):637–40.

[10] Wagner HD, Lourie O, Feldman Y, Tenne R. Stress-induced fragmentation of multiwall carbon nanotubes in a polymer matrix. Appl Phys Lett 1998;72(2):188–90.

[11] Walters DA, Ericson LM, Casavant MJ, Liu J, Colbert DT, Smith KA, et al. Elastic strain of freely suspended single-wall carbon nanotube ropes. Appl Phys Lett 1999;74(25):3803–5.

[12] Yu MF, Files BS, Arepalli S, Ruoff RS. Tensile loading of ropes of single wall carbon nanotubes and their mechanical properties. Phys Rev Lett 2000;84(24):5552–5.

[13] Xie SS, Li WZ, Pan ZW, Chang BH, Sun LF. Mechanical and physical properties on carbon nanotube. J Phys Chem Solids 2000;61(7):1153–8.

[14] Dai H. Nanotube growth and characterization: carbon nanotubes. Berlin: Springer; 2001.

[15] Iijima S. Helical microtubules of graphitic carbon. Nature 1991;354(6348):56–8.

[16] Guo T, Nikolaev P, Thess A, Colbert DT, Smalley RE. Catalytic growth of single-walled manotubes by laser vaporization. Chem Phys Lett 1995;243:49–54.

[17] Jose-Yacaman M, Miki-Yoshida M, Rendon L, Santiesteban JG. Catalytic growth of carbon microtubules with fullerene structure. Appl Phys Lett 1993;62(2):202–4.

[18] Chung YH, Jou S. Carbon nanotubes from catalytic pyrolysis of polypropylene. Mater Chem Phys 2005;92(1):256–9.

[19] Nakazawa S, Yokomori T, Mizomoto M. Flame synthesis of carbon nanotubes in a wall stagnation flow. Chem Phys Lett 2005;403(1–3):158–62.

[20] Zhang YF, Gamo MN, Xiao CY, Ando T. Liquid phase synthesis of carbon nanotubes. Physica B 2002;323(1–4):293–5.
[21] Andrews R, Jacques D, Qian DL, Rantell T. Multiwall carbon nanotubes: synthesis and application. Accounts Chem Res 2002;35(12):1008–17.
[22] Strong KL, Anderson DP, Lafdi K, Kuhn JN. Purification process for single-wall carbon nanotubes. Carbon 2003;41(8):1477–88.
[23] Bougrine A, Naji A, Ghanbaja J, Billaud D. Purification and structural characterization of single-walled carbon nanotubes. Synthrtic Metals 1999;103:2480–1.
[24] Hernadi K, Siska A, Thien-Nga L, Forro L, Kiricsi I. Reactivity of different kinds of carbon during oxidative purification of catalytically prepared carbon nanotubes. Solid State Ionics 2001;141:203–9.
[25] Shen K, Curran S, Xu HF, Rogelj S, Jiang YB, Dewald J, et al. Single-walled carbon nanotube purification, pelletization, and surfactant-assisted dispersion: a combined TEM and resonant micro-Raman spectroscopy study. J Phys Chem B 2005;109(10):4455–63.
[26] Fang HT, Liu CG, Chang L, Feng L, Min L, Cheng HM. Purification of single-wall carbon nanotubes by electrochemical oxidation. Chem Mater 2004;16(26):5744–50.
[27] Ko CJ, Lee CY, Ko FH, Chen HL, Chu TC. Highly efficient microwave-assisted purification of multiwalled carbon nanotubes. Microelectron Eng 2004;73-4:570–7.
[28] Journet C, Bernier P. Production of carbon nanotubes. Appl Phys a-Mater 1998;67(1):1–9.
[29] Colomer JF, Stephan C, Lefrant S, Tendeloo GV, Willems I, Kónya Z, et al. Large-scale synthesis of single-wall carbon nanotubes by catalytic chemical vapor deposition (CCVD) method. Chem Phys Lett 2000;317(1–2):83–9.
[30] Couteau E, Hernadi K, Seo JW, Thien-Nga L, Miko C, Gaal R, et al. CVD synthesis of high-purity multiwalled carbon nanotubes using CaCO3 catalyst support for large-scale production. Chem Phys Lett 2003;378(1-2):9–17.
[31] Kathyayini H, Nagaraju N, Fonseca A, Nagy JB. Catalytic activity of Fe, Co and Fe/Co supported on Ca and Mg oxides, hydroxides and carbonates in the synthesis of carbon nanotubes. J Mol Catal a-Chem 2004;223(1-2):129–36.
[32] Kong J, Cassell AM, Dai H. Chemical vapor deposition of methane for single-walled carbon nanotubes. Chem Phys Lett 1998;292:567–74.
[33] Zhao NQ, He CN, Jiang ZY, Li JJ, Li YD. Fabrication and growth mechanism of carbon nanotubes by catalytic chemical vapor deposition. Mater Lett 2006;60(2):159–63.
[34] Qian WZ, Liu T, Wei F, Wang ZW, Li YD. Enhanced production of carbon nanotubes: combination of catalyst reduction and methane decomposition. Appl Catal a-Gen 2004;258(1):121–4.
[35] Li YL, Kinloch IA, Shaffer MS, Geng JF, Johnson B, Windle AH. Synthesis of single-walled carbon nanotubes by a fluidized-bed method. Chem Phys Lett 2004;384(1-3):98–102.
[36] Qian WZ, Wei F, Wang ZW, Liu T, Yu H, Luo GH, et al. Production of carbon nanotubes in a packed bed and a fluidized bed. Aiche J 2003;49(3):619–25.
[37] Corrias M, Caussat B, Ayral A, Durand J, Kihn Y, Kalck P, et al. Carbon nanotubes produced by fluidized bed catalytic CVD: first approach of the process. Chem Eng Sci 2003;58(19):4475–82.
[38] Weidenkaff A, Ebbinghaus SG, Mauron P, Reller A, Zhang Y, Zuttel A. Metal nanoparticles for the production of carbon nanotube composite materials by decomposition of different carbon sources. Mat Sci Eng C-Bio S 2002;19(1-2):119–23.
[39] Nasibulin AG, Moisala A, Brown DP, Jiang H, Kauppinen EI. A novel aerosol method for single walled carbon nanotube synthesis. Chem Phys Lett 2005;402(1–3):227–32.
[40] Moisala A, Nasibulin AG, Brown DP, Jiang H, Khriachtchev L, Kauppinen EI. Single-walled carbon nanotube synthesis using ferrocene and iron pentacarbonyl in a laminar flow reactor. Chem Eng Sci 2006;61(13):4393–402.

[41] Chen Y, Sun Z, Li YN, Tay BK. Optimization of carbon nanotube powder growth using low pressure floating catalytic chemical vapor deposition. Mater Chem Phys 2006;98(2-3):256–60.
[42] Hiraoka T, Bandow S, Shinohara H, Iijima S. Control on the diameter of single-walled carbon nanotubes by changing the pressure in floating catalyst CVD. Carbon 2006;44(9):1853–9.
[43] Zhou ZP, Wan DY, Bai Y, Dou XY, Song L, Zhou WY, et al. Ring formation from the direct floating catalytic chemical vapor deposition. Physica E 2006;33(1):24–7.
[44] Fan YY, Kaufmann A, Mukasyan A, Varma A. Single- and multi-wall carbon nanotubes produced using the floating catalyst method: synthesis, purification and hydrogen up-take. Carbon 2006;44(11): 2160–70.
[45] Teo KBK, Chhowalla M, Amaratunga GAJ, Milne WI, Pirio G, Legagneux P, et al. Characterization of plasma-enhanced chemical vapor deposition carbon nanotubes by Auger electron spectroscopy. J Vac Sci Technol B 2002;20(1):116–21.
[46] Chaisitsak S, Yamada A, Konagai M. Hot filament enhanced CVD synthesis of carbon nanotubes by using a carbon filament. Diam Relat Mater 2004;13(3):438–44.
[47] Lin CH, Lee SH, Hsu CM, Kuo CT. Comparisons on properties and growth mechanisms of carbon nanotubes fabricated by high-pressure and low-pressure plasma-enhanced chemical vapor deposition. Diam Relat Mater 2004;13(11-12):2147–51.
[48] Choi S, Lee S, Koh KH. In situ optical investigation of carbon nanotube growth in hot-filament chemical vapor deposition. Curr Appl Phys 2006;6S1:e38–42.
[49] Kim CD, Jang HS, Lee HR, Kim DH. Low temperature growth of carbon nanotubes in a magnetic field. Mater Lett 2007;61(10):2075–8.
[50] Bondi SN, Lackey WJ, Johnson RW, Wang X, Wang ZL. Laser assisted chemical vapor deposition synthesis of carbon nanotubes and their characterization. Carbon 2006;44(8):1393–403.
[51] Teo KB, Singh C, Chhowalla M, Milne WI. Catalytic synthesis of carbon nanotubes and nanofibers. Encyclopedia Nanosci Nanotechnol 2003;10(1).
[52] Suh DJ, Park T-J, Kim J-H, Kim K-L. Fast sol-gel synthetic route to high surface area alumina aerogels. Chem Mater 1997;9:1903–5.
[53] Ivanov V, Nagy J, Lambin P, Lucas A, Zhang X, Zhang X, et al. The study of carbon nanotubules produced by catalytic method. Chem Phys Lett 1994;223(4):329–35.
[54] Xu CB, Zhu J. One-step preparation of highly dispersed metal-supported catalysts by fluidized-bed MOCVD for carbon nanotube synthesis. Nanotechnology 2004;15(11):1671–81.
[55] Schwarz JA, Contescu C, Contescu A. Methods for preparation of catalytic materials. Chem Rev 1995;95(3):477–510.
[56] Nikolaev P, Bronikowski MJ, Bradley RK, Rohmund F, Colbert DT, Smith KA, et al. Gas-phase catalytic growth of single-walled carbon nanotubes from carbon monoxide. Chem Phys Lett 1999;313(1–2): 91–7.
[57] Gavillet J, Loiseau A, Journet C, Willaime F, Ducastelle F, Charlier JC. Root-growth mechanism for single-wall carbon nanotubes. Phys Rev Lett 2001;87(27).
[58] Kataura H, Kumazawa Y, Maniwa Y, Ohtsuka Y, Sen R, Suzuki S, et al. Diameter control of single-walled carbon nanotubes. Carbon 2000;38(11–12):1691–7.
[59] Li YL, Kinloch IA, Shaffer MSP, Singh C, Geng JF, Johnson BFG, et al. Growth of single-walled carbon nanotubes by the rapid heating of a supported catalyst. Chem Mater 2004;16(26):5637–43.
[60] Baker RTK, Harris PS, Thomas RB, Waite RJ. Formation of filamentous carbon from iron, cobalt and chromium catalyzed decomposition of acetylene. J Catal 1973;30(1):86–95.
[61] Baker RTK. Catalytic growth of carbon filaments. Carbon 1989;27(3):315–23.
[62] Alvarez L, Guillard T, Sauvajol JL, Flamant G, Laplaze D. Growth mechanisms and diameter evolution of single wall carbon nanotubes. Chem Phys Lett 2001;342(1–2):7–14.

[63] Gavillet J, Loiseau A, Ducastelle F, Thair S, Bernier P, Stephan O, et al. Microscopic mechanisms for the catalyst assisted growth of single-wall carbon nanotubes. Carbon 2002;40(10):1649−63.
[64] Jost O, Gorbunov AA, Moller J, Pompe W, Liu X, Georgi P, et al. Rate-limiting processes in the formation of single-wall carbon nanotubes: pointing the way to the nanotube formation mechanism. J Phys Chem B 2002;106(11):2875−83.
[65] Zhu WZ, Miser DE, Chan WG, Hajaligol MR. Characterization of multiwalled carbon nanotubes prepared by carbon arc cathode deposit. Mater Chem Phys 2003;82(3):638−47.
[66] See CH, Harris AT. A review of carbon nanotube synthesis via fluidized-bed chemical vapor deposition. Ind Eng Chem Res 2007;46(4):997−1012.
[67] Lacerda RG, Teh AS, Yang MH, Teo KBK, Rupesinghe NL, Dalal SH, et al. Growth of high-quality single-wall carbon nanotubes without amorphous carbon formation. Appl Phys Lett 2004;84(2): 269−71.
[68] Arepalli S, Nikolaev P, Gorelik O, Hadjiev VG, Bradlev HA, Holmes W, et al. Protocol for the characterization of single-wall carbon nanotube material quality. Carbon 2004;42(8−9):1783−91.
[69] Belin T, Epron F. Characterization methods of carbon nanotubes: a review. Mat Sci Eng B-Solid 2005; 119(2):105−18.
[70] Arepalli S, Freiman SW, Hooker SA, Migler KD. Measurement Issues in Single Wall Carbon Nanotubes; 2008.
[71] Singh P, Campidelli S, Giordani S, Bonifazi D, Bianco A, Prato M. Organic functionalisation and characterisation of single-walled carbon nanotubes. Chem Soc Rev 2009;38(8):2214−30.
[72] Dresselhaus MS, Dresselhaus G, Saito R, Jorio A. Raman spectroscopy of carbon nanotubes. Phys Rep 2005;409(2):47−99.
[73] Doorn SK, Heller DA, Barone PW, Usrey ML, Strano MS. Resonant Raman excitation profiles of individually dispersed single walled carbon nanotubes in solution. Appl Phys a-Mater 2004;78(8):1147−55.
[74] Fogden S, Howard CA, Heenan RK, Skipper NT, Shaffer MSP. Scalable method for the reductive dissolution, purification, and separation of single-walled carbon nanotubes. ACS Nano 2012;6(1):54−62.
[75] Kataura H, Kumazawa Y, Maniwa Y, Umezu I, Suzuki S, Ohtsuka Y, et al. Optical properties of single-wall carbon nanotubes. Synthetic Met 1999;103(1−3):2555−8.
[76] Fantini C, Jorio A, Souza M, Strano MS, Dresselhaus MS, Pimenta MA. Optical transition energies for carbon nanotubes from resonant Raman spectroscopy: environment and temperature effects. Phys Rev Lett 2004;93(14):147406.
[77] Heeg S, Malic E, Casiraghi C, Reich S. Quantitative composition of a single-walled carbon nanotube sample: Raman scattering versus photoluminescence. Phys Status Solidi B 2009;246(11-12):2740−3.
[78] Mevellec JY, Bergeret C, Cousseau J, Buisson JP, Ewels CP, Lefrant S. Tuning the Raman resonance behavior of single-walled carbon nanotubes via covalent functionalization. J Am Chem Soc 2011;133 (42):16938−46.
[79] Heller DA, Barone PW, Swanson JP, Mayrhofer RM, Strano MS. Using Raman spectroscopy to elucidate the aggregation state of single-walled carbon nanotubes. J Phys Chem B 2004;108(22):6905−9.
[80] Strano MS, Dyke CA, Usrey ML, Barone PW, Allen MJ, Shan HW, et al. Electronic structure control of single-walled carbon nanotube functionalization. Science 2003;301(5639):1519−22.
[81] Fantini C, Usrey ML, Strano MS. Investigation of electronic and vibrational properties of single-walled carbon nanotubes functionalized with diazonium salts. J Phys Chem C 2007;111(48):17941−6.
[82] Kim WJ, Usrey ML, Strano MS. Selective functionalization and free solution electrophoresis of single-walled carbon nanotubes: separate enrichment of metallic and semiconducting SWNT. Chem Mater 2007;19(7):1571−6.
[83] Lee JH, Paik U, Choi JY, Kim KK, Yoon SM, Lee J, et al. Dispersion stability of single-walled carbon nanotubes using Nafion in bisolvent. J Phys Chem C 2007;111(6):2477−83.

[84] Moore VC, Strano MS, Haroz EH, Hauge RH, Smalley RE, Schmidt J, et al. Individually suspended single-walled carbon nanotubes in various surfactants. Nano Lett 2003;3(10):1379–82.
[85] Bergin SD, Sun ZY, Streich P, Hamilton J, Coleman JN. New solvents for nanotubes: approaching the dispersibility of surfactants. J Phys Chem C 2010;114(1):231–7.
[86] Zhao B, Itkis ME, Niyogi S, Hu H, Zhang J, Haddon RC. Study of the extinction coefficients of single-walled carbon nanotubes and related carbon materials. J Phys Chem B 2004;108(24):8136–41.
[87] Bachilo SM, Strano MS, Kittrell C, Hauge RH, Smalley RE, Weisman RB. Structure-assigned optical spectra of single-walled carbon nanotubes. Science 2002;298(5602):2361–6.
[88] McDonald TJ, Blackburn JL, Metzger WK, Rumbles G, Heben MJ. Chiral-selective protection of single-walled carbon nanotube photoluminescence by surfactant selection. J Phys Chem C 2007;111 (48):17894–900.
[89] Cognet L, Tsyboulski DA, Rocha JDR, Doyle CD, Tour JM, Weisman RB. Stepwise quenching of exciton fluorescence in carbon nanotubes by single-molecule reactions. Science 2007;316 (5830):1465–8.
[90] Wang YH, Shan HW, Hauge RH, Pasquali M, Smalley RE. A highly selective, one-pot purification method for single-walled carbon nanotubes. J Phys Chem B 2007;111(6):1249–52.
[91] Sato Y, Yanagi K, Miyata Y, Suenaga K, Kataura H, Lijima S. Chiral-angle distribution for separated single-walled carbon nanotubes. Nano Lett 2008;8(10):3151–4.
[92] Vijayaraghavan A, Timmermans MY, Grigoras K, Nasibulin AG, Kauppinen EI, Krupke R. Imaging conduction pathways in carbon nanotube network transistors by voltage-contrast scanning electron microscopy. Nanotechnology 2011;22(26):265715.
[93] Lee Y, Jeon KS, Lim H, Shin HS, Jin SM, Byon HR, et al. Silencing of metallic single-walled carbon nanotubes via spontaneous hydrosilylation. Small 2009;5(12):1398–402.
[94] Balasubramanian K, Sordan R, Burghard M, Kern K. A selective electrochemical approach to carbon nanotube field-effect transistors. Nano Lett 2004;4(5):827–30.
[95] Kelly BT. Physics of graphite. London: Applied Science; 1981.
[96] Kelly A, Macmillan NH. Strong solids. Oxford: Oxford University Press; 1986.
[97] Bacon R. Growth, structure, and properties of graphite whiskers. J Appl Phys 1960;31(2):283–90.
[98] Overney G, Zhong W, Tomanek D. Structural rigidity and low-frequency vibrational-modes of long carbon tubules. Z Phys D Atom Mol Cl 1993;27(1):93–6.
[99] Lu JP. Elastic properties of single and multilayered nanotubes. J Phys Chem Solids 1997;58 (11):1649–52.
[100] Treacy MMJ, Ebbesen TW, Gibson JM. Exceptionally high Young's modulus observed for individual carbon nanotubes. Nature 1996;381(6584):678–80.
[101] Poncharal P, Wang ZL, Ugarte D, de Heer WA. Electrostatic deflections and electromechanical resonances of carbon nanotubes. Science 1999;283(5407):1513–16.
[102] Falvo MR, Clary GJ, Taylor RM, Chi V, Brooks FP, Washburn S, et al. Bending and buckling of carbon nanotubes under large strain. Nature 1997;389(6651):582–4.
[103] Salvetat JP, Kulik AJ, Bonard JM, Briggs GAD, Stockli T, Metenier K, et al. Elastic modulus of ordered and disordered multiwalled carbon nanotubes. Adv Mater 1999;11(2):161–5.
[104] Salvetat JP, Briggs GAD, Bonard JM, Bacsa RR, Kulik AJ, Stockli T, et al. Elastic and shear moduli of single-walled carbon nanotube ropes. Phys Rev Lett 1999;82(5):944–7.
[105] Miko C, Milas M, Seo JW, Couteau E, Barisic N, Gaal R, et al. Effect of electron irradiation on the electrical properties of fibers of aligned single-walled carbon nanotubes. Appl Phys Lett 2003;83 (22):4622–4.
[106] Kis A, Csanyi G, Salvetat JP, Lee TN, Couteau E, Kulik AJ, et al. Reinforcement of single-walled carbon nanotube bundles by intertube bridging. Nat Mater 2004;3(3):153–7.

[107] Bauhofer W, Kovacs JZ. A review and analysis of electrical percolation in carbon nanotube polymer composites. Compos Sci Technol 2009;69(10):1486–98.
[108] Chou TW, Gao LM, Thostenson ET, Zhang ZG, Byun JH. An assessment of the science and technology of carbon nanotube-based fibers and composites. Compos Sci Technol 2010;70(1):1–19.
[109] Gojny FH, Wichmann MHG, Kopke U, Fiedler B, Schulte K. Carbon nanotube-reinforced epoxy-compo sites: enhanced stiffness and fracture toughness at low nanotube content. Compos Sci Technol 2004;64(15):2363–71.
[110] Suhr J, Koratkar N, Keblinski P, Ajayan P. Viscoelasticity in carbon nanotube composites. Nat Mater 2005;4(2):134–7.
[111] Coleman JN, Khan U, Blau WJ, Gun'ko YK. Small but strong: a review of the mechanical properties of carbon nanotube-polymer composites. Carbon 2006;44(9):1624–52.
[112] Chae HG, Choi YH, Minus ML, Kumar S. Carbon nanotube reinforced small diameter polyacrylonitrile based carbon fiber. Compos Sci Technol 2009;69(3-4):406–13.
[113] Veedu VP, Cao AY, Li XS, Ma KG, Soldano C, Kar S, et al. Multifunctional composites using reinforced laminae with carbon-nanotube forests. Nat Mater 2006;5(6):457–62.
[114] Garcia EJ, Wardle BL, Hart AJ, Yamamoto N. Fabrication and multifunctional properties of a hybrid laminate with aligned carbon nanotubes grown In Situ. Compos Sci Technol 2008;68(9):2034–41.
[115] Bakshi SR, Agarwal A. An analysis of the factors affecting strengthening in carbon nanotube reinforced aluminum composites. Carbon 2011;49(2):533–44.
[116] Kashiwagi T, Du FM, Douglas JF, Winey KI, Harris RH, Shields JR. Nanoparticle networks reduce the flammability of polymer nanocomposites. Nat Mater 2005;4(12):928–33.
[117] Beigbeder A, Degee P, Conlan SL, Mutton RJ, Clare AS, Pettitt ME, et al. Preparation and characterisation of silicone-based coatings filled with carbon nanotubes and natural sepiolite and their application as marine fouling-release coatings. Biofouling 2008;24(4):291–302.
[118] Wu ZC, Chen ZH, Du X, Logan JM, Sippel J, Nikolou M, et al. Transparent, conductive carbon nanotube films. Science 2004;305(5688):1273–6.
[119] De S, Coleman JN. The effects of percolation in nanostructured transparent conductors. Mrs Bull 2011; 36(10):774–81.
[120] Cao Q, Rogers JA. Ultrathin films of single-walled carbon nanotubes for electronics and sensors: a review of fundamental and applied aspects. Adv Mater 2009;21(1):29–53.
[121] Ionescu AM, Riel H. Tunnel field-effect transistors as energy-efficient electronic switches. Nature 2011;479(7373):329–37.
[122] Tans SJ, Verschueren ARM, Dekker C. Room-temperature transistor based on a single carbon nanotube. Nature 1998;393(6680):49–52.
[123] Appenzeller J, Lin YM, Knoch J, Avouris P. Band-to-band tunneling in carbon nanotube field-effect transistors. Phys Rev Lett 2004;93(19):196805.
[124] Jensen K, Weldon J, Garcia H, Zettl A. Nanotube radio. Nano Lett 2007;7(11):3508–11.
[125] Franklin AD, Luisier M, Han SJ, Tulevski G, Breslin CM, Gignac L, et al. Sub-10 nm carbon nanotube transistor. Nano Lett 2012;12(2):758–62.
[126] Park H, Afzali A, Han SJ, Tulevski GS, Franklin AD, Tersoff J, et al. High-density integration of carbon nanotubes via chemical self-assembly. Nat Nanotechnol 2012;7(12):787–91.
[127] Cao Q, Kim HS, Pimparkar N, Kulkarni JP, Wang CJ, Shim M, et al. Medium-scale carbon nanotube thin-film integrated circuits on flexible plastic substrates. Nature 2008;454(7203):495–500.
[128] Sun DM, Timmermans MY, Tian Y, Nasibulin AG, Kauppinen EI, Kishimoto S, et al. Flexible high-performance carbon nanotube integrated circuits. Nat Nanotechnol 2011;6(3):156–61.
[129] McCarthy MA, Liu B, Donoghue EP, Kravchenko I, Kim DY, So F, et al. Low-voltage, low-power, organic light-emitting transistors for active matrix displays. Science 2011;332(6029):570–3.

[130] Chen PC, Fu Y, Aminirad R, Wang C, Zhang JL, Wang K, et al. Fully printed separated carbon nanotube thin film transistor circuits and its application in organic light emitting diode control. Nano Lett 2011;11(12):5301–8.
[131] Jung M, Kim J, Noh J, Lim N, Lim C, Lee G, et al. All-printed and roll-to-roll-printable 13.56-MHz-operated 1-bit RF tag on plastic foils. IEEE Trans Electron Devices 2010;57(3):571–80.
[132] Dai LM, Chang DW, Baek JB, Lu W. Carbon nanomaterials for advanced energy conversion and storage. Small 2012;8(8):1130–66.
[133] Kohler AR, Som C, Helland A, Gottschalk F. Studying the potential release of carbon nanotubes throughout the application life cycle. J Clean Prod 2008;16(8-9):927–37.
[134] Evanoff K, Khan J, Balandin AA, Magasinski A, Ready WJ, Fuller TF, et al. Towards ultrathick battery electrodes: aligned carbon nanotube—enabled architecture. Adv Mater 2012;24(4):533–7.
[135] Sotowa C, Origi G, Takeuchi M, Nishimura Y, Takeuchi K, Jang IY, et al. The reinforcing effect of combined carbon nanotubes and acetylene blacks on the positive electrode of lithium-ion batteries. Chemsuschem 2008;1(11):911–15.
[136] Gogotsi Y, Simon P. True performance metrics in electrochemical energy storage. Science 2011;334(6058):917–18.
[137] Izadi-Najafabadi A, Yasuda S, Kobashi K, Yamada T, Futaba DN, Hatori H, et al. Extracting the full potential of single-walled carbon nanotubes as durable supercapacitor electrodes operable at 4 V with high power and energy density. Adv Mater 2010;22(35):E235–41.
[138] De Volder MF, Tawfick SH, Baughman RH, Hart AJ. Carbon nanotubes: present and future commercial applications. Science 2013;339(6119):535–9.
[139] Matsumoto T, Komatsu T, Arai K, Yamazaki T, Kijima M, Shimizu H, et al. Reduction of Pt usage in fuel cell electrocatalysts with carbon nanotube electrodes. Chem Commun 2004;7:840–1.
[140] Gong KP, Du F, Xia ZH, Durstock M, Dai LM. Nitrogen-doped carbon nanotube arrays with high electrocatalytic activity for oxygen reduction. Science 2009;323(5915):760–4.
[141] Le Goff A, Artero V, Jousselme B, Tran PD, Guillet N, Metaye R, et al. From hydrogenases to noble metal-free catalytic nanomaterials for H-2 production and uptake. Science 2009;326(5958):1384–7.
[142] Lee JM, Park JS, Lee SH, Kim H, Yoo S, Kim SO. Selective electron- or hole-transport enhancement in bulk-heterojunction organic solar cells with N- or B-doped carbon nanotubes. Adv Mater 2011;23(5):629–33.
[143] Gabor NM, Zhong ZH, Bosnick K, Park J, McEuen PL. Extremely efficient multiple electron-hole pair generation in carbon nanotube photodiodes. Science 2009;325(5946):1367–71.
[144] Holt JK, Park HG, Wang YM, Stadermann M, Artyukhin AB, Grigoropoulos CP, et al. Fast mass transport through sub-2-nanometer carbon nanotubes. Science 2006;312(5776):1034–7.
[145] Corry B. Designing carbon nanotube membranes for efficient water desalination. J Phys Chem B 2008;112(5):1427–34.
[146] Heller DA, Baik S, Eurell TE, Strano MS. Single-walled carbon nanotube spectroscopy in live cells: towards long-term labels and optical sensors. Adv Mater 2005;17(23):2793–9.
[147] De La Zerda A, Zavaleta C, Keren S, Vaithilingam S, Bodapati S, Liu Z, et al. Carbon nanotubes as photoacoustic molecular imaging agents in living mice. Nat Nanotechnol 2008;3(9):557–62.
[148] Kam NWS, O'Connell M, Wisdom JA, Dai HJ. Carbon nanotubes as multifunctional biological transporters and near-infrared agents for selective cancer cell destruction. P Natl Acad Sci USA 2005;102(33):11600–5.
[149] Kurkina T, Vlandas A, Ahmad A, Kern K, Balasubramanian K. Label-free detection of few copies of DNA with carbon nanotube impedance biosensors. Angew Chem Int Edit 2011;50(16):3710–14.
[150] Heller DA, Jin H, Martinez BM, Patel D, Miller BM, Yeung TK, et al. Multimodal optical sensing and analyte specificity using single-walled carbon nanotubes. Nat Nanotechnol 2009;4(2):114–20.

[151] Snow ES, Perkins FK, Houser EJ, Badescu SC, Reinecke TL. Chemical detection with a single-walled carbon nanotube capacitor. Science 2005;307(5717):1942–5.

[152] Chen Z, Tabakman SM, Goodwin AP, Kattah MG, Daranciang D, Wang XR, et al. Protein microarrays with carbon nanotubes as multicolor Raman labels. Nat Biotechnol 2008;26(11):1285–92.

[153] Star A, Tu E, Niemann J, Gabriel JCP, Joiner CS, Valcke C. Label-free detection of DNA hybridization using carbon nanotube network field-effect transistors. P Natl Acad Sci USA 2006;103(4):921–6.

[154] Esser B, Schnorr JM, Swager TM. Selective detection of ethylene gas using carbon nanotube-based devices: utility in determination of fruit ripeness. Angew Chem Int Edit 2012;51(23):5752–6.

[155] Shi XH, von dem Bussche A, Hurt RH, Kane AB, Gao HJ. Cell entry of one-dimensional nanomaterials occurs by tip recognition and rotation. Nat Nanotechnol 2011;6(11):714–19.

[156] Hong SY, Tobias G, Al-Jamal KT, Ballesteros B, Ali-Boucetta H, Lozano-Perez S, et al. Filled and glycosylated carbon nanotubes for in vivo radioemitter localization and imaging. Nat Mater 2010;9(6): 485–90.

[157] Liu Z, Sun XM, Nakayama-Ratchford N, Dai HJ. Supramolecular chemistry on water-soluble carbon nanotubes for drug loading and delivery. Acs Nano 2007;1(1):50–6.

[158] Bianco A, Kostarelos K, Prato M. Making carbon nanotubes biocompatible and biodegradable. Chem Commun 2011;47(37):10182–8.

[159] Poland CA, Duffin R, Kinloch I, Maynard A, Wallace WAH, Seaton A, et al. Carbon nanotubes introduced into the abdominal cavity of mice show asbestos-like pathogenicity in a pilot study. Nat Nanotechnol 2008;3(7):423–8.

CHAPTER

CLASSICAL MOLECULAR DYNAMICS SIMULATIONS

3

3.1 INTRODUCTION

Ever since Iijima discovered carbon nanotubes (CNTs) in 1991 by using high-resolution transmission election microscopy [1] and, in particular, after they were successfully synthesized in 1993 by Bethune et al. [2], and Iijima and Ichihashi [3], CNTs have attracted considerable attention. CNTs are a new type of carbon material with very broad prospects [4–6] because of their unique structure and excellent physical and chemical properties. In particular, their excellent mechanical properties have aroused widespread interest in the scientific community, and progress [7–10] has been made in applying them to toughen materials. As the excellent mechanical properties of CNTs are likely to be related to their structure, much research attention in recent years has been paid to their structural properties and applications [7–17]. Recent studies have indicated that CNTs possess an exceptionally high elastic modulus [12,18] and can sustain large elastic and failure strains [19,20]. CNTs have thus been identified as a promising superfiber for nanotube-composite materials. As the excellent mechanical properties of CNTs are likely to be related to their structure, much research attention in recent years has been paid to their structural properties and applications.

Molecular dynamics (MD) simulation is by far the most popular tool for investigating the mechanical properties of CNTs. For example, Yakobson et al. [11] investigated axially compressed armchair single-walled carbon nanotubes (SWCNTs) using MD simulations and compared the solutions with those obtained from a simple continuum thin shell model. They observed that the continuum shell models can predict the buckling properties of SWCNTs satisfactorily provided that the mechanical parameters, such as the Young's modulus and effective wall thickness, are judiciously adopted. Wang et al. [12] used MD simulation to study the elastic response and buckling modes of SWCNTs under axial compression. Cornwell and Wille [21] used MD simulation to study the response of a set of armchair SWCNTs and calculated the Young's modulus of the tubes. Srivastava et al. [22] performed a MD simulation to examine the nanoplasticity of zigzag CNTs under compression. Recently, Liew et al. [13–17] studied the mechanical properties of armchair and zigzag SWCNTs and multiwalled carbon nanotubes (MWCNTs) composed of multiple armchair SWCNTs with different diameters. Wang et al. [12] applied MD simulation to study the elastic response and buckling modes of SWCNTs under axial compression.

3.2 COMPUTATIONAL MODEL

Compared to the first-principle quantum mechanics, classical MD simulation is an empirical method which is easy to implement in large systems (millions to billions of atoms). The classical

Mechanical Behaviors of Carbon Nanotubes. DOI: http://dx.doi.org/10.1016/B978-0-323-43137-8.00003-2

MD method was first proposed by Alder and Wainwright [23]. Up to now, MD simulation is the most popular tool for investigating the mechanical properties and structural deformation of CNTs. The basic concept of the method is to simulate the time evolution of a system. It describes a system's atomic-scale dynamics, in which atoms and molecules move while interacting with many of the other atoms and molecules in the vicinity. The method relies on a mathematical description of total energy of the system as a function of all of atomic coordinates. Atoms in the system are treated as point-like masses that interact with each other according to a given potential energy $E(\mathbf{r}_1, \mathbf{r}_2, \ldots, \mathbf{r}_N)$, where $\mathbf{r}_j$ is the vector position of the jth atom. The evolution is computed at every time-step, and the instantaneous location and velocity of each atom are then determined by solving Hamilton's classical equation of motion from Newton's second law.

$$\mathbf{F}_i = m_i \times \frac{d^2\mathbf{r}_i}{dt^2} = -\nabla_i E(\mathbf{r}_1, \mathbf{r}_2, \ldots, \mathbf{r}_N) \tag{3.1}$$

where m_i and $\mathbf{r}_i$ are the mass and spatial coordinates of the ith atom, respectively, E is the empirical potential for the system, and ∇ denotes the spatial gradient. Due to the small scale involved, explicit integration algorithms, such as the Verlet method [24] and other high-order methods, are commonly used to ensure a high order of accuracy.

The reliability of MD simulation depends on the use of an appropriate potential. Allinger and his colleagues [25–28] developed a molecular mechanics MM2 force field and an improved version known as the MM3 force field. These methods are designed for a broad class of problems in organic and inorganic systems. Abell [29] proposed the quantum-mechanical concept of bond order formalism for CNT characteristics. Using Morse-type potential, Abell showed that the degree of bonding universality could be well maintained in molecular modeling for similar elements. Tersoff [30,31] introduced this concept for the modeling of group four elements such as carbon, silicon, and germanium, and reasonably accurate results were reported. Nordlund et al. [32] modified the Tersoff potential such that the interlayer interaction is also considered. Brenner [33] made further improvements to the Tersoff potential by introducing additional terms into the bond order function. Compared with the Tersoff potential, Brenner potential shows robustness in the treatment of conjugacy, and it allows for the forming and breaking of the bond with the correct representation of the bond order. Brenner's potential has enjoyed success in the analysis of the formulation of fullerenes and their properties [34,35], surface patterning [35], indentation and friction at the nanoscale [36], and the energetics of nanotubes [35].

In the short-range Brenner potential, the system energy means the sum of the energy of each bond, and the energy of each bond is composed of a repulsive and attractive pair. Following Tersoff [30] and Brenner [33], the expression for the bonding energy between atoms i and j is

$$E_B(r_{ij}) = V_R(r_{ij}) - B_{ij} V_A(r_{ij}) \tag{3.2}$$

where r_{ij} is the distance between the pairs of nearest neighboring atoms i and j, and the function $V_R(r_{ij})$ represents the interatomic repulsion, such as the core–core interactions, the function $V_A(r_{ij})$ represents the attraction that results from the valence electrons, which are given by

$$V_R(r_{ij}) = \frac{D^{(e)}}{S-1} e^{-\sqrt{2S}\beta(r_{ij}-R^{(e)})} f_c(r_{ij}) \tag{3.3}$$

$$V_A(r_{ij}) = \frac{D^{(e)}S}{S-1} e^{-\sqrt{2/S}\beta(r_{ij}-R^{(e)})} f_c(r_{ij}) \tag{3.4}$$

f_c is a smooth cutoff function that limits the range of the potential, and is given by

$$f_c(r_{ij}) = \begin{cases} r_{ij} > R^{(2)} & r_{ij} < R^{(1)} \\ \frac{1}{2}\left\{1 + \cos\left[\frac{\pi(r_{ij} - R^{(1)})}{R^{(2)} - R^{(1)}}\right]\right\} & R^{(1)} < r_{ij} < R^{(2)} \\ 0 & r_{ij} > R^{(2)} \end{cases} \tag{3.5}$$

where $R^{(1)}$ and $R^{(2)}$ are the effective ranges of the cutoff function.

The parameter B_{ij} in Eq. (3.2) represents multibody coupling between the bond $i-j$ and the local environment of atom i, and is given by

$$B_{ij} = \left[1 + \sum_{k \neq i,j} G(\theta_{ijk}) f_c(r_{ik})\right]^{-\delta} \tag{3.6}$$

where θ_{ijk} is the angle between the bonds $i-j$ and $i-k$, and the angle function G is given by

$$G(\theta_{ijk}) = a_0 \left[1 + \frac{c_0^2}{d_0^2} - \frac{c_0^2}{d_0^2 + (1 + \cos(\theta_{ijk}))^2}\right] \tag{3.7}$$

For atoms i and j with different local environments, Brenner [33] suggested replacing B_{ij} in (Eq. 3.6) with

$$\overline{B}_{ij} = \frac{1}{2}(B_{ij} + B_{ji}) \tag{3.8}$$

These parameters $D^{(e)}$, S, β, $R^{(e)}$, δ, $R^{(1)}$, $R^{(2)}$, a_0, c_0, d_0 in the aforementioned equations can be determined by fitting with the known physical properties of carbon. Brenner gave two sets of parameters, which are listed in Table 3.1 [33].

Recently, a further revision of Brenner potential [37] called the second-generation reactive empirical bond order (REBO) potential, which is able to describe exactly the bonding structure and properties of graphite, diamond, and hydrocarbon molecules, is provided. In comparison with the Brenner potential, the second-generation potential includes both improved analytic functions for the

Table 3.1 Brenner Potential Parameters of Sets I and II [33]

Parameter	Set I	Set II
$D^{(e)}$	6.325 eV	6.0 eV
S	1.29	1.22
β	15 nm^{-1}	21 nm^{-1}
$R^{(e)}$	0.1315 nm	0.139 nm
$R^{(1)}$	0.17 nm	0.17 nm
$R^{(2)}$	0.2 nm	0.2 nm
δ	0.80469	0.5
a_0	0.011304	0.00020813
c_0	19	330
d_0	2.5	3.5

interatomic interactions and an extended fitting database, which results in a significantly better description of the bond length, energies, and force constants for hydrocarbon molecules and elastic properties, interstitial defect energies, and surface energy for diamonds.

The REBO potential is also expressed as Eq. (3.2). To tell the second-generation potential from Brenner potential, superscript symbols are used, such as V^R and V^A are expressed as the pair term of repulsive and attractive potential

$$V^R(r_{ij}) = f^c(r_{ij})(1 + Q/r_{ij})Ae^{-\alpha r_{ij}} \tag{3.9}$$

$$V^A(r_{ij}) = f^c(r_{ij}) \sum_{n=1,3} B_n e^{-\beta_n r_{ij}} \tag{3.10}$$

where Q is the effective charge in the screened coulomb potential, and $f^c(r_{ij})$ limits the range of the covalent interactions and is given by

$$f^c(r_{ij}) = \begin{cases} 1 & r_{ij} < D_{ij}^{\min} \\ \dfrac{1}{2}\left[1 + \cos\left(\dfrac{r_{ij} - D_{ij}^{\min}}{D_{ij}^{\max} - D_{ij}^{\min}}\right)\right] & D_{ij}^{\min} < r_{ij} < D_{ij}^{\max} \\ 0 & r_{ij} > D_{ij}^{\max} \end{cases} \tag{3.11}$$

where $D_{ij}^{\max} - D_{ij}^{\min}$ is the distance over which the function goes from one to zero. The parameters obtained through data-fitting [37] are listed in Table 3.2.

The bond-order function between atoms i and j is given by

$$\overline{B}_{ij} = \frac{1}{2}\left(B_{ij}^{\sigma-\pi} + B_{ji}^{\sigma-\pi}\right) + B_{ij}^{\pi} \tag{3.12}$$

where the functions $B_{ij}^{\sigma-\pi}$ and $B_{ji}^{\sigma-\pi}$ depend on the local coordination and bond angles for atoms i and j, respectively. The function B_{ij}^{π} can be written as

$$B_{ij}^{\pi} = \pi_{ij}^{RC} + B_{ij}^{DH} \tag{3.13}$$

The first term π_{ij}^{RC} in Eq. (3.13) depends on whether the bond has a radical character and is part of the conjugated system. The second term B_{ij}^{DH} depends on the torsion angle for the C–C bonds. For bonds that are not part of the conjugated system, and for planar double bonds, both terms are equal to zero. The term $B_{ij}^{\sigma-\pi}$ in Eq. (3.12) is given as

$$B_{ij}^{\sigma-\pi} = \left\{1 + \sum_{k \neq i,j} f_{ik}^c(r_{ik}) G_C\left[\cos(\theta_{ijk})\right]\right\}^{-1/2} \tag{3.14}$$

Table 3.2 Parameter Values for C–C Pair Terms [37]

$B_1 = 12388.79197798$ eV	$\beta_1 = 4.7204523127$ Å^{-1}	$Q = 0.3134602960833$ Å
$B_2 = 17.56740646509$ eV	$\beta_2 = 1.4332132499$ Å^{-1}	$A = 10953.544162170$ eV
$B_3 = 30.71493208065$ eV	$\beta_3 = 1.3826912506$ Å^{-1}	$\alpha = 4.7465390606595$ Å^{-1}
$D_{min} = 1.7$ Å	$D_{max} = 2.0$ Å	

An identical expression is given for $B_{ji}^{\sigma-\pi}$ by wrapping the i and j indices. The angle function $G_C\left[\cos(\theta_{ijk})\right]$ modulates the contribution that each nearest neighbor makes to the empirical bond-order according to the cosine of the angle of the bonds between atoms i and k, and i and j. To allow for both overcoordinated and undercoordinated atoms, a second spline function $\gamma_c[\cos(\theta)]$ was determined for angles less than 109.47 degrees that was coupled to the original angular function $G_C[\cos(\theta)]$. The term π_{ij}^{RC} in Eq. (3.13) represents the influence of radical energies and the π bond conjugation on the bond energies. It accurately describes radical structures and accounts for nonlocal conjugation effects such as those governing the different properties of the C–C bonds in graphite. It is taken as a tricubic spline. The detail on the term B_{ij}^{DH} that describes the torsion angle for the C–C bonds can be found in [37].

As mentioned above, the modified second-generation Brenner potential allows for covalent bond breaking and forming. Zhang et al. [38] adopted the modified second-generation Brenner potential to study the fracture of defected CNTs by atomistic and multiscale simulations. However, their numerical results revealed that the fracture stresses predicted by the second-generation Brenner potential is several times larger than those of quantum mechanics. This is due to the functional form of the cutoff function in the potential, which artificially magnifies the bond force for the distances between 1.7 and 2.0 Å and thus leads to a nonphysical failure mechanism. To overcome this problem, it is suggested [38,39] that this cutoff function be removed in the fracture analysis, but included in the C–C interactions only for those atom pairs that are less than 2.0 Å apart in the initial and undeformed configurations. Such an improvement makes the potential no longer able to handle bond formation, but it can provide accurate results for the fracture.

The short-range Brenner and REBO potentials do not model the nonbonding interactions between atoms. Hence, we computed the interspacing interaction between the SWCNTs in the bundles or multiwalled CNTs by using the long-range van der Waals potential which is described as

$$E_{vdw} = \begin{cases} 0 & r_{ij} < r'_s \\ c_{3,k}\left(r_{ij}-r_k\right)^3 + c_{2,k}\left(r_{ij}-r_k\right)^2 & r'_s < r_{ij} < r'_m \\ E^{IJ}\left(r_{ij}\right) & r'_m < r_{ij} < r'_b \end{cases} \tag{3.15}$$

where $c_{n,k}$ are cubic spline coefficients and $r'_s = 0.2$ nm, $r'_m = 0.32$ nm, and $r'_b = 1.0$ nm. E^{IJ} is defined as the Lennard–Jones potential, which is computed as [40]

$$E_{\mathrm{vdw}}(r_{ij}) = 4\varepsilon\left[\left(\frac{\sigma}{r_{ij}}\right)^{12} - \left(\frac{\sigma}{r_{ij}}\right)^{6}\right] \tag{3.16}$$

where r_{ij} is the distance between atoms i and j, $\sigma = 0.34$ nm and $\varepsilon = 2.8437$ meV. Eq. (3.16) reveals that the attenuation of the vdW interaction is quite rapid as the distance increases. It should be noted that our computational model will incorporate the van der Waals potential only if the covalent potential is zero, such that there is no artificial reaction barrier formed by the steep repulsive wall of the Lennard–Jones potential to prevent the nonbonded atoms from chemical reaction.

Therefore, the total potential stored in the whole system is a sum of the short-range covalent and long-range noncovalent interactions

$$E_{\mathrm{TOT}} = \sum_i \sum_{j>1} \left(E_{\mathrm{b}} + E_{\mathrm{vdw}}\right) \tag{3.17}$$

3.3 ELASTIC PROPERTIES OF CNTs

3.3.1 YOUNG'S MODULUS OF SINGLE-WALLED CNTs WITH IMPURITIES

The system strain energy here is calculated using the REBO potential, and the Young's modulus is obtained by applying the classical elasticity theory from

$$Y = -\frac{1}{V_0}\frac{\partial^2 E}{\partial \varepsilon^2}\Big|_{\varepsilon=0} \tag{3.18}$$

where the function E is the total strain energy of the system, ε is the strain, and V_0 is the equilibrium volume.

3.3.1.1 Model of SWCNTs with impurities

To construct armchair (5, 5) (10, 10) and zigzag (9, 0) (18, 0) SWCNTs, the atomic space coordinates in the basic unit are first determined. The structure of the SWCNTs can then be achieved by selecting carbon as the atom and coupling the bond properly. Similarly, the structure of BN-doped SWCNTs can be achieved by substituting a boron atom and a nitrogen atom for two adjacent carbon atoms. There is a deviation in the bond length due to the curvature effect of the tube wall. In general, it is found that the smaller the tube diameter, the greater the impact on the bond length. The initial bond lengths of the C—C, C—N, and C—B tubes are set to 1.42 Å, and the bond lengths of the B—N tubes are set to 1.45 Å. The calculated models are shown in Fig. 3.1, where Fig. 3.1C denotes a pure single-walled BNNT and Fig. 3.1D a pure SWCNT. The periodic structural units are then built by setting the basic unit of the SWCNTs in a cuboid crystal lattice; one end of the nanotubes is located at plane xy, and the other end is located on the opposite side plane, the axes of which are parallel to the crystal lattice axis z and pass the two centers of the planes that are face to face. To minimize the impact between CNTs, the lattice parameters, which are much larger than the diameter of the nanotubes, are set to $a = b = 100 \times 10^{-10}$ m, as shown in Fig. 3.2. Finally, the structure is optimized, the electron density distribution is determined by using density functional theory (DFT), and the elastic parameters are computed using the MD method.

3.3.1.2 Effect of impurities on single-walled CNTs

Fig. 3.3 shows the effect of various percentages of BN impurity (ξ) on the Young's modulus (Y) of armchair (5, 5) (10, 10) and zigzag (9, 0) (18, 0) SWCNTs. The results indicate that the Young's modulus of SWCNTs is generally very large, with the values for the armchair (5, 5) (10, 10) and zigzag (9, 0) (18, 0) SWCNTs without impurities being 948, 901 and 804, 860 GPa, respectively [41]. The Young's modulus of the (5, 5) SWCNTs is very close to the value of 0.971 TPa reported in [42]. When the ratio of BN impurity in the armchair (5, 5) (10, 10) SWCNTs is about 6.7%, the Young's modulus decreases by about 17%. However, further increases in the doping ratio do not cause the modulus to reduce further: it remains at about 800 and 760 GPa, respectively. The behavior of the zigzag SWCNTs is different. When the impurity ratio is about 7.5%, the Young's modulus decreases by about 40%, but rises with further increases in the doping ratio. When the impurity ratio is 37%, the Young's modulus decreases by only 17%. When the impurity ratio reaches 100% (i.e., boron–nitride nanotubes are formed), the Young's modulus is 780 and 835 GPa, which is 97% of that of the corresponding pure SWCNTs (these values are not shown in Fig. 3.3).

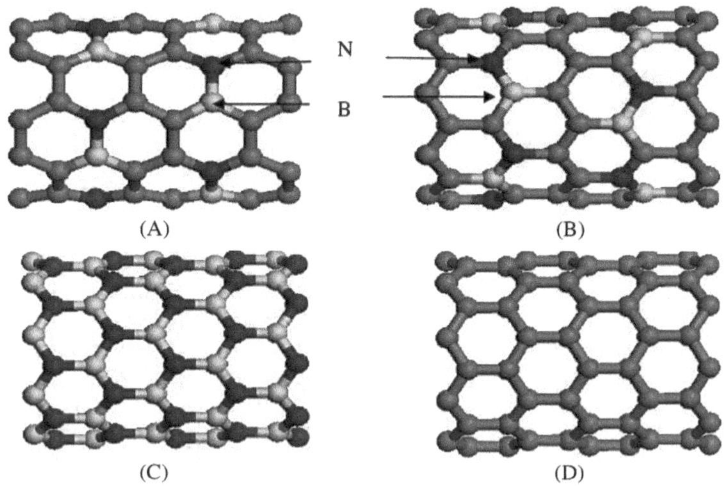

FIGURE 3.1

The atomic structures of SWCNTs doped with BN. (A) (5, 5) SWCNTs doped with BN; (B) (9, 0) SWCNTs doped with BN; (C) (9, 0) BNNT; (D) (9, 0) SWCNTs [41].

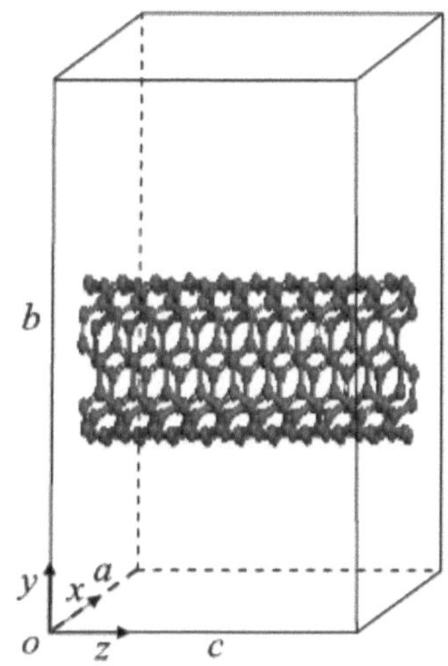

FIGURE 3.2

The calculation model of SWCNTs [41].

3.3.1.3 Analysis of the results

Because the effects of the different BN impurity ratios on the Young's modulus of the same types of SWCNTs are similar, only (5, 5) was chosen from among the armchair SWCNTs and (9, 0) from among the zigzag SWCNTs to analyze the effects of BN doping on the electron density of

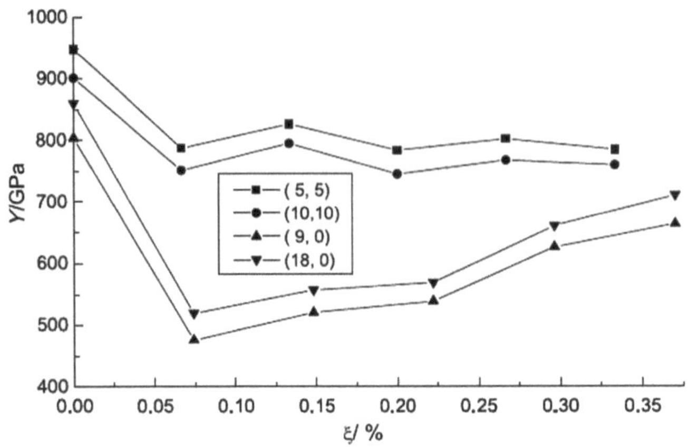

FIGURE 3.3

The effect on Young's modulus (*Y*) of the SWCNTs with different doping ratios of BN impurities [41].

SWCNTs. The comparison of the deformation electron density of SWCNTs with BN impurity and no impurity, which was derived using the local density approximation, is shown in Fig. 3.4.

Molecular orbitals are formed by an atomic orbital wave function through linear combination. Several atomic orbitals can be combined into several molecular orbitals. Half of the molecular orbitals are formed by two atomic orbitals with the same positive and negative symbols. The electronic probability density between the two nuclei increases, and the atomic energy becomes lower than the original atomic orbital energy. This is conducive to bonding and results in bonding molecular orbitals, such as the σ or π orbits. The other half of the molecular orbitals is formed by two atomic orbitals with different positive and negative symbols. The electronic probability density between the two nuclei decreases, and the atomic energy becomes higher than the original atomic orbital energy. This is not conducive to bonding, and thus these orbitals are known as antibonding molecular orbitals, such as the σ^* or π^* orbits. The π or π^* bonds in CNTs are considered to be a hybridization of π or π^* orbit bonding, which occurs in nanotubes when the π bond on the graphite layer degenerates into a large π bond on the carbon ring. Through the interaction and overlapping of adjacent large π bonds, a hybridization of the π bond and π^* bond is formed.

As shown in Fig. 3.4, it is clear that atom B reveals more, atom C is less exposed, and atom N has been inundated in an electronic cloud. Therefore, compared with atom C, atom N can obtain electrons from atom B more easily. It can also be seen that when BN impurity is distributed around the circumference of the nanotubes, the existence of each N or B atom mostly affects the one bond in the axial direction in the zigzag SWCNTs, but also, to a lesser extent, affects the two bonds in the nonaxial direction in the armchair SWCNTs. This indicates that in the presence of BN doping, the Young's modulus of zigzag SWCNTs decreases more than that of armchair SWCNTs. This phenomenon can be attributed to the change in the π orbital electron cloud that occurs when the π orbits bonded in the axial direction turn into antibonding π^* orbits, which decreases the electron transfer in the circumferential direction. When the ratio of BN impurity in the SWCNTs increases, the increasing number of electrons in the circumferential direction improves the connectivity of

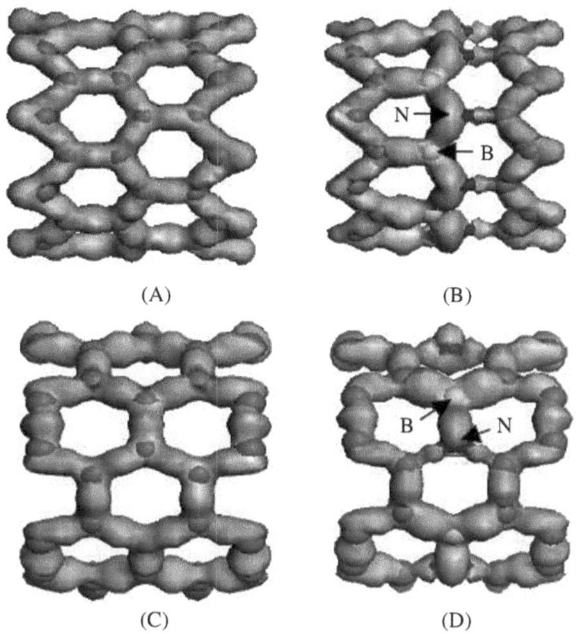

FIGURE 3.4

Deformation electron density comparison between SWCNTs with doped BN impurities and SWCNTs with no impurities: (A) No impurities in (9, 0); (B) BN impurities in (9, 0); (C) No impurities in (5, 5); (D) BN impurities in (5, 5) [41].

zigzag SWCNTs, and strengthens the circumference direction σ bond. However, because the circumferential electron cloud of armchair SWCNTs is not connective, the increasing electron cloud in the circumferential direction does not strengthen the circumferential σ bond. Generally speaking, the σ bond and π bond in CNTs are related to their mechanical and electric properties, respectively. Therefore, the Young's modulus of zigzag SWCNTs obviously increases with the percentage of BN impurity, whereas that of armchair SWCNTs tends to remains unchanged. These results concur with those shown in Fig. 3.3.

In performing data analysis on the Mulliken charges of SWCNTs doped with BN, it is find that the charge transfers only a little between B and C, but relatively more between B–N and N–C. For instance, in SWCNTs doped with 37% of BN, a 0.07*e* charge is transferred from B to C between B and C, a 0.17N charge is transferred from C to N between N and C, and a 0.55*e* charge is transferred from B to N between B and N. This means that each N obtains 0.72*e* and each B loses 0.62*e*. There are also small charge transfers among other atoms. This is caused by electronegativity: N > C > B, but the electronegativity within atom C is the same.

In BN-doped SWCNTs, polarity covalent bonds are formed in C–N and C–B, an electrovalent bond is formed in B–N, and a covalent bond is formed in C–C. Generally speaking, a mixed sp^2 and sp^3 covalent bond is formed in C–C close to N or B, and the other bonds are sp^2 covalent bonds. The

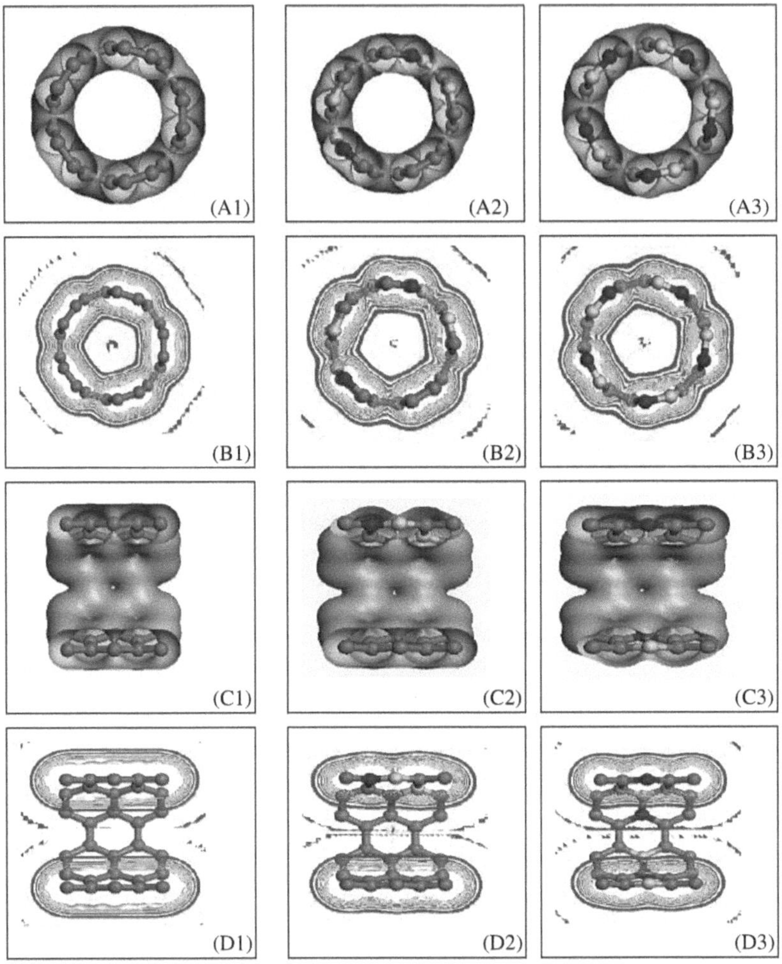

FIGURE 3.5

Electron density distribution cross-sections (A, B) and axes-sections (C, D) of (5, 5) SWCNTs and BN-doped SWCNTs [41].

electron state distribution in BN-doped SWCNTs is thus relatively intricate. The cross-sections (*A*, *B*) and axial sections (*C*, *D*) of the electron density distribution of (5, 5), (9, 0) SWCNTs and the corresponding BN-doped SWCNTs are shown in Figs. 3.5 and 3.6, respectively. In these figures, (A), (C) show the electron distributing isosurface shape, and (B), (D) show the section isoline structure. The left, middle, and right figures show the situation for no, partial, and full impurity at the circumference, respectively. The figures show that there are some differences in the atomic bonding properties of the different constructions. The electron density distribution changes greatly for the N atom in the B−N bond (Fig. 3.5B3), which means that approximate electrovalent bonds are formed.

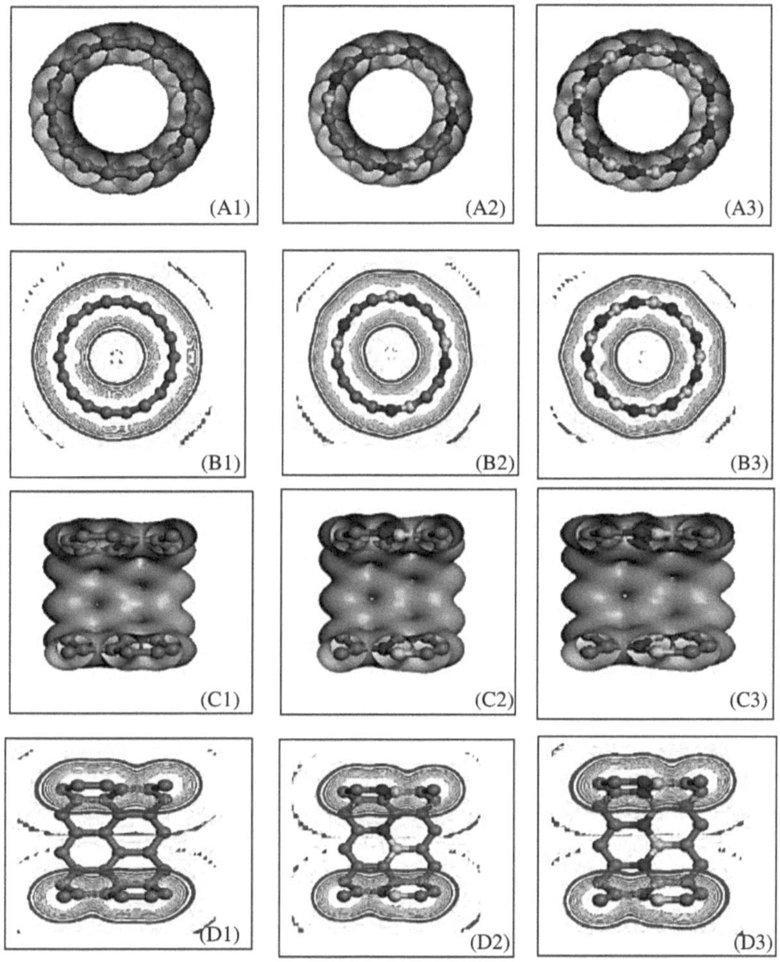

FIGURE 3.6

Electron density distribution cross-sections (A, B) and axes-sections (C, D) of (9, 0) SWCNTs and BN-doped SWCNTs [41].

In the N–C bond, the electron density distribution migrates slightly to N (Fig. 3.5B2), which means that a polarity covalent bond with a higher binding energy is formed. The electron migration in the CB bond can be attributed to the antibonding π^* orbits (see Fig. 3.5B2).

It is clearly shown in (B1) of Figs. 3.5 and 3.6 that because the electron density distribution among the C atoms overlaps, the symmetrical large ring orbits appear both inside and outside of the carbon ring. Parts (B2) and (B3) of the figures show that the electronic localization difference between B and N causes the distortion of the large ring orbits, especially for the lining orbits. Comparing Fig. 3.5D1 and Fig. 3.6D2, it can be observed that in the axial direction, the electric electron density distribution that results from the presence of large π bonds on the carbon ring in (5, 5) is clearly superior to that in (9, 0). Therefore, the electrical conductivity of the armchair nanotubes is better than that of the zigzag nanotubes in this direction. Figs. 3.5D2 and 3.6D2 show that the effects of BN impurity on the two types of nanotube are also different. The effects of BN on the inner and outer large ring orbits of (5, 5) are greater than the effects of the orbits in (9, 0) in these figures, whereas in Figs. 3.5B and 3.6B the effects of BN on (9, 0) are greater than those on (5, 5). These results indicate that the primary effect of BN impurity in armchair nanotubes occurs in the electron density distribution in the direction of the tube-axis, whereas in zigzag nanotubes it takes place in the circumferential direction. The results also indicate that the changes in the large π bond, which correlates with the electrical properties of nanotubes, also have an effect on their mechanical properties.

In CNTs, both the σ bond and the π bond can affect the Young's modulus, but the contribution to the modulus originates mainly from the σ bond. Because the electron cloud of the σ bond overlaps more than that of the π bond, the local level of the σ bond is higher than that of the π bond. The electron density distribution on the outer orbit is relatively close to a line. Because of the structural characteristics of the armchair SWCNTs, the connectivity of the circumferential electron density distribution is relatively poor, and thus the mechanical properties mainly depend on the electron density distribution in the axial direction. As shown in Fig. 3.5D1–D3, the electron density distribution in SWCNTs undergoes an obvious change when the SWCNTs are doped with BN, but this change does not increase with increases in the impurity ratio. Fig. 3.6D1–D3 show that the electron density distribution in the direction of the tube-axis does not change greatly when the tube is doped with BN, but the electron density distribution in the circumferential direction does change in the presence of impurity, eventually reverting to the state without impurity. Therefore, when the impurity ratio is relatively low, the structural stability falls because of the difference in electronic localization between atoms B, N, and C. As the impurity ratio increases, the whole electron density distribution becomes more uniform and symmetrical, which compensates for the difference in electronic localization. It is known from previous analyses that the compensatory effect of zigzag SWCNTs is superior to that of armchair SWCNTs. In conclusion, when the ratio of BN impurity in armchair and zigzag SWCNTs is low, their Young's modulus decreases. However, further increases in impurity do not give rise to a further decrease in the Young's modulus. The modulus of the armchair SWCNTs remains at a certain value and fluctuates around it, whereas that of the zigzag SWCNTs increases with further increases in the impurity ratio. At the point at which pure boron–nitride nanotubes are formed, the Young's modulus approaches that for pure SWCNTs.

3.3.2 EFFECTS OF VACANCY DEFECT RECONSTRUCTION ON THE ELASTIC PROPERTIES OF CNTs

It is nearly impossible to obtain perfect CNT samples. In the process of the growth of CNTs, a variety of structural defects inevitably occur, which can have a big impact on the mechanical properties of the CNTs. In some cases, even a small number of defects can fundamentally change the mechanical performance. So, it is of great practical significance to conduct systematic research into CNT defects. In recent years, there have been many reports of 5–77–5 and vacancy defects and other types of structural CNT defects [43–51]. However, most of the studies have concentrated on the impact of defects on the electrical properties of CNTs. How do structural defects affect their mechanical properties? At present, there are still few reports. Therefore, it is necessary to study the impact of defects on the mechanical properties of CNTs. A research has been conducted to help us to gain a more comprehensive understanding of the structure of good CNTs [52].

Vacancies are the most common defects in CNTs, and have drawn much research attention in recent years. Ajayan et al. [53] studied (10, 10) SWCNTs and confirmed that an ideal single vacancy containing three hanging bonds is unstable. He stated that two of the three dangling bonds tend to form a new C–C bond, and that the structure becomes a pentagonal ring coupled with one dangling bond, which forms a so-called 5–1DB defect. Lu and Pan [44] carried out a systematic calculation of single vacancies in SWCNTs and obtained the relationship between the vacancy formation energies and radii of different chiral SWCNTs. Krasheninnikov et al. [45] confirmed that vacancy defects are probable when using the Ar-ion process for low irradiation on (10, 10) SWCNTs, and stated that the process will be an instrument to control the structural defects of CNTs. In the early study of graphite structure, people found that there are usually two types of vacancies in graphite: single vacancy and di-vacancy [54]. As CNTs can be understood as curled graphite substrates, MD method is used to optimize these two kinds of initial structure defects at different temperatures and study the effect of different defect ratios on the Young's moduli of SWCNTs. First, the reconstruction of single and di-vacancy structures in SWCNTs is optimized at different temperatures. Then, the Young's moduli of SWCNTs with different single vacancy defect ratios are calculated. Finally, the results are analyzed according to the law of electron cloud coupling between the hanging bonds in the vacancies and the reconstruction features of the vacancy defects. Based on the analysis, the following conclusions can be derived. The reconstruction of vacancy defects can be carried out by way of not only temperature but also the defect ratios in SWCNTs because of the activity of dangling bonds and their space. Vacancy defects will bring about a decrease in the Young's modulus, but their reconstruction will be an important factor in stabilizing the modulus. SWCNTs with vacancy defects can be energy suppliers by controlling vacancy reconstruction, which provides opportunities for designing new nanoelectromechanical system devices [55,56].

3.3.2.1 Model and methods

The atomic space coordinates are first calculated according to the construction of armchair (5, 5) and (10, 10) SWCNTs. The structure of the SWCNTs can then be achieved by selecting carbon as the atom and coupling the bond properly (as shown in Fig. 3.7A). The structure of single and di-vacancy defects can be achieved by removing one carbon atom and two adjacent carbon atoms, respectively, from the perfect SWCNTs (as shown in Fig. 3.7B and C). Periodic structural units are then built by setting the obtained SWCNTs in a cuboid crystal lattice. One end of the nanotube is located on plane xy and the other end is located on the opposite plane, the axes of which are parallel to the crystal lattice axis z and pass the two centers of the planes that are face to face. To minimize the impact between CNTs, the lattice parameters, which are much larger than the diameter of the nanotubes, are set to $a = b = 100 \times 10^{-10}$ m. Finally, the structures are optimized at different temperatures, their electron density distribution is calculated using DFT, and the elastic parameters of SWCNTs with different single vacancy defect ratios are calculated using the MD method [52].

DFT is a quantum mechanical theory used in physics and chemistry to investigate the electronic structure of many-body systems, in particular atoms, molecules, and the condensed phases. DFT

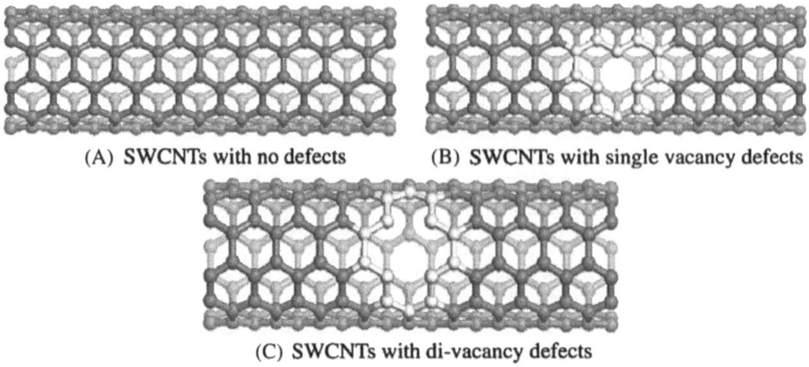

FIGURE 3.7

Atomic structures of (5, 5) SWCNTs with initial vacancy defects [52].

has been very popular for calculations in solid-state physics since the 1970s. In many cases, the results of DFT calculations for solid-state systems agreed quite satisfactorily with experimental data [57–59].

3.3.2.2 Vacancy defect reconstructions in single-walled CNTs

Fig. 3.8A and B are the optimized structures of (5, 5) single-walled CNTs with single and di-vacancy defects at different temperatures. The results indicate that when the temperature is below 2400K, the optimized single vacancy shows few changes, and there is no reconstruction. However, when the temperature is equal to or higher than 2400K, the optimized single vacancy shows obvious reconstruction. The structure becomes a pentagonal ring coupled with one dangling bond, which forms a so-called 5–1DB defect (Fig. 3.8A—2400K). The reconstruction of a di-vacancy is similar to that of a single vacancy. However, the di-vacancy has a higher reconstruction temperature and becomes an octagonal ring pinched into two pentagonal rings when the temperature is about or higher than 2600K. We call this structure a 5–8–5 defect (Fig. 3.8B—2600K). In Fig. 3.8, it is evident that the reconstruction configurations for single vacancy and di-vacancy remain unchanged when the temperature for (1) single vacancy is beyond 2500K and (2) di-vacancy is beyond 2700K. Therefore, the reconstruction configurations for them beyond these temperatures will not be presented.

In CNTs, the reconstruction of different types of vacancy defects takes place at different temperatures. The main reason is that when the vacancy defect is formed, it leads to a hanging bond around the vacancy (see the atoms with hanging bonds pointed by arrow in Fig. 3.8). If these dangling bonds need to be reformed into a new C–C bond, their activity must be enhanced to a certain extent. Temperature is a physical parameter that reflects the average kinetic energy of the microparticles: the higher is the temperature, the quicker are the microparticles and the greater is their average kinetic energy. Therefore, by increasing the temperature of the system, the electrons constituting a hanging bond can get more energy and the activity of the hanging bond will be enhanced. When the temperature is below a certain point, the activity of the hanging bond is below rebonding activity, so the transformation by optimizing the structure is very small, and no

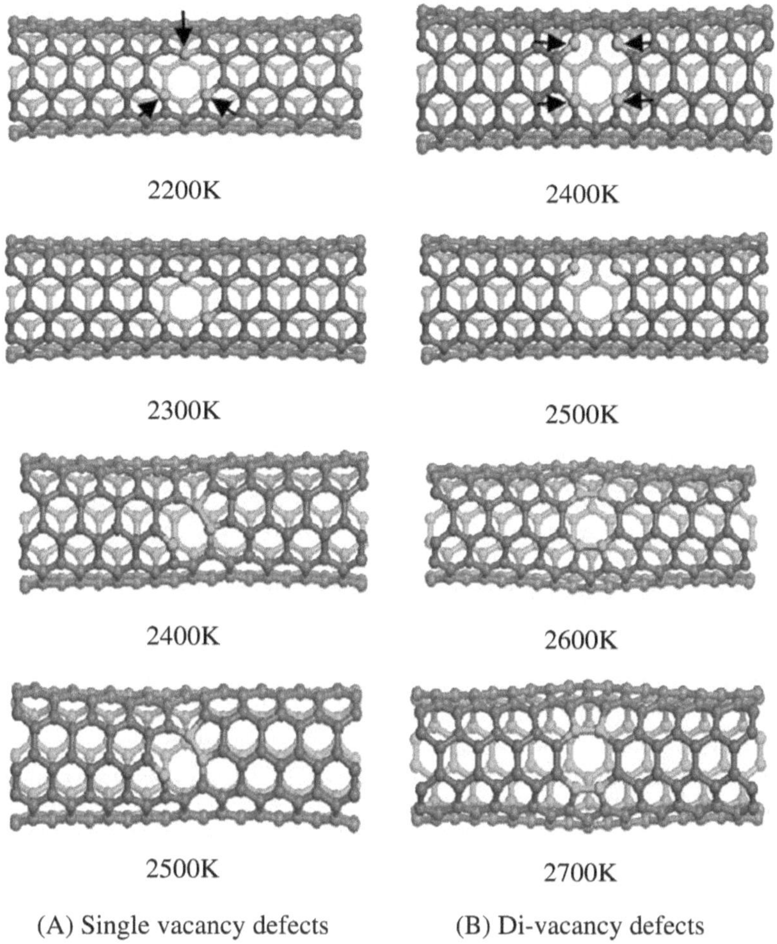

FIGURE 3.8

Optimized structure of (5, 5) SWCNTs with single and di-vacancy defects at different temperatures [52].

reconstruction happens. How single vacancies and di-vacancies are reconstructed and the temperatures required for the reconstruction primarily depend on the distance between the hanging bonds and surrounding environment. In general, the smaller the distance between the hanging bonds and the freer the electrons composing the hanging bonds, the lower the temperature necessary for reconstruction is. Thus, it can be seen from Fig. 3.8 that the single vacancy is easily reconstituted into a 5−1DB defect and the di-vacancy into a 5−8−5 defect. The reconstitution temperature of the di-vacancy is higher than that of the single vacancy, as the new 5−8−5 structure does not have a hanging bond; its acquired energy is required for four hanging bonds, close to two 5−1DB defects. Of course, the reconstitution temperature of di-vacancy defects also depends on the diameters of

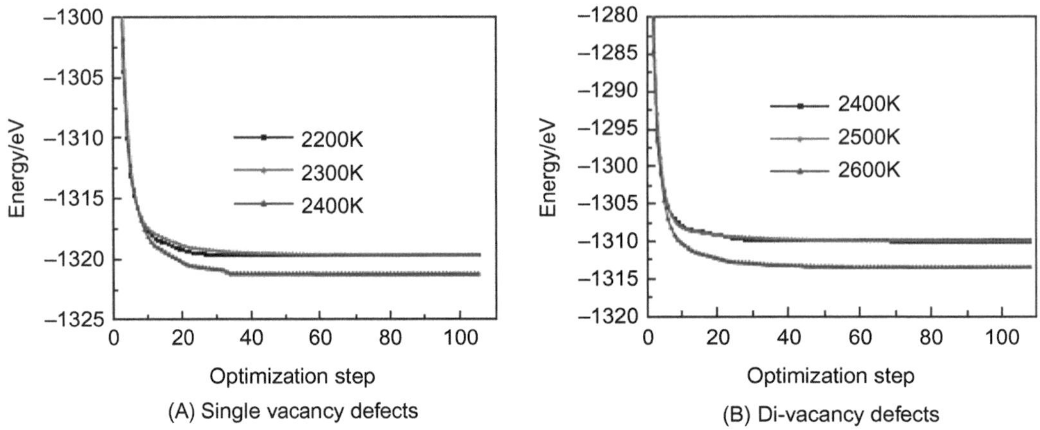

FIGURE 3.9

Potential energy of optimized (5, 5) SWCNTs with single and di-vacancy defects at different temperatures [52].

tube and orientation of the defects. It can be forecasted that because of the curvature of the tube wall, either different diameters or a different orientation will lead to small changes in the distance between the hanging bonds and greater changes in the reconstitution temperature. In addition, the reconstruction of the defects can further release the potential energy of the system to make the structure more compact, which helps to improve the stability of the structure. Fig. 3.9 presents the optimization energy curves of the aforementioned structures. When the temperature is less than that required for reconstruction, it can be seen from the figure that the optimized single and di-vacancy structures have different stable potential energies, and that despite the temperature increase, the potential energies remain unchanged. These are -1319.6 and -1309.9 eV, respectively. The potential energy of the di-vacancy is higher than that of the single vacancy when the number of hanging bonds in the di-vacancy is greater. When the temperature reaches the degree necessary for reconstruction, the potential energies at stability for the single and di-vacancy structures fall suddenly to -1321.2 and -1313.5 eV, respectively, which indicates that reconstruction happens, energy is released, and the structure becomes more stable. From the point of view of quantum mechanics, there is a barrier between the vacancy states before and after reconstruction. Only when the hanging bonds get sufficient energy can they overcome the barrier and bond to form a more stable structure. The higher reconstitution temperature for the di-vacancy compared to that for the single vacancy indicates that its barrier is higher. Using the reconstitution temperature (T) and the potential energy before and after reconstruction (E_{p1}, E_{p2}), the following results can be calculated: vacancy barrier $\Delta\varepsilon = k_B T$ (where the Boltzmann constant $k_B = 1.38 \times 10^{-23}$ J/K), the potential energy difference between the pre- and postreconstruction stages $\Delta E_p = E_{p2} - E_{p1}$, and the releasable energy in the process of the reconstruction $\Delta E = \varepsilon + \Delta E_p$. The results are shown in Table 3.3. The greater decline in potential energy indicates that di-vacancy reconstruction greatly contributes to improving the stability of the structure.

Fig. 3.10 shows the deformation electron densities and their isolines on sections of CNTs with single and di-vacancy defects before and after reconstruction. Therein, the left

Table 3.3 Energy (eV) of (5, 5) SWCNTs With Vacancy Defects [52]

Type of Vacancy	T (K)	E_{p1}	E_{p2}	$\Delta\varepsilon$	ΔE_p	ΔE
Single vacancy	2400	−1319.6	−1309.9	0.207	1.6	1.807
Di-vacancy	2600	−1321.2	−1313.5	0.224	3.6	3.824

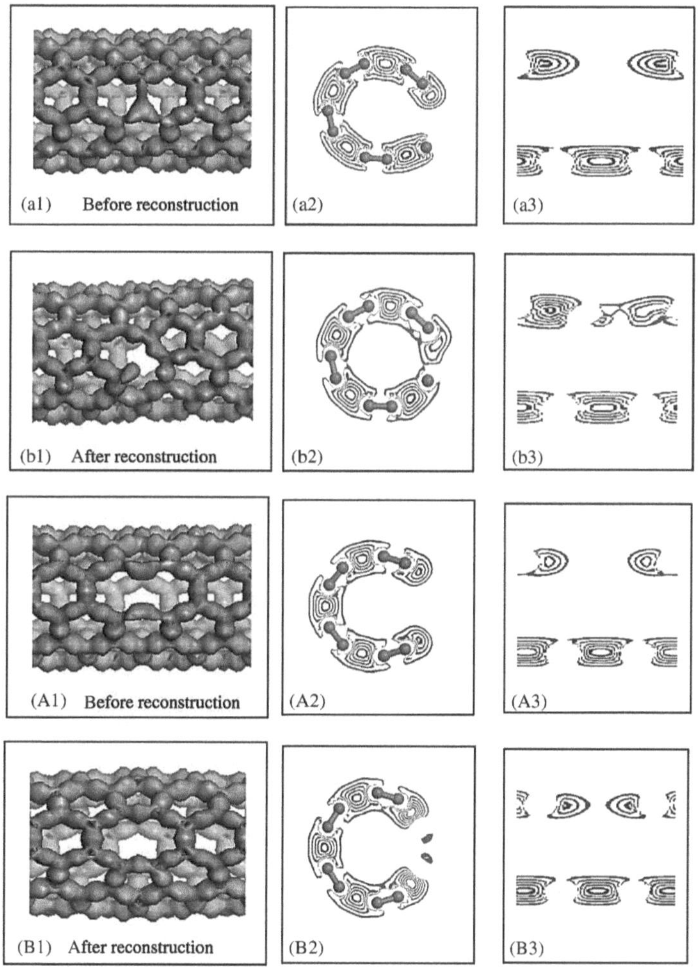

FIGURE 3.10

Deformation electron density (left) and cross-section (middle) and axial section isoline structure (right) of (5, 5) SWCNTs with single and di-vacancy defects [52].

figures (Fig. 3.10a(A) and b(B)) represent the distribution of deformation electron densities, and the middle and right figures represent the distribution of their cross-section and axial section isolines, respectively. The electron density is the same as the online equivalent, and the deformation electron density gradually becomes smaller on the vertical isolines from the inside outward. The isoline intervals reflect the local electronic conditions, in that the greater is the interval, the weaker is the localization and the freer the electrons, and vice versa. By comparing Fig. 3.10a1 and b1, it is clear that a new bond has formed between two dangling bonds in the nonaxial direction, and that the electron density distribution composing the bond is close to that of other C–C bonds of nondefective CNTs. This indicates the formation of a new C–C bond. However, the situation for Fig. 3.10A1 and A2 is somewhat different, as the formation of new C–C bonds takes place between two pairs of axial dangling bonds. Comparing Fig. 3.10a2(A2) and b2(B2), the electrons of the atomic outer layers around the defects on the cross-section transfer to the defects after the reconstruction, and the distribution of the electron density takes on a connecting trend. From Fig. 3.10a3(A3) and b3(B3), we can also see that the axial section resembles the transition circumstances of the cross-section, with narrowing of the electron density center making the spatial distribution region uniform and improving the stability of the structure. In particular, the di-vacancy makes a relatively great contribution in the axial direction.

Molecular orbitals are formed by an atomic orbital wave function through a linear combination. Several atomic orbitals can be combined into several molecular orbitals. Half of a molecular orbital is formed by two atomic orbitals with the same positive and negative symbols. The electronic probability density between the two nuclei increases, and the atomic energy becomes lower than the original atomic orbital energy. This is conducive to bonding and results in bonding molecular orbitals, such as the σ or π orbits. The other half of a molecular orbital is formed by two atomic orbitals with different positive and negative symbols. The electronic probability density between the two nuclei decreases, and the atomic energy becomes higher than the original atomic orbital energy. This is not conducive to bonding, and thus these kinds of orbitals are known as antibonding molecular orbitals such as the σ^* or π^* orbits. The CNTs under study are composed of a number of carbon hexagonal rings, and their structure can thus be considered to be a rolled-up graphene sheet in which the localized strong σ bond and delocalized π bond are formed by combining each carbon atom with the three adjacent carbon atoms through sp^2 hybrid orbital bonding. The σ and π bonds are located, respectively on and perpendicular to the tube wall. The transformation of vacancy defects in CNTs can be understood as follows. Because of the vacancy, class π bonds near the defects on the tube wall are destroyed, creating much local stress at the defects. The formation of new C–C bonds can effectively eliminate part of the stress, resulting in the spontaneous transformation of the vacancy into a 5–1DB or 5–8–5 defect.

From Fig. 3.10a1, we can see the electronic distribution that appears among the three hanging bonds; however, the electrons are little bound by their C atomic kernel, their energy is great, the mid-density between the two nuclei is small, and the bonds are weak. This indicates that σ^* or π^* bonds that are related to conductivity are mainly taken on. Comparing Fig. 3.10a3 and b3, we can see that the isoline interval around the defects becomes denser after reconstruction. Therefore, after reconstruction, the partial σ^* or π^* bonds have primarily been transformed into σ or π bonds, which make a larger contribution to the elasticity. From Fig. 3.10B1 (compared with Fig. 3.10A1), we find that some axial C atoms near the defect are more revealed, and that other C atoms are less exposed or are inundated in an electronic cloud. This is because partial electrons recombined to form new C–C bonds come from the transfer of axial *p* electrons that form π bonds. There are two recombined bonds in a 5–8–5 defect and only one in a 5–1DB defect, and thus more electrons are required in the 5–8–5 defect.

3.3.2.3 Effect of single vacancy defect ratio on Young's modulus of single-walled CNTs

Fig. 3.11 shows the effect of various percentages of vacancy defects (ξ) on the Young's modulus (Y) of armchair (5, 5) and (10, 10) SWCNTs at 298 K. The results indicate that the Young's modulus of SWCNTs is generally very large, with the values for the armchair (5, 5) and (10, 10) SWCNTs with no defects being 948 and 901 GPa, respectively, at room temperature. The Young's modulus of the (5, 5) SWCNTs is very close to the value of 0.971 TPa reported in [42]. With an increase in the single vacancy defect ratio in SWCNTs, the Young's moduli will decrease. However, when the vacancy defect achieves a certain ratio, such as $\xi = 2.5\%$ and $\xi = 5.83\%$ for (5, 5) and $\xi = 2.2\%$ and $\xi = 5.86\%$ for (10, 10) SWCNTs, there is a sudden slowdown in the curves of the Young's moduli versus the vacancy defect radio and the platform phenomenon emerges.

A vacancy will result in hanging bonds near the carbon atoms and a relaxed structure, leading ultimately to a decrease in the elastic moduli. The platform phenomenon in this decrease means that the influence of the defects on the system structure will be reduced, and the elasticity of the system will become stable for the following two reasons. First, normally, when the vacancy rate is smaller, mainly single vacancies are present in the system, and di-vacancies seldom appear. In contrast, when the rate becomes larger, the space between two adjacent single vacancy defects will become smaller. When the space is reduced to zero (namely, a di-vacancy), the boundaries of the defects will be partly superimposed, which leads to a great reduction in the number of hanging bonds. From the structural characteristics of the defects, we know that two single vacancies produce six hanging bonds, whereas one di-vacancy produces only four hanging bonds. Therefore, when the defect rate is small, compared with the influence of the two separate single vacancies, that of the di-vacancy on the CNT is smaller. However, the main reason that the system elasticity tends to become stable should be that reconstruction has taken place in some of the vacancies. In the

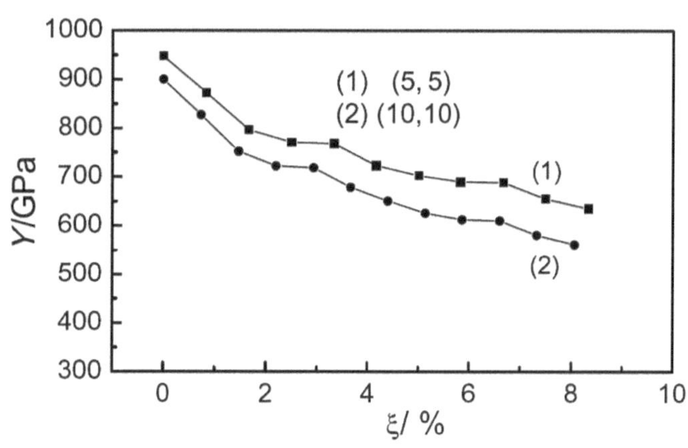

FIGURE 3.11

Effect on the Young's modulus (Y) of SWCNTs of different vacancy defect ratios [52].

abovementioned investigation, we consider only the impact of temperature on the reconstruction of one defect in a CNT. However, for CNTs that have a higher vacancy rate, the reconstruction of defects still occurs. When the vacancy rate increases gradually, the distance between the defects decreases naturally, and most of the hanging bonds are created. When the distance is narrowed to a certain pitch, the bonds can easily interact with each other. The activity of the large number of hanging bonds should follow a distribution similar to Maxwell speed distribution in thermodynamics. As a result, even at a temperature lower than the reconstruction temperature required for single vacancy defects, when the defects reach a certain percentage, there are always some highly active hanging bonds, and some of them are neighbors, so then the bonds will combine with each other to form 5−1DB or 5−8−5 defects. Then, the energy of the system will begin to be released, and the released energy can also activate more hanging bonds with lower activity to allow reconstruction to take place continuously within a certain range. In this way, the structure will be stable for the time being, which results in a sudden delay in the decline of the Young's moduli, and subsequently, the plateau effect. In addition, the energy brought by the reconstruction is enough to activate the hanging bond with lower activity to bond (see Table 3.3, $\Delta E >> \Delta\varepsilon$), and a source releasing energy can be generated by controlling the temperature at a location. The activation effect spreads further, from the near to the distant vacancy, creating a new reconstruction and releasing energy continuously. As a result, the magnitude and continuous time releasing energy can be adjusted by the defect rate, and the defect rate can be controlled by using the Ar-ion process for low irradiation on SWCNTs [60]. Therefore, an SWCNT with vacancy defects is expected to be an energy supplier by controlling vacancy reconstruction.

3.3.3 YOUNG'S MODULI OF SINGLE-WALLED CNTs WITH GRAFTS

3.3.3.1 Model of SWCNTs with grafts

As shown in Fig. 3.2, the atomic space coordinates in the basic unit are first calculated according to the construction of the armchair (5, 5) (10, 10) and zigzag (9, 0) (18, 0) SWCNTs (see Tables 3.4 and 3.5). The structure of the SWCNTs can then be achieved by selecting carbon as the atom and coupling the bond properly. Usually, there are two basic types of CNT ends: open and closed. Open-ended CNTs are considered to be more active, and are often selected to practically enhance composite materials. Therefore, open-ended CNTs are considered here, and 2−8 amine functional groups are linked symmetrically at both ends of the CNTs. In Fig. 3.12, the functional group atom marked with an arrow is grafted to a carbon atom at one end such that the C−C bonds at the grafts are parallel to the tube axes of the CNTs. The calculated model is shown in Fig. 3.12. The DMol3 software is used to optimize the structure and calculate the related electric properties. The periodic structural units are then built by setting the basic unit of the SWCNTs in a cuboid crystal lattice. One end of the nanotubes is located on plane xy and the other end is located on the opposite plane, the axes of which are parallel to the crystal lattice axis z and pass the two centers of the planes, which are face to face. To minimize the impact between the CNTs, the lattice parameters, which are much larger than the diameter of the nanotubes, are set to $a = b = 100 \times 10^{-10}$ m, as shown in Fig. 3.2. Once the structure is optimized, the electron density distribution is calculated by using the DFT. Then, the amine functional groups are cut off from the optimized structures. Finally, the elastic parameters are calculated by using the MD method.

Table 3.4 Atom Coordinates (nm) in the Basic Cell of (5, 5) SWCNTs [66]

Atomic Order Number	x	y	z	Atomic Order Number	x	y	z
1	5.311295	5.138886	0.00000	11	5.170208	5.295335	0.122976
2	4.964107	5.338977	0.00000	12	4.771717	5.253141	0.122976
3	4.666522	5.070614	0.00000	13	4.688705	4.861114	0.122976
4	4.829792	4.704665	0.00000	14	5.035893	4.661023	0.122976
5	5.228283	4.746859	0.00000	15	5.333478	4.929386	0.122976
6	5.340872	5.000000	0.00000	16	5.275771	5.20036	0.122976
7	5.105335	5.324189	0.00000	17	4.894665	5.324189	0.122976
8	4.724229	5.200360	0.00000	18	4.659128	5.000000	0.122976
9	4.724229	4.799640	0.00000	19	4.894665	4.675811	0.122976
10	5.105335	4.675811	0.00000	20	5.275771	4.79964	0.122976

Table 3.5 Atom Coordinates (nm) in the Basic Cell of (9, 0) SWCNTs [66]

Atomic Order Number	x	y	z	Atomic Order Number	x	y	z
1	5.354094	5.000000	0.00000	15	4.728748	4.772393	0.71000
2	5.271252	5.227607	0.00000	16	4.938512	4.651285	0.71000
3	5.061488	5.348715	0.00000	17	5.177047	4.693345	0.71000
4	4.822953	5.306655	0.00000	18	5.33274	4.878893	0.71000
5	4.66726	5.121107	0.00000	19	5.33274	5.121107	0.71000
6	4.66726	4.878893	0.00000	20	5.177047	5.306655	0.71000
7	4.822953	4.693345	0.00000	21	4.938512	5.348715	0.213000
8	5.061488	4.651285	0.00000	22	4.728748	5.227607	0.213000
9	5.271252	4.772393	0.00000	23	4.645906	5.000000	0.213000
10	5.33274	5.121107	0.00000	24	4.728748	4.772393	0.213000
11	5.177047	5.306655	0.71000	25	4.938512	4.651285	0.213000
12	4.938512	5.348715	0.71000	26	5.177047	4.693345	0.213000
13	4.728748	5.227607	0.71000	27	5.33274	4.878893	0.213000
14	4.645906	5.000000	0.71000				

To accurately determine the Young's modulus of CNT, its wall thickness of different diameters should take on different values, however, it was reported that the differences are generally very small [61–65]. In this study, the effects of grafted amines on the CNT are studied where the CNT wall thickness of 0.34 nm is used in the simulation.

3.3.3.2 Effect of amine grafts on SWCNTs

Fig. 3.13 shows the effect of different numbers of grafted amines on the Young's moduli of two hexagonal units of tube-length armchair (5, 5) (10, 10) and zigzag (9, 0) (18, 0) SWCNTs. The

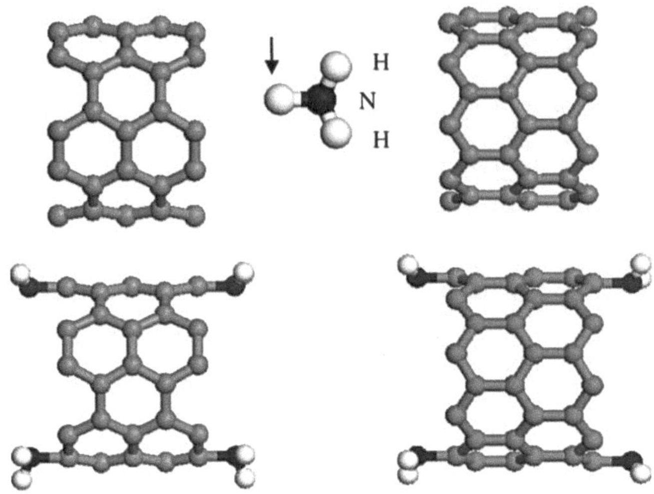

FIGURE 3.12

Atomic structures of (5, 5) and (9, 0) SWCNTs with different graft numbers [66].

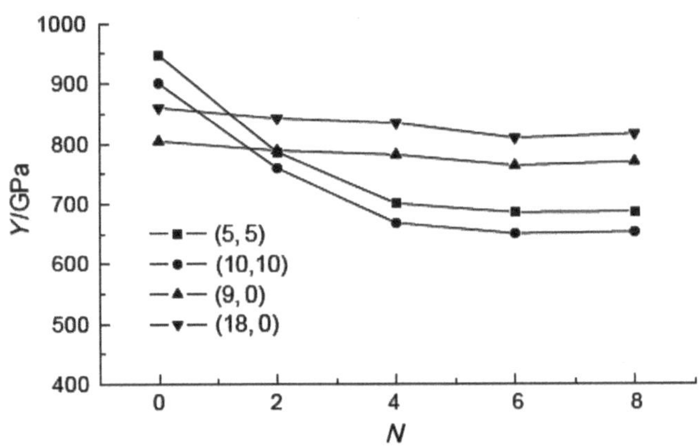

FIGURE 3.13

The effect of different graft numbers (*N*) on the Young's modulus (*Y*) of SWCNTs [66].

results indicate that the Young's moduli of the SWCNTs are generally very large, with the values for the armchair (5, 5) (10, 10) and zigzag (9, 0) (18, 0) SWCNTs without grafts being 948, 901 and 804, 860 GPa, respectively [66]. The Young's moduli of the (5, 5) SWCNTs are very close to the value of 0.94 TPa reported in [64]. In general, the effect of tube length on the Young's modulus of CNT is small. For the comparison purposes, the CNTs in this study are all set in the same cuboids crystal lattice (see Fig. 3.2), i.e., the CNTs are of infinite lengths. It was found that the effect of tube diameter on the elastic properties of CNT are size-dependent [65,67], but the Young's modulus of CNT is not a monotone continuous variation with the tube diameter [68]. It is known that the armchair CNT and zigzag CNT are different in their structure. Their basic hexagon units are perpendicular to each other; therefore, the size-dependent Young's modulus should be different. It was reported that the Young's modulus increases with the decrease of the tube radius [69,70]; but opposite trends were also reported [65]. It indicates that a further investigation of the Young's modulus is worth carrying out.

When the zigzag SWCNTs are grafted with amine, their Young's moduli decrease slightly, but an increase in the graft number does not cause any further reductions in the moduli, which basically remain at about 95% of those of the zigzag SWCNTs without grafts. The armchair SWCNTs behave somewhat differently when grafted with amine, with a marked drop in the Young's moduli. When the graft number is 2 or 4, the Young's moduli drop by 17% and 26%, respectively, but the decrease then slows until the moduli remain at a certain value with further increases in the graft number. When the graft number reaches 6 or 8, the Young's moduli are about 72% of those of the armchair SWCNTs without grafts.

SWCNTs can be classified as zigzag, armchair, or chiral types according to their chiral angle. The chiral angle of zigzag nanotubes is 0 degree, whereas that of armchair nanotubes is 30 degrees. The chiral angle of chiral SWCNTs is greater than 0 degree and less than 30 degrees. Therefore, the foregoing results indicate that the effects of amine grafts on the elastic properties of SWCNTs are related to their chiral angle.

3.3.3.3 Analysis of the results

As the effects of amine grafting on the elastic modulus follow a similar law for the same types of CNTs, only (5, 5) and (9, 0) CNTs are selected for analysis. Figs. 3.14 and 3.15 show the deformation of the electronic distribution and density sections of the isolines as calculated by DFT theory by simulating (5, 5) and (9, 0) SWCNTs in a 0 to 8−graft model. The electron density is same as the online equivalent, and the deformation of the electron density gradually becomes smaller on the vertical isolines from the inside outward. The isoline intervals reflect the local electronic conditions, in that the greater the interval, the weaker the localization and the freer the electrons, and vice versa.

As shown in Fig. 3.14, when the armchair SWCNTs are grafted, the outer electron density of the CNTs decreases and their inner isoline interval become obviously denser as the number of grafts increases, which indicates an increase in the localization of the s electron. The inner s electronic isoline interval tends to remain stable, which shows that the electronic shackles from the atomic nucleus become stabilized to some extent when enhanced. The outer-layer electrons are the main cause of the covalent bond. The stronger the shackles of the nucleus on the s electron, the less likely the possibility of electronic stimulation from $2s$ to $2p$. Consequently, the sp hybrid probability declines and the outer-layer electrons and number of bonds decrease, which results in structural stability and a decline in the elastic modulus. At the same time, it is observed for 1−4 amines that the nonlocalized outer large π bond electrons gradually flow to the grafting area at the ends of the nanotubes. This indicates that both the conductivity and the bonding electrons, which maintain the stability of the structure, have decreased. The resulting elastic moduli reflect a decline in structural stability. As the number of grafts increases further, the electron density tends toward a stable state and the elastic moduli remain the same. However, the situation for the zigzag SWCNTs

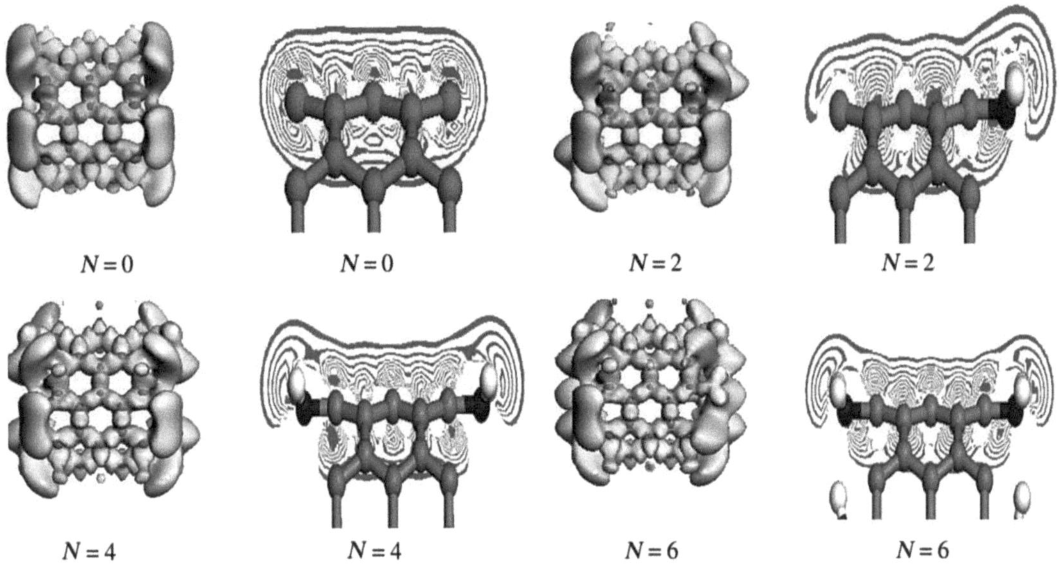

FIGURE 3.14

The deformation electron density (left) and isoline structure (right) of (5, 5) SWCNTs with different graft numbers (N) [66].

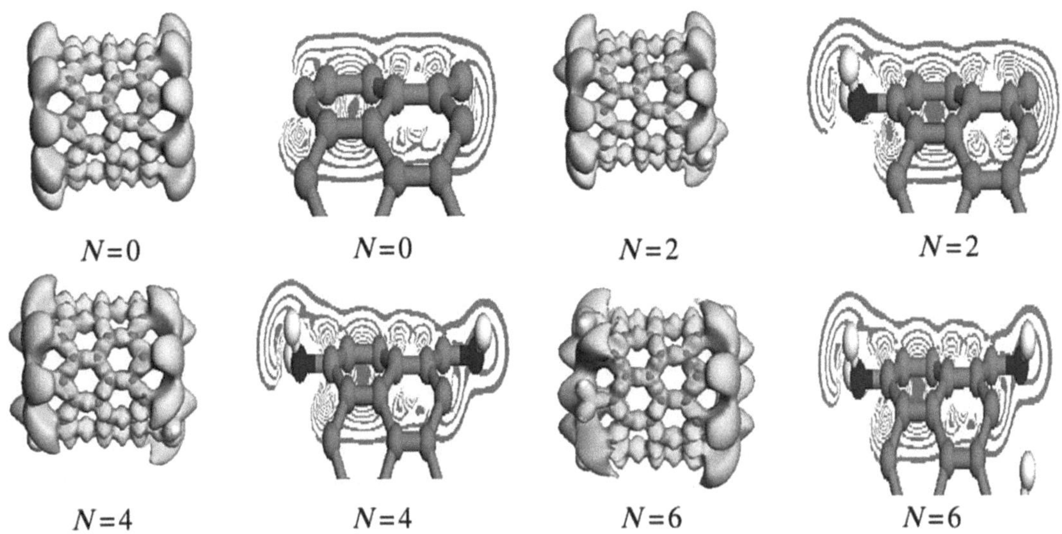

FIGURE 3.15

The deformation electron density (left) and isoline structure (right) of (9, 0) SWCNTs with different graft numbers (N) [66].

is different. It can be seen from Fig. 3.15, which describes the zigzag SWCNTs, that both the electron density of the nonlocalized outer layer of the large π bond, which is related to the conductivity, and the electron density of the r bond within the domain change little as the quantity of grafts changes. The isoline intervals also show no obvious change when the graft quantity changes. Fig. 3.14 and 3.15 show that the deformation electron density follows a similar change law to the elastic modulus (Fig. 3.13).

As shown in Tables 3.6 and 3.7, the C–C bond length of the SWCNTs changes as a result of grafting. The C–C covalent bond length is closely related to s and p components in the hybrid orbital. The reduction of the p component and the increase of the s component result in a shortening of the bond length. This is because the space configuration of the s orbital determines that its electronic cloud is close to the atomic nucleus and its orbital radius is smaller, and the space configuration of the p orbital determines that its electronic cloud is further away from the atomic nucleus and its orbital radius is larger. There are usually three hybrid forms of carbon atoms, namely, sp^3, sp^2, and sp. In a hybrid orbital, if the s component increases, then the orbital energy will be reduced and the effective radius decreased, and the covalent bond length will be shortened accordingly. The CNTs under study are composed of a number of carbon hexagon rings, and their structure can thus be considered as a rolled-up graphene sheet where the localized strong σ bond and delocalized π bond are formed by combining each carbon atom with the three adjacent carbon atoms through sp^2 hybrid orbital bonding.

Table 3.6 The C–C Bond-Length (nm) of the Optimized (5, 5) Nanotubes With Grafts [66]

	Center		Port Circumferential Direction	
Graft Number	Circumferential Direction	Axial Direction	Further from Graft	Nearer to Graft
0	0.1469	0.1468	0.1448	0.1448
2	0.1467	0.1482	0.1446	0.1567
4	0.1465	0.1493	0.1447	0.1570
6	0.1466	0.1497	0.1447	0.1572
8	0.1465	0.1497	0.1448	0.1573

Table 3.7 The C–C Bond-Length (nm) of the Optimized (9, 0) Nanotubes With Grafts [66]

	Center		Port Circumferential Direction	
Graft Number	Circumferential Direction	Axial Direction	Further from Graft	Nearer to Graft
0	0.1443	0.1504	0.1405	0.1405
2	0.1442	0.1502	0.1406	0.1418
4	0.1442	0.1505	0.1407	0.1426
6	0.1444	0.1501	0.1408	0.1424
8	0.1443	0.1506	0.1411	0.1427

Tables 3.6 and 3.7 show that the C–C bond length in the center of the optimized zigzag (9, 0) SWCNTs changes little in the circumferential and axial directions when the graft number increases. This indicates that the stability of the structure does not change, and thus that the Young's moduli undergo almost no change as the graft number increases. The armchair (5, 5) SWCNTs, in contrast, behave totally differently. When their ends are grafted with amines, the C–C bond-length in the center is shortened slightly in the circumferential direction, and the impact of the graft number on the bond length is small. In contrast, the C–C bond length in the axial direction is significantly elongated, and the impact of the graft number on the bond length is large. When the nanotubes are grafted with 0–4 amines, the bond length is elongated, but when the graft number reaches 6, the bond length basically remains unchanged. In general, the greater the bond-length difference in the circumferential and axial directions, the greater the decline in structural stability and the smaller the elastic modulus. This may be the main reason for the difference in the elastic moduli of the armchair and zigzag SWCNTs of the same tube diameter without grafts. It is clear from Tables 3.6 and 3.7 that the C–C bond-length changes in the grafted armchair SWCNTs are almost the same as the elastic modulus changes. The bond-length changes can be attributed to a certain number of p electron-bonding transfers. The p electrons in the center of the grafted armchair SWCNTs transfer from the circumferential direction to the axial direction and the grafted ends. This results in a reduction in the p component and an increase in the s component in the circumferential direction, and an increase in the p component and reduction in the s component in the axial direction in the bonding capacity of the hybrid orbital. In CNTs, the σ bond and π bond can be built by electron coupling in the outer layer of carbon atoms. Both the σ bond and the π bond can affect the Young's modulus, but the main contribution comes from the σ bond. Because the stability of the σ bond is greater, it is impossible for its electron-bonding to undergo an obvious shift. The mobility of the π electron bond is greater, which is directly related to the conductivity. Consequently, electron transfers in the circumferential direction can be divided into two types. In the first type, p electrons form a σ bond that transfers in the axial direction to the nearest neighbor, but the transfer number has a saturation property. This means that when the number of grafts reaches 6, the number of p electrons in the axial direction no longer increases. In the second type, more p electrons form and the bond is transferred to the distant ports where the grafts are. In this case, the bond-length changes are not very large, and the shorter the bond length, the smaller the distance between the atoms and the greater the energy absorbed from the splitting of the covalent bond. Consequently, if the bond energy is high, then the bond is strongly stable and the elastic modulus is large. The elastic moduli of amine-grafted armchair SWCNTs thus depend primarily on the axial bond length in the center. The axial bond-length changes in the center of armchair SWCNTs with different numbers of amine grafts have a strong relation with the corresponding changes in the Young's modulus.

To further analyze the impact of grafts on the Young's moduli of SWCNTs, the total deformation potential energy of the optimized structure of SWCNTs with different numbers of grafts is calculated. The calculated results for (5, 5) and (9, 0) SWCNTs with 0–8 grafts at their ports are shown in Figs. 3.16 and 3.17. It can be seen that the potential energy is about 2710 kJ/mol after the structure of the (9, 0) SWCNTs has been optimized to a stable state, and that the value does not show any clear correlation with the graft number. The behavior of (5, 5) SWCNTs is rather different, with a stability potential energy value when no grafts are present of 2650 kJ/mol. This value rises to 2715 kJ/mol with a 2–amine graft and 2748 kJ/mol with a 4–amine graft, and then remains

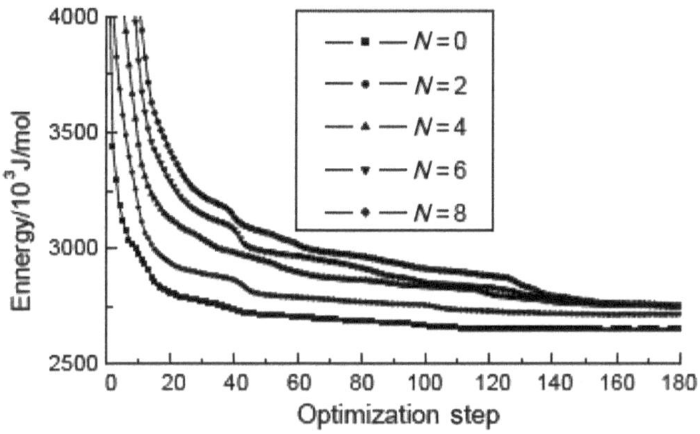

FIGURE 3.16

The deformation potential energy of (5, 5) SWCNTs with different graft [66].

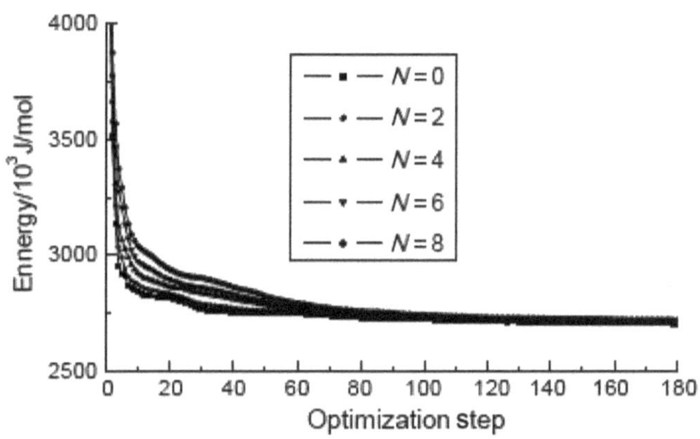

FIGURE 3.17

The deformation potential energy of (9, 0) SWCNTs with different graft numbers (*N*) [66].

at about 2750 kJ/mol for any further increases in graft number. These changes basically concur with those shown in Fig. 3.13.

The total deformation potential energy of a molecular structure is calculated using an analytical function that is a sum of a number of individual energy terms. At its simplest level, this function includes bond stretching, valence angle bending, torsion, and nonbonded interaction terms, all of which determine the energetic penalty of the structure based upon deviations from idealized equilibrium geometry. For instance, the bond term is a summation of all of the bonds in the structure, in which the energy of each bond is evaluated based on how far it is deformed from its equilibrium value. The total deformation potential energy being different from the aforementioned optimized structures indicates that the deviations from the idealized equilibrium geometry are different. The deformation of a structure creates equilibrium space variety among its atoms. When the distance between a pair of atoms is greater, an attraction will appear between them. When the distance is small, the electronic cloud repulsion between the atoms will work quickly in accordance with Pauli's exclusion principle. The equilibrium spacing between atoms near the origin equilibrium position is fixed by two opposite effects. According to linear elastic theory, the net force between the atoms is proportional to the deviations between the actual and idealized equilibrium spacing, and its potential role is proportional to the square of the deviations; the greater the structural deviations, the higher the potential energy. This suggests that the looser its intermolecular combining, the more easily the structure will be deformed and the smaller its Young's modulus will be. The total deformation potential energy variabilities of the optimized armchair (5, 5) and zigzag (9, 0) SWCNTs with different graft numbers have a good relation with their corresponding Young's modulus variations.

3.4 STRUCTURAL STABILITY AND BUCKLING OF CNTs

3.4.1 BUCKLING OF SWCNTs AND MWCNTs

This study is to provide a comprehensive study on the buckling behavior of perfectly structured single-walled and multiwalled CNTs under compressive deformation. In this study, the optimum diameter of an SWCNT was computed at its maximum buckling load, and the effect of the number of layers in an MWCNT on its properties during buckling was investigated [16]. The spontaneous plastic collapse of the nanotubes from the simulation is in qualitative agreement with the experimental observations of Lourie et al. [71] and the simulated observations of Srivastava et al. [22] using the quantum generalized tight binding method.

In this study, the buckling behavior of single-walled and multiwalled CNTs were simulated by solving the equations of motions using the Gear's predictor-corrector algorithm [72]. The axial compression of perfectly structured single-walled and multiwalled CNTs is achieved by applying a rate of 20 and 10 m/s, respectively, at both ends. At the same time, the atoms at both ends of the nanotube are kept transparent to the interatomic forces. The end atoms are then moved inwardly along the axis by small steps, followed by a conjugate gradient minimization method whilst keeping the end atoms fixed.

3.4.1.1 Single-walled carbon nanotubes

To characterize the buckling behavior of SWCNTs, a (8, 0) SWCNT is used in the simulation. The nanotube has a length $l = 43$ Å and diameter $d = 6.3$ Å. Each time-step used in this simulation is equivalent to 1 fs and the simulations are allowed to run for 20,000 time-steps.

Fig. 3.18 shows the plot of strain energy per atom, which is determined as the difference in total energy per atom of the strained and unstrained CNT, against strain, which is defined as the ratio of elongation/original length for the (8, 0) CNTs after it is compressed axially using the Brenner's second-generation empirical potential function [37]. In addition, for comparison, the similar strain energy calculated for (8, 0) CNT using quantum generalized tight-binding MD (GTBMD) [22] is also shown. At the elastic regime, both Brenner's second-generation empirical analytical potential function and the GTBMD simulation agree well (only with a slight difference). Furthermore,

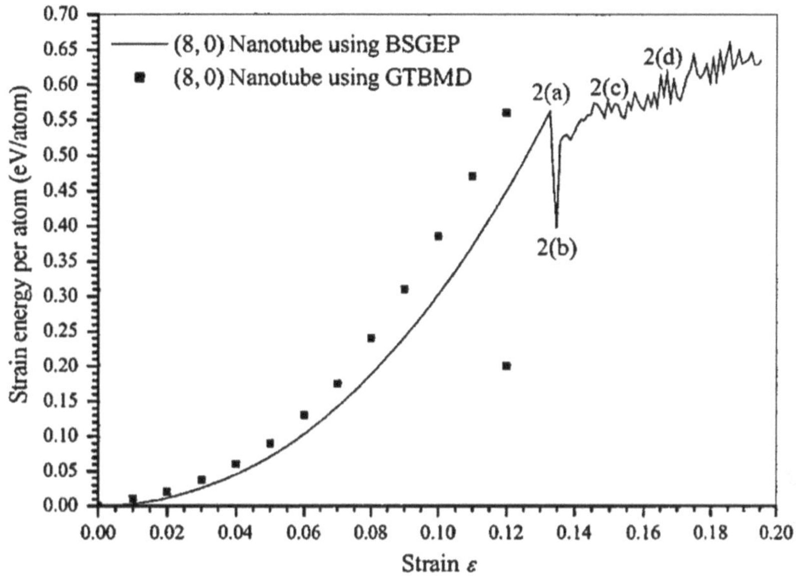

FIGURE 3.18

Computed strain energy per atom for (8, 0) single-walled carbon nanotubes using Brenner's "second-generation" empirical potential (BSGFP) [37] and (8, 0) single-walled carbon nanotube using GTBMD [22] (Liew et al. [16]).

Brenner's second-generation empirical analytical potential function shows that the (8, 0) CNT can be compressed up to a strain $\varepsilon = 0.13$ before buckling; while the GTBMD simulation for the same nanotube shows that buckling starts at strain $\varepsilon = 0.12$. Hence, generally both the GTBMD and Brenner's second-generation empirical analytical potential agree well with each other, validating the accuracy of the latter.

Fig. 3.18 depicts that the (8, 0) SWCNT undergoes elastic deformation before collapsing catastrophically at a certain critical strain of $\varepsilon = 0.13$, resulting in a 30% spontaneous drop of strain energy per atom to 0.4 eV/atom. The nature of the spontaneous plastic collapse of single-walled nanotubes reported in this paper is in qualitative agreement with the experimental observation of Lourie et al. [71] and the simulated observation of Srivastava et al. [22] using the GTBMD scheme. The calculated critical stress $\sigma_{cr} = 149$ GPa for the (8, 0) SWCNT in this work is also close to the simulation work of Srivastava et al. [22] who obtained $\sigma_{cr} = 153$ GPa. The (8, 0) CNT is subjected to acute morphological changes, hence, higher strains are found in the (8, 0) CNT, particularly around the kinks.

In Fig. 3.19 , the shading indicates the strain energy per atom, equally spaced from below 0.25 eV (lightest) to above 1.25 eV (darkest). According to Fig. 3.19A, at $\varepsilon = 0.13$ the compressed (8, 0) CNT maintains its cylindrical shape before spontaneously deforming into a symmetric three-fin pattern as shown in Fig. 3.19B leading to a plastic collapse and a net release of energy. This phenomenon has also been discussed by Srivastava et al. [22] The highly symmetric morphological shape is intrinsic to the mechanics of a perfect, undisturbed CNT as reported [11,22]. Furthermore, higher strain energy per atom is found along the edges of the fin. At $\varepsilon = 0.15$, the nanotube

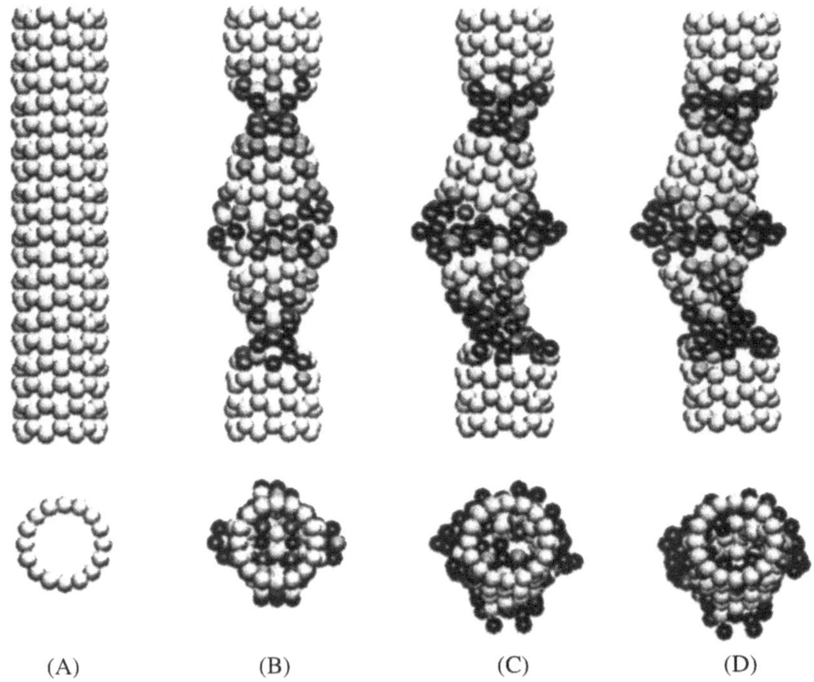

FIGURE 3.19

Morphological changes for (8, 0) single-walled nanotube showing high strains concentrated at the kinks. The (8, 0) nanotube at $\varepsilon = 0.13$ (A) spontaneously collapse into a three-fin pattern while maintaining its straight axis (B). At $\varepsilon = 0.15$, the nanotube buckles sideways (C) before further reduction in length at $\varepsilon = 0.17$ (D) [16].

continues to deform with the pinches thinning while buckling sideways, as shown in Fig. 3.19C. It is observed that high strain energy per atom is located across the pinch as the nanotube is being compressed. Further increase in strain beyond $\varepsilon = 0.17$ causes the nanotube to continue to buckle sideways with further reduction in length l, as shown in Fig. 3.19D, where high strain energy per atom is concentrated at the kinks.

A length/diameter (l/d) ratio of 7.7:1 is used to illustrate its relationship with the buckling loads Pcr of selected zigzag and armchair SWCNTs listed in Table 3.8. From the table, it is observed that as the diameters d increase, the buckling loads P_{cr} also increase steadily until the optimum diameters of $d = 11.76$ Å and $d = 20.36$ Å for the zigzag and armchair nanotubes, respectively, were attained. Beyond these optimum diameters d, the buckling loads P_{cr} start to decrease instead. It is, however, important to note that for any l/d ratio, there is an optimum diameter.

For clarity, Fig. 3.20 shows the plot of buckling load P_{cr} against diameter d for zigzag SWCNTs. As the diameter d of the zigzag nanotube increases, there is a rapid increase in its buckling load P_{cr}. However, as the diameter d reaches an optimum value $d = 11.76$ Å, any further increment beyond that will cause the buckling load P_{cr} to decrease slowly to a steady value of about 1.10×10^{-7} N. It is evident that the length l of the CNTs will affect the critical strain ε_{cr}.

Table 3.8 Buckling Loads for Selected Zigzag and Armchair Single-Walled Carbon Nanotubes [16]

Zigzag						Armchair					
(*n*, *m*)	Diameter *d* (Å)	P_{cr} ($\times 10^{-7}$ N)	(*n*, *m*)	Diameter *d* (Å)	P_{cr} ($\times 10^{-7}$ N)	(*n*, *m*)	Diameter *d* (Å)	P_{cr} ($\times 10^{-7}$ N)	(*n*, *m*)	Diameter *d* (Å)	P_{cr} ($\times 10^{-7}$ N)
(4, 0)	3.13	0.72	(20, 0)	15.67	1.25	(4, 4)	5.43	0.87	(20, 20)	27.15	1.19
(5, 0)	3.92	0.94	(21, 0)	16.46	1.24	(5, 5)	6.79	0.88	(21, 21)	28.50	1.17
(6, 0)	4.70	1.01	(22, 0)	17.24	1.23	(6, 6)	8.14	0.90	(22, 22)	29.86	1.16
(7, 0)	5.49	1.09	(23, 0)	18.02	1.21	(7, 7)	9.50	0.90	(23, 23)	31.22	1.16
(8, 0)	6.27	1.10	(24, 0)	18.81	1.18	(8, 8)	10.86	0.91	(24, 24)	32.58	1.15
(9, 0)	7.05	1.14	(25, 0)	19.59	1.16	(9, 9)	12.22	1.03	(25, 25)	33.93	1.14
(10, 0)	7.84	1.19	(26, 0)	20.38	1.15	(10, 10)	13.57	1.04	(26, 26)	35.29	1.12
(11, 0)	8.62	1.20	(27, 0)	21.16	1.14	(11, 11)	14.93	1.12	(27, 27)	36.65	1.11
(12, 0)	9.40	1.22	(28, 0)	21.94	1.14	(12, 12)	16.29	1.13	(28, 28)	38.01	1.11
(13, 0)	10.19	1.25	(29, 0)	22.73	1.13	(13, 13)	17.65	1.15	(29, 29)	39.36	1.10
(14, 0)	10.97	1.27	(30, 0)	23.51	1.12	(14, 14)	19.00	1.16	(30, 30)	40.72	1.09
(15, 0)	11.76	1.30	(35, 0)	27.43	1.11	(15, 15)	20.36	1.16	(32, 32)	43.44	1.09
(16, 0)	12.54	1.29	(40, 0)	31.35	1.10	(16, 16)	21.72	1.18	(34, 34)	46.15	1.09
(17, 0)	13.32	1.27	(45, 0)	35.27	1.10	(17, 17)	23.08	1.19	(35, 35)	47.51	1.09
(18, 0)	14.11	1.26	(50, 0)	39.18	1.09	(18, 18)	24.43	1.19	(37, 37)	50.22	1.09
(19, 0)	14.89	1.25	(55, 0)	43.10	1.09	(19, 19)	25.79	1.20	(40, 40)	54.29	1.09

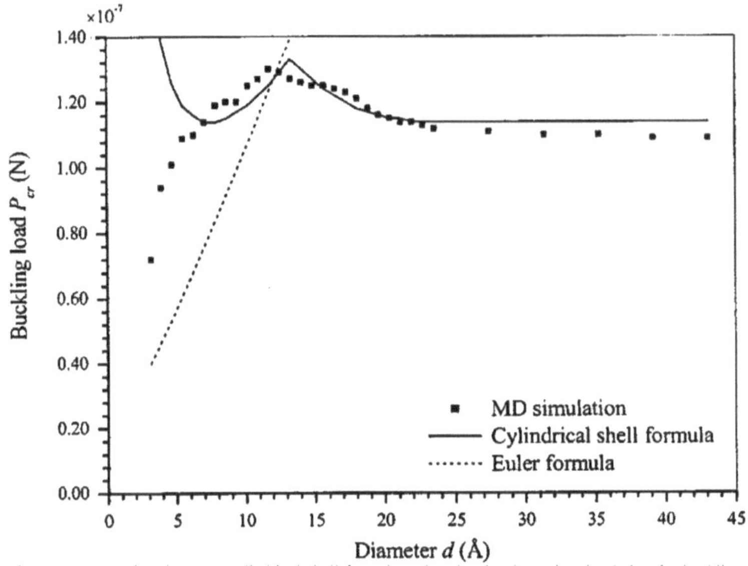

FIGURE 3.20

Comparison between cylindrical shell formula and molecular dynamics simulation for buckling loads P_{cr} of various zigzag single-walled nanotubes with different diameters [16].

It is worthy of note that the classical shell theory can be used for the local buckling analysis of nanotubes. For a layer of cylindrical shell with length l, radius r, thickness t, Young's modulus E, Poisson's ratio ν, and m and $2n$ longitudinal and circumferential wave numbers, the critical stress for the buckling of the cylindrical shell is obtained as [73]

$$\sigma_{\mathrm{cr}} = \frac{Et^3}{12(1-\nu^2)}\left\{\frac{\left[n^2+\left(m^2\pi^2r^2/l^2\right)\right]^2}{r^2\left(m^2\pi^2r^2/l^2\right)}\right\} + \frac{E\left(m^2\pi^2r^2/l^2\right)}{\left[n^2+\left(m^2\pi^2r^2/l^2\right)\right]^2} \tag{3.19}$$

In predicting the buckling load of nanotubes by using the cylindrical shell model, almost all previous literatures adopted the interlayer separation of graphite, i.e., $t = 3.4$ Å as the representative thickness of single-walled nanotubes. However, the buckling loads obtained from Eq. (3.19) are much larger than our MD simulated results if $t = 3.4$ Å is used. Here, we took the diameter of carbon atom (1.54 Å) as the thickness of the nanotube, as shown in Fig. 3.19A. The Young modulus $E = 1.28$ TPa is directly extracted from the experimental results of Wong et al. [74] for the calculation of buckling loads using Eq. (3.19). The buckling load is not sensitive to the value of Poisson's ratio, thus in this study, it is taken as $\nu = 0.25$. Using the cylindrical shell formula in Eq. (3.19) and MD, buckling loads are computed for various zigzig single-walled nanotubes, as shown in Fig. 3.20. It is observed from Fig. 3.20 that the two sets of results are in good agreement with diameter d ranging from 5.49 to 43.1 Å or larger. In spite of this, the results obtained from the cylindrical shell formula (Eq. 3.19) for diameters d smaller than 4.7 Å do not agree well with that of the MD simulation. This is because the nanotube behaves more similarly to a rod than a

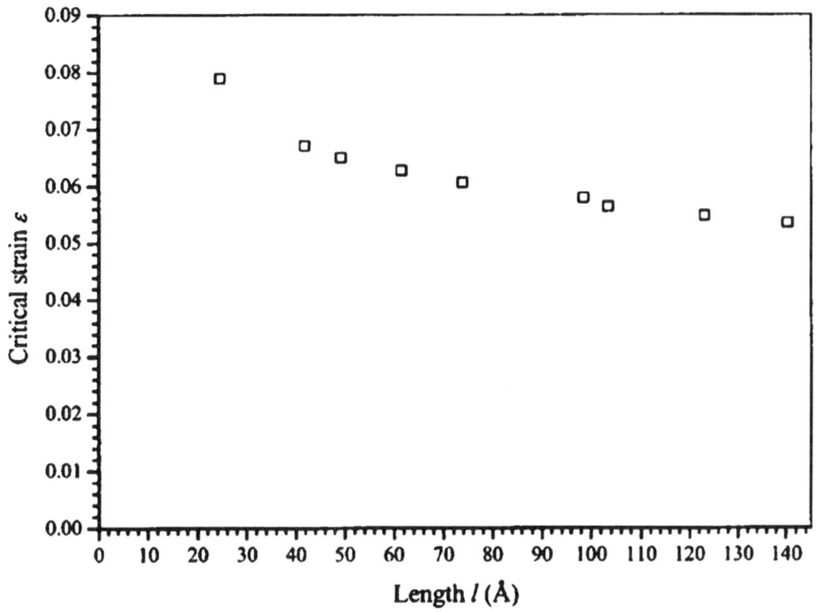

FIGURE 3.21

Critical strains ε_{cr} for (10, 10) single-walled carbon nanotubes with diameter of 13.57 Å at different lengths l [16].

cylindrical shell with thickness/radius (t/r) ratio larger than 0.66. In view of this, the buckling loads of zigzag nanotubes with fixed ends are also calculated using Euler's formula [75] and are presented in Fig. 3.20,

$$P_{cr} = \frac{4\pi^2 EI}{l^2} \tag{3.20}$$

where I is the moment of inertia. For the fixed length/diameter (l/d) ratio, the buckling loads obtained from Euler's formula for nanotubes with fixed ends are proportional to the diameter d of nanotubes, as shown in Fig. 3.20. It is obvious from the trend that the Euler formula is more reasonable than the cylindrical shell model in the estimation of the buckling loads of nanotubes with (t/r) ratio larger than 0.66.

Fig. 3.21 plots the various critical strains ε_{cr} of a (10, 10) SWCNT at different lengths l from 24.6 to 140.0 Å. This plot shows that as the length l increases, the critical strain ε_{cr} decreases. This indicates that a larger l/d ratio results in a lower critical strain ε_{cr}.

3.4.1.2 Multiwalled carbon nanotubes

Comparisons are also made between MWCNTs to determine the effect of the number of layers on the properties of the multiwalled nanotubes. Three configurations of multiwalled nanotubes are considered in the simulation: the first is a two-walled (5, 5) and (10, 10) nanotube; the second is a three-walled (5, 5), (10, 10), and (15, 15) nanotube; and the last is a four-walled (5, 5), (10, 10), (15, 15), and (20, 20) nanotube. The lengths of all three multiwalled nanotubes are $l = 60$ Å. And

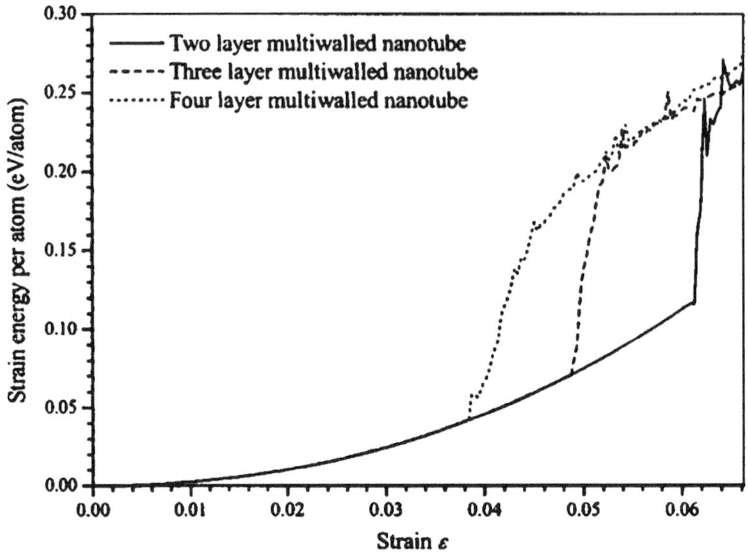

FIGURE 3.22

Strain energy per atom for two layers ((5, 5) and (10, 10) with diameter of 13.57 Å), three layers ((5, 5), (10, 10), and (15, 15) with diameter of 20.36 Å), and four layers ((5, 5), (10, 10), (15, 15), and (20, 20) with diameter of 27.15 Å) multiwalled carbon nanotubes [16].

their diameters d are 13.52, 20.36, and 27.15 Å for the two-, three-, and four-walled nanotubes, respectively.

It is worth noting that the van der Waals potential is nonzero only if the covalent potential is zero, such that there is no artificial reaction barrier formed by the steep repulsive wall of the Lennard–Jones 12–6 potential to prevent nonbonded atoms from chemical reactions. The three multiwalled nanotubes are compressed axially using the same method as the single-walled nanotubes. Each time-step is equivalent to 1 fs and there are a total of 20,000 time-steps, which corresponds approximately to $\varepsilon = 0.066$. The resulting plots are shown in Fig. 3.22. For multiwalled nanotubes with m layers, the total strain energy is

$$E_s = \sum_{i}^{m} E_i \tag{3.21}$$

Fig. 3.22 depicts that the two-walled CNT manages to keep its elasticity for the longest with the largest critical strain at $\varepsilon = 0.06$, whilst the four-walled CNT has the lowest critical strain at $\varepsilon = 0.038$. Instead of a spontaneous decrease of strain energy upon buckling as evidenced in the SWCNTs, the strain energy per atom increases until all the layers in the MWCNTs are fully buckled.

The sudden increase in strain energy per atom for each layer is due to the spontaneous increase in potential energy of each atom for each layer as depicted in Fig. 3.23. From the plot, it is observed that the middle layer has the largest increment of potential energy of each atom for each layer followed by the inner layer and then the outer layer. Being sandwiched between the outer and

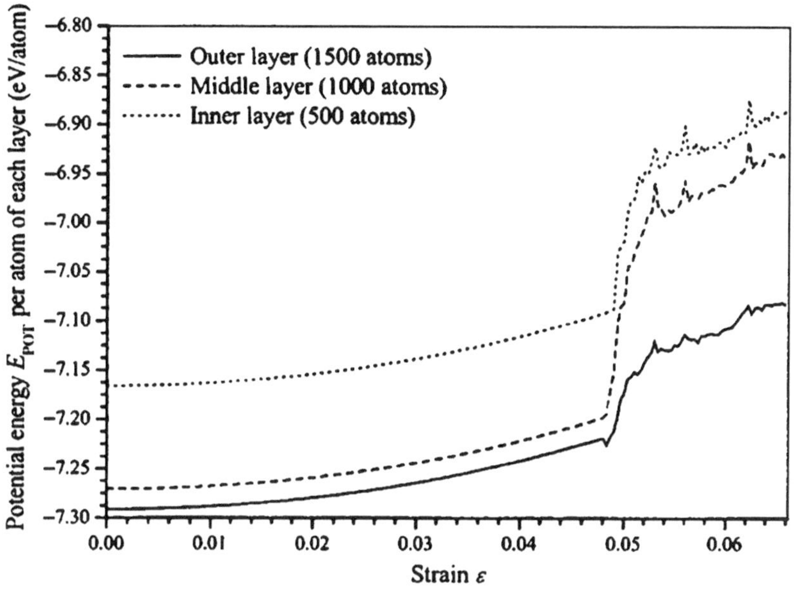

FIGURE 3.23

Potential energy per atom for each layer for (5, 5), (10, 10), and (15, 15) three-walled carbon nanotube with diameter 20.36 Å [16].

inner layer, when the three-walled CNT starts to buckle, the atoms at the middle layer will have more nearest neighbors than those of the outer and inner layer. This causes the formation of more chemical bonds to form, from sp^2 to sp^3 configurations. Thus, it leads to a larger increase in the empirical bond order function and consequently the potential energy. Hence the middle layer has the highest strain energy per atom for each layer after buckling, as seen in Fig. 3.24.

Fig. 3.25 depicts the distances of a selective pair of atoms from different layers at $\varepsilon = 0.047$ (before buckling) and $\varepsilon = 0.055$ (after buckling) of the three-walled CNT. It is observed that at $\varepsilon = 0.047$, the distances between the atoms from different layers are approximately 3.4 Å apart. However, at $\varepsilon = 0.055$, the distances between these atoms are reduced to values less than 1.6 Å. The decrease in distance allows the formation of sp^3 bonds that result in a rise in the potential energy.

To further explain the phenomenon of the increase in strain energy per atom upon buckling as observed in Fig. 3.22, the plot of the strain energy per atom for each layer of the three-walled nanotube against the strain ε is presented in Fig. 3.24. Before buckling, the three layers have the same strain energy per atom for each layer. However, at $\varepsilon = 0.0484$ the outer layer starts to buckle first, followed by the middle layer at $\varepsilon = 0.0486$, and then inner layer at $\varepsilon = 0.0493$ (see the inset of Fig. 3.24 for amplification).

Fig. 3.26 shows the three-dimensional and cross-sectional three-walled nanotube at different strains. It is evident from Fig. 3.26A that the outer layer starts to deform first into a ring pattern at $\varepsilon = 0.0484$, while the middle and inner layers are undeformed and their strain energies remain low. However, further compression at $\varepsilon = 0.0486$ causes the middle layer to deform too, resulting in a

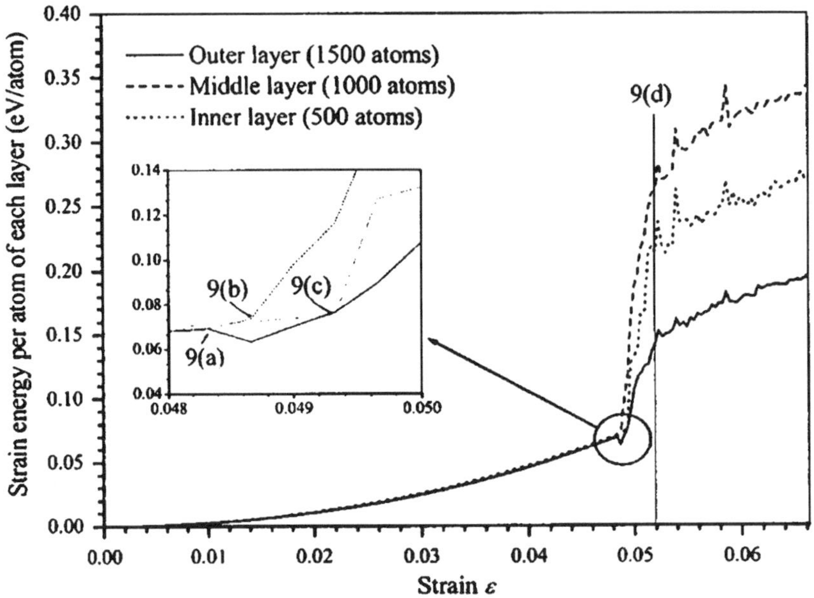

FIGURE 3.24

Strain energy per atom of each layer for (5, 5), (10, 10), and (15, 15) three-walled carbon nanotube with diameter 20.36 Å [16].

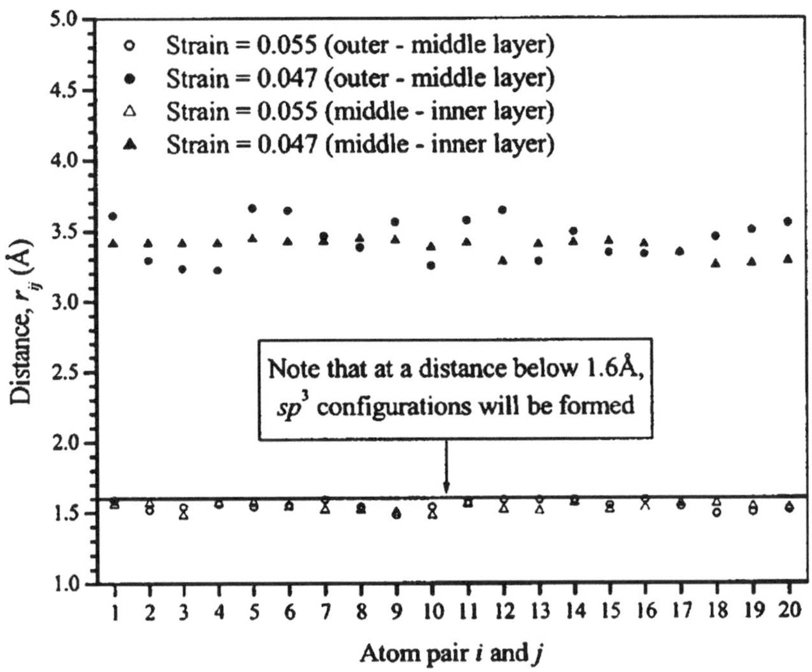

FIGURE 3.25

Distances between atoms from different layers before buckling at $\varepsilon = 0.047$ and after buckling at $\varepsilon = 0.055$ for the three-walled carbon nanotube [16].

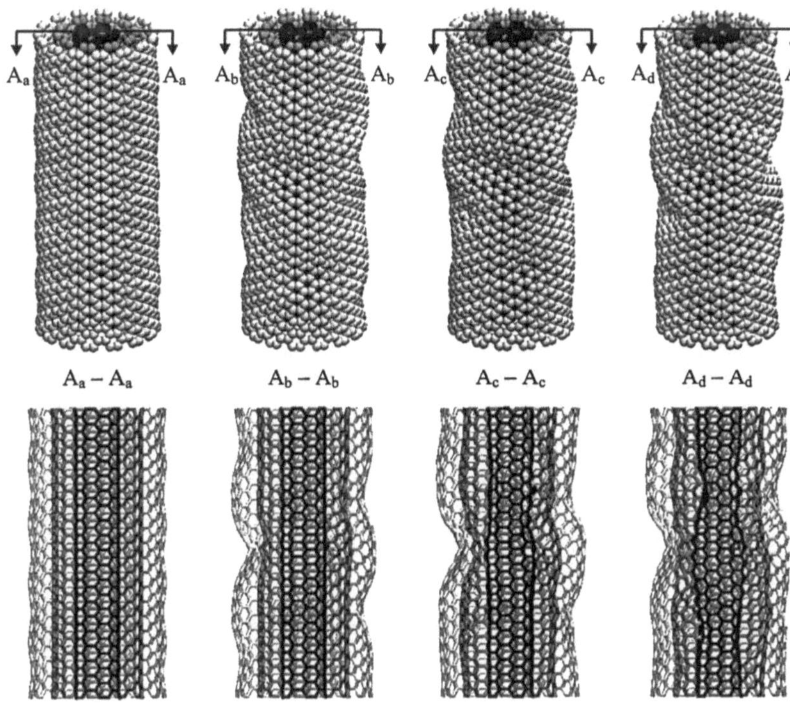

FIGURE 3.26

Three-dimensional and cross-sectional view of morphological changes for three-walled (5, 5), (10, 10), and (15, 15) carbon nanotube with diameter 20.4 Å. At $\varepsilon = 0.0484$ the outer layer deforms into a ''ring'' pattern (A). Further compression at $\varepsilon = 0.0486$ causes the middle later to deform too (B). At $\varepsilon = 0.0493$, the inner layer also buckles (C) and at $\varepsilon = 0.052$, all the three layers are deformed (D) [16].

sudden increase in the strain energy per atom because of the increase in potential energy due to the atoms in the middle layer. Similarly at this time, the inner layer maintains the elasticity of the nanotube temporarily, as shortly after at $\varepsilon = 0.0493$ the inner layer also starts to buckle, therefore increasing the overall potential energy and hence further increasing the strain energy per atom further. After $\varepsilon = 0.052$, all the layers are deformed, hence there is no further abrupt increase in potential energy due to any sudden addition of nearest neighbors.

The force−deformation plot in Fig. 3.27 shows the buckling load P_{cr} that each layer is able to withstand. Being a layer with a wider diameter d, the outer layer of the three-walled CNT has the highest bucking load P_{cr} at about 1.18×10^{-7} N. However, it also tends to buckle fractionally earlier than the other two layers after deforming $\Delta l = 2.9$ Å. The buckling loads P_{cr} for the middle and inner layers are lower at 0.88×10^{-7} N and 0.42×10^{-7} N, respectively.

According to Fig. 3.28, the buckling load P_{cr} increases as the number of layers in a multiwalled nanotube increases. The four-walled nanotube buckles first at $\Delta l = 2.3$ Å even though it has the

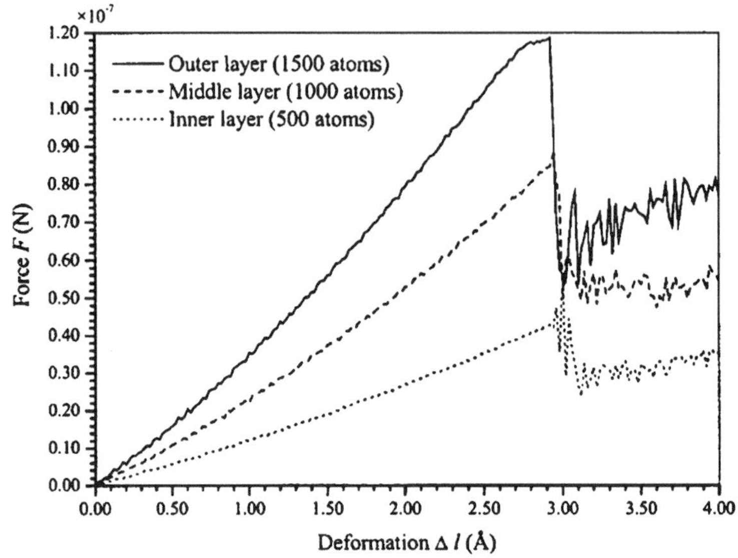

FIGURE 3.27

Force-deformation curves of each layer for (5, 5), (10, 10), and (15, 150) three-walled carbon nanotube with diameter 20.4 Å [16].

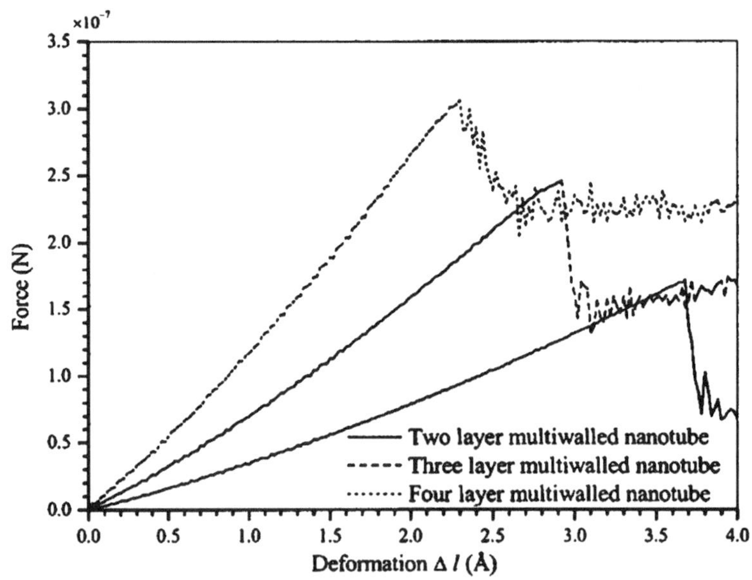

FIGURE 3.28

Force-deformation curves for two layers ((5, 5) and (10, 10) with diameter of 13.57 Å), three layers ((5, 5), (10, 10), and (15, 15) with diameter of 20.36 Å), and four layers ((5, 5), (10, 10), (15, 15), and (20, 20) with diameter of 27.15 Å) multiwalled carbon nanotubes [16].

Table 3.9 Buckling Loads for Selected Two and Three-Walled Carbon Nanotubes [16]

Two-walled		Three-walled	
(n_1, m_1) (n_2, m_2)	P_{cr} ($\times 10^{-7}$N)	(n_1, m_1) (n_2, m_2) (n_3, m_3)	P_{cr} ($\times 10^{-7}$ N)
(5, 5) (10, 10)	1.72	(5, 5) (10, 10) (15, 15)	2.46
(10, 10) (15, 15)	2.02	(10, 10) (15, 15) (20, 20)	2.73
(15, 15) (20, 20)	2.09	(15, 15) (20, 20) (25, 25)	2.81
(20, 20) (25, 25)	2.10	(20, 20) (25, 25) (30, 30)	3.05

highest buckling load P_{cr} at 3.2×10^{-7} N; while the three-walled nanotube buckles at $\Delta l = 2.9$ Å with $P_{cr} = 2.5 \times 10^{-7}$ N. And finally, the two-walled nanotube buckles last at $\Delta l = 3.7$ Å but with the lowest buckling load $P_{cr} = 1.7 \times 10^{-7}$ N. Hence the more layers a multiwalled nanotube has, the higher the buckling load P_{cr} is required.

The buckling loads P_{cr} for two and three-walled CNTs are tabulated in Table 3.9. It is observed from the table that generally MWCNTs have larger buckling loads P_{cr}. For instance, the (10, 10), (15, 15) two-walled CNT has a higher buckling load $P_{cr} = 2.02 \times 10^{-7}$ N compared to the (5, 5), (10, 10) two-walled CNT that has a buckling load $P_{cr} = 1.72 \times 10^{-7}$ N (see Table 3.9). Moreover, Table 3.9 also verifies that as the number of layers in a CNT increases, its buckling load P_{cr} increases too. According to Table 3.8, the buckling load P_{cr} for a (5, 5) SWCNT is 0.88×10^{-7} N. However, if an extra outer layer is added to form a (5, 5), (10, 10) two-walled CNT, its buckling load P_{cr} increases to 1.72×10^{-7} N as seen in Table 3.9. If another layer is added to form a (5, 5), (10, 10), and (15, 15) three-walled CNT, the buckling load P_{cr} increases to 2.46×10^{-7} N.

3.4.2 STRUCTURAL STABILITY OF A COAXIAL CNTs INSIDE A BORON–NITRIDE NANOTUBE

Coaxial C@BNNT nanocables consisting of armchair and zigzag CNT cores and BNNT sheaths are optimized by the DFT method [76]. The structural stability is investigated by analyzing the parametric variation in the tube radius, energy, and deformation electron density. The interwall distance in the most favorable nanocables is about 0.35 nm for armchair C@BNNTs and from 0.33 to 0.36 nm for zigzag C@BNNTs. The best armchair and zigzag C@BNNTs are C(5, 5)@BN(10, 10) and C(9, 0)@BN(m, 0) ($m = 17$, 18), respectively. The interwall interaction between CNTs and BNNTs is a typical van der Waals interaction.

3.4.2.1 Modeling method

Using C(5, 5) or C(9, 0) as the inner tube and BN(n, n) or BN(m, 0) as the outer tube, coaxial armchair C(5, 5)@BN(n, n) and zigzag C(9, 0)@BN(m, 0) nanotubes are constructed. The inner tube and outer tube are of the same tube type and possess the same repeat units. The armchair and zigzag C@BNNTs comprise eight and four basic repeat units, respectively, and the corresponding tube lengths are 1.968 and 1.704 nm. To achieve electron saturation, hydrogen atoms are used to link the two ends of the inner and outer nanotubes. Corresponding to the different diameters of the tubes, the values $n = 7-15$ for the BN (n, n) nanotubes and $m = 14-22$ for the BN (m, 0) nanotubes are selected.

First, the structure of the C@BNNT, its inner tube, and its outer tube are optimized. The energy and electron density are then calculated. Several representative structures of optimized C(5, 5)@BN(n, n) and C(9, 0)@BN(m, 0) structures are shown in Fig. 3.29. Energy calculation is used to quantitatively analyze the internal interaction of the structures, determine the rules governing their interaction, and elucidate the interaction mechanism. By calculating the deformation electron density, the bonding state can be intuitively understood, the bonding characterization can be analyzed, and the interaction mechanism between CNTs and BNNTs is revealed from another angle.

The DMol3 software is used to optimize the structure and calculate the related electrical properties [77,78]. The function PW91 [79] belonging to the generalized gradient approximation is used to process the exchange correlation potential energy, the effective core potential approach is used to process the core electron, and the orbit population parameter smearing is set at 0.0005 a.u. By using the universal force field (UFF) [80,81], several kinds of energy of the optimized structures can also be calculated. The potential energy of the UFF is expressed as the sum of bonded and nonbonded interactions.

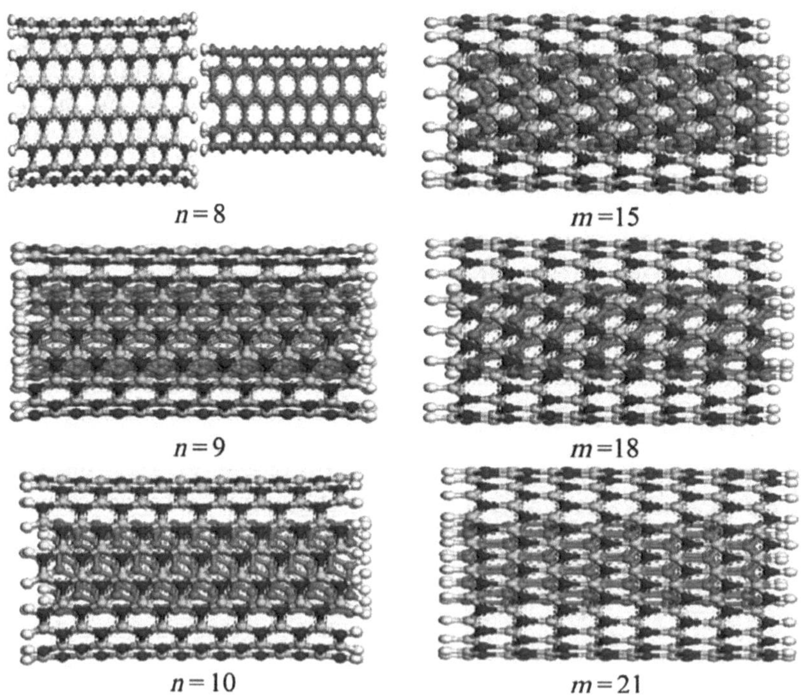

FIGURE 3.29

Structure of the optimized C(5, 5)@BN(n, n) and C(9, 0)@BN(m, 0) [76].

$$E = E_R + E_\theta + E_\tau + E_\omega + E_{vdw} + E_{el} \quad (3.22)$$

The bonded interactions include the bond stretching E_R, the angle bending E_θ, the dihedral angle torsion E_τ, and the inversion terms E_ω, respectively. The nonbonded interactions include the van der Waals interaction E_{vdw} and the electrostatic interaction E_{el}. The UFF and the parameters for C, B, and N are described in detail elsewhere in the literature [80–82].

3.4.2.2 Structural parameters

The radii of the inner and outer nanotubes for the optimized C(5, 5)@BN(n, n) (n = 9–15) are shown in Fig. 3.30 and those for C(9, 0)@BN(m, 0) (m = 15–21) are shown in Fig. 3.31. R_r is the radius of the C(5, 5) or BN(n, n) in the C(5, 5)@BN(n, n) structure, and R_f is the radius of the free C(5, 5) or BN(n, n). The radial increment $\delta R = R_r - R_f$ and the space between the inner and outer nanotubes ($\Delta R = R_r$(BNNT)$-R_r$(CNT)) is the difference between the radii of the inner and outer nanotubes. The full computational details for radius of the coaxial nanotubes are given in Table 3.10.

When $n = 9$, the radii of the outer BN(9, 9) for C(5, 5)@BN(9, 9) and of the free BN(9, 9) are 0.63600 and 0.62142 nm, respectively, the former being larger than the latter by 0.01458 nm. The radii of the inner C(5, 5) for C(5, 5)@BN(9, 9) and the free C(5, 5) are 0.33368 and 0.34094 nm, the former being less than the latter by 0.00726 nm. The space between the inner and outer nanotubes is 0.34892 nm. It is obvious that the radius of C(5, 5) decreases and that of BN(9, 9) increases when they are combined into C(5, 5)@BN(9, 9). This indicates the occurrence of greater repulsive interactions between the inner CNT and the outer BNNT. When $n = 10$, the radii of the outer BN(10, 10) for C(5, 5)@BN(10, 10) and of the free BN(10, 10) are 0.68967 and 0.69003 nm, the former being only slightly smaller than the latter by 0.00036 nm. The radius of the inner C(5, 5) for C(5, 5)@BN(10, 10) is 0.34075 nm, which is also slightly less than that of the free C(5, 5) by 0.000194 nm. The space between the inner and outer nanotubes is 0.34892 nm. It is obvious that

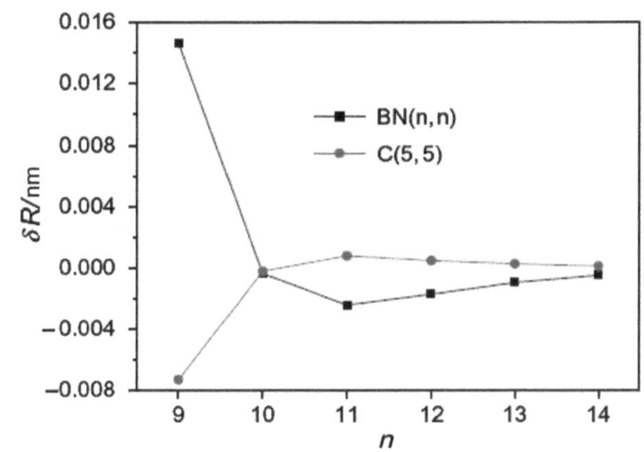

FIGURE 3.30

Radial increment (nm) versus parameter n for the inner and outer nanotubes of the optimized C(5, 5)@ BN (n, n) [76].

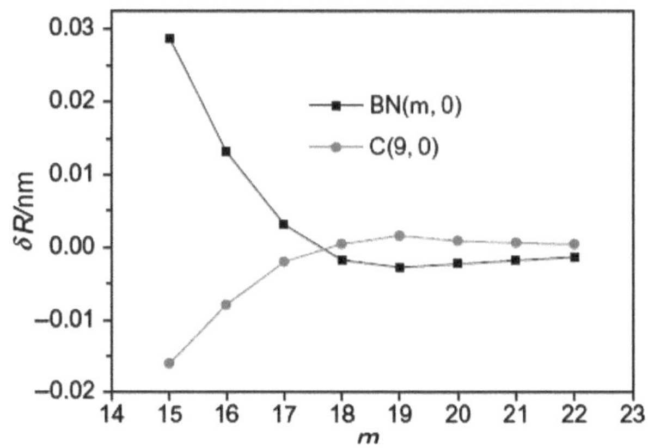

FIGURE 3.31

Radial increment (nm) versus parameter *m* for the inner and outer nanotubes of the optimized C(9, 0)@ BN(*m*, 0) [76].

Table 3.10 Radii (nm) of the Inner and Outer Nanotubes for the Optimized C(5, 5)@BN(*n*, *n*) and C(9, 0)@BN(*m*, 0) [76]

	BN(*n*, *n*)			C(5, 5)			
n	R_f	R_r	δR	R_f	R_r	δR	ΔR
9	0.62142	0.63600	0.01458	0.34094	0.33368	−0.00726	0.30233
10	0.69003	0.68967	−0.00036	0.34094	0.34075	−0.00019	0.34892
11	0.75863	0.75624	−0.00239	0.34094	0.34175	0.00081	0.41449
12	0.82712	0.82543	−0.00169	0.34094	0.34143	0.00048	0.48400
13	0.89578	0.89484	−0.00094	0.34094	0.34121	0.00027	0.55362
14	0.96428	0.96382	−0.00047	0.34094	0.34107	0.00012	0.62275
	BN(*m*, 0)			C(9, 5)			
m	R_f	R_r	δR	R_f	R_r	δR	ΔR
15	0.59842	0.62725	0.02883	0.35409	0.33806	−0.01603	0.28919
16	0.63808	0.6513	0.01322	0.35409	0.34614	−0.00795	0.30515
17	0.67781	0.68094	0.00312	0.35409	0.35206	−0.00203	0.32888
18	0.71744	0.71566	−0.00179	0.35409	0.35451	0.00042	0.36115
19	0.75344	0.75066	−0.00278	0.35409	0.35567	0.00158	0.39499
20	0.79265	0.79034	−0.00231	0.35409	0.35496	0.00087	0.43537
21	0.83636	0.83453	−0.00184	0.35409	0.3547	0.00060	0.47983
22	0.87612	0.87475	−0.00136	0.35409	0.35452	0.00043	0.52023

the radii of the inner and outer nanotubes are very close to those of the free structures. This indicates that the interaction between the inner CNT and the outer BNNT approaches zero. When $n = 11$, the radii of the outer BN(11, 11) for C(5, 5)@BN(11, 11) and the free BN(11, 11) are 0.75624 and 0.75863 nm, the former being smaller than the latter by 0.00239 nm. The radius of the inner C(5, 5) for C(5, 5)@BN(10, 10) is 0.34175 nm, which is larger than that of the free C(5, 5) by 0.000805 nm. The space between the inner and outer nanotubes (ΔR) is 0.41449 nm. It is obvious that the radius of C(5, 5) increases and that of BN(11, 11) decreases when they are combined into C(5, 5)@BN(11, 11). This again indicates greater attraction interactions between the inner CNT and the outer BNNT. As the parameter n increases, this attraction interaction decreases gradually and tends to zero. Consequently, C(5, 5)@BN(10, 10) is a relatively stable armchair structure.

The situation for the zigzag C(9, 0)@BN(m, 0) ($m = 15-21$) nanotubes (see Fig. 3.31) is similar to that for the armchair C(5, 5)@BN(n, n) already described. As parameter m increases, the interactions between the inner CNT and the outer BNNT switch from repulsion to attraction, although the transition is smoother. The corresponding transition chiral parameter m is about 17–18 and the space between the inner and outer nanotubes (ΔR) is 0.33–0.36 nm. Thus, the most stable zigzag structures are C(9, 0)@BN(17, 0) and C(9, 0)@BN(18, 0). Nanotubes are classified as zigzag, armchair, or chiral types according to their chiral angle. The chiral angle of zigzag nanotubes is 0 degree, whereas that of armchair nanotubes is 30 degrees. The chiral angle of chiral nanotubes is within 0 and 30 degrees. Comparing the foregoing armchair C(5, 5)@BN(n, n) with zigzag C(9, 0) @BN(m, 0) (see Figs. 3.30 and 3.31), it is found that the interaction between the inner and outer tube for armchair C(5, 5)@BN(10, 10) is much closer to zero than that for zigzag C(9, 0)@BN(17, 0) or C(9, 0)@ BN(18, 0). Therefore, the optimal interwall distances in C@BNNTs are related to their chiralities. The greater the chiral angle is, the smaller the interwall interaction of combination structure is. It is obvious that the armchair C@BNNTs achieve a more stable combination structure than the zigzag case.

In addition, the radii of the outer BN(10, 10) in optimal C(5, 5)@BN(10, 10) and the free BN (10, 10) are 0.68967 and 0.69003 nm; the former being slightly smaller than the latter by 0.00036 nm. Thus it is concluded that the BNNT in optimal C@BNNT is hardly influenced by the combination structure. Similar to the radii of BNNT, the structural parameter of CNTs in optimal C@BN is barely affected by the coating BNNTs. They display the same fundamental characteristics as CNTs in the free space. The BNNT around the CNTs effectively helps to protect them and makes them more stable, such as oxidation degradation of CNTs is reduced by coating with BNNTs, and the thermal stability of C@BNNTs is far superior to CNTs [83–86].

To explain the interactions between the inner CNT and the outer BNNT thoroughly, the radial distribution function (RDF) of the C(5, 5) and BN(n, n) nanotubes in the C(5, 5)@BN(n, n) ($n = 9$, 10, and 11) structure are selected for analysis (see Fig. 3.32 and Table 3.11) [22]. In the RDF of nanotubes, the peak locations (abscissa values) correspond to the first neighbor, second neighbor, and other neighbor distances between the atoms of the hexagonal atomic ring on the tube wall. Fig. 3.32 (Table 3.12) shows first and second neighbor distances for the C(5, 5) in C(5, 5)@BN(10, 10) to be 0.141 and 0.243 nm and those for the BN(10, 10) in C(5, 5)@BN(10, 10) to be 0.145 and 0.249 nm, respectively. Compared with these data in C(5, 5)@BN(10, 10), Fig. 3.32A shows that first and second neighbor distances for the C(5, 5) in C(5, 5)@BN(9, 9) are reduced to 0.139 and 0.239 nm, and the other neighbor distances also decrease. However, Fig. 3.32B shows the opposite effect for the BN(9, 9). Its first and second neighbor distances increase to 0.147 and 0.253 nm,

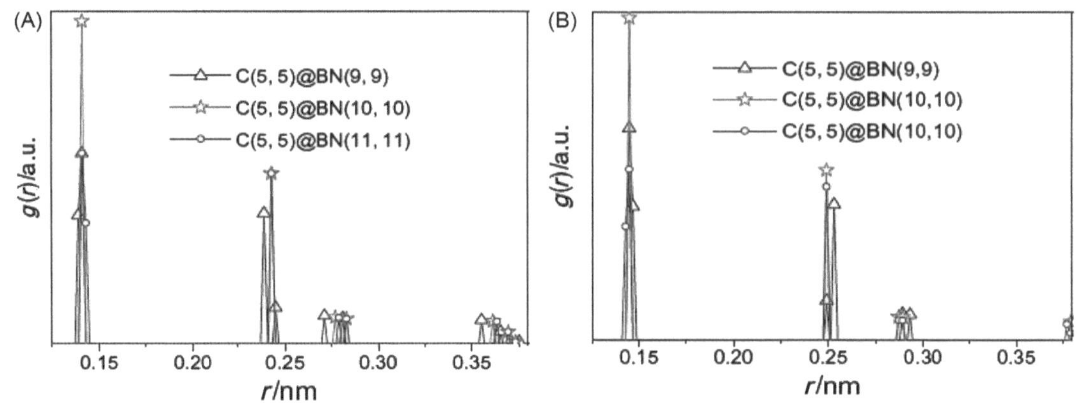

FIGURE 3.32

RDF of the C(5, 5) and BN(*n*, *n*) in C(5, 5)@BN(*n*, *n*) (A) C(5, 5) (B) BN(*n*, *n*) [76].

Table 3.11 Peak Locations (nm) of RDF of the C(5, 5) and BN(*n*, *n*) in C(5, 5)@BN(*n*, *n*) [76]

Nanotubes	*n*	Highest Peak	Subpeak
C(5, 5)	9	0.139	0.239
	10	0.141	0.243
	11	0.143	0.243
BN(*n*, *n*)	9	0.147	0.253
	10	0.145	0.249
	11	0.143	0.249

respectively. It is clear in this case that larger mutual interactions occur between the C(5, 5) and the BN(9, 9). Fig. 3.32A and B also reveal that the first neighbor distances of the C(5, 5) increase to 0.143 nm and that of BN(11, 11) are reduced to 0.143 nm in the optimized C(5, 5)@BN(11, 11), but the second and other neighbor distances only change slightly. This indicates a weak attractive interaction between the C(5, 5) and BN(11, 11). These results concur with those shown in Fig. 3.30.

3.4.2.3 Binding energy

The stability of C@BNNT is determined by the interaction force between the inner and outer nanotubes and the deformation caused by the force. The interaction energy E_i and its van der Waals energy E_{iv} between the inner and outer nanotubes and the binding energy E_b of each C@BNNT are calculated, as shown in Fig. 3.33. Tables 3.12 and 3.13 show the energy of C@BNNT and its inner CNT and outer BNNT. Their total potential energies are denoted by $E_{@}$, E_C, and E_{BN}, respectively, and their van der Waals energies are denoted by $E_{@v}$, E_{Cv}, and E_{BNv}, respectively. E_{Cf} and E_{BNf} represent the total potential energy of the inner CNT and the outer BNNT in their free state. E_C and E_{BN} are obtained by calculating the single point energy of the inner CNT and the outer BNNT in a C@BNNT. The interaction energy E_i ($E_i = E_{@} - E_C - E_{BN}$) between the inner and outer nanotubes is calculated. The binding energy E_b ($E_b = E_{@} - E_{Cf} - E_{BNf}$) of each C@BNNT is also

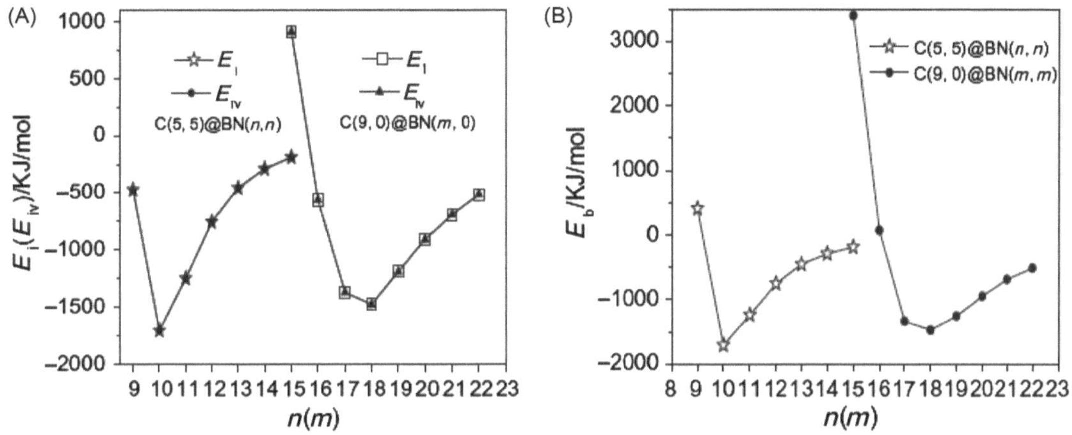

FIGURE 3.33

Interaction energy E_i (A) and binding energy E_b (B) for different $n(m)$ for C(5, 5)@BN(n, n) and C(9, 0)@BN (m, 0) [76].

calculated, in which the E_{Cf} of C(5, 5) and C(9, 0) is 6754.7044 and 5756.2738 KJ/mol, respectively. E_{iv} ($E_{iv} = E_{@v} - E_{Cv} - E_{BNv}$) denotes the van der Waals interaction between the inner and outer nanotubes, and is a component of the interaction energy E_i.

Fig. 3.33 shows that E_b is a positive value only when $n = 9$, $m = 16$. This indicates that the interaction force between the inner CNT and outer BNNT is one of repulsion. However, the interaction energy remains slightly negative. Only when $m(n)$ is further reduced does a positive E_b occur. This is probably because the single point energy of the inner CNT and the outer BNNT in a C@BNNT is influenced by the H bond at either ends of the actual deformation, which means that the computed values of E_C and E_{BN} are slightly larger than their actual values. For a C@BNNT with $n \geq 10$, $m \geq 17$, E_b and E_i are always negative. This indicates that the interaction between the inner and outer nanotubes is one of attraction. Fig. 3.33 shows the curves for E_i and E_b with different $n(m)$ for C(5, 5)@BN(n, n) and C(9, 0)@BN(m, 0). It can be clearly seen from the figure that the greatest attraction is displayed at $n = 10$ for C(5, 5)@B(n, n) and at $m = 17-18$ for C(9, 0)@BN(m, 0). At this point, the space between the tube walls for the armchair and zigzag C@BNNTs are about 0.35 and 0.33–0.36 nm, respectively. These values are just within the distance between the layers of 0.33–0.36 nm for graphite and h-BN. When $n \geq 11$ and $m \geq 19$, the attraction between the inner CNT and the outer BNNT begins to decline. As E_b is the energy change when the free CNT and free BNNT are bound together as a C@BNNT system, $n = 9$ and $m = 15$ and 16 result in endothermic processes, and the rest are exothermic processes. The heat of the exothermic process is highest when $n = 10$ and $m = 17-18$. This indicates that the systems with $n = 10$ and $m = 17-18$ are most stable. The difference between E_i and E_b reflects the energy loss caused by the deformation of the CNT and BNNT when the free CNT and free BNNT are bound together as a C@BNNT. Fig. 3.33 shows that the maximal energy loss takes place in the C@BNNT, where $n = 9$ and $m = 15$ and 16. This is because there is greater repulsion between the inner and outer nanotubes, which leads to greater deformation and results in a larger energy loss.

Table 3.12 Energy of C(5, 5)@BN(*n*, *n*) and Its Inner C(5, 5) and Outer BN(*n*, *n*) (KJ/mol) [76]

n	$E_@$	E_C	E_{BN}	E_i	E_{BNf}	E_b	$E_{@v}$	E_{Cv}	E_{BNv}	E_{iv}
9	13141.3599	7060.7654	6554.1584	−473.5639	5978.2527	408.4028	4958.5919	2962.2490	2469.9077	−473.5647
10	11227.7768	6755.2610	6177.2600	−1704.7442	6175.8915	−1702.8191	4124.2066	2412.3867	3416.5648	−1704.7449
11	11930.7525	6758.0083	6421.5398	−1248.7956	6409.8503	−1233.8022	4932.3540	2354.0673	3827.0826	−1248.7959
12	12678.7635	6755.9761	6678.0435	−755.2561	6672.8308	−748.7717	5725.5299	2364.1495	4116.6354	−755.2549
13	13259.3362	6755.0949	6960.0436	−455.8023	6958.5380	−453.9062	6338.3555	2377.3792	4416.7803	−455.8039
14	13729.1222	6754.8079	7263.0619	−288.7476	7262.5076	−288.0898	6831.7000	2386.1948	4734.2555	−288.7502
15	14147.9122	6754.7701	7581.5968	−188.4547	7581.2117	−188.0039	7260.2253	2386.3369	5062.3437	−188.4553

Table 3.13 Energy of C(9, 0)@BN(*m*, 0) and Its Inner C(9, 0) and Outer BN(*m*, 0) (KJ/mol) [76]

m	$E_@$	E_C	E_{BN}	E_i	E_{BNf}	E_b	$E_{@v}$	E_{Cv}	E_{BNv}	E_{iv}
15	14466.7376	6775.0002	6783.2829	908.4545	5312.3118	3398.152	5602.8929	3053.4566	1640.9844	908.4519
16	10782.8325	5990.191	5355.8065	−563.165	4954.4244	72.1343	4110.6857	2504.7563	2169.094	−563.1646
17	9450.1524	5773.1001	5048.7748	−1371.7225	5026.822	−1332.9434	3484.2938	2203.1192	2652.8955	−1371.7209
18	9395.5699	5757.1002	5115.2682	−1476.7985	5110.5014	−1471.2053	3582.4326	2083.3956	2975.834	−1476.797
19	9761.4202	5759.2165	5191.5173	−1189.3136	5258.5779	−1253.4315	4205.3893	2065.6473	3329.6584	−1189.9164
20	10198.1675	5758.4794	5348.434	−908.7459	5389.3492	−947.4555	4656.3946	2065.8981	3499.2452	−908.7487
21	10485.9898	5757.5615	5419.4814	−691.0531	5414.5755	−684.8595	4820.8316	2070.1659	3441.7242	−691.0585
22	10773.8162	5756.9206	5532.3457	−515.4501	5529.9226	−512.3802	5139.7949	2075.9928	3579.2546	−515.4525

It should be noted that when $n = 8$ (not shown in the Fig. 3.33), the C(5, 5)@BN(n, n) structure cannot be maintained because of the powerful repulsion between the inner and outer tubes. The optimal structure occurs with a separation along the axial direction (see Fig. 3.29). When $n \geq 11$or $m \geq 19$, the distances between the inner and outer nanotubes increase, and the deformation and energy loss decrease rapidly. In the formation of the C(5, 5)@BN(n, n) and C(9, 0)@BN(m, 0) systems, the energy change derives mainly from the interaction force between molecules, and the effect of deformation on the energy change is slight. A comparison of E_{i} and E_{iv} in Fig. 3.33 shows that the difference is very small. It can thus be concluded that the interaction between the inner CNT and outer BNNT is a typical van der Waals interaction between molecules.

3.4.2.4 Electronic structure and bonding model

To further understand the interaction between the CNT and BNNT in a C@BNNT, the deformation electron density and its isoline structure are calculated. As the effects of different values of the parameter $n(m)$ on the electronic distribution of C(5, 5)@BN(n, n) and C(9, 0)@BN(m, 0) follow a similar law, only C(5, 5)@BN(n, n) is used in the analysis. Fig. 3.34 shows the deformation electron density and its section isoline as calculated by DFT by simulating a C(5, 5)@BN(n, n) model for $n = 9$, 10, and 11. The electron density is the same as the online equivalent, and gradually becomes smaller on the vertical isolines from the inside outward. The isoline intervals reflect the local electronic conditions, in that the greater the interval, the weaker the localization and the more free the electrons, and vice versa.

As shown in Fig. 3.34, in the outer tube BN(n, n)NT, the outer-layer electrons are obviously partial to the N atom in the B–N bond. However, in the inner CNT, the electrons are more concentrated at the center of the C–C bond. The outer-layer electrons are the main cause of the covalent bond. If the outer-layer electrons spend most of their time within the nuclei of the two atoms, then the covalent bond is placed in the bonding orbital, in which case the distribution shape of the electron density is small at either end and large in the middle. In contrast, if the electrons spend most of their time outside the nuclei of the two atoms, then the covalent bond is placed in the antibonding orbital, in which case the distribution shape of the electron density is large at either end and small in the middle. This is because in the bonding orbital there is an increase in electron density between the nuclei, whereas in the antibonding orbital there is a decrease in electron density. Placing an electron in the bonding orbital stabilizes the molecule because it remains in between the two nuclei. Conversely, placing electrons in the antibonding orbital cause a decrease in stability [87]. The deformation electron density plot in Fig. 3.34 shows that C(5, 5)@BN(10, 10)NT has a good bonding electronic distribution that is small at either end and large in the middle. The electrons are uniformly distributed along the atoms in rings. The shape when $n = 11$ is similar to that when $n = 10$, but is different when $n = 9$. This is mainly because a part of the bond of the inner CNT shows an antibonding effect. The isoline structure in Fig. 3.34 shows some regular change as the parameter n increases. For example, when $n = 9$, because of the repulsive interactions between the inner CNT and the outer BNNT, some of the outer electrons tend toward the inside. Consequently, the isoline interval is smaller and a partial antibonding effect may occur. When $n = 10$, the repulsive interactions disappear, the outer layer regains its electron density, the isoline interval is medium-sized, and it is difficult for the antibonding effect to occur. The electron distribution also has a good bonding characterization. Thus, the structural stability of C(5, 5)@BN(10, 10) is superior to that of C(5, 5)@BN(9, 9). When $n = 11$, due to the attraction between the CNT and the BNNT, the electronic localization of the inner layer may be

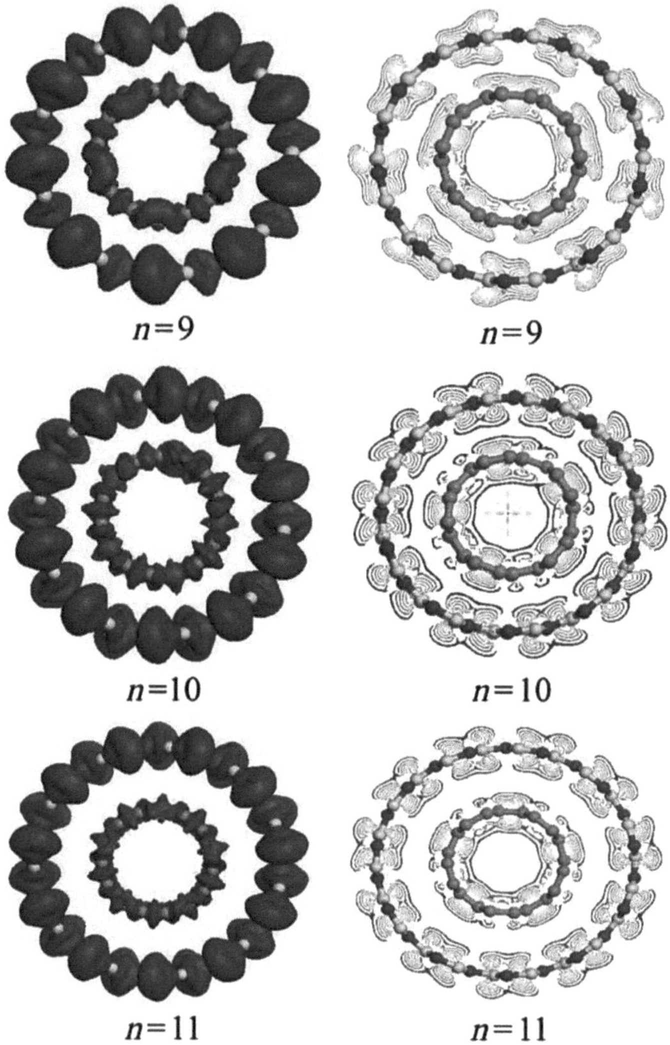

FIGURE 3.34

Deformation electron density (left) and isoline structure (right) of C(5, 5)@BN(n, n) in cross-section [76].

weakened, some s electrons become p electrons due to excitation, and partial electronic delocalization of the outer layer occurs, which results in an increase in electron mobility. Consequently, the outside electron density for $n = 11$ decreases and the isoline interval for the inner layer become somewhat sparse. Of course, as this effect is small, the electron density distribution is still similar to that when $n = 10$. In general, the numbers of nonlocalized outer-layer electrons, which are related to the conductivity, increase, and the number of bonding electrons, which maintain the stability of the structure, naturally decrease. Thus, the structural stability of C(5, 5)@BN(10, 10) is superior to that of C(5, 5)@BN(11, 11).

Fig. 3.34 clearly shows that bonding electronic distributions are only formed between B and N or C and C, and no bond is formed between B and C or N and C. The distribution state remains basically unchanged despite changes in parameter n. This indicates that there is no bond mix between the CNT and the BNNT in a C@BNNT, and the interactions between the CNT and BNNT are only van der Waals forces between the molecules. This result concurs with the conclusion drawn from the energy analysis.

3.5 BUCKLING OF CNTs BUNDLES

In the MD simulations of CNT bundles, the analysis was carried out without considering temperature control. Axial tension and compression of the perfectly structured CNT bundles were achieved by applying a constant rate of 20 m/s at both ends. At the beginning of each simulation, the CNT bundle was allowed to equilibrate before proceeding into the "production phase" where the trajectories of the atoms, energies, and forces were extracted and later analyzed. The end atoms were then shifted either inwardly (compression) or outwardly (tension) along the axis by small steps and the whole CNT bundle was subsequently relaxed by a conjugate gradient minimization method, while keeping the end atoms fixed. This process repeats until the CNT bundle breaks. Each simulation had a total of 25,000 time-steps, and each simulation time-step is equivalent to 1 fs.

The algorithm was implemented in parallel by a domain decomposition method, i.e., dividing a CNT bundle lengthwise across individual processors of the computer. A Message-Passing Interface library [88] was used in the parallelization for communication of data. The MD simulations reported here were run on parallel platforms using four processors on SGI Origin 3400, to reduce computation time significantly.

To obtain an insight into the buckling behavior of the CNT bundles, two sets of MD simulations were carried out for each size of CNT bundle, i.e., with and without the long-range nonbonding van der Waals interactions. Fig. 3.35 shows the plot of the buckling load P_{cr} against the CNT bundle size. From the plot, we can see a linear relationship between the buckling load P_{cr} and the CNT bundle size, with a bigger CNT bundle size having a larger buckling load P_{cr}. For instance, a CNT bundle with four SWCNTs has a buckling load of $P_{cr} = 0.527 \times 10^{-6}$ N, while a CNT bundle with 25 SWCNTs has a much larger buckling load at $P_{cr} = 3.30 \times 10^{-6}$ N. For comparison, a (10, 10) SWCNT is included in the plot, which reveals that the addition of more SWCNTs to a CNT bundle enhances its rigidity because the total cross-sectional area increases, and hence greater force is needed to buckle the bundle, leading to an increase in its buckling load P_{cr}.

Fig. 3.35 also shows the buckling loads P_{cr} of the CNT bundles without considering the long-range nonbonding van der Waals interactions. When the long-range interactions are not considered, the buckling loads P_{cr} of the bundles drop by a small percentage. The buckling loads P_{cr} of the various CNT bundles with and without consideration of the long-range nonbonding van der Waals interactions are listed in Table 3.14.

The inclusion of long-range nonbonding van der Waals interactions has a strengthening effect and thereby increases the buckling loads P_{cr} of the various CNT bundles by about 3.8–5.2%. The

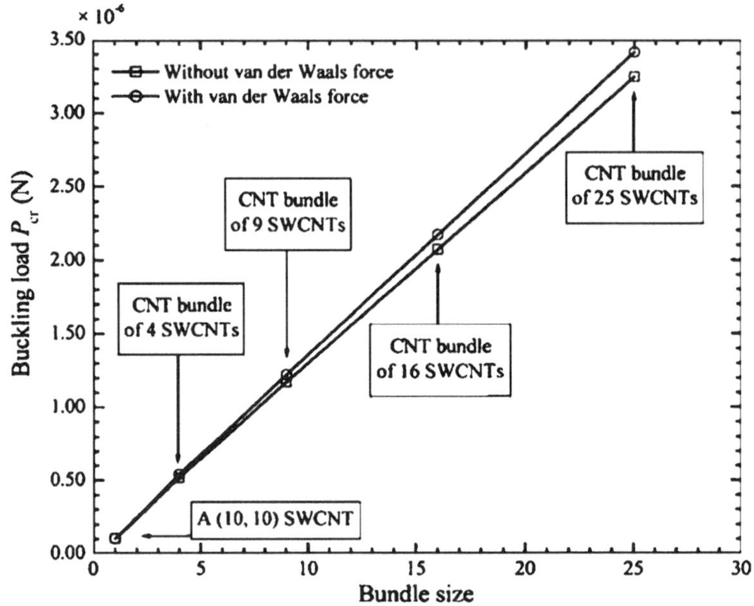

FIGURE 3.35

Buckling loads P_{cr} of various sizes of CNT bundles with and without van der Waal interactions [14].

Table 3.14 Buckling Loads P_{cr} of Different Sizes of CNT Bundles With and Without van der Waals Interactions and Their Difference (in Percentage) [14]

	Buckling Load P_{cr}		
CNT Bundle Size	Without van der Waals ($\times 10^{-6}$ N)	With van der Waals ($\times 10^{-6}$ N)	Differences in Percentage (%)
1	0.130	0.136	4.6
4	0.520	0.540	3.8
9	1.170	1.220	4.3
16	2.070	2.170	4.8
25	3.240	3.410	5.2

importance of van der Waals interactions increases as the size of the CNT bundle increases. Fig. 3.36 shows the total energy due to van der Waal interactions for different sizes of CNT bundles. The biggest CNT bundle of 25 SWCNTs has the largest van der Waals potential because of the larger interspacial gaps, while the CNT bundle of four SWCNTs has the lowest van der Waals potential. Hence, greater van der Waals interactions due to larger interspacial gaps contribute to a strengthening effect, which results in an improvement in the buckling load P_{cr} of CNT bundles.

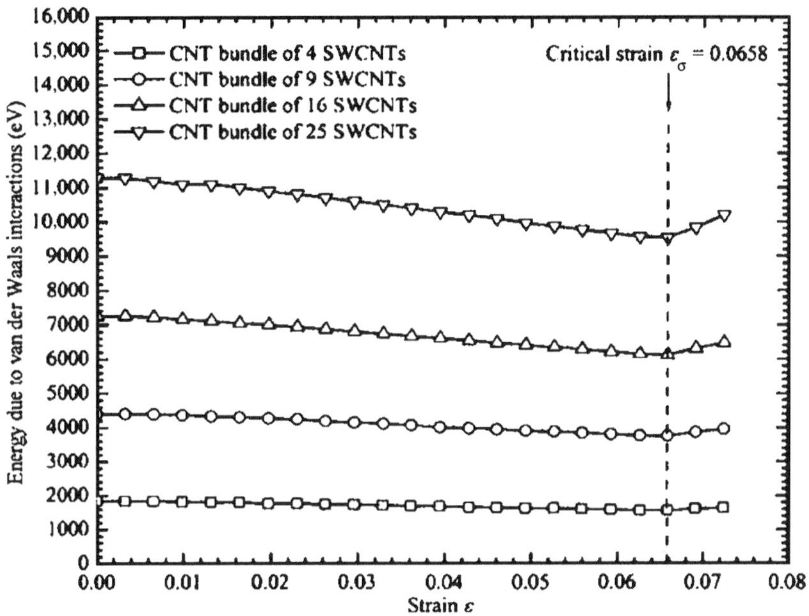

FIGURE 3.36

Energy due to van der Waals interactions at different strains ε for different sizes of CNT bundles [14].

It is also evident in Table 3.14 that even without taking into account van der Waals interactions, the inclusion of more SWCNTs in a CNT bundle already has a large influence on its buckling load P_{cr}. For example, the buckling load P_{cr} of a CNT bundle with 25 SWCNTs is five times larger that of a CNT with four SWCNTs without considering van der Waals interactions. Hence, even though the van der Waal interactions play an important role in strengthening the CNT bundle, the addition of extra SWCNTs more significantly contributes to the overall strength of the bundle.

From Fig. 3.35, we can also see that the buckling load P_{cr} of a CNT bundle is proportional to that of an SWCNT. For instance, the buckling load P_{cr} of a (10, 10) SWCNT is about 0.136×10^{-6} N (see Table 3.14) and a CNT bundle of four (10, 10) SWCNTs has a buckling load P_{cr} four times higher than that of an SWCNT. Hence, it would be useful to determine for any size of CNT bundle whether the buckling load P_{cr}^{bundle} is the product of the buckling load P_{cr}^{swcnt} of a similar SWCNT and the total number of SWCNTs in the CNT bundle. Therefore,

$$P_{cr}^{bundle} = nP_{cr}^{swcnt} \tag{3.23}$$

where n is the total number of SWCNTs in a CNT bundle, P_{cr}^{bundle} is the buckling load of a CNT bundle and P_{cr}^{swcnt} is the buckling load of an SWCNT.

The critical strains ε_{cr} at which the CNT bundles buckle were also investigated (see Table 3.15). Being a strengthening mechanism, the van der Waals interactions delay the collapse of the CNT bundle by creating weak intertube bonds due to the intertube attractive forces. Despite the

Table 3.15 Critical Strains ε_{cr} of Different Sizes of CNT Bundles With and Without van der Waals Interactions and Their Difference (In Percentage) [14]

	Critical Strains ε_{cr}		
CNT Bundle Size	Without van der Waals ($\times 10^{-6}$ N)	With van der Waals ($\times 10^{-6}$ N)	Differences in Percentage (%)
1	0.0625	0.0658	5.3
4	0.0625	0.0658	5.3
9	0.0625	0.0658	5.3
16	0.0625	0.0658	5.3
25	0.0625	0.0658	5.3

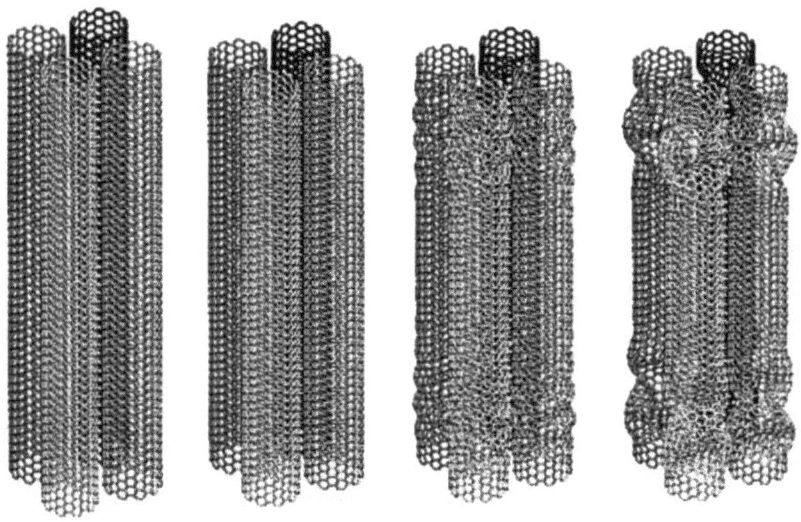

FIGURE 3.37

Morphological structural deformation of a CNT bundle of four (10, 10) SWCNTs at different strains ε [14].

different sizes of CNT bundles, they all had the same critical strains of $\varepsilon_{cr} = 0.0658$ and $\varepsilon_{cr} = 0.0625$ for those with and without long-range van der Waals interactions.

The morphological deformation of the CNT bundles when compressed was also investigated. During compression, CNT bundles undergo structural deformation. Fig. 3.37 shows the morphological changes at various stages of the compression process for the CNT bundle consisting of four SWCNTs. At strain $\varepsilon = 0.00$, the CNT bundle remains at its original length as shown in Fig. 3.37A. Further compression until $\varepsilon = 0.046$ reveals an overall decrease in length l_t of the CNT bundle (Fig. 3.37B). At $\varepsilon = 0.066$, each SWCNT in the CNT bundle begins to buckle and spontaneously collapse into "ring" patterns as revealed in Fig. 3.37C. At $\varepsilon = 0.082$, the whole CNT bundle is fully buckled (Fig. 3.37D).

In summary, the mechanical properties, in particular the critical buckling load P_{cr} and critical strain ε_{cr}, of different sizes of CNT bundles, were studied using MD simulations. We found that the buckling loads P_{cr} of CNT bundles are much higher than that of an SWCNT. At the same time, the bigger the CNT bundle, the higher the buckling load P_{cr}. Additionally, the critical strains ε_{cr} of the CNT bundles are not affected by their sizes. It was also deduced that the van der Waals interactions play a role in strengthening the overall rigidity of the CNT bundle. Depending on the size of the CNT bundle, the van der Waals interactions improve the buckling load P_{cr} by more than 3.8%, and the critical strain ε_{cr} by about 5.3%. The buckling load P_{cr} of a CNT bundle was also found to be directly proportionally to that of an SWCNT.

Fig. 3.38 shows the schematic of two CNT bundles used in the MD simulations. The CNT bundle consisted of three or seven individual SWCNTs vertically aligned next to one another. In the MD simulations, various sizes of CNT bundles were used. These sizes were determined by the size of the individual SWCNTs in the bundle.

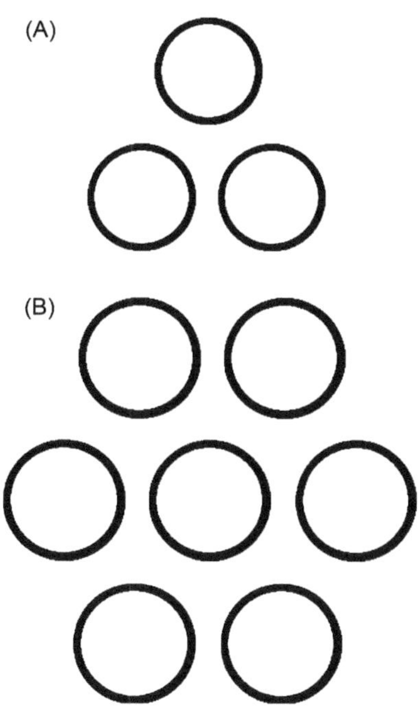

FIGURE 3.38

Schematic cross-sectional view of: (A) a CNT bundle of three SWCNTs; and (B) a CNT bundle of seven SWCNTs. The SWCNTs in the CNT bundles are all equally spaced [15].

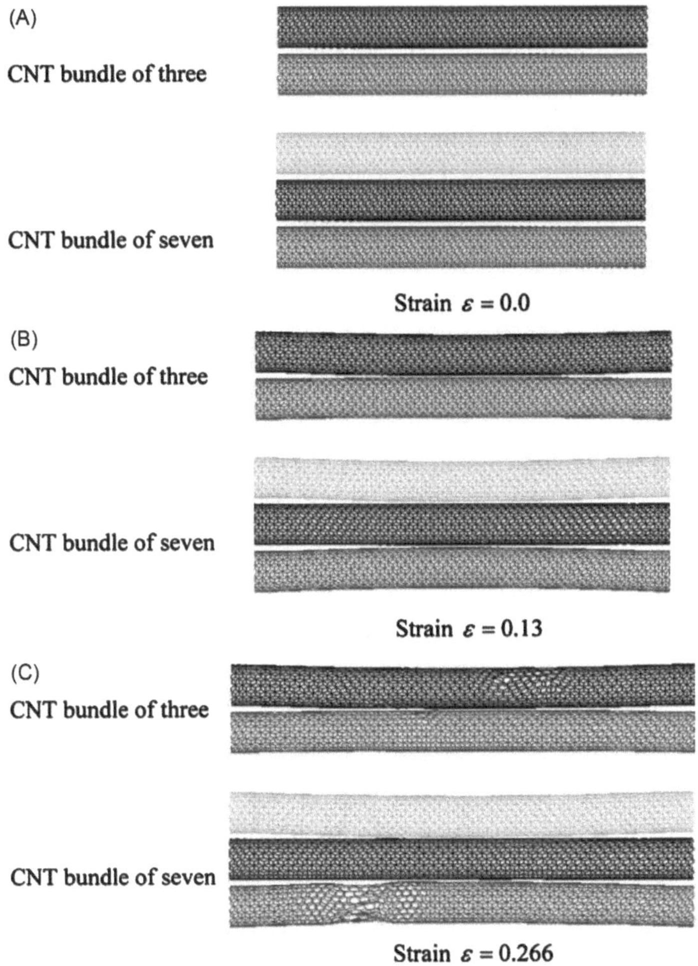

FIGURE 3.39

Morphological structural deformation of CNT bundle of three (10, 10) SWCNTs and seven (10, 10) SWCNTs at different strains during tensile loading [15].

3.5.1 CNT BUNDLES UNDER AXIAL TENSION

In the MD simulations of CNT bundles under axial tension, altogether there were four sizes of SWCNTs used, namely (5, 5), (7, 7), (10, 10), and (12, 12) SWCNTs. The interspacial gap between the individual SWCNTs in the CNT bundle was fixed at 0.34 nm.

During tensile loading, the CNT bundles undergo structural deformation as shown in Fig. 3.39. Fig. 3.39A shows that there was no structural deformation in both CNT bundles of three and seven (10, 10) SWCNTs at strain $\varepsilon = 0.0$. When the CNT bundles are further subjected to tensile force as seen in Fig. 3.39B, and at strain $\varepsilon = 0.130$, the CNT bundles tend to shrink their overall diameter

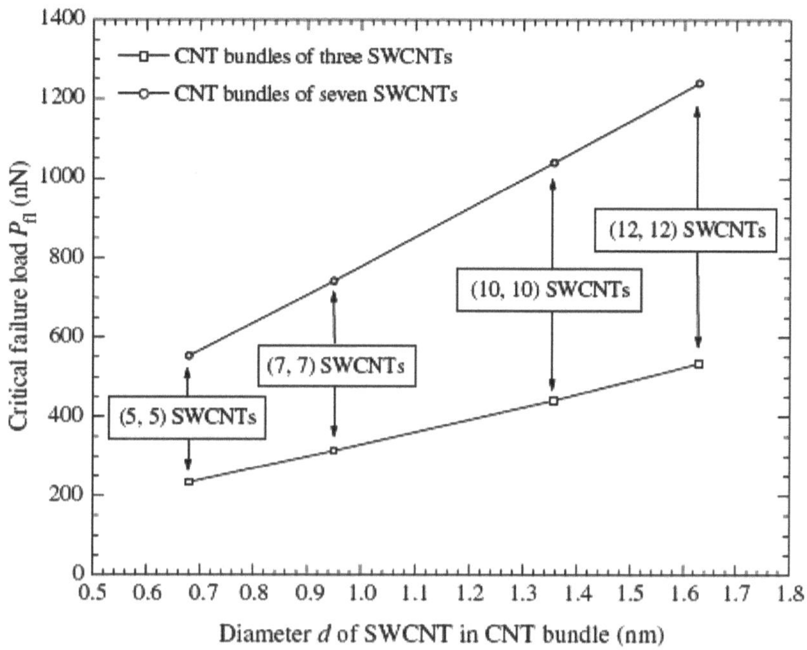

FIGURE 3.40

Critical failure loads P_{fl} of various sizes of CNT bundles of three and seven SWCNTs [15].

along their lengths. At strain $\varepsilon = 0.266$, both CNT bundles begin to fracture before they completely disintegrate.

Fig. 3.40 depicts the failure loads P_{fl} of various sizes of the CNT bundles of three and seven SWCNTs, respectively. The plots clearly demonstrate that CNT bundles with larger SWCNTs exhibit higher failure loads P_{fl}. From the plots and Table 3.16, it is also evident that the average failure load P_{fl}^{avg} of each SWCNT in a CNT bundle of three is very close to that of a CNT bundle of seven. It can therefore be deduced that the critical failure loads P_{fl} for CNT bundles are directly proportional to the average critical failure load P_{fl}^{avg} of each SWCNT in the CNT bundle. This information allows simple computation of CNT bundles of up to hundreds of SWCNTs without having to perform computationally expensive MD simulations with millions of atoms. For instance, the average critical failure loads P_{fl}^{avg} of a CNT bundle of three and seven (10, 10) SWCNTs are 147.1 and 148.1 nN, respectively (according to Table 3.16). Hence, the critical failure load P_{fl} of a CNT bundle of twenty (10, 10) SWCNTs is approximated to be 294.2 nN.

The figure also shows that the critical failure loads P_{fl} follow an almost linear relationship with respect to the diameters d of each SWCNT in the CNT bundles. This enables the determination of critical failure loads P_{fl} of other sizes of CNT bundles by simply extrapolating the data.

The critical strains ε_{cr} at which these CNT bundles fail are also presented in Table 3.16. From the table, it can be observed that the critical strains ε_{cr} are relatively insensitive to the various CNT bundles, despite the difference in their size and total number of SWCNTs in the CNT bundles.

Table 3.16 Failure Loads P_{fl} and Critical Strains ε_{cr} of Various Configurations of CNT Bundles [15]

	Failure Load P_{fl} (nN)	Average Failure Load P_{fl}^{avg} of Each SWCNT (nN)	Critical Strain ε_{cr}
CNT bundle of three (5, 5) SWCNTs	233.5	77.8	0.269
CNT bundle of three (7, 7) SWCNTs	313.0	104.3	0.269
CNT bundle of three (10, 10) SWCNTs	441.2	147.1	0.266
CNT bundle of three (12, 12) SWCNTs	532.7	177.6	0.256
CNT bundle of seven (5, 5) SWCNTs	551.6	78.8	0.266
CNT bundle of seven (7, 7) SWCNTs	741.3	105.9	0.266
CNT bundle of seven (10, 10) SWCNTs	103.7	148.1	0.266
CNT bundle of seven (12, 12) SWCNTs	123.6	176.6	0.243

Although the critical strains ε_{cr} are generally within 0.266–0.269, further studies are needed to understand why they drop to 0.243 and 0.256 for CNT bundles of three and seven (12, 12) SWCNTs, respectively.

3.5.2 CNT BUNDLES UNDER AXIAL COMPRESSION

When compressed, the CNT bundles will undergo structural deformation. Fig. 3.41 shows the morphological changes at various stages of the compression process of CNT bundles of three and seven (10, 10) SWCNTs. Fig. 3.41A shows the CNT bundles at strain $\varepsilon = 0.0$, where there is no deformation. Further compression until strain $\varepsilon = 0.04435$ will cause the CNT bundle of three to spontaneously collapse into a three-fin pattern, while the CNT bundle of seven is bent slightly outwards from the core SWCNT. At strain $\varepsilon = 0.05010$, the CNT bundle of three continues to deform. However, the CNT bundle of seven begins to buckle forming three-fin patterns in each of its SWCNTs. Finally at strain $\varepsilon = 0.1536$, the CNT bundle of three buckles sideways at the same direction, while the CNT bundle of seven buckles outwards from its core SWCNT forming a "bird-cage" pattern.

Investigations were carried out on the buckling properties of CNT bundles. Altogether, seven sizes of CNT bundles were used. They are namely (5, 5), (7, 7), (10, 10), (12, 12), (15, 15), (18, 18), and (20, 20). The plot in Fig. 3.42 shows the critical buckling loads P_{cr} for the different sizes of CNT bundles of three and seven. From the plot, it can be seen that as the size of the individual SWCNT in a CNT bundle increases, its buckling load P_{cr} also increases exponentially. Also, the critical buckling load P_{cr} of the CNT bundles of three tends to reach a maximum value of $P_{cr} \approx 300$ nN despite further increases in the diameters d.

However, the CNT bundles with seven SWCNTs have critical buckling loads P_{cr} of about two to three times higher than that of CNT bundles of three SWCNTs. As the diameters d of the SWCNTs in the CNT bundle of seven increase, the difference in critical buckling loads P_{cr} between the CNT bundles of three and seven SWCNTs also increases. For instance, at diameter $d = 0.68$ nm, the critical buckling load P_{cr} of the CNT bundle of seven (5, 5) SWCNTs is about 2.1

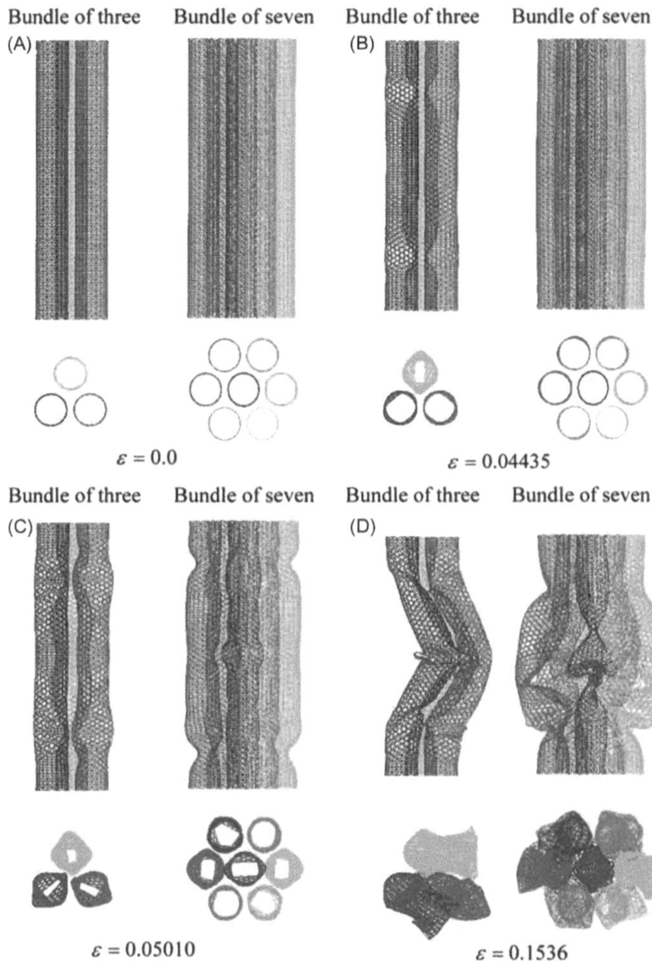

FIGURE 3.41

Morphological structural deformation of CNT bundle of three (10, 10) SWCNTs and seven (10, 10) SWCNTs at different strains during compressive loading [15].

times larger. However, when the diameter increases to $d = 2.44$ nm, the critical buckling load P_{cr} of the CNT bundle of seven (18, 18) SWCNTs is about 2.4 times larger.

To understand how the intertube van der Waals interactions affect the compressive properties of CNT bundles, two sets of MD simulations were carried out: (1) with long-range intertube van der Waals interactions; and (2) without long-range intertube van der Waals interactions. It is noteworthy that intratube van der Waals interactions were omitted since they are not crucial to the calculations [15]. Fig. 3.43 depicts the buckling loads P_{cr} for the various sizes of CNT bundles, with (filled circles) and without (open circles) consideration of the intertube van der Waals interactions.

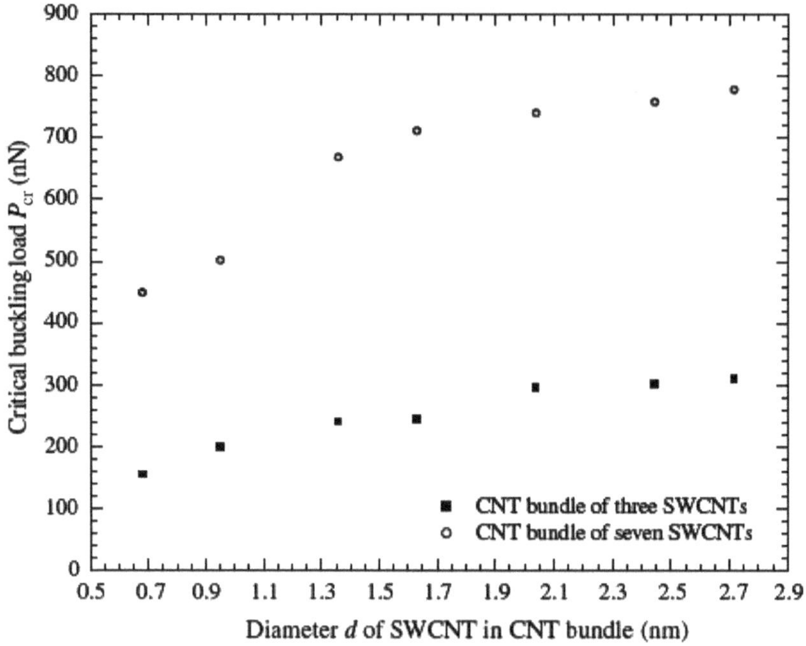

FIGURE 3.42

Plot of critical buckling loads P_{cr} of CNT bundles of three and seven as a function of the diameter d of SWCNT [15].

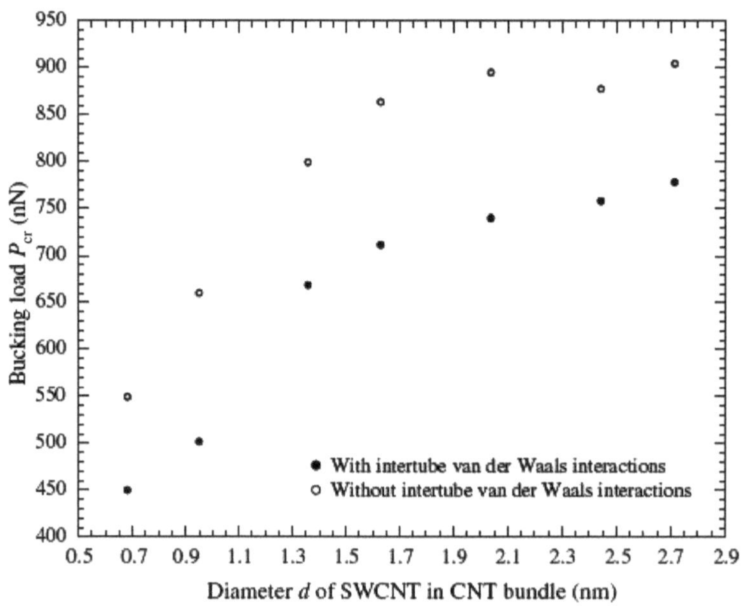

FIGURE 3.43

Buckling loads P_{cr} of various sizes of CNT bundles with and without long-range intertube van der Waals interactions [15].

From the plot, it is revealed that despite having the same number of SWCNTs in the bundles, the buckling loads P_{cr} will increase exponentially as the sizes of the individual SWCNTs in the CNT bundles increase. Even though the increasing trends of the buckling loads P_{cr} are similar, it is revealed that higher buckling loads P_{cr} can be achieved when the intertube van der Waals interactions were not considered. This is in agreement with the findings of Lu [42] who revealed that CNT bundles tend to have lower elastic moduli compared to individual SWCNT due to the weak intertube van der Waals interactions that make the CNT bundles weaker, since the individual SWCNTs can easily rotate and slide with respect to one another.

Apart from the above, similar MD simulations were performed at different interspacial gaps, where the long-range intertube van der Waals interactions can be found. This is to determine if the size of the interspacial gaps plays a role in the buckling properties of CNT bundles. In this work, a total of three interspacial gaps were used, i.e., 0.2, 0.34, and 0.51 nm. The interspacial gaps were not chosen smaller than 0.2 nm since according to Eq. (3.15), intertube covalent bonds will form when $r_{ij} < 0.2$ nm. Fig. 3.44 shows the plot of buckling loads P_{cr} for different sizes of CNT bundles at various interspacial gaps. When the long-range intertube van der Waals interactions were not considered, the SWCNTs in the CNT bundles act as individual SWCNTs. The buckling loads P_{cr} for the different sizes of CNT bundles at different interspacial gaps are similar. They have an increasing exponential trend until the diameter d of the SWCNT in the CNT bundle reaches 2.04 nm, thereafter any further increase in diameter d will not result in any increase in its buckling load P_{cr}.

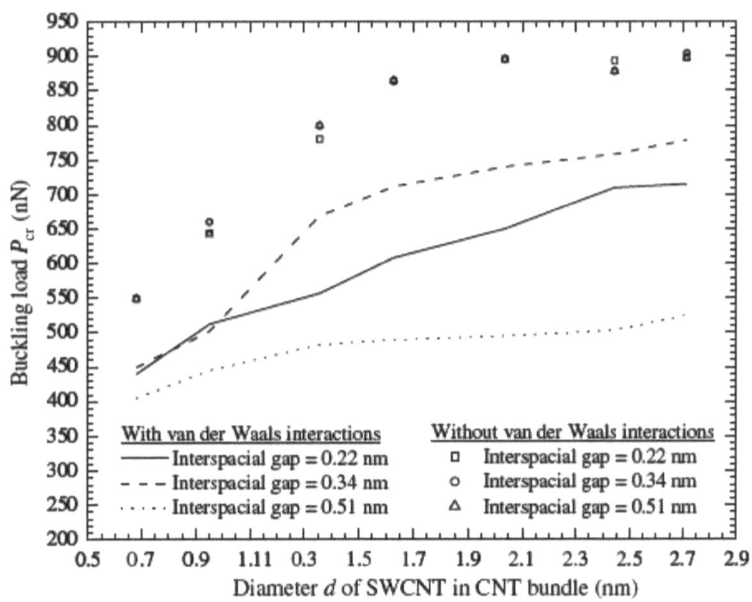

FIGURE 3.44

Buckling loads P_{cr} of various sizes of CNT bundles at different interspacial gaps with and without long-range intertube van der Waals interactions [15].

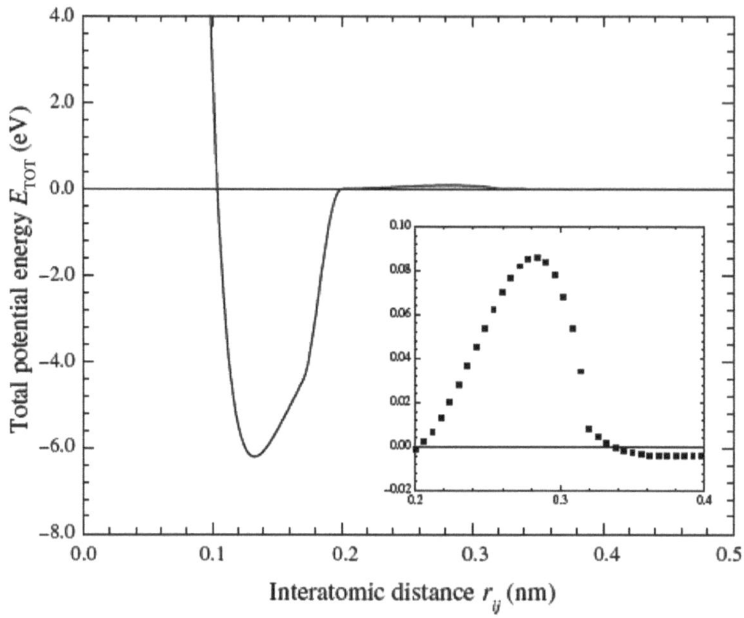

FIGURE 3.45

Plot of total potential energy E_{TOT} of a pair of carbon atoms as a function of interatomic distance r_{ij}. Inset shows the amplification for 0.2 nm $\leq r_{ij} <$ 0.4 nm [15].

However, when the long-range intertube van der Waals interactions were considered in the computation, the buckling loads P_{cr} for the different sizes of CNT bundles were not the same. At an interspacial gap of 0.22 nm, the buckling load P_{cr} of the CNT bundle will be higher than that with an interspacial gap of 0.51 nm. This is due to the possibility of the formation of covalent bonds between individual SWCNTs when the CNT bundle is compressed, leading to a reduction in the interspacial gaps to less than 0.2 nm. Additionally, stronger intertube van der Waals interactions will be present than for a CNT bundle with an interspacial gap of 0.51 nm. The repulsive long-range intertube van der Waals interactions in the CNT bundle with an interspacial gap of 0.22 nm also explains why it has a lower buckling load P_{cr} than a CNT bundle with an interspacial gap of 0.34 nm (see inset of Fig. 3.45).

The critical strains ε_{cr} of the various CNT bundles were also investigated. Fig. 3.46 shows that, in general, as the diameters d of the SWCNTs in the CNT bundle increase, the critical strain ε_{cr} will decrease. Furthermore, the critical strains ε_{cr} of CNT bundles of seven SWCNTs is generally higher than that of CNT bundles of three SWCNTs. For example, the critical strain ε_{cr} of the CNT bundle of seven (10, 10) SWCNTs is $\varepsilon_{cr} = 0.05010$, while the critical strain ε_{cr} of the CNT bundle of three (10, 10) SWCNTs is lower at $\varepsilon_{cr} = 0.04435$. This is due to the existence of more SWCNTs in the CNT bundle of seven which gives rise to higher strength and thus higher resistance to buckling. However, as the diameter d increases, the difference between the critical strains ε_{cr} of both the CNT bundles of seven and three SWCNTs becomes smaller.

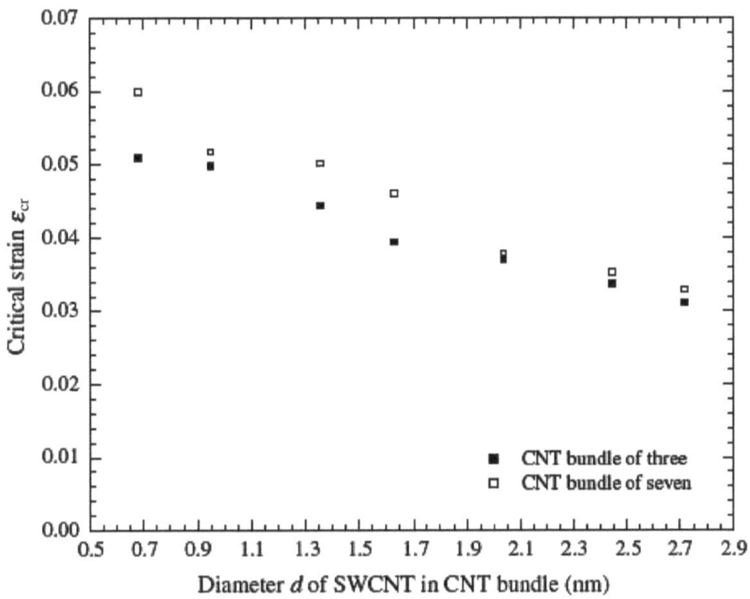

FIGURE 3.46

Critical strains ε_{cr} of various sizes of CNT bundles of three and seven [15].

3.5.3 TWISTING EFFECTS OF CNTs BUNDLES

MD simulations were performed on twisted CNT bundles, which comprise seven (5, 5) SWCNTs that are spaced 0.34 nm apart, under axial compression and tension. A twisting load was applied to six of the (5, 5) SWCNTs that surround a core SWCNT to form a twisted CNT bundle. A schematic cross-sectional view of the CNT bundle under study, which comprises six (5, 5) SWCNTs that surround a core (5, 5) SWCNT, is shown in Fig. 3.47.

The SWCNTs in the CNT bundle are spaced 0.34 nm apart, and the length of the bundle is $l_t = 12.3$ nm. The diameter of each (5, 5) SWCNT is $d_t = 0.68$ nm, and each CNT bundle contains 7000 atoms. A twisting load is applied to the six SWCNTs that surround core SWCNT. The twist angle θ_t is quantified as the relative twist between the two ends of the CNT bundle. A total of eight cases with different twisting angles θ_t of 0, 15, 30, 60, 75, 90, 100, and 120 degrees were studied.

MD simulations were carried out on twisted CNT bundles under axial compressive and tensile forces. The axial compressive and tensile loads were achieved by applying a rate of 50 and 20 m/s, respectively, to both ends of the CNT bundles. The atoms at both ends of the CNT bundles were made transparent to the interatomic forces and were then moved either inwardly (compression) or outwardly (tension) along the axis in small steps followed by a conjugate gradient minimization method while the end atoms were kept fixed.

When the six surrounding SWCNTs were twisted, the overall lengths of these SWCNTs were maintained at 12.3 nm. As a result, the atoms in these SWCNTs move further apart from one another, stretching the C–C bonds between them. This effect is nevertheless insignificant as

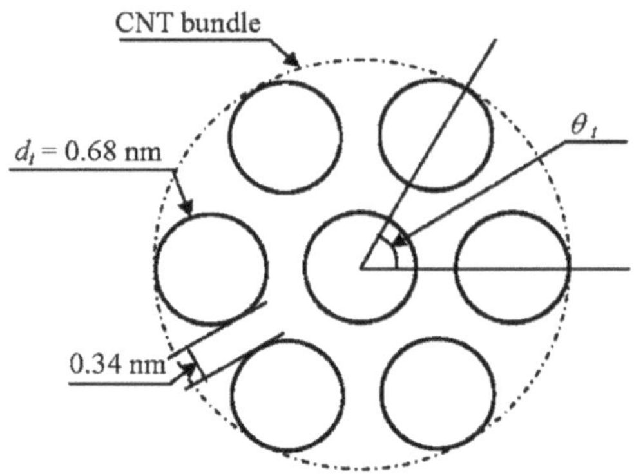

FIGURE 3.47

Schematic cross-sectional diagram of a CNT bundle of six (5, 5) SWCNTs surrounding a core (5, 5) SWCNT, where the diameter and length of each SWCNT are $d_t = 0.68$ nm and $l_t = 12.3$ nm, respectively [100].

depicted in Fig. 3.48, which shows the average binding energy per atom of a surrounding SWCNT twisted at different angles θ_t. The plot shows that as the twist angle θ_t increases, the average binding energy per atom becomes weaker because the atoms are shifted further apart from one another. Despite the extension of the C–C bond lengths, the maximum difference in the average binding energy per atom when compared to a nontwisted SWCNT is only about 0.6%. This difference is small and is hence considered to be negligible in this study.

3.5.3.1 Twisted CNT bundles under axial compression

When a twisted CNT bundle is under axial compression, the six surrounding SWCNTs will expand and contract radially around the midlength along the center axis of the CNT bundle. To quantify the radial expansions and contractions and the morphological changes of the CNT bundle, the total intertube van der Waals force acting on the core SWCNT (F_{vdw}^{core}) of each CNT bundle twisted at various angles (i.e., $\theta_t = 0$, 30, 60, and 90 degrees) and its deformation Δl_t (i.e., defined as $\Delta l_t = (l_t - l)/l_t$, where l is the current length of the CNT bundle), were computed and plotted in Fig. 3.49. The graph shows that F_{vdw}^{core} undergoes a cyclic pattern during compression. When the six surrounding SWCNTs expand radially, F_{vdw}^{core} becomes weaker as shown by the "valleys" of the plots in Fig. 3.49. However, when the six surrounding SWCNTs contract radially, F_{vdw}^{core} grows stronger, as shown by the "peaks" of the plots in Fig. 3.49.

A twisted CNT bundle is expected to have a lower critical buckling load P_{cr} and critical strain ε_{cr} than a nontwisted CNT bundle due to the curved geometries of the six surrounding SWCNTs that are formed by the twisting. Fig. 3.50 depicts the plot of the critical buckling load P_{cr} against the angle of twist θ_t in the CNT bundle. As the twist angle θ_t increases from 0 to 60 degrees, the critical buckling load P_{cr} decreases by about 5% from 325 to 308 nN. A further increase in the twist angle θ_t from 60 to 120 degrees, however, results in a larger drop of about 42% in the critical buckling load P_{cr} from 325 to 188 nN. This is because as the twist angle θ_t grows, the compressive force at each end of the surrounding SWNCT becomes more misaligned, which causes the surrounding SWNCT to buckle more easily and at a lower load P_{cr}.

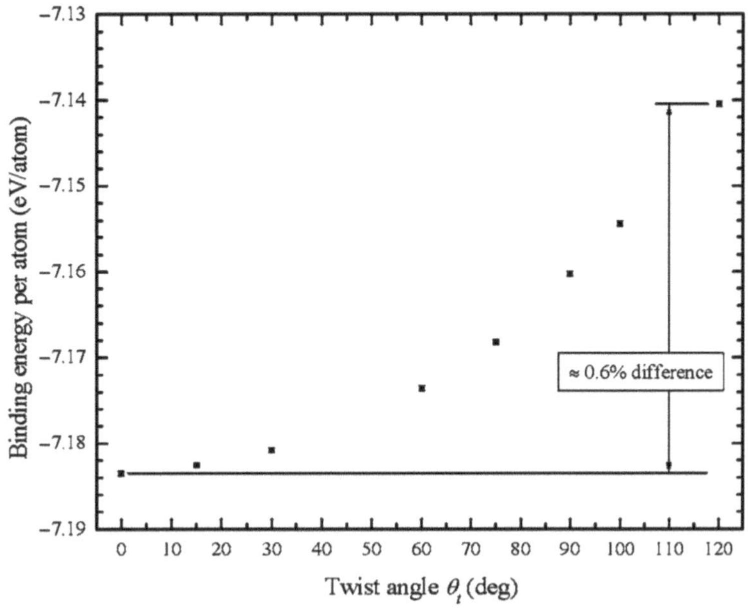

FIGURE 3.48

Binding energy per atom of a surrounding SWCNT at different twist angles θ_t [100].

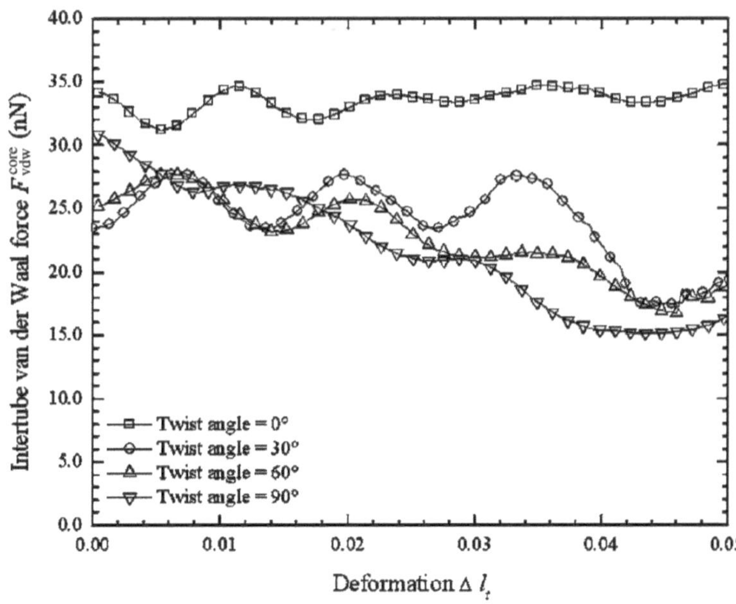

FIGURE 3.49

Intertube van der Waals force acting on the core SWCNTs (F_{vdw}^{core}) of CNT bundles twisted at various angles θ_t during axial compression [100].

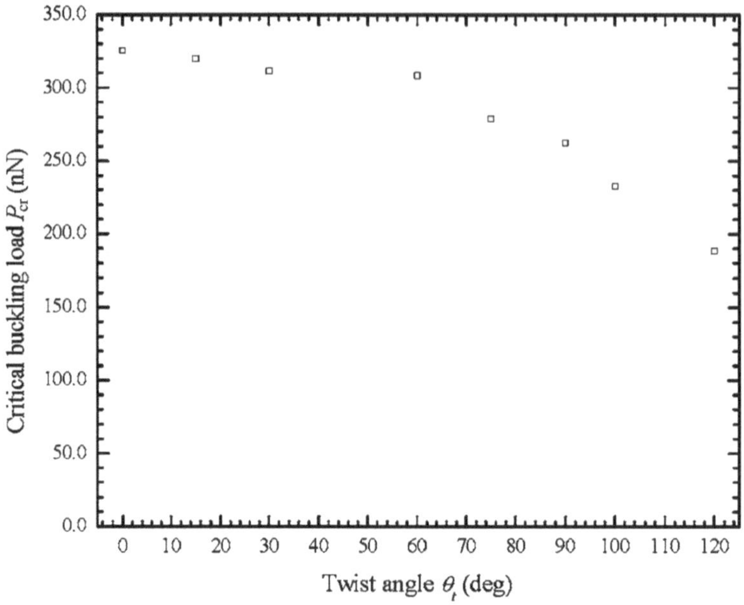

FIGURE 3.50

Critical buckling load P_{cr} of CNT bundles twisted at different angles θ_t [100].

Additionally, it is also interesting to note that the six surrounding SWCNTs will buckle earlier than the core SWCNT due to their curved geometries. Fig. 3.51 shows the morphological changes of a 60 degree-twisted CNT bundle at different deformations Δl_t. In the initial stage when $\Delta l_t = 0$, the whole CNT bundle is relaxed, and the helical angle θ_{he} that is formed due to the twisting of the six surrounding SWCNTs is about 77 degrees. Upon further compression, up to when the deformation $\Delta l_t = 0.048$ (see Fig. 3.51B), the six surrounding SWCNTs buckle outwards from the center axis fractionally earlier and form a "bird-cage" pattern, while the core SWCNT remains unbuckled. The fully buckled CNT at $\Delta l_t = 0.06$ is shown in Fig. 3.51C.

To gain further insight into the axial load P that each SWCNT experiences in the twisted CNT bundle during compressive loading, the axial force F of each individual SWCNT in the CNT bundle is plotted against its deformation Δl_t as in Fig. 3.52. As the axial force F that acts on the six surrounding SWCNTs is the same, therefore only the axial force F that acts on one surrounding SWCNT and the core SWCNT is plotted. It is evident from the plot that the surrounding SWCNT buckles slightly earlier than the core SWCNT and at a lower critical buckling load Pcr. This is because since the surrounding SWCNT is geometrically twisted, therefore it is more susceptible to buckling than the core SWCNT.

3.5.3.2 Twisted CNT bundles under axial tension

In contrast to wire ropes, which provide better structural reliability and are mainly used in applications in which the tensile load is dominant, twisted CNT bundles tend to underperform when

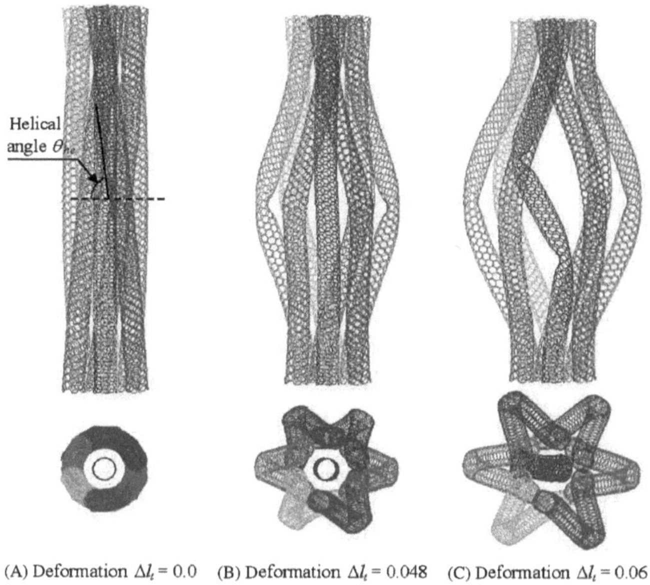

FIGURE 3.51

Morphological changes of a 60 degree-twisted CNT bundle due to a compressive load at different deformations Δl_t [100].

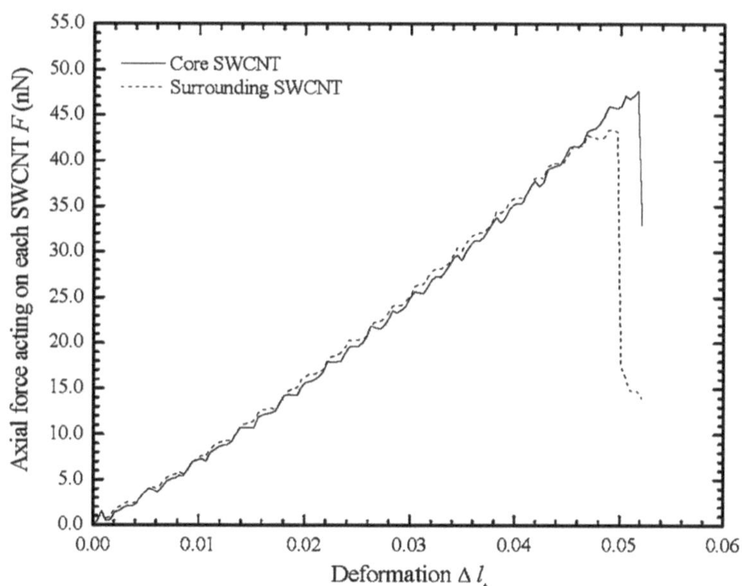

FIGURE 3.52

Axial force F acting on the core SWCNT and surrounding SWCNT of a 60 degree-twisted CNT bundle during axial compression [100].

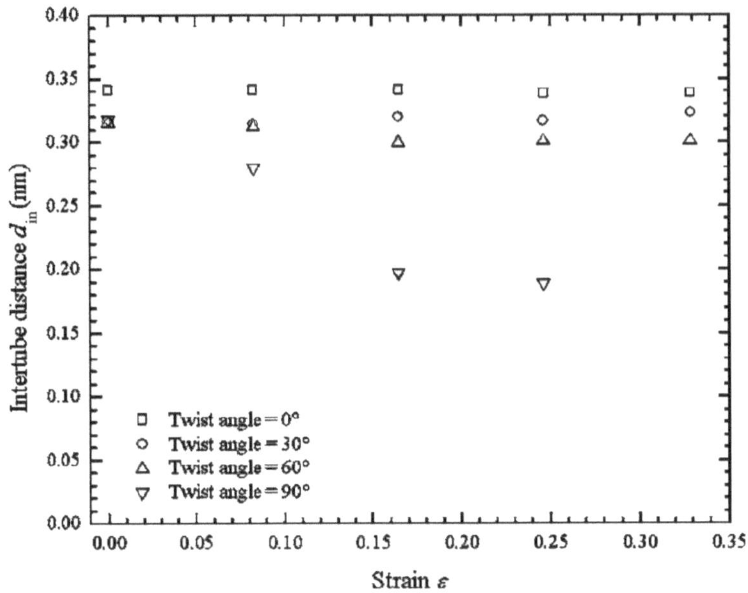

FIGURE 3.53

Intertube distance d_{in} for CNT bundles twisted at different angles θ_t at different strains ε [100].

compared with nontwisted CNT bundles subjected to an axial tensile load. As a result of the twisting, the six surrounding SWCNTs are brought closer to each other under a helical angle, and junctions are formed between the neighboring SWCNTs. During axial tensile loading, the six surrounding SWCNTs were pulled even closer to the core SWCNT, which induced a reduction in the intertube distance d_{in} between individual SWCNTs. The plot in Fig. 3.53 shows the relationship between the intertube distance d_{in} and strain ε, which is computed as $\varepsilon = (l - l_t)/l_t$, where l and l_t are the current and initial length of the CNT bundle, respectively. From the plot, it is evident that the intertube distance d_{in} decreases as the strain ε increases for CNT bundles with a large twist angle θ_t. This is because when the six surrounding SWCNTs are twisted at a larger angle θ_t, they are more closely intertwined and are geometrically curved more prominently. Consequently when a tensile load is applied to both ends of the CNT bundles, the curved geometry of each of the six individual surrounding SWCNTs tends to straighten out, which leads to a reduction in the intertube distance d_{in}.

This phenomenon can also be confirmed by the plot of F_{vdw}^{core} (total intertube van der Waals force F_{vdw}^{core} acting on the core SWCNT) at different strains ε, as shown in Fig. 3.54. When a tensile load is applied to the CNT bundle, F_{vdw}^{core} increases until it reaches a maximum value. This indicates a decrease in the intertube distance d_{in} between the surrounding six SWCNTs and the core SWCNT. F_{vdw}^{core} then slowly decreases, which is not a result of an increase in the intertube distance between the six surrounding SWCNTs and the core SWCNT, but is an indication of an increase in the repulsive energy, which is explained further below.

In contrast to the theory of wire ropes in a continuum domain, twisted CNT bundles do not exhibit better tensile properties, as shown in Fig. 3.55. When the twisting angle θ_t is greater than 75 degrees,

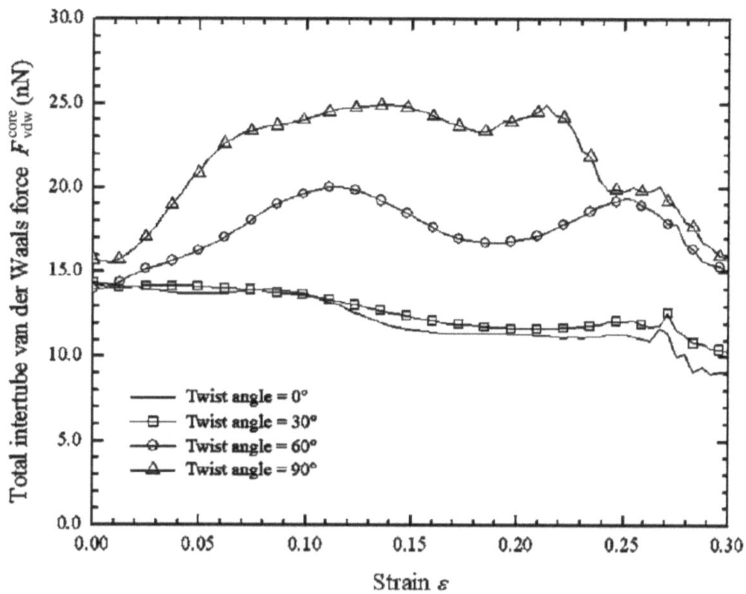

FIGURE 3.54

Total intertube van der Waals force acting on the core SWCNT (F_{vdw}^{core}) of each CNT bundle (twisted at angles $\theta_t = 0$, 30, 60, and 90 degrees) during axial tension [100].

the failure load P_{fl} decreases sharply. An investigation into the cause of this was carried out, and it was found that when a CNT bundle with a twisting angle θ_t that is greater than 75 degrees is subjected to axial tension, the intertube distance d_{in} between the seven SWCNTs decreases to less than 0.2 nm. This results in an increase in the repulsive energy between the seven SWCNTs since the contact distance is less than 0.34 nm. Consequently, these CNT bundles tend to fail at the junctions at lower loads and strains. This is in qualitative agreement with previous work [89–92], in which it was found that the dangling C–C bonds that were formed between SWCNTs and which were brought close together were not strong enough to withstand bending and axial forces.

The total intertube van der Waals force F_{vdw}^{total} and intertube repulsive force F_{pre}^{total} around the midlength of the CNTs that has been twisted at 0 and 90 degrees were extracted from the MD simulations and are depicted in Figs. 3.56 and 3.57, respectively. The midlength of the CNT bundle was chosen for the depictions, as this is where pairs of the twisted surrounding SWCNTs meet each other at a helical angle and form a junction where they overlap. As the strain ε becomes larger, F_{vdw}^{total} in the 90 degree-twisted CNT bundle increases due to the decreasing of intertube distance d_{in} (see Fig. 3.53). However, F_{vdw}^{total} in the nontwisted CNT bundle weakens steadily as the strain ε increases due to the elongation of the CNT bundle during tension, as there is no change in the intertube distance d_{in}.

Throughout the whole tensile process, the nontwisted CNT bundle has no F_{pre}^{total} (intertube repulsive force) due to the large intertube spacing d_{in}, as shown in Fig. 3.57. Likewise, there is no F_{pre}^{total} present in the 90 degree-twisted CNT bundle that is twisted at 90 degrees up to $\varepsilon = 0.185$. This is because prior to $\varepsilon = 0.185$, the intertube distance d_{in} is greater than 0.2 nm, and, according to

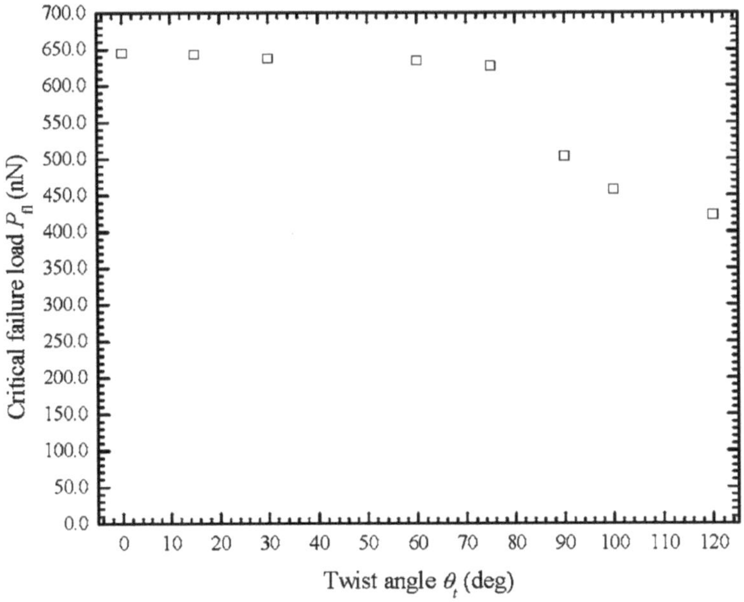

FIGURE 3.55

Critical failure load P_{fl} of CNT bundles twisted at different angles θ_t due to axial tensile force [100].

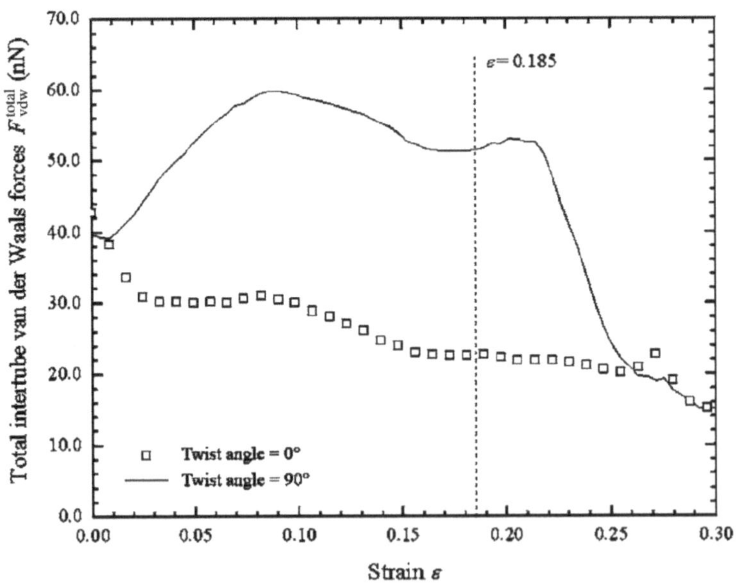

FIGURE 3.56

Total intertube van der Waals force F^{total}_{vdw} around the midlength of CNT bundles twisted at 0 and 90 degrees [100].

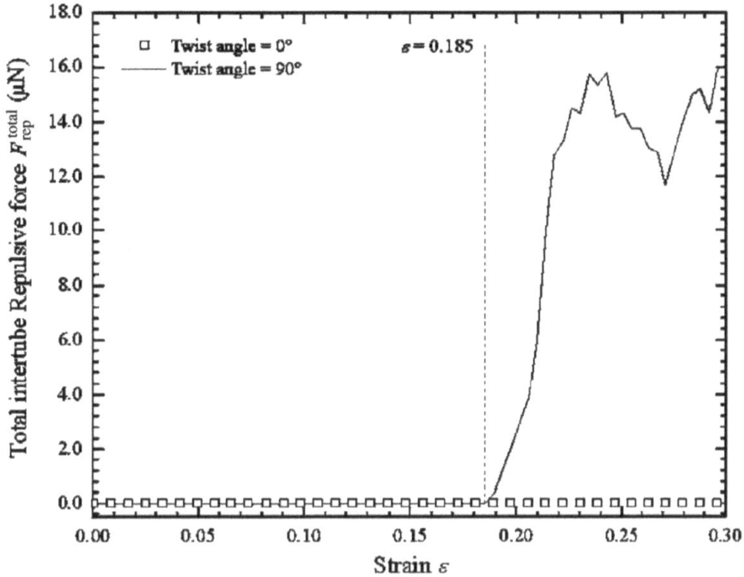

FIGURE 3.57

Total intertube repulsive force F_{rep}^{total} around the midlength of a nontwisted CNT bundle and a 90 degree-twisted CNT bundle [100].

Eq. (3.15), the repulsive force F_{rep} is only computed when the distance between two atoms is less than 0.2 nm. However, after $\varepsilon > 0.185$, the intertube distance d_{in} for the 90 degree-twisted CNT bundle reduces to less than 0.2 nm (see Fig. 3.53), which results in a sudden increase in F_{pre}^{total} and a decrease in F_{vdw}^{total} (total intertube van der Waals force), as shown in Fig. 3.56. It is worth noting that F_{vdw}^{total} and F_{pre}^{total} act radially on neighboring SWCNTs. As a result of the sudden surge in F_{pre}^{total}, the 90 degree-twisted CNT bundle is weakened severely, and therefore only a low tensile load is needed to damage it. The damage often occurs at the junction areas (see Fig. 3.58).

As a consequence of the sudden increase in F_{pre}^{total} at the junction areas at which the individual SWCNTs overlap, CNT bundles that are twisted at an angle $\theta_t > 75$ degrees will collapse at lower critical strains ε_{cr}, as shown in Fig. 3.59. The plot shows that the critical strains remain at around $\varepsilon_{cr} \approx 0.35$ up to a twisting angle θ_t of 75 degrees, and decrease significantly thereafter to $\varepsilon_{cr} \approx 0.18$.

Fig. 3.60 shows the cross-sectional view of two CNT bundles with a twisting angle of 0 and 120 degrees, respectively, at different strains. It is evident from this figure that the 120 degree-twisted CNT bundle suffers from a large amount of distortion of the individual SWCNTs compared with the nontwisted CNT bundle. The SWCNTs in the 120 degree-twisted CNT bundle lose their cylindrical shapes at high strains, which results in a dislocation of the hexagonal arrangement of the C atoms. The distortion of individual SWCNTs in the CNT bundle is manifested by the compressive van der Waals force F_{vdW} and the repulsive force F_{rep} between the seven SWCNTs.

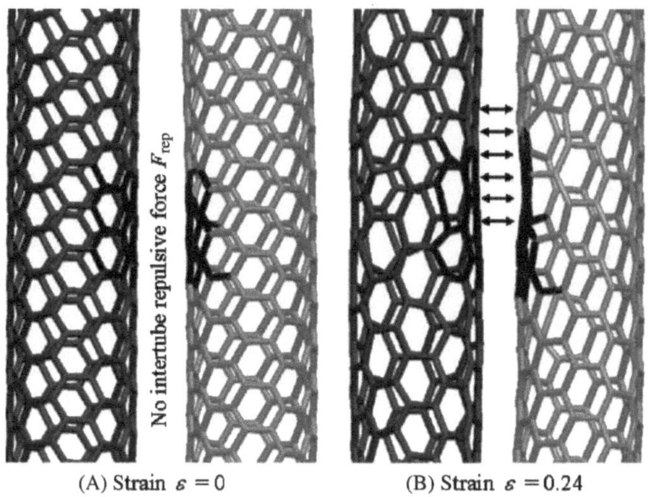

FIGURE 3.58

Enlarged pictorial view of the core SWCNT (dark gray) and a surrounding SWCNT (light gray) of a 90 degree-twisted at (A) strain $\varepsilon = 0$, where the intertube distance d_{in} is about 0.32 nm; (B) strain $\varepsilon = 0.24$, where the intertube distance d_{in} is about 0.19 nm. The junction area at which the two SWCNTs overlap is shown in black. The double-headed arrows represent the intertube repulsive force F_{rep} [100].

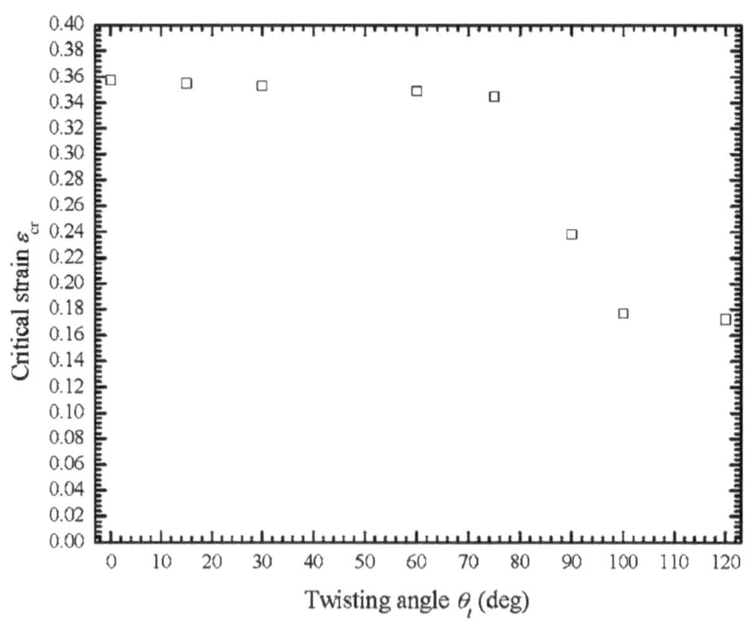

FIGURE 3.59

Critical strain ε_{cr} of CNT bundles twisted at different angles θ_t due to axial tensile force [100].

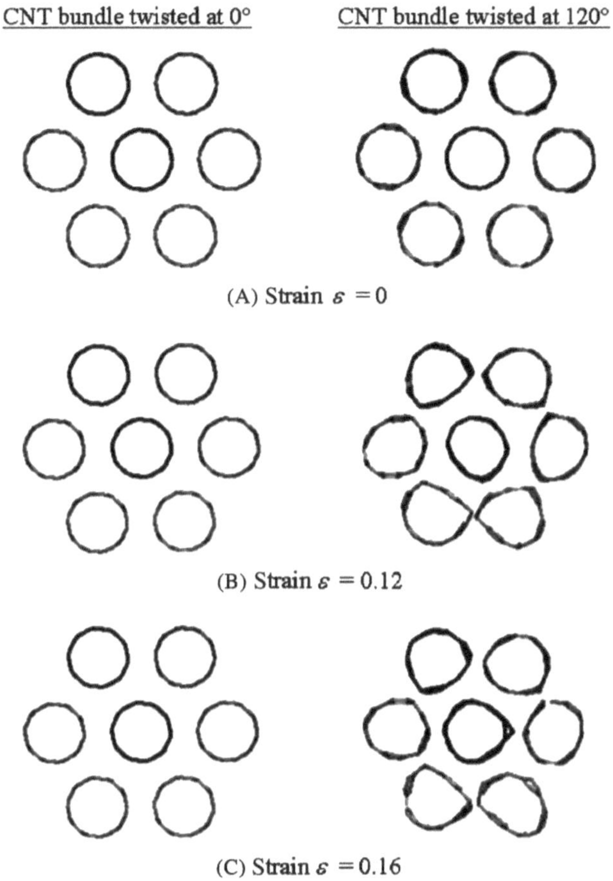

FIGURE 3.60

Cross-sectional view of CNT bundles twisted at 0 and 120 degrees at different strains ε [100].

3.6 FRACTURE OF CNTs

Usually, Young's modulus is assumed to be constant. It can be determined from the second derivative of potential function by fitting the potential energy into a cubic polynomial function. For CNTs, Young's modulus is dependent on strain and treated as constant if the strain is small. In general, we can obtain the derivative of a potential function in a smooth stress–strain curve in the elastic region, but not for the elasto-plastic-failure region. This is due to large fluctuation of stresses. Thus it is difficult to determine some mechanical properties, such as the tensile strength, elastic strain limit, maximum strain, by fitting the potential energy into a smooth stress–strain curve.

In order to describe the complete stress–strain relationship from the elastic deformation to rapture, the stress–strain curves of single and multiwalled CNTs under tension deformation are obtained in our simulations by applying a rate of 20 m/s to both ends. The atoms at the end of the CNTs are then moved outwardly along the axis by small steps, followed by a conjugate gradient minimization method while keeping them fixed in plane.

In the microcanonical ensemble MD simulations of the CNTs, the strain is computed by $\varepsilon = (L - L_0)/L_0$ in which L_0 and L being the initial and current length of the CNTs, respectively, and the stress is given by $\sigma = F/S$. The axial force F is obtained by summing the interatomic force for the atoms at the end of CNTs and the cross-sectional area is $S = \pi dh$, where d is the diameter of the CNT and h the thickness of the CNT that is taken as $h = 0.335$ nm for (10, 10) single-walled CNT, $h = 0.670$nm for (5, 5) and (10, 10) two-walled CNT, $h = 1.005$ nm for (5, 5), (10, 10), and (15, 15) three-walled CNT, and $h = 1.34$ nm for (5, 5), (10, 10), (15, 15), and (20, 20) four-walled CNT.

In order to validate the present MD simulation, the stress–strain responses are determined for a (12, 12) single-walled CNT with a length-to-diameter ratio $L/D = 9.1$. Our MD simulation is first compared with the results of Belytschko et al. [64] as depicted in Fig. 3.61. It is observed that our stress–strain curve deviates slightly from the results in [64]; nevertheless, they are in reasonable agreement. In the second comparison study listed in Table 3.17, our MD simulations of Young's modulus of various single-walled CNTs are in close agreement with the ab initio results of Kudin et al. [93]. With the above comparison studies, we therefore confirm the accuracy of our MD simulation.

The elastoplastic responses of the CNTs are simulated and plotted in Fig. 3.62A–H for the length-to-diameter ratio of 4.5 and 9.1. It is evident that when $0 \le \varepsilon \le \varepsilon_P$, the stress–strain relationship follows the Hooke's law, i.e., $\sigma = E\varepsilon$. Here, ε_P corresponds to point **P** in Fig. 3.62A–H, and is the proportional strain limit. It is observed that as the strain increases, the stress–strain curve behaves nonlinearly with a decreasing gradient until it reaches the yield strain ε_s, which corresponds to yield point S in Fig. 3.62A–H. Using the regression analysis, the stress–strain curve is represented by $\sigma = (A\varepsilon + B)\varepsilon$ where $\varepsilon_p \le \varepsilon \le \varepsilon_s$. The values of A and B are determined by fitting

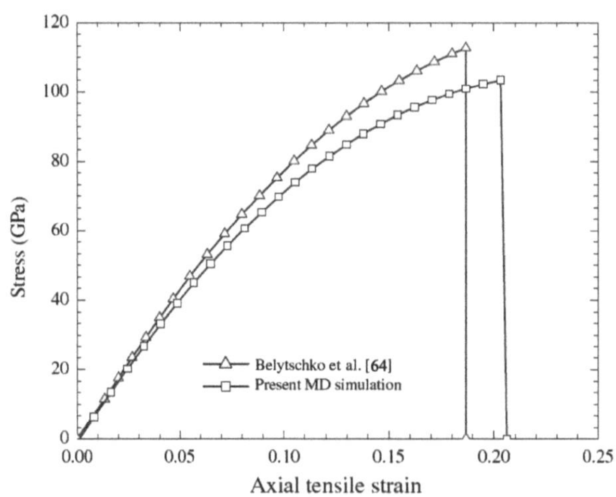

FIGURE 3.61

Stress–strain curve of a single-walled CNT (12, 12) under axial tension as compared with results from Ref. [64] (Liew et al. [13]).

Table 3.17 Young's Modulus of Various CNTs Computed by MD (This Work) and Ab Initio Simulations [13]

	Young's Modulus	
CNT	**MD**	**Ab Initio [93]**
(4, 4)	55.3	56.4
(7, 0)	55.7	56.3
(7, 7)	56.0	56.5
(12, 0)	56.4	55.2

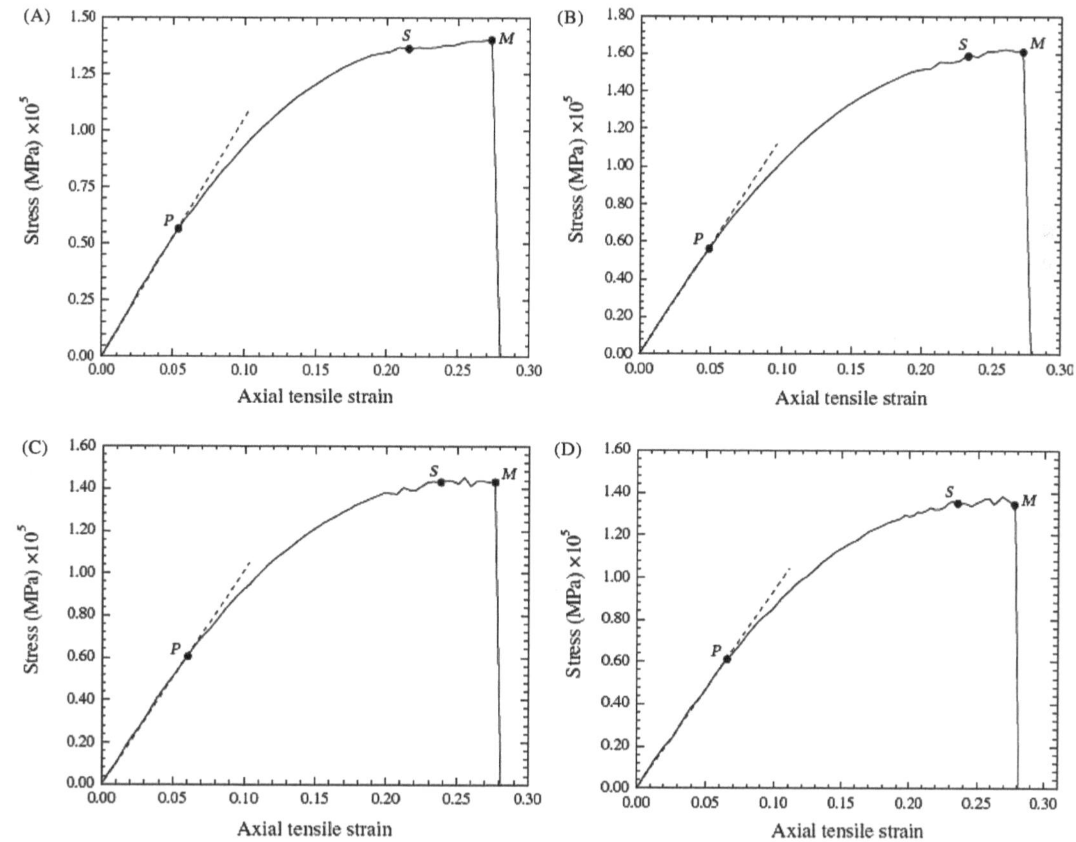

FIGURE 3.62

(A) The transverse elastic–plastic response of a single-walled CNT (10, 10) under axial tension with $L/D = 4.5$. (B) The transverse elastic–plastic response of a two-walled (5, 5) and (10, 10) CNT under axial tension with $L/D = 4.5$. (C) The transverse elastic–plastic response of a three-walled (5, 5), (10, 10), and (15, 15) CNT under axial tension with $L/D = 4.5$. (D) The transverse elastic–plastic response of a four-walled (5, 5), (10, 10), (15, 15), and (20, 20) CNT under axial tension with $L/D = 4.5$. (E) The transverse elastic–plastic response of a single-walled CNT (10, 10) under axial tension with $L/D = 9.1$. (F) The transverse elastic–plastic response of a two-walled (5, 5) and (10, 10) CNT under axial tension with $L/D = 9.1$. (G) The transverse elastic–plastic response of a three-walled (5, 5), (10, 10), and (15, 15) CNT under axial tension with $L/D = 9.1$. (H) The transverse elastic–plastic response of a four-walled (5, 5), (10, 10), (15, 15), and (20, 20) CNT under axial tension with $L/D = 9.1$ [13].

the stress–strain relationship to a smooth curve again using the regression analysis, as shown in Table 3.18 for various CNTs. At the yield point, we have $d\sigma/d\varepsilon = 0$, and thus the yield strain limit ε_s becomes $\varepsilon_s = -B/2A$. Beyond the yield point S, considerable plastic deformation occurs without a noticeable increase in the stress; at the strain limit of $\varepsilon_s \leq \varepsilon \leq \varepsilon_h$, the stress can be approximately expressed as $\sigma = \sigma_s$. This ductile behavior can be observed in our animated MD simulations, showing wave propagation from the center to the two ends of the CNTs and then back to the center. This phenomenon of wave propagation displays the rearrangement of configuration of the CNTs for strain release via the formation of the Stone–Wales defects.

Table 3.18 lists the computed Young's modulus, Poisson's ratio, yield strength, tensile strength, elastic strain limit, and maximum strain for selected single to four-walled armchair CNTs with the length-to-diameter ratios of the CNTs equal to 4.5 and 9.1. It is observed that the properties are dependent on the number of layers of the CNT. Nevertheless, there are small differences in the properties of CNTs for the length-to-diameter ratios of 4.5 and 9.1. It is evident that the effect of geometrical size on the mechanical properties of CNTs is relatively small and, with the same number of layers, we find that the mechanical properties of CNTs can be treated as the same for all length-to-diameter ratios.

Fig. 3.63 shows the formation of (5–77–5) defects due to a 90 degrees rotation of a bond during the plastic deformation. The initial (5–77–5) defect is formed at a strain of 24%. As the strain increases, more (5–77–5) defects are generated along the circumferential direction, as shown in Fig. 3.63. When the strain reaches the maximum strain point M at 0.256, two bonds are broken leading to two holes, as shown in Fig. 3.64. This observation is in close agreement with the quantum mechanical simulated results of Dumitrica et al. [94], which showed that failures occurred at strain above 0.240 for armchair CNTs. As the strain increases, more bonds are broken and the holes become larger until the CNT fractures. The brittle fracture process is displayed in Fig. 3.64A–D for the outermost layer of the four-walled CNT. Upon reaching the maximum strain ε_m, the CNTs rupture very quickly and completely fracture within a small period of time, which is displayed in Fig. 3.62A–H. Fig. 3.65 shows the entire failure process of the four-walled CNT of (5, 5), (10, 10), (15, 15), and (20, 20). When strain is at 0.263, bond breaking initially takes place at the outermost layer (first layer) with all the other three layers remaining undamaged. The phenomenon whereby the outermost layer of multiwalled CNT is first damaged is consistent with the experimental result of Yu et al. [95]. As the strain increases to 0.268, the outermost layer is almost totally fractured, while one bond in the second layer begins to break, and the third and innermost layers remain undamaged. However, a bond is observed to have broken in the third layer as the strain increases to 0.275. At this time, the third layer starts fracturing, even though no bond breaking is found in the innermost layer (fourth layer). When strain is at 0.279, a bond starts to break in the innermost layer and the three outer layers are all completely damaged. Eventually, the entire four-walled CNT disintegrates when the strain is at 0.28. The above exemplifies that multiwalled CNTs can sustain elongations up to 28%, which is in good agreement with the experimental result of Bozovic et al. [96].

3.7 THERMAL STABILITY OF CNTs

To study how the atomic structures of CNTs change and their stability at various temperatures, we have incorporated the Berendsen thermostat [97] into the MD simulation. This method requires the CNT to be coupled to an external heat bath, which is fixed at a desired temperature. The bath acts

Table 3.18 Material Properties of Selected Single and Multiwalled CNTs [13]

	L/D	Young's Modulus (TPa)	*A* (TPa)	*B* (TPa)	Proportional Strength (MPa)	Yield Strength (MPa)	Tensile Strength (MPa)	Proportional Strain Limit	Elastic Strain Limit	Maximum Strain
Single-walled CNT of (10, 10)	4.5	1.043	−2.625	1.211	6.103×10^4	1.369×10^5	1.404×10^5	0.0585	0.231	0.280
	9.1	1.031	−2.522	1.190	6.271×10^4	1.421×10^5	1.485×10^5	0.0594	0.236	0.279
Two-walled CNT of (5, 5) and (10, 10)	4.5	1.161	−2.543	1.259	7.231×10^4	1.614×10^5	1.624×10^5	0.0627	0.247	0.279
	9.1	1.175	−2.810	1.362	7.287×10^4	1.633×10^5	1.684×10^5	0.0621	0.242	0.281
Three-walled CNT of (5, 5), (10, 10) and (15, 15)	4.5	1.000	−2.358	1.160	6.068×10^4	1.430×10^5	1.434×10^5	0.0605	0.238	0.281
	9.1	0.972	−2.275	1.120	5.645×10^4	1.381×10^5	1.414×10^5	0.0611	0.246	0.282
Four-walled CNT of (5, 5), (10, 10), (15, 15) and (20, 20)	4.5	0.932	−2.234	1.103	6.075×10^4	1.343×10^5	1.382×10^5	0.0654	0.235	0.281
	9.1	0.872	−2.132	1.023	5.784×10^4	1.278×10^5	1.327×10^5	0.0633	0.241	0.280

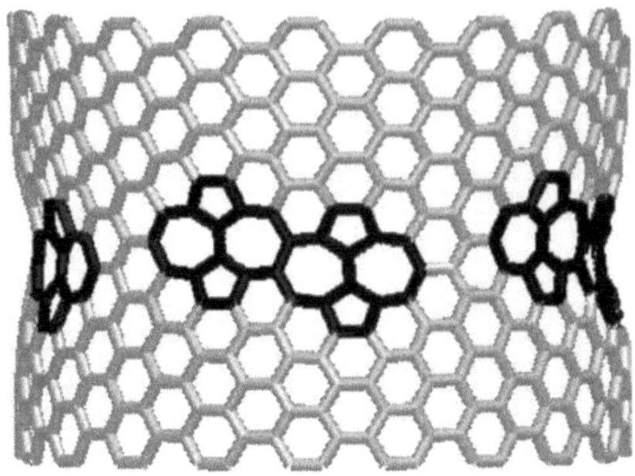

FIGURE 3.63

The formation of (5–77–5) defects in the outermost layer of a four-walled (5, 5), (10, 10), (15, 15), and (20, 20) CNT under axial tension during the plastic deformation [13].

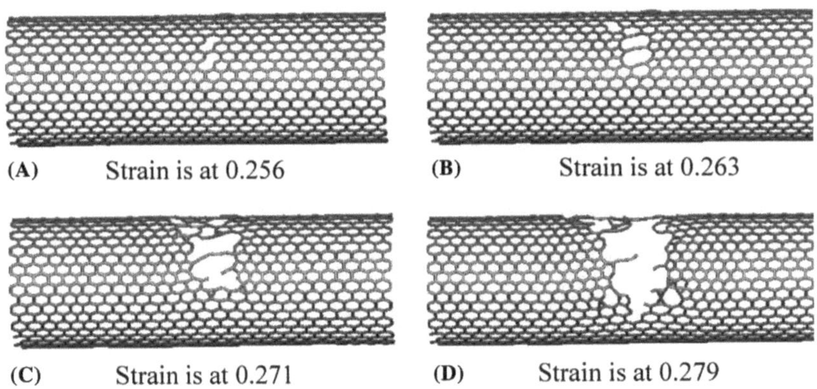

FIGURE 3.64

The brittle fracture process of the outermost layer of four-walled CNTs. (A) Formation of initial fracture. (B) Fracture propagation in circumferential direction. (C) Fracture propagation in the axial and circumferential direction. (D) The CNT is almost completely broken [13].

as a source of thermal energy, supplying or removing heat from the CNT as appropriate. The velocities of the atoms are scaled at each time-step, such that the rate of temperature change is proportional to the difference in temperature between the bath and the CNT

$$\frac{dT(t)}{t} = \frac{1}{\tau}[T_{\text{bath}} - T(t)] \tag{3.24}$$

where τ is the coupling parameter whose magnitude determines how tightly the bath and the CNT are coupled together. A larger τ will give a weaker coupling, while a smaller τ will give a stronger

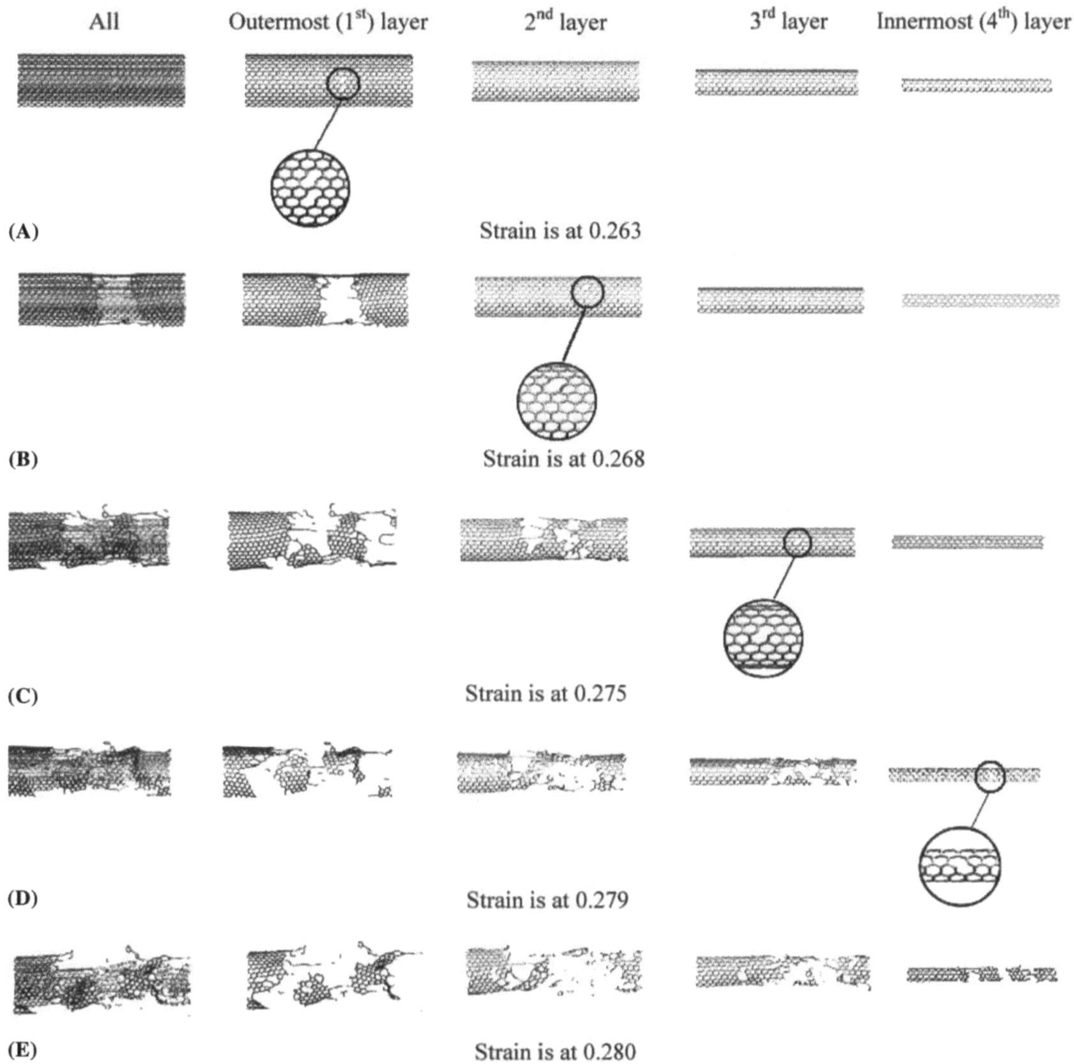

FIGURE 3.65

The brittle fracture process of the four-walled (5, 5), (10, 10), (15, 15), and (20, 20) CNT with $L/D = 9.1$. (A) Bond breaking first takes place in the outermost layer when the strain is at 0.263. (B) The second layer starts to break when the strain is at 0.268. (C) The third layer starts to break when the strain is at 0.275. (D) The innermost layer starts to break when the strain is at 0.279. (E) The entire four-walled CNT completely fractures when the strain is at 0.280 [13].

coupling. This method gives an exponential decay/growth of the system toward the desired temperature. The change in temperature between successive time-steps is

$$\Delta T = \frac{\delta t}{\tau}[T_{\text{bath}} - T(t)] \tag{3.25}$$

and therefore, the scaling factor λ for the velocities is

$$\lambda = \sqrt{1 + \frac{\delta t}{\tau}\left[\frac{T_{\text{bath}}}{T(t)} - 1\right]} \tag{3.26}$$

From Eq. (3.26), it can be seen that if the coupling parameter τ equals to the time-step δt, then the Berendsen scheme is equivalent to the simple ad hoc velocity scaling method. A coupling constant τ of approximately 0.4 ps has been suggested [97] as a suitable value to use when the time-step is 1 fs, thereby giving $\delta t/\tau \approx 0.0025$.

To apply the Berendsen thermostat [97] to the study of CNTs, the atoms are first given random initial velocities from a uniform velocity distribution on an interval $[-1, +1]$. These velocities are scaled so that the total linear momentum is zero. Therefore,

$$v_{i\alpha}^{new} = v_{i\alpha}^{old} - \frac{1}{N}\sum_{i}^{N} v_{i\alpha}^{old} \tag{3.27}$$

where α represents the x, y, and z directions. These velocities are then rescaled to achieve the desired temperature T by using the scaling factor in Eq. (3.26) for all directions.

From these initially assigned velocities, the Maxwell velocity distribution should develop during equilibration. The Maxwell–Boltzmann law [98] is used to evaluate the Maxwell velocity distribution analytically, and is defined as

$$f(\mathbf{v})d\mathbf{v} = \frac{N(\mathbf{v})d\mathbf{v}}{N} \tag{3.28}$$

which measures the fraction of N atoms having velocities within an interval $d\mathbf{v}$ of a particular value $\mathbf{v}$. From Eq. (3.28), the Maxwell velocity distribution can then be evaluated analytically as

$$f(v_\alpha)dv_\alpha = \sqrt{\frac{m}{2\pi k_B T}}\exp\left(-\frac{m{v_\alpha}^2}{2k_B T}\right)dv_\alpha \tag{3.29}$$

The distribution is Gaussian with standard deviation $\sigma_{sd} = (k_B T/m)^{1/2}$ about the mean value $\langle V_\alpha \rangle = 0$. Fig. 3.66 shows the Maxwell velocity distribution of a (10, 10) single-walled CNT exposed to various temperatures T at 100K, 200K, 300K, and 400K employing the Berendsen thermostat [97]. As the temperature T increases, the standard deviation α_{sd} approaches to uniformity.

To verify if the Maxwell velocity distribution has developed, the velocity distribution $f(v_\alpha)$ is time averaged over a number of time-steps and compared to the Maxwell velocity distribution at the same kinetic energy. Fig. 3.67 shows the comparison of the (10, 10) single-walled CNT at temperature $T = 300$K. The plot shows good agreement and signifies that an equilibration state is achieved.

During the simulation, it is instructive to follow the development of the Maxwell distribution and, hence the approach to equilibrium, by monitoring the time behavior of the kinetic portion of Boltzmann's celebrated H-function [98] expressed as

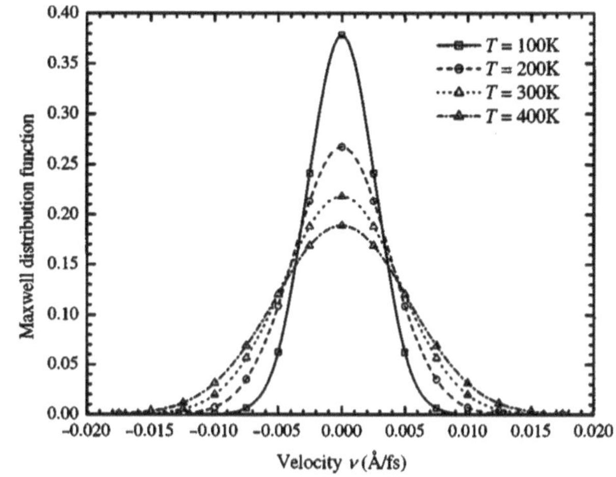

FIGURE 3.66

Effect of temperature on the distribution of the velocity of (10, 10) single-walled CNT, as computed from the Maxwell velocity distribution in Eq. (3.29) (Liew et al. [13]).

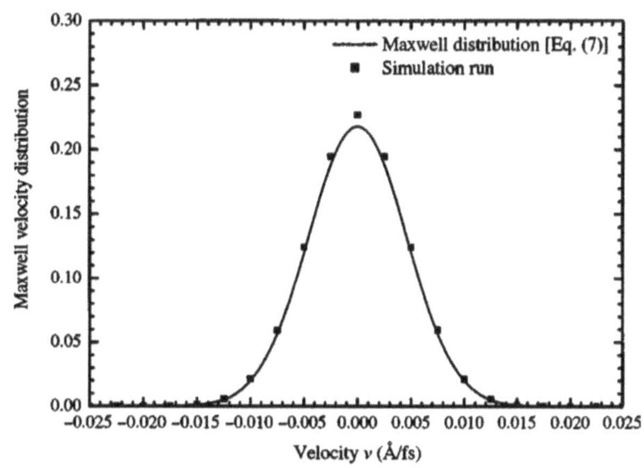

FIGURE 3.67

The velocity distribution averaged over 100,000 time-steps of the run (point) compared to the Maxwell distribution (line) for the (10, 10) single-walled CNT at temperature $T = 300\text{K}$ [17].

$$H_\alpha(t) = \int_{-\infty}^{+\infty} f(v_\alpha)\ln(v_\alpha)\, d(v_\alpha) \tag{3.30}$$

Fig. 3.68 shows the instantaneous H-function computed during the equilibration of the (10, 10) single-walled CNT at temperature $T = 300\text{K}$. In the figure, the horizontal line at $H = -0.00505$ is the value of the H-function for the Maxwell velocity distribution at the kinetic energy of the

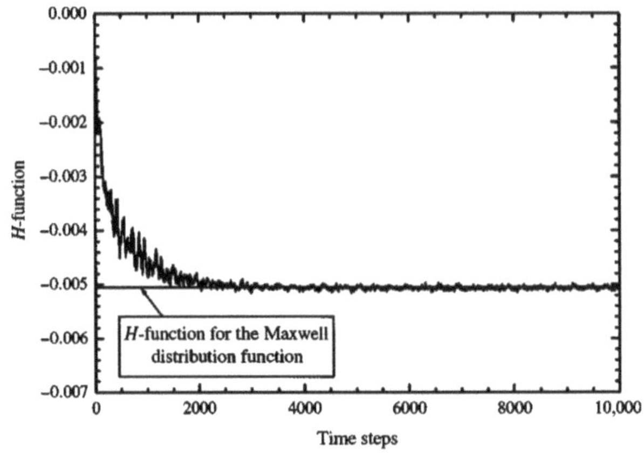

FIGURE 3.68

Approach to equilibrium for the (10, 10) single-walled CNT at temperature $T = 300$K based on H-function [17].

simulation. After about 3000 time-steps, the instantaneous H-function converges to the Maxwell's value of the H-function, thus indicating that Maxwell distribution develops within about 3000 time-steps. Above 3000 time-steps, the instantaneous H-function fluctuates along the Maxwell's value of the H-function, and hence the system is considered to be in equilibrium.

At equilibration, the potential energy E_{POT} and kinetic energy E_K fluctuate about a steady value. Fig. 3.69 shows that the (10, 10) single-walled CNT achieves equilibration at temperature $T = 300$K. The kinetic energy E_K fluctuates around 9.0 eV, while the potential energy E_{POT} fluctuates around −7230 eV over 4000 time-steps. The plot also indicates the conservation of the total energy. During the simulation, the total energy E_{TOT} has a maximum deviation of 0.033% from its average value of −7220.92 eV.

Both open-ended and close-capped single-walled CNTs are investigated to determine the influence of the caps on the CNTs. All the single-walled CNTs are made to adhere to the Maxwell velocity distribution and the H-function to confirm their equilibrium state.

3.7.1 CLOSE-CAPPED SINGLE-WALLED CNTs

Three different lengths of (5, 5) close-capped single-walled CNTs are investigated. The lengths L are 122, 183, and 244 Å. Fig. 3.70 shows that a 122 Å-long (5, 5) close-capped single-walled CNT remains undeformed at temperatures T from 300K up to 2000K. At $T = 2000$K, the close-capped single-walled CNT begins to deform slightly at its body and disintegrates completely at $T = 3489$K.

A similar simulation is carried out on a 183 Å-long (5, 5) close-capped single-walled CNT as depicted in Fig. 3.71. The longer close-capped single-walled CNT begins to deform structurally at temperature $T = 2000$K before breaking down completely at a lower temperature $T = 3333$K.

Similarly atomic structural deformation is evident for the 244 Å-long (5, 5) close-capped single-walled CNT before collapsing at $T = 2946$K.

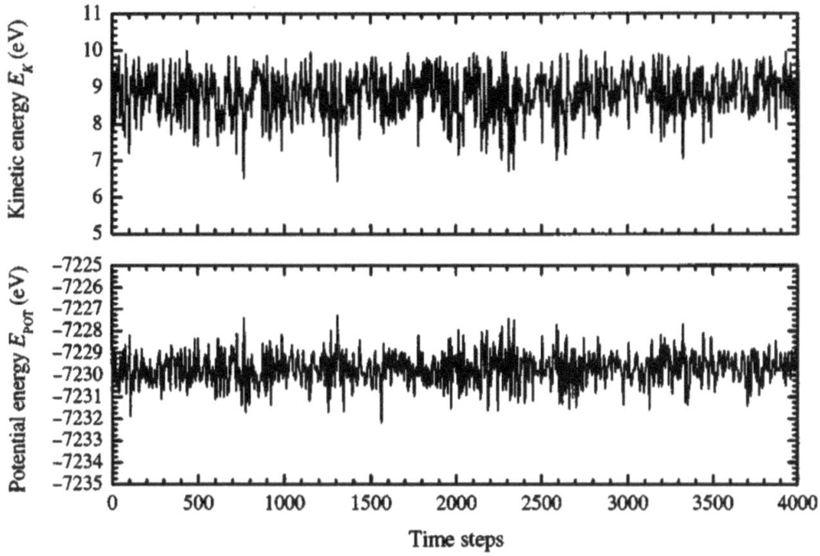

FIGURE 3.69

Plot of kinetic energy and potential energy for the (10, 10) single-walled CNT at temperature $T = 300\text{K}$ after equilibration [17].

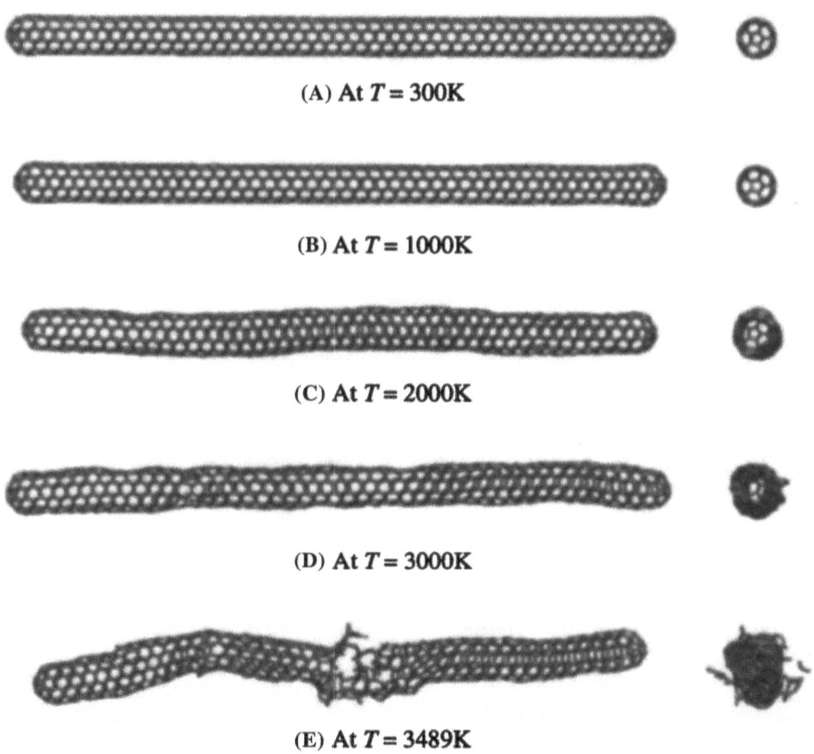

FIGURE 3.70

The relaxed structures of close-capped (5, 5) single-walled CNT ($L = 122$ Å) at various temperatures [17].

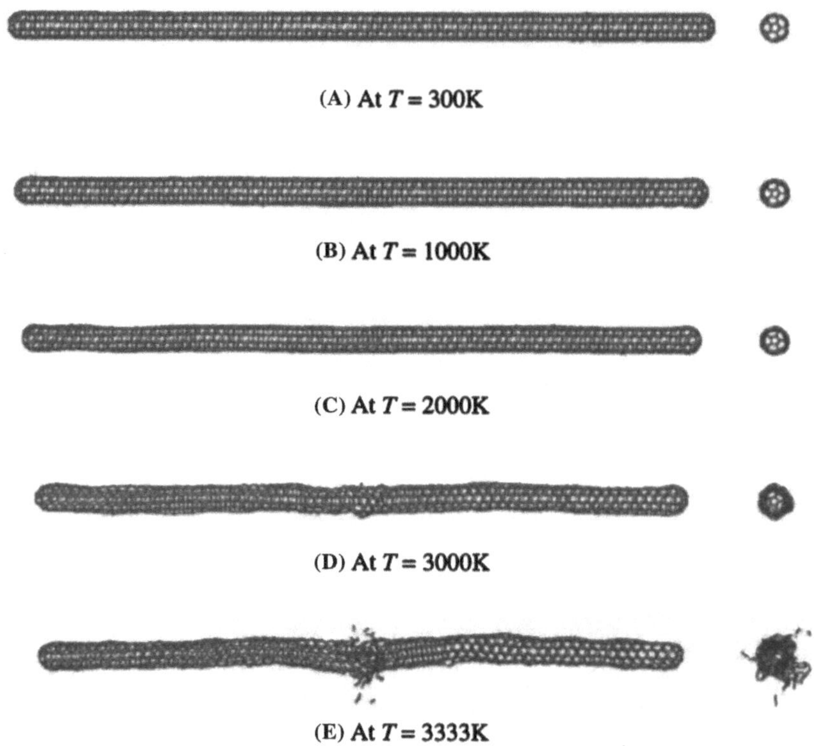

FIGURE 3.71

The relaxed structures of close-capped (5, 5) single-walled CNT ($L = 183$ Å) at various temperatures [17].

3.7.2 OPEN-ENDED SINGLE-WALLED CNTs

As a comparison to close-capped (5, 5) single-walled CNTs, different lengths ($L = 122$, 183, and 244 Å) of open-ended (5, 5) single-walled CNTs are also investigated. The relaxed CNTs at various temperatures are shown in Figs. 3.72 and 3.73.

It can be seen in Fig. 3.72 that the ends of the 122 Å-long open-ended (5, 5) single-walled CNT begins to flange slightly at temperature $T = 300$K. As the temperature increases to $T = 1000$K, the open-ended (5, 5) single-walled CNT begins to deform structurally. And when the temperature increases to $T = 3000$K, the CNT bends before disintegrating at $T = 3582$K. The bending-induced thermal contraction causes the end-to-end distance to shrink, and agrees with the findings of Schelling et al. [99].

Deformation at the ends of the 183 Å-long open-ended (5, 5) single-walled CNT is apparent at temperature $T = 1000$K, as shown in Fig. 3.73. The CNT starts to bend from $T = 2000–3000$K, causing a reduction of the end-to-end distance. At $T = 3337$K, the 183 Å-long open-ended (5, 5) single-walled CNT can no longer withstand the thermal load and disintegrates.

A 244 Å-long open-ended (5, 5) single-walled CNT is also modeled and simulated as thermal load is applied to it. The results show that it collapses after it reaches a temperature of $T = 2798$K.

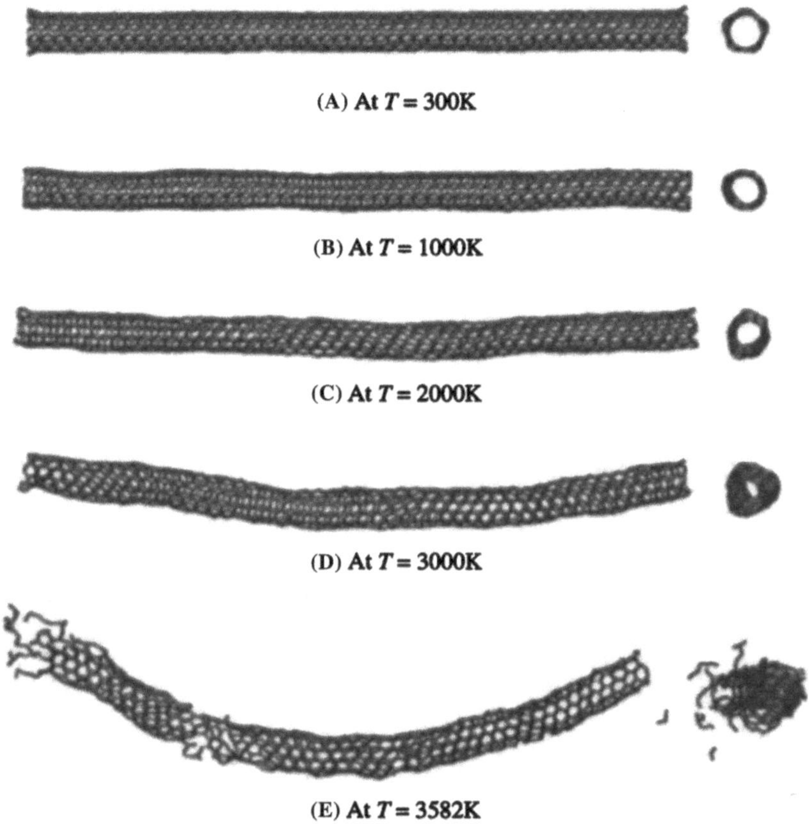

FIGURE 3.72

The relaxed structures of open-ended (5, 5) single-walled CNT ($L = 122$ Å) at various temperatures [17].

From Figs. 3.70–3.73, we can observe that the atomic structural deformation differs between close-capped and open-ended (5, 5) single-walled CNTs. For open-ended CNTs, the end atoms begin to deform first, probably due to unsatisfied bonds at the ends of the CNTs to satisfy their bonds with bonding to other carbon atoms, while the body of the close-capped CNTs deform first when subjected to thermal loads. However, the maximum critical temperatures both the open-ended and close-capped CNTs can withstand before breaking down are less than 5% different (see Fig. 3.74). Hence for the ease of modeling, further open-ended (10, 10) and (12, 12) single-walled CNTs are investigated.

The response to temperature for the open-ended (10, 10) and (12, 12) single-walled CNT is included in Fig. 3.74. From the plots, all the single-walled CNTs tend to have lower thermal stability as their lengths L increase. For instance at length $L = 122$ Å, the (10, 10) single-walled CNT can resist up to $T = 3350K$ before collapsing. However, the maximum temperature a 244 Å-long (10, 10) single-walled CNT can withstand is lower, i.e., $T = 2798K$.

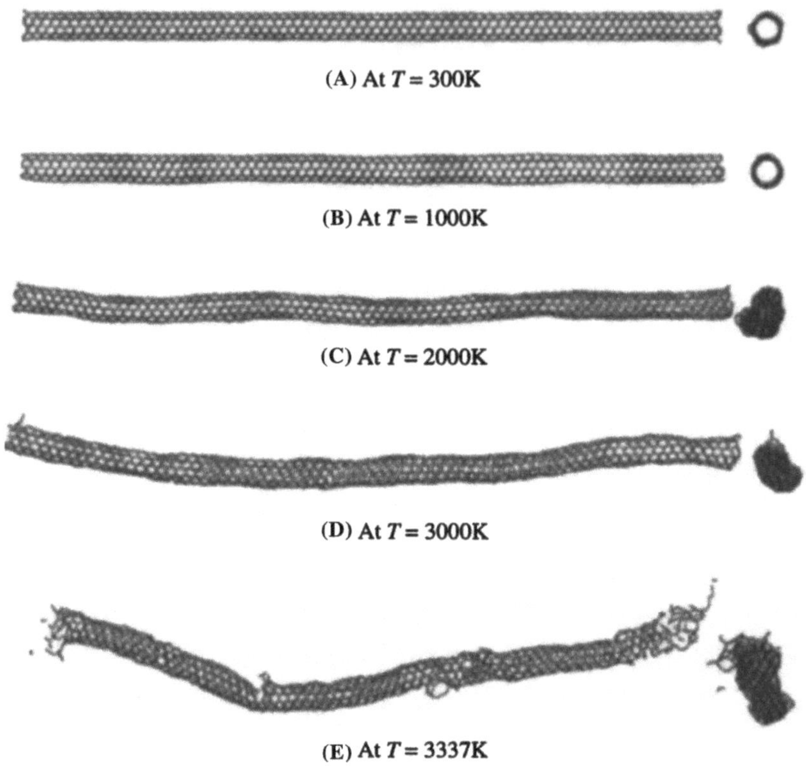

FIGURE 3.73

The relaxed structures of open-ended (5, 5) single-walled CNT ($L = 183$ Å) at various temperatures [17].

Additionally, as the diameter d_t of the CNTs increases, its ability to withstand thermal loads is lower. This is exemplified by the maximum temperature $T = 2868$K the (12, 12) single-walled CNT can withstand as compared to the lower maximum temperature $T = 3337$K for the (5, 5) single-walled CNT when both lengths $L = 183$ Å.

According to Fig. 3.74, it is evident that the length L of the single-walled CNT affects its response to temperature T. Fig. 3.75 explains the cause of this phenomena using the (10, 10) single-walled CNT as an example.

Comparing to the shorter (10, 10) single-walled CNT that is shown in Fig. 3.74, it can be deduced that longer single-walled CNTs tend to break down at a lower temperature than shorter single-walled CNTs at longer lengths. This can be explained by the plot in Fig. 3.75. Although both lengths of (10, 10) single-walled CNT have similar kinetic energy E_K per atom at a certain temperature T, the potential energy E_{POT} per atom is different. The longer (10, 10) single-walled CNT has lower potential energy E_{POT} per atom compared to the shorter (10, 10) single-walled CNT at all temperatures T. When subjected to heat, the chemical bonds from the longer single-walled CNT break down more easily, because each atom from the longer (10, 10) single-walled CNT has lower potential energy E_{POT}. Consequently, the longer (10, 10) single-walled CNT collapses at a lower temperature T than the shorter (10, 10) single-walled CNT.

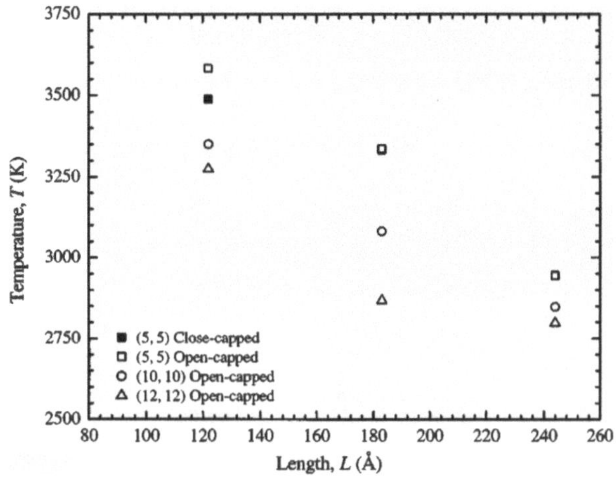

FIGURE 3.74

Plot depicting the maximum temperature T various single-walled CNT of different lengths can resist before collapsing [17].

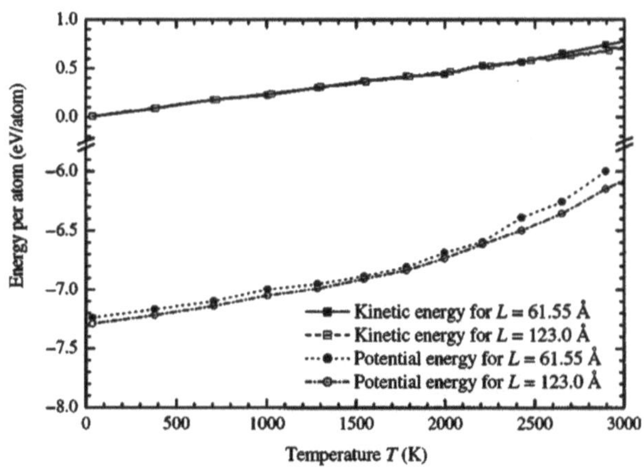

FIGURE 3.75

Plot depicting the kinetic energies E_K and potential energies E_{POT} for the short ($L = 61.55$ Å) and long ($L = 123.0$ Å) (10, 10) single-walled CNTs at various temperatures T [17].

3.7.3 OPEN-ENDED MULTIWALLED CNTs

The response of multiwalled CNTs with different geometries at different temperatures T are also investigated. In this study, two multiwalled CNTs, namely, the (5, 5) and (10, 10), and (10, 10) and (15, 15) multiwalled CNTs with the same lengths $L = 61.55$ Å are considered. The relaxed multiwalled CNTs at various temperatures T are shown in Fig. 3.76 and 3.77. All of the multiwalled CNTs adhere to the Maxwell velocity distribution and the H-function to confirm their equilibrium state.

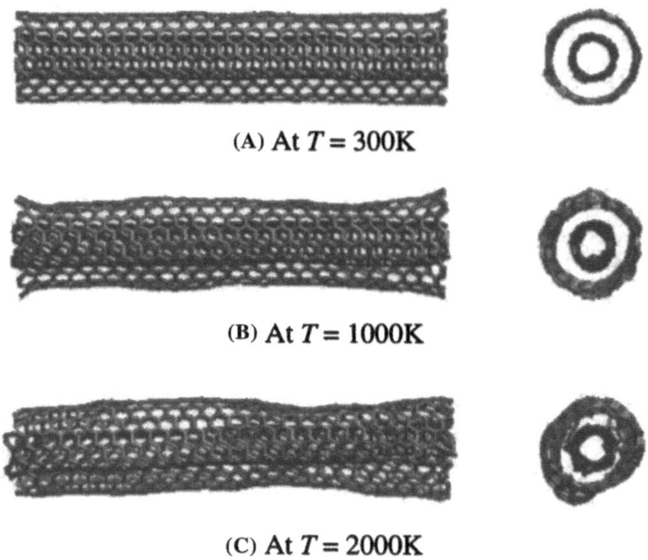

FIGURE 3.76

The relaxed structures of (5, 5) and (10, 10) multiwalled CNT at various temperatures [17].

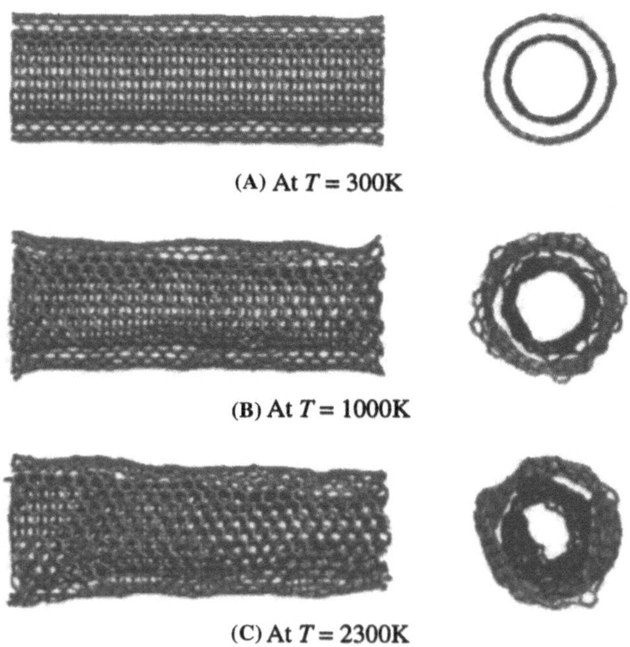

FIGURE 3.77

The relaxed structures of (10, 10) and (15, 15) multiwalled CNT at various temperatures [17].

Fig. 3.76 depicts the morphological changes of the (5, 5) and (10, 10) multiwalled CNT when subjected to an increasing temperature T from 0K to 3000K. At $T = 300$K, the multiwalled CNT is similar to that of an ideal structure. However, at $T = 1000$K, both layers start to deform. Significant flanges at both ends of the outer layer of the multiwalled CNT are also observed, and the ends of the inner layer start to close together. Before the multiwalled CNT collapses after $T = 2000$K, more deformation occurs at both layers. Due to the high temperature, the atoms at both layers vibrate more and collide with each other resulting in the disintegration of the multiwalled CNT.

When exposed to an increasing temperature T, the second (10, 10) and (15, 15) multiwalled CNT deforms as shown in Fig. 3.77. Like all of the multiwalled CNTs, the (10, 10) and (15, 15) multiwalled CNT maintains its overall geometry at $T = 300$K. Flanges at both ends of the multiwalled CNT are formed at $T = 1000$K until $T = 2300$K, when the ends of both layers begin to close. On further increase of temperature T, the bonds break and the multiwalled CNT eventually collapses.

REFERENCES

[1] Iijima S. Helical microtubules of graphitic carbon. Nature 1991;354:56–8.

[2] Bethune D, Klang C, De Vries M, Gorman G, Savoy R, Vazquez J, et al. Cobalt-catalysed growth of carbon nanotubes with single-atomic-layer walls. Nature 1993;363(6430):605–7.

[3] Iijima S, Ichihashi T. Single-shell carbon nanotubes of 1-Nm diameter. Nature 1993;363:603–5.

[4] Goldsmith BR, Coroneus JG, Khalap VR, Kane AA, Weiss GA, Collins PG. Conductance-controlled point functionalization of single-walled carbon nanotubes. Science 2007;315:77–81.

[5] Zaric S, Ostojic GN, Kono J, Shaver J, Moore VC, Strano MS, et al. Optical signatures of the Aharonov-Bohm phase in single-walled carbon nanotubes. Science 2004;304:1129–31.

[6] Barreiro A, Rurali R, Hernandez ER, Moser J, Pichler T, Forro L, et al. Subnanometer motion of cargoes driven by thermal gradients along carbon nanotubes. Science 2008;320:775–8.

[7] Koziol K, Vilatela J, Moisala A, Motta M, Cunniff P, Sennett M, et al. High-performance carbon nanotube fiber. Science 2007;318:1892–5.

[8] Meyer JC, Paillet M, Roth S. Single-molecule torsional pendulum. Science 2005;309:1539–41.

[9] Sun L, Banhart F, Krasheninnikov AV, Rodriguez-Manzo JA, Terrones M, Ajayan PM. Carbon nanotubes as high-pressure cylinders and nanoextruders. Science 2006;312:1199–202.

[10] Wang XY, Wang X. Torsional buckling of multi-walled carbon nanotubes subjected to torsional loads. J Reinf Plast Compos 2007;26:479–94.

[11] Yakobson BI, Brabec CJ, Bernholc J. Nanomechanics of carbon tubes: instabilities beyond linear response. Phys Rev Lett 1996;76:2511–14.

[12] Wang Y, Wang XX, Ni XG, Wu HA. Simulation of the elastic response and the bucIrding modes of single-walled carbon nanotubes. Comput Mater Sci 2005;32:141–6.

[13] Liew KM, He XQ, Wong CH. On the study of elastic and plastic properties of multi-walled carbon nanotubes under axial tension using molecular dynamics simulation. Acta Mater 2004;52:2521–7.

[14] Liew KM, Wong CH, Tan MJ. Buckling properties of carbon nanotube bundles. Appl Phys Lett 2005;87:041901.

[15] Liew KM, Wong CH, Tan MJ. Tensile and compressive properties of carbon nanotube bundles. Acta Mater 2006;54:225–31.

[16] Liew KM, Wong CH, He XQ, Tan MJ, Meguid SA. Nanomechanics of single and multiwalled carbon nanotubes. Phys Rev B 2004;69:115429.

[17] Liew KM, Wong CH, He XQ, Tan MJ. Thermal stability of single and multi-walled carbon nanotubes. Phys Rev B 2005;71:075424.
[18] Li CY, Chou TW. Elastic moduli of multi-walled carbon nanotubes and the effect of van der Waals forces. Compos Sci Technol 2003;63:1517–24.
[19] Wei CY, Cho KJ, Srivastava D. Tensile strength of carbon nanotubes under realistic temperature and strain rate. Phys Rev B 2003;67.
[20] Liu P, Zhang YW, Lu C, Lam KY. Tensile and bending properties of double-walled carbon nanotubes. J Phys D-Appl Phys 2004;37:2358–63.
[21] Cornwell C, Wille L. Elastic properties of single-walled carbon nanotubes in compression. Solid State Commun 1997;101:555–8.
[22] Srivastava D, Menon M, Cho K. Nanoplasticity of single-wall carbon nanotubes under uniaxial compression. Phys Rev Lett 1999;83:2973.
[23] Alder BJ, Wainwright TE. Phase transition for a hard sphere system. J Chem Phys 1957;27:1208–9.
[24] Verlet L. Computer experiments on classical fluids. I. Thermodynamical properties of Lennard-Jones molecules. Phys Rev 1967;159:98.
[25] Allinger NL. Conformational-analysis .130. MM2—hydrocarbon force-field utilizing V1 and V2 torsional terms. J Am Chem Soc 1977;99:8127–34.
[26] Allinger NL, Yuh YH, Lii JH. Molecular mechanics—the MM3 force-field for hydrocarbons .1. J Am Chem Soc 1989;111:8551–66.
[27] Lii JH, Allinger NL. Molecular mechanics—the MM3 force-field for hydrocarbons .2. Vibrational frequencies and thermodynamics. J Am Chem Soc 1989;111:8566–75.
[28] Lii JH, Allinger NL. Molecular mechanics—the MM3 force-field for hydrocarbons .3. The vanderwaals potentials and crystal data for aliphatic and aromatic-hydrocarbons. J Am Chem Soc 1989;111:8576–82.
[29] Abell GC. Empirical chemical pseudopotential theory of molecular and metallic bonding. Phys Rev B 1985;31:6184–96.
[30] Tersoff J. New empirical-approach for the structure and energy of covalent systems. Phys Rev B 1988;37:6991–7000.
[31] Tersoff J. Empirical interatomic potential for carbon, with applications to amorphous-carbon. Phys Rev Lett 1988;61:2879–82.
[32] Nordlund K, Keinonen J, Mattila T. Formation of ion irradiation induced small-scale defects on graphite surfaces. Phys Rev Lett 1996;77:699–702.
[33] Brenner DW. Empirical potential for hydrocarbons for use in simulating the chemical vapor-deposition of diamond films. Phys Rev B 1990;42:9458–71.
[34] Robertson DH, Brenner DW, Mintmire JW. Energetics of nanoscale graphitic tubules. Phys Rev B 1992;45:12592–5.
[35] Robertson DH, Brenner DW, White CT. On the way to fullerenes—molecular-dynamics study of the curling and closure of graphitic ribbons. J Phys Chem 1992;96:6133–5.
[36] Harrison JA, White CT, Colton RJ, Brenner DW. Investigation of the atomic-scale friction and energy-dissipation in diamond using molecular-dynamics. Thin Solid Films 1995;260:205–11.
[37] Brenner DW, Shenderova OA, Harrison JA, Stuart SJ, Ni B, Sinnott SB. A second-generation reactive empirical bond order (REBO) potential energy expression for hydrocarbons. J Phys-Conden Matter 2002;14:783–802.
[38] Zhang SL, Mielke SL, Khare R, Troya D, Ruoff RS, Schatz GC, et al. Mechanics of defects in carbon nanotubes: atomistic and multiscale simulations. Phys Rev B 2005;71:115403.
[39] Shenderova OA, Brenner DW, Omeltchenko A, Su X, Yang LH. Atomistic modeling of the fracture of polycrystalline diamond. Phys Rev B 2000;61:3877–88.

[40] Zhbanov AI, Pogorelov EG, Chang YC. Van der Waals interaction between two crossed carbon nanotubes. Acs Nano 2010;4:5937–45.
[41] Yuan JH, Liew KM. Effects of boron nitride impurities on the elastic properties of carbon nanotubes. Nanotechnology 2008;19:445703.
[42] Lu JP. Elastic properties of carbon nanotubes and nanoropes. Phys Rev Lett 1997;79:1297.
[43] Charlier J-C, Ebbesen T, Lambin P. Structural and electronic properties of pentagon-heptagon pair defects in carbon nanotubes. Phys Rev B 1996;53:11108.
[44] Lu AJ, Pan BC. Nature of single vacancy in achiral carbon nanotubes. Phys Rev Lett 2004;92.
[45] Krasheninnikov A, Nordlund K, Sirviö M, Salonen E, Keinonen J. Formation of ion-irradiation-induced atomic-scale defects on walls of carbon nanotubes. Phys Rev B 2001;63:245405.
[46] Krasheninnikov A, Nordlund K. Stability of irradiation-induced point defects on walls of carbon nanotubes. J Vacuum Sci Technol B 2002;20:728–33.
[47] Krasheninnikov AV, Elesin VF. The effect of interstitial clusters and vacancies on the scanning tunneling microscopy image of graphite. Surface Sci 2000;454:519–24.
[48] Terrones M, Terrones H, Banhart F, Charlier JC, Ajayan PM. Coalescence of single-walled carbon nanotubes. Science 2000;288:1226–9.
[49] Choi HJ, Ihm J, Louie SG, Cohen ML. Defects, quasibound states, and quantum conductance in metallic carbon nanotubes. Phys Rev Lett 2000;84:2917–20.
[50] Hansson A, Paulsson M, Stafstrom S. Effect of bending and vacancies on the conductance of carbon nanotubes. Phys Rev B 2000;62:7639–44.
[51] Rossato J, Baierle RJ, Fazzio A, Mota R. Vacancy formation process in carbon nanotubes: first-principles approach. Nano Lett 2005;5:197–200.
[52] Yuan JH, Liew KM. Effects of vacancy defect reconstruction on the elastic properties of carbon nanotubes. Carbon NY 2009;47:1526–33.
[53] Ajayan P, Ravikumar V, Charlier J-C. Surface reconstructions and dimensional changes in single-walled carbon nanotubes. Phys Rev Lett 1998;81:1437.
[54] Kelly BT. Physics of graphite. London: Applied Science Publishers; 1981.
[55] Shim SB, Imboden M, Mohanty P. Synchronized oscillation in coupled nanomechanical oscillators. Science 2007;316:95–9.
[56] Wang XD, Song JH, Liu J, Wang ZL. Direct-current nanogenerator driven by ultrasonic waves. Science 2007;316:102–5.
[57] Tomasio SD, Walsh TR. Atomistic modelling of the interaction between peptides and carbon nanotubes. Mol Phys 2007;105:221–9.
[58] Ye LH, Liu BG, Wang DS, Han R. Ab initio phonon dispersions of single-wall carbon nanotubes. Phys Rev B 2004;69:235409.
[59] Lazar P, Redinger J, Podloucky R. Density functional theory applied to VN/TiN multilayers. Phys Rev B 2007;76:174112–21.
[60] Krasheninnikov AV, Nordlund K, Sirvio M, Salonen E, Keinonen J. Formation of ion-irradiation-induced atomic-scale defects on walls of carbon nanotubes. Phys Rev B 2001;63:245405–11.
[61] Zhang ZQ, Liu B, Chen YL, Jiang H, Hwang KC, Huang Y. Mechanical properties of functionalized carbon nanotubes. Nanotechnology 2008;19:395702.
[62] Haskins RW, Maier RS, Ebeling RM, Marsh CP, Majure DL, Bednar AJ, et al. Tight-binding molecular dynamics study of the role of defects on carbon nanotube moduli and failure. J Chem Phys 2007;127:074708.
[63] Shen LX, Li J. Equilibrium structure and strain energy of single-walled carbon nanotubes. Phys Rev B 2005;71:165427.

[64] Belytschko T, Xiao SP, Schatz GC, Ruoff RS. Atomistic simulations of nanotube fracture. Phys Rev B 2002;65:235430.
[65] Wang LF, Zheng QS, Liu JZ, Jiang Q. Size dependence of the thin-shell model for carbon nanotubes. Phys Rev Lett 2005;95:105501.
[66] Yuan JH, Liew KM. The effects of grafted amine groups on the elastic properties of single-walled carbon nanotubes. Carbon N Y 2009;47:713–21.
[67] Chang TC, Geng JY, Guo XM. Chirality- and size-dependent elastic properties of single-walled carbon nanotubes. Appl Phys Lett 2005;87:251929.
[68] Zhang LY, Peng YJ, Jin QH, Wang YF, Li BH, Ding DT. Elastic properties of single-wall nanotubes. Acta Physica Sinica 2006;55:4193–6.
[69] Popov VN, Van Doren VE, Balkanski M. Elastic properties of crystals of single-walled carbon nanotubes. Solid State Commun 2000;114:395–9.
[70] Xiao SP, Hou WY. Studies of size effects on carbon nanotubes' mechanical properties by using different potential functions. Fullerenes Nanotubes Carbon Nanostruct 2006;14:9–16.
[71] Lourie O, Cox D, Wagner H. Buckling and collapse of embedded carbon nanotubes. Phys Rev Lett 1998;81:1638.
[72] Gear CW. Numerical initial value problems in ordinary differential equations. Englewood Cliffs, NJ: Prentice Hall PTR; 1971.
[73] Allen HG, Bulson PS. Background to buckling. New York, NY: McGraw, Hill; 1980.
[74] Wong EW, Sheehan PE, Lieber CM. Nanobeam mechanics: elasticity, strength, and toughness of nanorods and nanotubes. Science 1997;277:1971–5.
[75] Timoshenko SP, Gere JM. Mechanics of materials (Van Nostrand Reinhold Company, New York) 1973.
[76] Yuan JH, Liew KM. Structural stability of a coaxial carbon nanotube inside a boron–nitride nanotube. Carbon N Y 2011;49:677–83.
[77] Delley B. An all-electron numerical method for solving the local density functional for polyatomic molecules. J Chem Phys 1990;92:508–17.
[78] Delley B. From molecules to solids with the DMol(3) approach. J Chem Phys 2000;113:7756–64.
[79] Perdew JP, Wang Y. Accurate and simple analytic representation of the electron-gas correlation energy. Phys Rev B 1992;45:13244.
[80] Rappé AK, Casewit CJ, Colwell K, Goddard Iii W, Skiff W. UFF, a full periodic table force field for molecular mechanics and molecular dynamics simulations. J Am Chem Soc 1992;114:10024–35.
[81] Rappe A, Colwell K, Casewit C. Application of a universal force field to metal complexes. Inorg Chem 1993;32:3438–50.
[82] Casewit C, Colwell K, Rappe A. Application of a universal force field to main group compounds. J Am Chem Soc 1992;114:10046–53.
[83] Chen LL, Ye HH, Gogotsi Y, McNallan MJ. Carbothermal synthesis of boron nitride coatings on silicon carbide. J Am Ceramic Soc 2003;86:1830–7.
[84] Tang CC, Bando Y. Effect of BN coatings on oxidation resistance and field emission of SiC nanowires. Appl Phys Lett 2003;83:659–61.
[85] Gomathi A, Harika MR, Rao CNR. Urea route to coat inorganic nanowires, carbon fibers and nanotubes by boron nitride. Mater Sci Eng a-Struct Mater Propert Microstruct Process 2008;476:29–33.
[86] Das M, Basu AK, Ghatak S, Joshi AG. Carbothermal synthesis of boron nitride coating on PAN carbon fiber. J Eur Ceramic Soc 2009;29:2129–34.
[87] Smith DW. The antibonding effect. J Chem Educ 2000;77:780–4.
[88] Gropp W, Lusk E, Skjellum A, Using MPI. Portable parallel programming with the message-passing interface. Cambridge, MA: MIT Press; 1999.

[89] Hertel T, Walkup RE, Avouris P. Deformation of carbon nanotubes by surface van der Waals forces. Phys Rev B 1998;58:13870.
[90] Avouris P, Hertel T, Martel R, Schmidt T, Shea H, Walkup R. Carbon nanotubes: nanomechanics, manipulation, and electronic devices. Appl Surf Sci 1999;141:201–9.
[91] Yoon Y-G, Mazzoni MS, Choi HJ, Ihm J, Louie SG. Structural deformation and intertube conductance of crossed carbon nanotube junctions. Phys Rev Lett 2001;86:688.
[92] Ma XL, Wang HT, Yang W. Tribological behavior of aligned single-walled carbon nanotubes. J Eng Mater Technol-Transact Asme 2004;126:258–64.
[93] Kudin KN, Scuseria GE, Yakobson BI. C2F, BN, and C nanoshell elasticity from ab initio computations. Phys Rev B 2001;64:235406.
[94] Dumitrica T, Belytschko T, Yakobson BI. Bond-breaking bifurcation states in carbon nanotube fracture. J Chem Phys 2003;118:9485–8.
[95] Yu MF, Lourie O, Dyer MJ, Moloni K, Kelly TF, Ruoff RS. Strength and breaking mechanism of multiwalled carbon nanotubes under tensile load. Science 2000;287:637–40.
[96] Bozovic D, Bockrath M, Hafner JH, Lieber CM, Park H, Tinkham M. Plastic deformations in mechanically strained single-walled carbon nanotubes. Phys Rev B 2003;67.
[97] Berendsen HJ, Postma JPM, van Gunsteren WF, DiNola A, Haak J. Molecular dynamics with coupling to an external bath. J Chem Phys 1984;81:3684–90.
[98] Chapman S, Cowling TG. The mathematical theory of non-uniform gases: an account of the kinetic theory of viscosity, thermal conduction and diffusion in gases. Cambridge: Cambridge University Press; 1970.
[99] Schelling PK, Keblinski R. Thermal expansion of carbon structures. Phys Rev B 2003;68:035425.
[100] Liew KM, Wong CH, Tan MJ. Twisting effects of carbon nanotube bundles subjected to axial compression and tension. J Appl Phys 2006;99:114312.

CHAPTER

ATOMISTIC-CONTINUUM THEORY 4

4.1 INTRODUCTION

Although atomistic simulation is an effective method to truly capture physical behaviors and simulate mechanical properties of carbon nanotubes (CNTs), it has a high price of computational resources. Some equivalent continuum models [1–8] have also been developed, and have proved to be very efficient from the computational point of view. These continuum-based methods are much faster than atomistic simulations for systems of engineering interest, which makes them attractive. Hybrid approaches [9–12] possess both the advantages of atomistic simulations and continuum methods, i.e., incorporating the atomic information and largely reducing the degrees of freedom of the whole system. These methods recently have become promising and significant in engineering applications owing to their high performance in the treatment of large-scale systems. Tadmor et al. [9,10] analyzed complex Bravais crystals using a FEM framework incorporating interatomic interaction. Shenoy et al. [11] proposed a quasicontinuum method by mixed atomistic simulations and continuum methods which gives rise to a reduction of atomistic degrees of freedom. Liew et al. [12] examined the elastic modulus and fracturing of single-walled carbon nanotubes (SWCNTs) using an atomistic-based continuum method and found a good agreement with the results of previous experiments and atomistic simulations.

One new hybrid approach in the study of CNTs can be traced back to the continuum method incorporating interatomic potentials [13], in which the constitutive model is constructed based on the Cauchy–Born rule. Due to the lack of centrosymmetry of the hexagonal lattice structures in CNTs, Zhang et al. [13–15] introduced an internal relaxation to ensure the minimum energy of the atomic system. They recommended a combination of interatomic potentials with a continuum model and studied linear elastic modulus fracture nucleation. However, Arroyo and Belytschko [16–18] argued that this application of the standard Cauchy–Born rule may be inadequate. Essentially, SWCNTs are curved crystalline sheets with a one-atom thickness. Therefore, the curved effect should not be neglected. Arroyo and Belytschko's numerical simulation demonstrated that the constitutive relationship constructed based on the Cauchy–Born rule fails to truly describe buckling behaviors of CNTs because of the neglect of the bending effect. Sunyk and Steinmann [19] noted that deformations of the underlying crystal must be sufficiently homogeneous when the standard Cauchy–Born rule is applied. Later, Guo et al. [20] and Wang et al. [21] proposed a higher-order Cauchy–Born rule for analysis of elastic properties of SWCNTs. In the higher-order gradient theory, the second deformation gradient can accurately describe the bending effect which makes the results much more close to the actual case. Soon afterwards, Liew and Sun [22–24] developed a mesh-free computational framework based on the higher-order deformation gradient continuum theory to simulate the buckling behaviors of SWCNTs under various typical loadings.

Mechanical Behaviors of Carbon Nanotubes. DOI: http://dx.doi.org/10.1016/B978-0-323-43137-8.00004-4

Here we model CNTs in the theoretical framework of the higher-order deformation gradient continuum theory, in which the strain energy density depends not only on the first-order but also the second-order deformation gradients. Subsequently, based on the established constitutive model, an approximate numerical technique should be adopted to implement the numerical simulation. It is noteworthy that since the second-order deformation gradient involved in the theory requires C^1-continuity [25,26] of interpolation of displacements, it is a challenge to establish the elements and construct the interpolation function using the traditional numerical tools such as FEM. Recently, mesh-free methods [22–24,27–34] as a newly category of promising numerical approaches, due to their superiority of independence of meshes, have been applied to taking order with engineering application problems.

4.1.1 OVERVIEW OF MESH-FREE METHODS

Mesh-free methods can be traced back to the 1970s, but they did not receive much attention from researchers until recently. In 1977, Lucy [35] and Gingold and Monaghan [36] proposed the smooth particle hydrodynamics (SPH) method and successfully applied it in the astrophysical field. Subsequently, Johnson and Beissel [37] improved the accuracy of the SPH method using a normalized smoothing function algorithm which can calculate accurately the normal strain rates for conditions of constant strain rates. Swegle et al. [38], Dyka et al. [39], and Chen et al. [40] pointed out the possible reasons for instability of the SPH method and gave an improvement. Vignjevic et al. [41] presented an approach for the treatment of zero-energy modes in the SPH method. Monaghan [42–44] concluded the main characteristic of the SPH method. Although the SPH originated for the purpose of studying astrophysical problems, it has been widely applied in high velocity impact, penetration, underwater explosion, shock wave, solids, and fluids [45–55].

Nayroles et al. [56–58] proposed a diffuse element method (DEM) by introducing moving least square (MLS) [59] into the Galerkin method in 1992. Afterwards, Belytschko et al. [60] modified the DEM by preserving the omitted terms in the derivatives of the interpolations in the DEM and thus proposed an element-free Galerkin (EFG) method. They introduced the Lagrangian multiplier method for the imposition of essential boundary conditions since the trial function lacks Kronecker delta property. This proposed EFG method costs much more computational resource than SPH but has a better computational stability. Chung and Belytschko et al. [61–64] gave an error estimate for the EFG method and conducted thorough research on the trial function and numerical integration. The EFG method has overcome the shortcoming of FEM which needs remeshing when simulating the moving discontinuity problems. In addition, the use of MLS can easily construct a shape function with C^1-continuity. The above superior properties allowed the EFG method to be successfully applied in thin plates and shells [65,66], dynamic and static fracture and crack propagation problems [67–77], water flows and impact [78–80], and phase interface [81].

Inspired by the theory of wavelets, Liu et al. [82,83] proposed the reproducing kernel particle method, by introducing a correction function into the kernel function, to overcome the problems of the lack of consistency in the SPH method in 1995. Chen et al. [84] proposed a strain smoothing stabilization for nodal integration to eliminate spatial instability. This method has been widely employed in numerical analysis such as rubber hyperelasticity [85,86], acoustics [87], structural dynamics [83], fluid dynamics [88,89], stress concentration [90], and large deformation problems [85,91–94].

In 1996, Duarte and Oden [95] constructed a partition of unity using the MLS functions to develop an meshless method, termed h-p clouds. Garcia et al. [96] and Mendonca et al. [97] applied this h-p clouds method to investigate the bending for thick plate and Timoshenko beam problems. Later, Zhu et al. [98–100] and Atluri et al. [101–103] proposed the local boundary integral equation method and meshless local Petrov-Galerkin approach based on the MLS approximation. The main advantage of these methods was the simplification of integration process.

In addition, there have been also many other mesh-free methods in the literature, i.e., finite point method (FPM) [104], finite spheres method (FSM) [105–107], free-mesh method (FMM) [108], material point method (MPM) [109], particle-in-cell (PIC) method [110,111], and point interpolation method (PIM) [112,113].

Note that the imposition of essential boundary conditions is a crucial point in mesh-free methods since most of the mesh-free methods lack the Kronecker delta property. It needs a special treatment in the imposition of essential boundary conditions in contrast to mesh-based methods. Some appropriate techniques have been introduced to impose the essential boundary conditions, such as penalty function method [114], Lagrange multiplier method [69], and modified variational principle [115], etc. In recent years, a novel interpolation, Kriging interpolation, has attracted much attention as it inherently possesses the delta property. This superior property makes it capable of enforcing the essential boundary conditions directly as FEM.

4.1.2 ADVANTAGES AND DISADVANTAGES OF MESH-FREE METHODS

Mesh-free methods have been proved to have many particular advantages. One distinct advantage is that they are well-suited to the treatment of discontinuities that do not coincide with the original mesh lines, e.g., the growth of cracks with arbitrary and complex paths and the simulation of phase transformation. Mesh-free methods have been successfully applied to the study of crack propagation [67,69,116] and the treatment of discontinuity [81]. In the FEM, the most viable technique for handling these complications is to remesh the domain of the structure at every computational step during the simulation. This remeshing process avoids severe mesh distortion and keeps the mesh line coincident with any discontinuity. However, the remeshing technique requires the projection of field variables between the meshes in successive stages of the problem, which leads to projection error and reduces the accuracy of the numerical simulation. In mesh-free methods, the shape functions are developed using the sophisticated interpolating strategy of field variables at a global level, the approximate solution can be constructed entirely in terms of a set of nodes, and the discrete equations can be subsequently developed without the need for meshes. The burdensome work of meshing and remeshing that is associated with the FEM can thus be avoided.

Most of the mesh-free methods are constructed by embedding a highly smooth window function with a certain range of support sizes and reproducing the polynomials exactly. This methodology leads to higher-order polynomial approximation or a highly smooth shape function with higher-order continuity, which alleviates many of the numerical ill-conditions encountered in the FEM. Mesh-free methods have been successfully applied to the simulation of material with the effects of the strain gradient and higher-order deformation gradient [26].

Moreover, mesh-free methods can be viewed as nonlocal approximations, as every evaluated point in the domain is actually covered by the multiple shape functions of the node, as schematically demonstrated in Fig. 4.1. This property may smooth the discontinuous fields and significantly delay

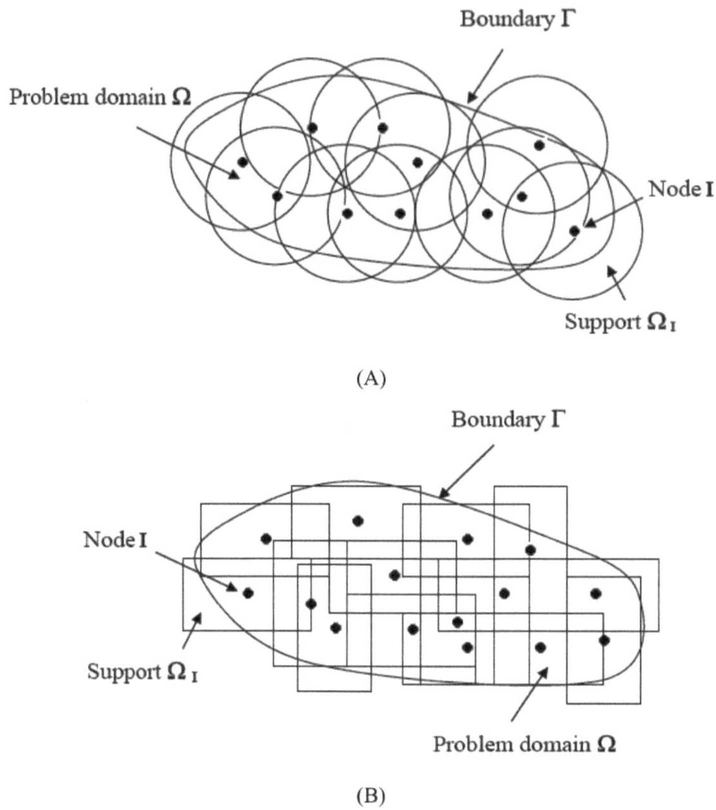

FIGURE 4.1

Mesh-free nodes collocated on the problem domain with the compact supports: (A) circular compact supports; (B) rectangular compact supports.

the mesh distortion in large deformation computations. It has been found that mesh-free methods can achieve much higher precision levels than the FEM for large deformation problems [85,86,117].

There are several key issues that need to be pointed out here. An important point is the imposition of essential boundary conditions in mesh-free methods. As the mesh-free shape functions generally lack the Kronecker delta property, the approximation generated by the MLS method does not pass through the nodal parameter values. Thus, the essential boundary conditions cannot be enforced as in the FEM. An appropriate technique is needed to impose the essential boundary conditions. Moreover, in the FEM, the integrals can be carried out on a series of discretized elements. However, the concept of elements disappears in mesh-free methods, and the evaluation of the integrals requires the use of certain techniques. The integration method is always an important factor that affects the stability and precision of numerical results. Furthermore, irregular geometrical shapes and irregular boundaries often have an effect on the performance of mesh-free methods and the accuracy of computational results.

4.2 CAUCHY–BORN RULE

The Cauchy–Born rule [9] is a fundamental kinematic assumption for linking the deformation of the lattice vectors of crystal to that of a continuum deformation field, and it has been applied as a multiscale approach to deal with continuum problems from the viewpoint of the crystal lattice structures. It describes the deformation of crystal lattice vector as:

$$\mathbf{r} = \mathbf{F} \cdot \mathbf{R} \tag{4.1}$$

where $\mathbf{r}$ and $\mathbf{R}$ are the deformed and undeformed crystal lattice vectors (see Fig. 4.2), and $\mathbf{F}$ is the deformation gradient. Arroyo and Belytschko [16–18] argued that this direct application of the standard Cauchy–Born rule may be inaccurate. The application requires sufficiently homogenous deformations mainly because the deformation gradient tensor only maps a tangent vector of the curved surface. A SWCNT is a curved sheet with only one atomic thickness in reality which indicates that the bending effect must be considered. Their computational results also revealed that the bending stiffness based on the standard Cauchy–Born rule is zero. Thus, it cannot truly capture the physical behaviors of CNTs such as buckling behaviors. The approximation can be enhanced by considering the effect of the second-order deformation gradient, and this extended rule is called the higher-order Cauchy–Born rule in the studies of Guo et al. [20] and Wang et al. [21].

From classical nonlinear continuum mechanics [19], the deformed lattice vector $\mathbf{r}$ can be expressed as

$$\mathbf{r} = \int_{\mathbf{0}}^{\mathbf{R}} \mathbf{F}(\mathbf{s}) d\mathbf{s} \tag{4.2}$$

Taking Taylor's expansion of the deformation field at $\mathbf{S} = \mathbf{0}$, and retaining up to the second-order term, the approximated deformed lattice vector can be derived as

$$\mathbf{r} \approx \mathbf{F}(\mathbf{0}) \cdot \mathbf{R} + \frac{1}{2}\mathbf{G}(\mathbf{0}):(\mathbf{R} \otimes \mathbf{R}) \tag{4.3}$$

where $\mathbf{F}$ and $\mathbf{G}$ denote the first- and second-order deformation gradients. With the involvement of the second deformation gradient, the accuracy of approximation for the deformed lattice vectors is largely enhanced. In particular, the second-order deformation gradient can accurately describe the

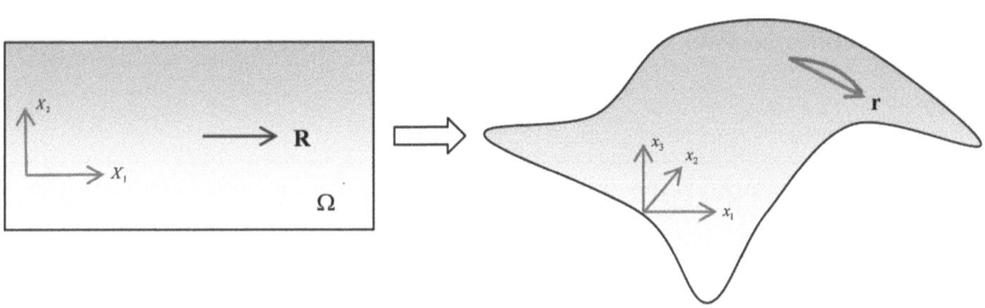

FIGURE 4.2

A planar sheet is transformed to a spatial curved surface. With this transformation, the vector in the original reference configuration is deformed as a curved line that corresponds to a spatial vector.

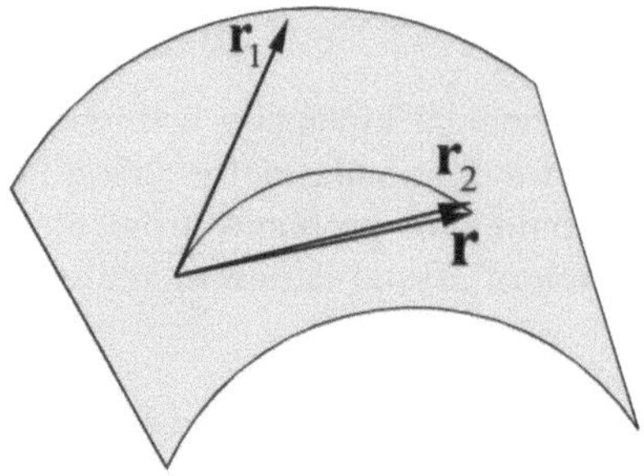

FIGURE 4.3

The mapping deformation vectors $\mathbf{r}_1$ and $\mathbf{r}_2$ denote deformation vectors obtained by the standard and higher-order Cauchy—Born rule. It reveals that the latter is much more close to the reality $\mathbf{r}$ than the former.

bending stiffness, which makes the results much more reasonable and capable to display the buckling behaviors. Deformation vector $\mathbf{r}_1$ obtained by the standard Cauchy—Born rule is linearly dependent on the original vector $\mathbf{R}$ and thus stands on the tangent plane of the current configuration. The involvement of the second-order deformation gradient in the higher-order Cauchy—Born rule makes $\mathbf{r}_1$ turn towards $\mathbf{r}_2$, which is much more approximate to the actual case $\mathbf{r}$ (shown in Fig. 4.3).

The simple Bravais lattice possesses the centrosymmetry and the Cauchy—Born rule can be applied directly. Unfortunately, the hexagonal lattice structure in the CNTs is a Bravais multilattice which lacks the centrosymmetry [13—15]. A simple method to solve this problem is decomposing the hexagonal lattice into two simple Bravais sublattices. Under the same deformation, a rigid body translation exists between the two triangular sublattices. Therefore, an internal relaxation η shown in Fig. 4.4 is required to introduce to maintain the minimum potential of atomic system. Eq. (4.3) should be modified as:

$$\mathbf{r} = \mathbf{F}\cdot(\mathbf{R} + \eta) + \frac{1}{2}\mathbf{G}:\left[(\mathbf{R} + \eta) \otimes (\mathbf{R} + \eta)\right] \tag{4.4}$$

Moreover, if we employ an equilibrium graphite sheet as the reference configuration, several additional optimization parameters need be added due to the differences of the bond length and bond angle in the equilibrium graphite sheet and the undeformed CNT.

4.3 ATOMISTIC-CONTINUUM THEORY

Based on the higher-order Cauchy—Born rule, an atomistic-continuum theory is derived to simulate the mechanical behaviors of CNTs. In the higher-order deformation gradient theoretical scheme, the strain energy is not only dependent on the first-order deformation gradient, the contribution of the second-order deformation gradient should also be considered. The transformation mapping a three-dimensional curved surface from a two-dimensional planar sheet is described using the higher-order gradient theory. The original planar sheet $\mathbf{X} = \mathbf{X}(X_1, X_2)$ and formed curved

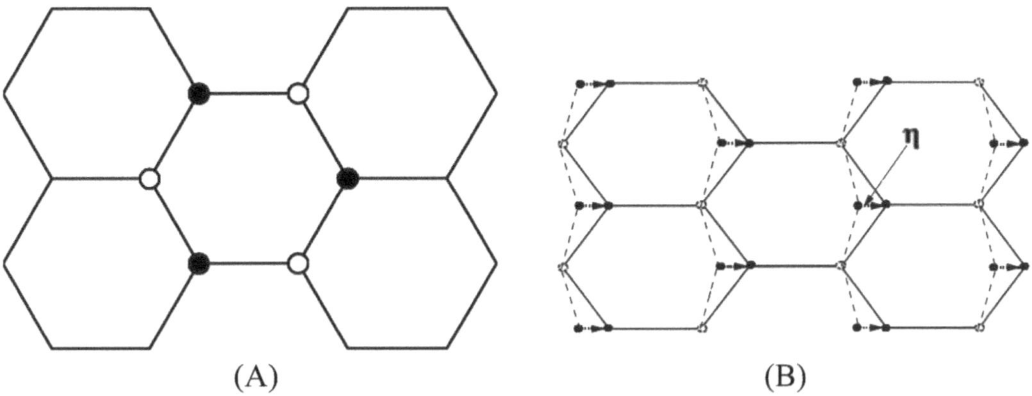

FIGURE 4.4

A hexagonal lattice is decomposed into two simple Bravais sublattices (triangles marked by the open and solid circles) to satisfy the centrosymmetry (A). Under the same deformation, a rigid body translation exists between the two triangular lattice structures. A shift vector η between the two sublattices is added to maintain the minimum potential of atomic structure (B).

surface $\mathbf{x} = \mathbf{x}(x_1, x_2, x_3)$ are selected as the reference and current configurations, respectively. Thus, the first- and second-order deformation gradients tensors at point in Eq. (4.4) can be calculated by:

$$\mathbf{F} = \frac{\partial \mathbf{x}}{\partial \mathbf{X}}, \quad \mathbf{G} = \frac{\partial^2 \mathbf{x}}{\partial \mathbf{X}^2} \tag{4.5}$$

Correspondingly, the first- and second-order stress tensors are [19]:

$$\mathbf{P} = \frac{\partial \omega}{\partial \mathbf{F}}, \quad \mathbf{Q} = \frac{\partial \omega}{\partial \mathbf{G}} \tag{4.6}$$

where ω is the strain energy density and generally determined by the empirical potential in the study of carbon nanostructures in numerical calculation.

The tangential moduli are given by

$$\mathbf{M}_{\mathbf{FF}} = \frac{\partial^2 \omega}{\partial \mathbf{F} \otimes \partial \mathbf{F}}, \quad \mathbf{M}_{\mathbf{FG}} = \frac{\partial^2 \omega}{\partial \mathbf{F} \otimes \partial \mathbf{G}}, \quad \mathbf{M}_{\mathbf{GF}} = \frac{\partial^2 \omega}{\partial \mathbf{G} \otimes \partial \mathbf{F}}, \quad \mathbf{M}_{\mathbf{GG}} = \frac{\partial^2 \omega}{\partial \mathbf{G} \otimes \partial \mathbf{G}} \tag{4.7}$$

The higher-order equilibrium equation and Neumann-type boundary conditions are given by:

$$\mathbf{Div}(\mathbf{P} - \mathbf{DivQ}) + \mathbf{b}_0 = \mathbf{0} \tag{4.8}$$

$$(\mathbf{P} - \mathbf{DivQ}) \cdot \mathbf{N} - K\mathbf{Q}:(\mathbf{N} \otimes \mathbf{N}) - \nabla_{\mathbf{X}}^{T}(\mathbf{Q} \cdot \mathbf{N}):\mathbf{I} = \mathbf{t}_0^{P} \tag{4.9}$$

$$\mathbf{Q}:(\mathbf{N} \otimes \mathbf{N}) = \mathbf{t}_0^{Q} \tag{4.10}$$

$$K = -\nabla_{\mathbf{X}}^{T}\mathbf{N}:\mathbf{I} \tag{4.11}$$

where $\mathbf{b}_0$ is the body force, $\mathbf{t}_0^P$ and $\mathbf{t}_0^Q$ are the first- and second-order stress tractions on the surface $\partial\Omega$, $\mathbf{N}$ is the outward normal.

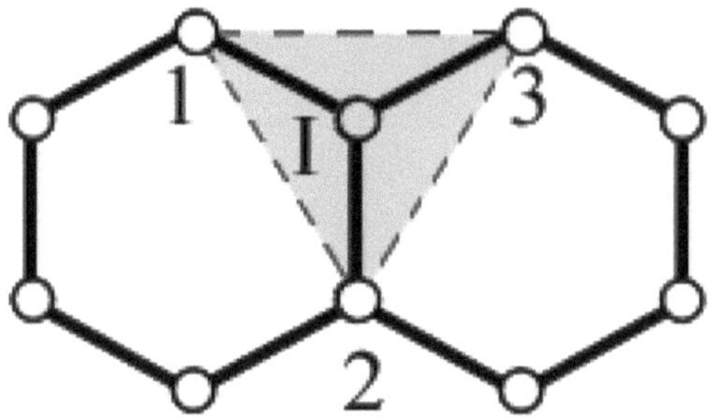

FIGURE 4.5

A representative cell described in the blue triangle. The central carbon atom *I* is connected to three neighboring carbon atoms.

As shown in the hexagonal cellular structure of carbon nanostructures, each carbon atom is connected to three neighboring carbon atoms and we select this atomic structure as the representative cell shown in Fig. 4.5. The chirality is determined by the orientation of this representative cell. The empirical potential for atom *I* is expressed as:

$$V_I = \frac{1}{2}\sum_{J=1,3} V_{IJ}(\mathbf{r}_{I1}, \mathbf{r}_{I2}, \mathbf{r}_{I3}) \tag{4.12}$$

Since the deformed bond length is a function of the first- and second-order gradient tensors, as well as the inner shift vector shown in Eq. (4.4), the strain energy of atom *I* can be expressed as:

$$W_I(\mathbf{F}, \mathbf{G}, \boldsymbol{\eta}) = \frac{1}{2}\sum_{J=1}^{3} V_{IJ}(r_{I2}, r_{I2}, r_{I3}) = V_I(\mathbf{F}, \mathbf{G}, \boldsymbol{\eta}) \tag{4.13}$$

The area strain energy density can be calculated by:

$$\omega_I(\mathbf{F}, \mathbf{G}, \boldsymbol{\eta}) = \nu_I(\mathbf{F}, \mathbf{G}, \boldsymbol{\eta}) = \frac{V_I(\mathbf{F}, \mathbf{G}, \boldsymbol{\eta})}{\Omega_I} \tag{4.14}$$

where Ω_I is the average area per atom being $3\sqrt{3}r_0^2/4$. r_0 is the C−C bond length in the initial equilibrium graphite sheet. It is noteworthy that the contribution of the inner shift vector should not be neglected. For a given deformed system, the strain energy should be minimized with respect to the inner shift

$$\frac{\partial \nu_I}{\partial \boldsymbol{\eta}} = \mathbf{0} \tag{4.15}$$

Therefore, the strain energy density is merely a function of **F** and **G**

$$\omega_I = \omega_I[\mathbf{F}, \mathbf{G}, \boldsymbol{\eta}(\mathbf{F}, \mathbf{G})] \tag{4.16}$$

Consequently, the first- and second-order stress tensors are calculated as

$$\mathbf{P} = \frac{\partial \omega_I}{\partial \mathbf{F}} = \frac{\partial \nu_I(\mathbf{F}, \mathbf{G}, \boldsymbol{\eta}(\mathbf{F}, \mathbf{G}))}{\partial \mathbf{F}} = \frac{\partial \nu_I}{\partial \mathbf{F}} + \frac{\partial \nu_I}{\partial \boldsymbol{\eta}}\frac{\partial \boldsymbol{\eta}}{\partial \mathbf{F}} = \frac{\partial \nu_I}{\partial \mathbf{F}} \tag{4.17}$$

$$\mathbf{Q} = \frac{\partial \omega_I}{\partial \mathbf{G}} = \frac{\partial \nu_I(\mathbf{F}, \mathbf{G}, \boldsymbol{\eta}(\mathbf{F}, \mathbf{G}))}{\partial \mathbf{G}} = \frac{\partial \nu_I}{\partial \mathbf{G}} + \frac{\partial \nu_I}{\partial \boldsymbol{\eta}}\frac{\partial \boldsymbol{\eta}}{\partial \mathbf{G}} = \frac{\partial \nu_I}{\partial \mathbf{G}} \tag{4.18}$$

Eqs. (3.40) and (3.41) have clearly demonstrated that the stress tensors are a linear function of the deformation gradient tensors.

The tangential moduli **M** is evaluated by

$$\mathbf{M}_{\mathbf{FF}} = \frac{\partial^2 \omega_I}{\partial \mathbf{F} \otimes \partial \mathbf{F}} = \frac{\partial^2 \nu_I}{\partial \mathbf{F} \otimes \partial \mathbf{F}} + \frac{\partial^2 \nu_I}{\partial \mathbf{F} \otimes \partial \boldsymbol{\eta}} \frac{\partial \boldsymbol{\eta}_I}{\partial \mathbf{F}} \tag{4.19}$$

$$\mathbf{M}_{\mathbf{FG}} = \frac{\partial^2 \omega_I}{\partial \mathbf{F} \otimes \partial \mathbf{G}} = \frac{\partial^2 \nu_I}{\partial \mathbf{F} \otimes \partial \mathbf{G}} + \frac{\partial^2 \nu_I}{\partial \mathbf{F} \otimes \partial \boldsymbol{\eta}} \frac{\partial \boldsymbol{\eta}_I}{\partial \mathbf{G}} \tag{4.20}$$

$$\mathbf{M}_{\mathbf{GF}} = \frac{\partial^2 \omega_I}{\partial \mathbf{G} \otimes \partial \mathbf{F}} = \frac{\partial^2 \nu_I}{\partial \mathbf{G} \otimes \partial \mathbf{F}} + \frac{\partial^2 \nu_I}{\partial \mathbf{G} \otimes \partial \boldsymbol{\eta}} \frac{\partial \boldsymbol{\eta}_I}{\partial \mathbf{F}} \tag{4.21}$$

$$\mathbf{M}_{\mathbf{GG}} = \frac{\partial^2 \omega_I}{\partial \mathbf{G} \otimes \partial \mathbf{G}} = \frac{\partial^2 \nu_I}{\partial \mathbf{G} \otimes \partial \mathbf{G}} + \frac{\partial^2 \nu_I}{\partial \mathbf{G} \otimes \partial \boldsymbol{\eta}} \frac{\partial \boldsymbol{\eta}_I}{\partial \mathbf{G}} \tag{4.22}$$

Differentiating Eq. (3.38) with respect to the first- and second-order deformation gradient vectors, we have

$$\frac{\partial^2 \nu_I}{\partial \boldsymbol{\eta} \otimes \partial \mathbf{F}} = \frac{\partial^2 \nu_I}{\partial \boldsymbol{\eta} \otimes \partial \mathbf{F}} + \frac{\partial^2 \nu_I}{\partial \boldsymbol{\eta} \otimes \partial \boldsymbol{\eta}} \frac{\partial \boldsymbol{\eta}_I}{\partial \mathbf{F}} = \mathbf{0} \tag{4.23}$$

$$\frac{\partial^2 \nu_I}{\partial \boldsymbol{\eta} \otimes \partial \mathbf{G}} = \frac{\partial^2 \nu_I}{\partial \boldsymbol{\eta} \otimes \partial \mathbf{G}} + \frac{\partial^2 \nu_I}{\partial \boldsymbol{\eta} \otimes \partial \boldsymbol{\eta}} \frac{\partial \boldsymbol{\eta}_I}{\partial \mathbf{G}} = \mathbf{0} \tag{4.24}$$

Thus,

$$\frac{\partial \boldsymbol{\eta}_I}{\partial \mathbf{F}} = -\left[\frac{\partial^2 \nu_I}{\partial \boldsymbol{\eta} \otimes \partial \boldsymbol{\eta}}\right]^{-1} \frac{\partial^2 \nu_I}{\partial \boldsymbol{\eta} \otimes \partial \mathbf{F}} \tag{4.25}$$

$$\frac{\partial \boldsymbol{\eta}_I}{\partial \mathbf{G}} = -\left[\frac{\partial^2 \nu_I}{\partial \boldsymbol{\eta} \otimes \partial \boldsymbol{\eta}}\right]^{-1} \frac{\partial^2 \nu_I}{\partial \boldsymbol{\eta} \otimes \partial \mathbf{G}} \tag{4.26}$$

Eqs. (4.17)–(4.22) can be further written as

$$\mathbf{M}_{\mathbf{FF}} = \frac{\partial^2 \omega_I}{\partial \mathbf{F} \otimes \partial \mathbf{F}} = \frac{\partial^2 \nu_I}{\partial \mathbf{F} \otimes \partial \mathbf{F}} - \frac{\partial^2 \nu_I}{\partial \mathbf{F} \otimes \partial \boldsymbol{\eta}} \left[\frac{\partial^2 \nu_I}{\partial \boldsymbol{\eta} \otimes \partial \boldsymbol{\eta}}\right]^{-1} \frac{\partial^2 \nu_I}{\partial \boldsymbol{\eta} \otimes \partial \mathbf{F}} \tag{4.27}$$

$$\mathbf{M}_{\mathbf{FG}} = \frac{\partial^2 \omega_I}{\partial \mathbf{F} \otimes \partial \mathbf{G}} = \frac{\partial^2 \nu_I}{\partial \mathbf{F} \otimes \partial \mathbf{G}} - \frac{\partial^2 \nu_I}{\partial \mathbf{F} \otimes \partial \boldsymbol{\eta}} \left[\frac{\partial^2 \nu_I}{\partial \boldsymbol{\eta} \otimes \partial \boldsymbol{\eta}}\right]^{-1} \frac{\partial^2 \nu_I}{\partial \boldsymbol{\eta} \otimes \partial \mathbf{G}} \tag{4.28}$$

$$\mathbf{M}_{\mathbf{GF}} = \frac{\partial^2 \omega_I}{\partial \mathbf{G} \otimes \partial \mathbf{F}} = \frac{\partial^2 \nu_I}{\partial \mathbf{G} \otimes \partial \mathbf{F}} - \frac{\partial^2 \nu_I}{\partial \mathbf{G} \otimes \partial \boldsymbol{\eta}} \left[\frac{\partial^2 \nu_I}{\partial \boldsymbol{\eta} \otimes \partial \boldsymbol{\eta}}\right]^{-1} \frac{\partial^2 \nu_I}{\partial \boldsymbol{\eta} \otimes \partial \mathbf{F}} \tag{4.29}$$

$$\mathbf{M}_{\mathbf{GG}} = \frac{\partial^2 \omega_I}{\partial \mathbf{G} \otimes \partial \mathbf{G}} = \frac{\partial^2 \nu_I}{\partial \mathbf{G} \otimes \partial \mathbf{G}} - \frac{\partial^2 \nu_I}{\partial \mathbf{G} \otimes \partial \boldsymbol{\eta}} \left[\frac{\partial^2 \nu_I}{\partial \boldsymbol{\eta} \otimes \partial \boldsymbol{\eta}}\right]^{-1} \frac{\partial^2 \nu_I}{\partial \boldsymbol{\eta} \otimes \partial \mathbf{G}} \tag{4.30}$$

Matrices **grad** and $\boldsymbol{\sigma}$ are defined as a combination of the first- and second-order deformation gradients and stress tensors:

$$\mathbf{grad} = [\mathbf{F} \quad \mathbf{G}]^T = [F_{11}, F_{12}, F_{21}, F_{22}, F_{31}, F_{32}, G_{111}, G_{112}, G_{122}, G_{211}, G_{212}, G_{222}, G_{311}, G_{312}, G_{322}]^T \tag{4.31}$$

$$\boldsymbol{\sigma} = [\mathbf{P} \quad \mathbf{Q}]^T = [P_{11}, P_{12}, P_{21}, P_{22}, P_{31}, P_{32}, Q_{111}, Q_{112}, Q_{122}, Q_{211}, Q_{212}, Q_{222}, Q_{311}, Q_{312}, Q_{322}]^T \tag{4.32}$$

The coordinate of central carbon atom is set as the original point, and the other atoms for the representative cell in the reference configuration and current configuration are denoted as

$$\mathbf{R}_{cell} = (X_1^1, X_2^1, X_1^2, X_2^2, X_1^3, X_2^3)^T \tag{4.33}$$

$$\mathbf{r}_{cell} = (x_1^1, x_2^1, x_3^1, x_1^2, x_2^2, x_3^2, x_1^3, x_2^3, x_3^3)^T \tag{4.34}$$

Consequently, the stress vector can be written as

$$\boldsymbol{\sigma} = \left[\frac{\partial v_I}{\partial \mathbf{grad}}\right]_{15\times1} = \left[\frac{\partial \mathbf{r}_{cell}}{\partial \mathbf{grad}}\right]_{15\times9} \cdot \left[\frac{\partial v_I}{\partial \mathbf{r}_{cell}}\right]_{9\times1} \tag{4.35}$$

The tangential moduli is

$$\begin{aligned}
\mathbf{M} &= \frac{\partial^2 v_I}{\partial \mathbf{grad} \otimes \partial \mathbf{grad}} - \frac{\partial^2 v_I}{\partial \mathbf{grad} \otimes \partial \boldsymbol{\eta}} \left[\frac{\partial^2 v_I}{\partial \boldsymbol{\eta} \otimes \partial \boldsymbol{\eta}}\right]^{-1} \frac{\partial^2 v_I}{\partial \boldsymbol{\eta} \otimes \partial \mathbf{grad}} \\
&= \left[\frac{\partial \mathbf{r}_{\mathrm{cell}}}{\partial \mathbf{grad}}\right]_{15\times9} \cdot \left[\frac{\partial^2 v_I}{\partial \mathbf{r}_{\mathrm{cell}}{}^2}\right]_{9\times9} \cdot \left[\frac{\partial \mathbf{r}_{\mathrm{cell}}}{\partial \mathbf{grad}}\right]_{15\times9}^T \\
&\quad - \left[\frac{\partial \mathbf{r}_{\mathrm{cell}}}{\partial \mathbf{grad}}\right]_{15\times9} \cdot \left[\frac{\partial^2 v_I}{\partial \mathbf{r}_{\mathrm{cell}}{}^2}\right]_{9\times9} \cdot \left[\frac{\partial \mathbf{r}_{\mathrm{cell}}}{\partial \boldsymbol{\eta}}\right]_{2\times9}^T \cdot \left[\frac{\partial^2 v_I}{\partial \boldsymbol{\eta} \otimes \partial \boldsymbol{\eta}}\right]_{2\times2}^{-1} \\
&\quad \cdot \left[\frac{\partial \mathbf{r}_{cell}}{\partial \boldsymbol{\eta}}\right]_{2\times9} \cdot \left[\frac{\partial^2 v_I}{\partial \mathbf{r}_{cell} \otimes \partial \mathbf{r}_{cell}}\right]_{9\times9} \cdot \left[\frac{\partial \mathbf{r}_{cell}}{\partial \mathbf{grad}}\right]_{15\times9}^T
\end{aligned} \tag{4.36}$$

$\left[\frac{\partial^2 v_I}{\partial \boldsymbol{\eta} \otimes \partial \boldsymbol{\eta}}\right]_{2\times2}^{-1}$ in Eq. (4.36) can be obtained by

$$\left[\frac{\partial^2 v_I}{\partial \boldsymbol{\eta} \otimes \partial \boldsymbol{\eta}}\right]_{2\times2} = \left[\frac{\partial \mathbf{r}_{cell}}{\partial \boldsymbol{\eta}}\right]_{2\times9} \cdot \left[\frac{\partial^2 v_I}{\partial \mathbf{r}_{cell} \otimes \partial \mathbf{r}_{cell}}\right]_{9\times9} \cdot \left[\frac{\partial \mathbf{r}_{cell}}{\partial \boldsymbol{\eta}}\right]_{2\times9}^T \tag{4.37}$$

$\left[\frac{\partial \mathbf{r}_{cell}}{\partial \mathbf{grad}}\right]_{15\times9}$ and $\left[\frac{\partial \mathbf{r}_{\mathrm{cell}}}{\partial \boldsymbol{\eta}}\right]_{2\times9}$, $\left[\frac{\partial v_I}{\partial \mathbf{r}_{cell}}\right]_{9\times1}$ and $\left[\frac{\partial^2 v_I}{\partial \mathbf{r}_{\mathrm{cell}}{}^2}\right]_{9\times9}$ in Eqs. (4.35)–(4.37) can be determined by the higher-order Cauchy–Born rule Eq. (4.4) and empirical potential Eq. (3.3), respectively. Consequently, the stress vector and tangential moduli can be obtained by being given a deformation description. The developed higher-order deformation gradient theoretical scheme will be used to evaluate the structural and elastic properties of SWCNCs. The whole procedure has been actualized by the developed Fortran program. The constitutive model established based on the higher-order Cauchy–Born rule incorporates the atomic interaction, thus it can accurately display the nonlinear responses of carbon nanostructures under a given deformation description **F** and **G**. The buckling and postbuckling behaviors of carbon nanostructures will be discussed in detail in the later numerical simulations.

4.4 STRUCTURAL AND ELASTIC PROPERTIES OF SWCNTs

4.4.1 TRANSFORMATION OF SWCNTs

An undeformed SWCNT can be considered as the rolling up of an initial equilibrium graphite sheet into a hollow cylindrical structure. It is noteworthy that the atomic structure of an initial equilibrium graphite sheet is different from that of an initial equilibrium SWCNT. All of the C–C bonds

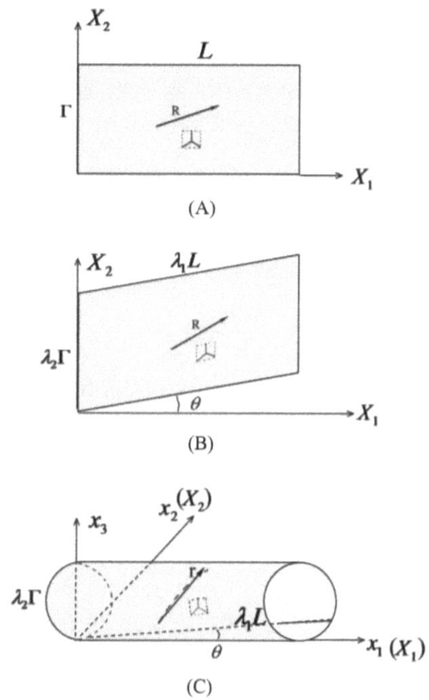

FIGURE 4.6

A SWCNT is formed by rolling an initial equilibrium graphite sheet into a cylindrical shape in two steps. This transforming process can be described by three structural parameters λ_1, λ_2, and θ, corresponding to the longitudinal and circumferential stretches, and shear strain, respectively (from (A) to (B)), and rigidly mapping a hollow cylinder from (B) to (C)

are the same in the initial equilibrium graphite sheet, while there are slight differences in C–C bonds in the counterpart SWCNT. Therefore, three geometrical parameters, λ_1, λ_2, and θ, corresponding to the uniform longitudinal, circumferential stretches, and the skew of any surface longitudinal line, respectively, are introduced to describe this shifting process. The three geometrical parameters can be determined by minimizing the strain energy of the representative cell to obtain an initial equilibrium configuration. The initial equilibrium graphite sheet and the optimized equilibrium SWCNT are selected as the reference configuration and the current configuration, respectively. The deformation process mapped from Fig. 4.6A–C can then be written:

$$x_1 = X_1 \lambda_1 \cos\theta \tag{4.38}$$

$$x_2 = \frac{\Gamma\lambda_2}{2\pi} \sin\left[2\pi\left(\frac{X_2}{\Gamma} + \frac{X_1\lambda_1 \sin\theta}{\Gamma\lambda_2}\right)\right] \tag{4.39}$$

$$x_3 = \frac{\Gamma\lambda_2}{2\pi} - \frac{\Gamma\lambda_2}{2\pi} \cos\left[2\pi\left(\frac{X_2}{\Gamma} + \frac{X_1\lambda_1 \sin\theta}{\Gamma\lambda_2}\right)\right] \tag{4.40}$$

where $\mathbf{x} = \mathbf{x}(x_1, x_2, x_3)$ and $\mathbf{X} = \mathbf{X}(X_1, X_2)$ are the current configuration and reference configuration, respectively. The first- and second-order deformation gradients $\mathbf{F}$ and $\mathbf{G}$ can be exactly calculated from the deformation process of SWCNTs in Eqs. (4.38)–(4.40).

When the three parameters λ_1, λ_2, and θ are introduced, the first- and second deformation gradients are determined by the three parameters and Eq. (4.14) can be further rewritten as:

$$\nu_I[\mathbf{F}, \mathbf{G}, \boldsymbol{\eta}] = \nu_I(\lambda_1, \lambda_2, \theta, \eta_1, \eta_2) \tag{4.41}$$

Denote the unknown vector as:

$$\boldsymbol{\lambda} = (\lambda_1, \lambda_2, \theta, \eta_1, \eta_2)^T \tag{4.42}$$

Structural parameters can be determined by minimizing the strain energy of the representative cell:

$$\frac{\partial \nu_I}{\partial \boldsymbol{\lambda}} = \mathbf{0} \tag{4.43}$$

The Taylor expansion of ν_I around an initial guess of $\boldsymbol{\lambda}_0$ for the equilibrium state is

$$\nu_I(\boldsymbol{\lambda}) = \nu_I(\boldsymbol{\lambda}_0) + \frac{\partial \nu_I(\boldsymbol{\lambda})}{\partial \boldsymbol{\lambda}}|_{\boldsymbol{\lambda}=\boldsymbol{\lambda}_0} \cdot \Delta\boldsymbol{\lambda} + \frac{1}{2}\Delta\boldsymbol{\lambda} \cdot \frac{\partial^2 \nu_I(\boldsymbol{\lambda})}{\partial \boldsymbol{\lambda}^2}|_{\boldsymbol{\lambda}=\boldsymbol{\lambda}_0} \cdot \Delta\boldsymbol{\lambda} + o(\Delta\boldsymbol{\lambda}^2) \tag{4.44}$$

Substitution of Eq. (4.44) into (4.43) yields the following governing equation

$$\frac{\partial^2 \nu_I}{\partial \boldsymbol{\lambda}^2} \cdot \Delta\boldsymbol{\lambda} = -\frac{\partial \nu_I}{\partial \boldsymbol{\lambda}} \tag{4.45}$$

The increment of structural vector $\Delta\boldsymbol{\lambda}$ can be determined by iteratively solving Eq. (4.45) until the right term reaches zero.

The aforementioned strain energy density ω_0 in Eq. (4.14) is defined as the area strain energy density, while the volume strain energy density is required in order to calculate the elastic and structural properties. Nevertheless, there is no agreed thickness concept for CNTs. Scholars have suggested an interlayer distance of approximately 0.33 ~ 0.34 nm for CNTs. In many publications [118,119], the interlayer distance of graphite sheet is chosen as 0.334 nm, hence we choose it as the thickness. Consequently, the volume strain energy density is calculated by:

$$\hat{\omega}_0 = \omega_0 / t \tag{4.46}$$

where t is the thickness.

Since geometrical parameters λ_1 and λ_2 correspond to the uniform axial and circumferential stretches, respectively, the axial and circumferential strains are calculated by:

$$\varepsilon_1 = \lambda_1 - 1, \quad \varepsilon_2 = \lambda_2 - 1. \tag{4.47}$$

Thus, axial and circumferential stresses are determined by:

$$\sigma_1 = \frac{\partial \hat{\omega}_0}{\partial \varepsilon_1} = \frac{\partial \hat{\omega}_0}{\partial \lambda_1}, \quad \sigma_2 = \frac{\partial \hat{\omega}_0}{\partial \varepsilon_2} = \frac{\partial \hat{\omega}_0}{\partial \lambda_2}. \tag{4.48}$$

In case of hydrostatic pressure, CNTs deform uniformly along the circumference. The radial strains of the CNTs equal the circumferential strains, i.e., $\varepsilon_r = \varepsilon_2$. Moreover, the hydrostatic pressure p has a relationship with σ_2 as

$$p 2R = \sigma_2 2t \tag{4.49}$$

Young's moduli along the axial and circumferential directions of CNTs E_1 and E_2 can be obtained by:

$$E_1 = \frac{\partial^2 \hat{\omega}_0}{\partial \lambda_1^2} - \frac{\left(\dfrac{\partial^2 \hat{\omega}_0}{\partial \lambda_1 \partial \lambda_2}\right)^2}{\dfrac{\partial^2 \hat{\omega}_0}{\partial \lambda_2^2}}, \quad E_2 = \frac{\partial^2 \hat{\omega}_0}{\partial \lambda_2^2} - \frac{\left(\dfrac{\partial^2 \hat{\omega}_0}{\partial \lambda_2 \partial \lambda_1}\right)^2}{\dfrac{\partial^2 \hat{\omega}_0}{\partial \lambda_1^2}} \tag{4.50}$$

Using Eq. (4.49), the radial Young's modulus can be calculated as

$$E_r = E_2 t / R \tag{4.51}$$

The angle θ is the shear strain at the surface and thus the shear stress at the cross-section is:

$$\tau = \frac{\partial \hat{\omega}_0}{\partial \theta} \tag{4.52}$$

The shear modulus is calculated by:

$$G = \frac{\partial^2 \hat{\omega}_0}{\partial \theta^2} \tag{4.53}$$

The Poisson's ratio is calculated by:

$$\nu_{12} = \frac{\partial^2 \hat{\omega}_0}{\partial \lambda_2 \partial \lambda_1} \Big/ \frac{\partial^2 \hat{\omega}_0}{\partial \lambda_1^2}, \quad \nu_{21} = \frac{\partial^2 \hat{\omega}_0}{\partial \lambda_1 \partial \lambda_2} \Big/ \frac{\partial^2 \hat{\omega}_0}{\partial \lambda_2^2} \tag{4.54}$$

Moreover, we have the relationship

$$\frac{\nu_{12}}{E_1} = \frac{\nu_{21}}{E_2} \tag{4.55}$$

Similar to the calculation of tangential modulus, it is noteworthy that the contribution of the inner shift vector should not be neglected when calculating elastic constants. Structural parameters of CNTs can be obtained by iteratively solving the incremental governing equation Eq. (4.45), and elastic properties are the values of Eqs. (4.50)–(4.54) at the initial equilibrium state of CNTs.

4.4.2 STRUCTURAL PARAMETERS

With the solution of Eq. (4.45), the structural parameters of the initial undeformed SWCNT can be obtained. Here, the second set of parameters [120] for the Brenner potential is employed. In the calculation, we need to present a number as the initial bond length for the representative cell in the reference graphite sheet (Fig. 4.7A). Essentially, we can choose any number as the initial optimization value, and the choice of this value has no effect on the final optimized results. Moreover, by setting a very large tube radius, we can obtain the bond length in the initial equilibrium graphite sheet. For the second set of parameters of the Brenner potential, our computation indicates that the bond length in the initial equilibrium graphite sheet is 0.14507 nm which is used as the initial value of the bond lengths in the reference configuration. With this choice, the evaluated values of λ_1, λ_2, and θ are, respectively, the axial and circumferential stretches and twisting angle of the tube relative to the initial equilibrium graphite sheet.

Table 4.1 shows the evaluated values of λ_1, λ_2, and θ for chiral (9, 6), armchair (5, 5), and zigzag (10, 0) SWCNTs. It is noted that $\theta = 0$ for the armchair and zigzag SWCNTs, whereas a nonzero

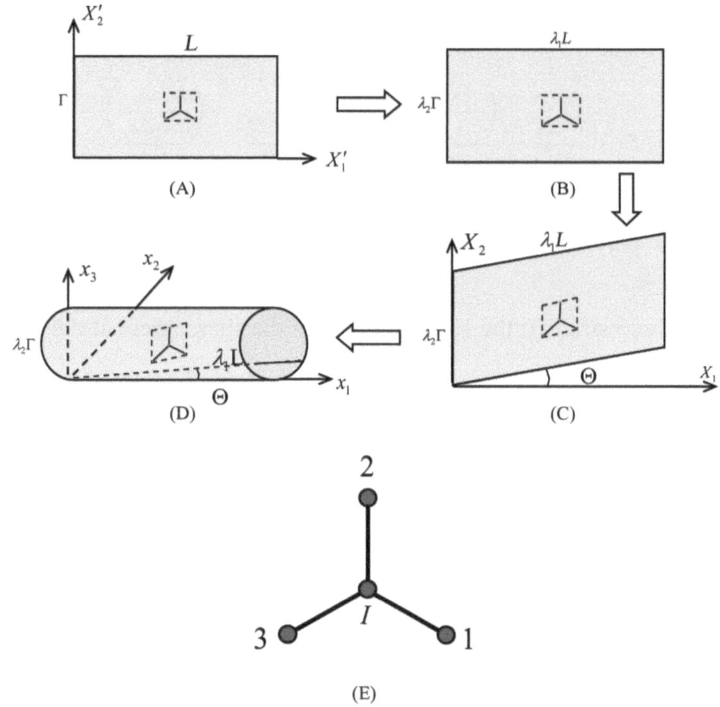

FIGURE 4.7

(A)–(D): an SWCNT is formed by rolling an equilibrium graphite sheet into a cylindrical shape through three steps; the change of the representative cell is also depicted. (E): the markers for the representative cell.

Table 4.1 Optimization Values for λ_1, λ_2, and θ [24]

SWCNT	λ_1	λ_2	θ (°)
(9, 6) Chiral	1.00100	0.99411	0.08554
(5, 5) Armchair	1.00307	0.98606	0
(10, 0) Zigzag	0.99820	0.99346	0

twisting angle θ occurs for the chiral CNT. This is because the representative cells in the armchair and zigzag CNTs are axisymmetric with respect to the axial and circumferential directions of the tube. The evaluated bond lengths and tube radii are listed in Table 4.2, along with the calculated atomic energy. Table 4.3 gives the results obtained by Chandraseker and Mukherjee [121]. The present results for bond lengths and atomic energy display excellent agreement with those of Chandraseker and Mukherjee [121]. However, the tube radius shows a slight discrepancy, and this may be because of a difference in the radius definitions in the two methods. (9, 6), (5, 5), and (10, 0) SWCNTs belong to tubes of small radius. The present computations show that the approximation with the first- and second-order deformation gradient is precise enough even for tubes of small radius.

Table 4.2 Bond Lengths, Tube Radius, and Atomic Energy in the Undeformed SWCNTs Obtained With the Present Method [24]

SWCNT	r_1 (nm)	r_2 (nm)	r_3 (nm)	R (nm)	V (eV)
(9, 6) Chiral	0.14520	0.14516	0.14533	0.51986	−7.33606
(5, 5) Armchair	0.14567	0.14532	0.14532	0.34150	−7.28637
(10, 0) Zigzag	0.14518	0.14544	0.14544	0.39729	−7.30785

Table 4.3 Bond Lengths, Tube Radius, and Atomic Energy in the Undeformed SWCNT in Ref. [121]

SWCNT	r_1 (nm)	r_2 (nm)	r_3 (nm)	R (nm)	V (eV)
(9, 6) Chiral	0.1452	0.1452	0.1453	0.5235	−7.3359
(5, 5) Armchair	0.1457	0.1453	0.1453	0.3490	−7.2844
(10, 0) Zigzag	0.1452	0.1454	0.1454	0.4039	−7.3076

4.4.3 ELASTIC PROPERTIES

Computations have been carried out for different types of CNTs. In the results analysis, we use ($3n$, $2n$) CNTs as an example to illustrate the properties of chiral CNTs. In all of the calculations, the first- and second-order derivatives are exactly computed. Fig. 4.8 shows the variation of the axial and circumferential Young's moduli with the tube radius. With an increasing tube radius, the axial and circumferential Young's moduli trend to the same constant that corresponds to the Young's modulus of the graphite sheet. The present estimated value is 0.7049 TPa, which is very close to that predicted by Arroyo and Belytschko [18]. When the tube radius is larger than 0.5 nm, the moduli are very close to those of the graphite sheet, and the maximum difference is less than 2%. Moreover, we also notice a special phenomenon: the axial and circumferential Young's moduli of the CNTs are generally less than those of the graphite sheet, but the circumferential Young's moduli of the zigzag CNTs are slightly larger than those of the graphite sheet. This indicates that the chirality of CNTs has an effect on their properties, but this effect is very slight when the tube radius is larger than 0.5 nm. Plotted in Fig. 4.9 is the radial Young's modulus versus the tube radius. The radial Young's modulus has a large dependence on the tube radius, because it is equal to the circumferential Young's modulus divided by the tube radius. Fig. 4.10 shows the variation of Poisson's ratio ν_{21} with the tube radius. Similar to the axial and circumferential Young's moduli, ν_{21} trends to the Poisson's ratio of the graphite sheet (the present estimated value is 0.4123). When the tube radius is larger than 0.5 nm, the Poisson's ratio is very close to 0.4123. Fig. 4.11 shows the dependence of the shear modulus on the tube radius. When the tube radius is larger than 0.5 nm, the shear modulus is very close to that of the graphite sheet. The present estimated shear modulus of the graphite sheet is approximately 0.25 TPa. In conclusion, the elastic properties of CNTs have a dependence on the chirality and the tube radius, but when the tube radius is larger

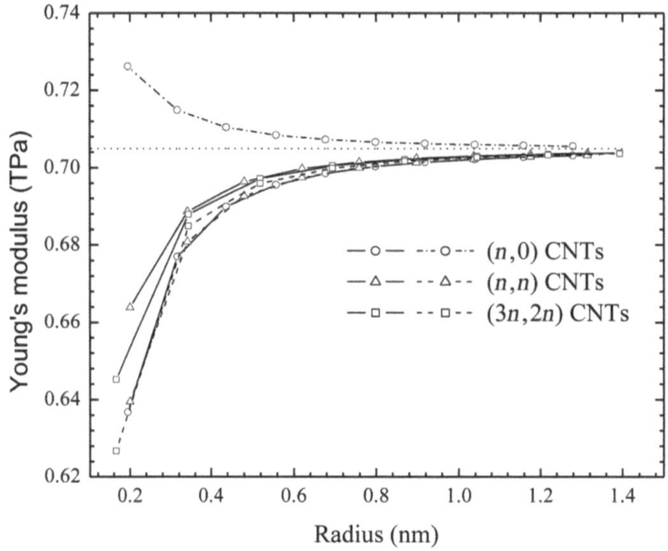

FIGURE 4.8

Axial and circumferential Young's moduli versus tube radius for (*n*, 0), (*n*, *n*), and (3*n*, 2*n*) CNTs: solid line for axial Young's modulus and dashed line for circumferential Young's modulus [24].

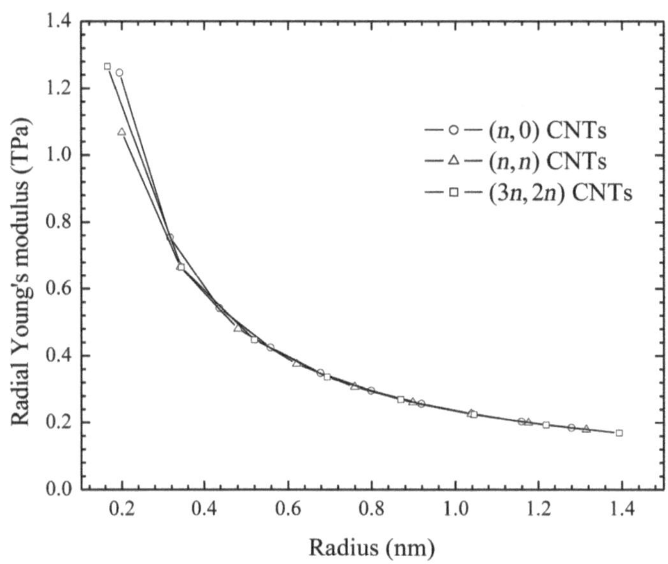

FIGURE 4.9

Radial Young's modulus versus tube radius for (*n*, 0), (*n*, *n*), and (3*n*, 2*n*) CNTs [24].

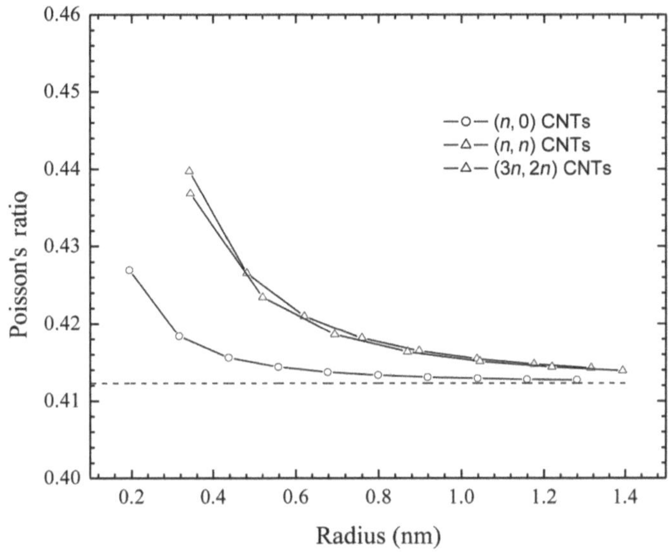

FIGURE 4.10

Poisson's ratio versus tube radius for (*n*, 0), (*n*, *n*), and (3*n*, 2*n*) CNTs [24].

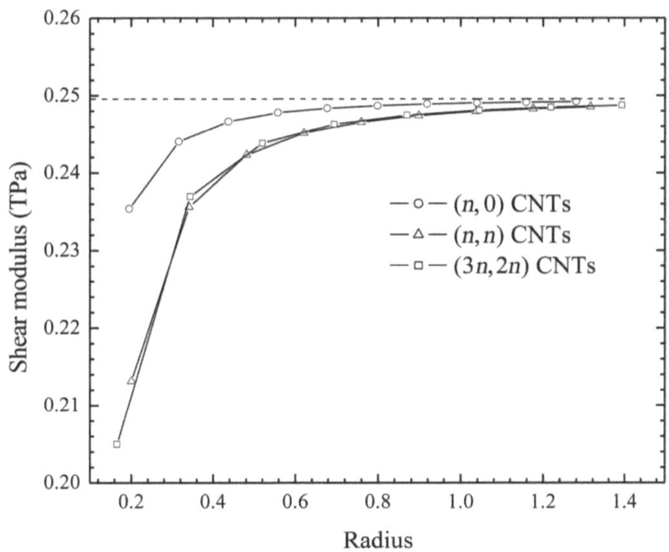

FIGURE 4.11

Shear Young's modulus versus tube radius for (*n*, 0), (*n*, *n*), and (3*n*, 2*n*) CNTs [24].

than 0.5 nm, the axial and circumferential Young's moduli, Poisson's ratio, and shear modulus are pretty close to those of the graphite sheet. Therefore, CNTs can be viewed as a curved surface on which the elastic properties at the tangent direct are almost the same with the graphite sheet.

4.4.4 PRESSURE–RADIAL STRAIN CURVE

The above method can be viewed as an analytical-numerical method, in which the first- and second-order derivatives of the strain energy density are calculated exactly, and Newton's method is used only to determine the minimum point of energy. Moreover, by varying the values of λ_1, λ_2, and θ, and then minimizing the energy density, we can calculate the mechanical response of CNTs under axial tension or compression (varying λ_1), hydrostatic pressure (varying λ_2), and twisting (varying θ). This method can also be used to study the interaction of different loading conditions, e.g., the twisting behavior of different chiralities of CNTs under axial or radial compression. Here, the hydrostatic pressure–radial strain and axial strain–stress curves are presented as two examples.

The axial strain–stress curves of SWCNTs can be obtained by varying the value of λ_1. For a given λ_1 that corresponds to an axial strain, the energy minimization results in the values of λ_2 and θ and the axial stress can be calculated using Eq. (4.48). By varying λ_1, the axial strain–stress can be obtained. Plotted in Fig. 4.12 is the axial strain–stress curve of a (15, 0) SWCNT.

The pressure–radial strain curve of SWCNTs under hydrostatic pressure can be obtained by varying the value of λ_2. For a given λ_2 that corresponds to a radial compressive strain, energy minimization gives the values of λ_1 and θ, and the hydrostatic pressure can be calculated as $p = \frac{\partial W_0}{R \partial \lambda_2}$.

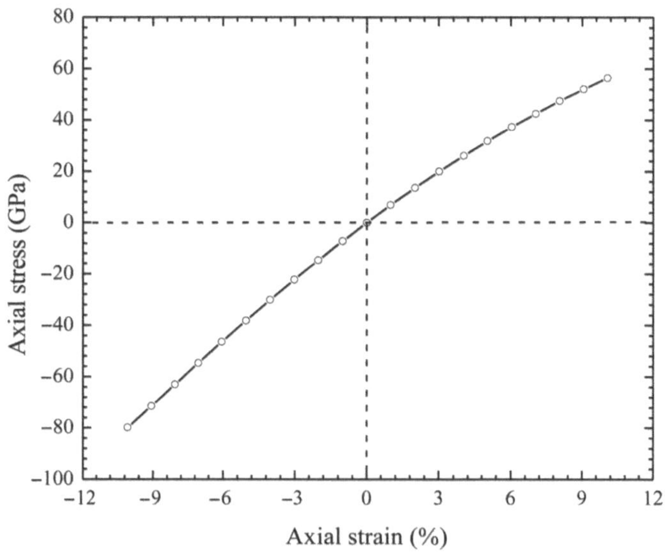

FIGURE 4.12

Axial strain–stress curve obtained with a semianalytical method [24].

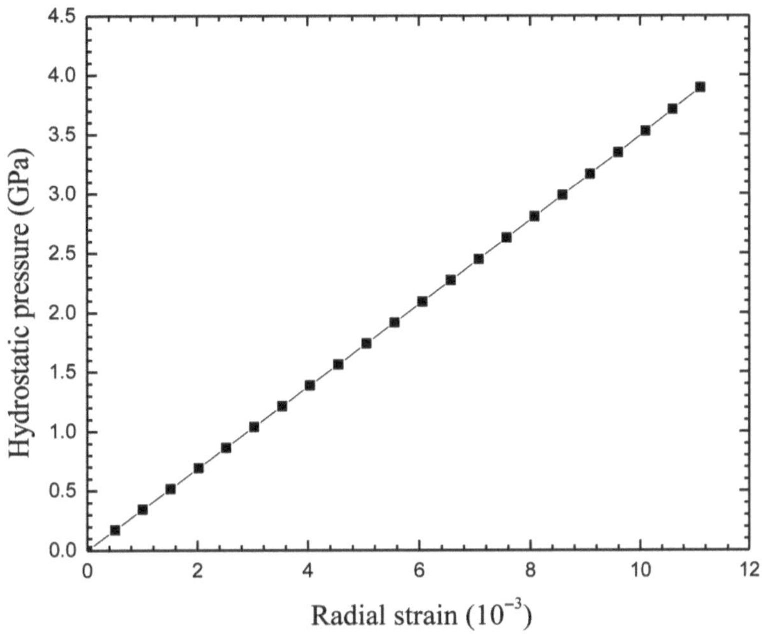

FIGURE 4.13

Hydrostatic pressure–radial strain curve obtained with a semianalytical method [24].

By varying λ_2, the curve of the hydrostatic pressure versus the radial strain $(\lambda_2 - 1)$ can be obtained. Plotted in Fig. 4.13 is the pressure–radial strain curve of a (10, 10) SWCNT.

4.5 MESH-FREE COMPUTATIONAL FRAMEWORK

Before deriving a mesh-free formulation for SWCNTs, we need to discuss the establishment of the appropriate reference configuration and the determination of the structure of the representative cell in the reference configuration. In the previous analysis, we use three geometric parameters λ_1, λ_2, and θ to define a transformation from a planar graphite sheet to an undeformed SWCNT. The computations show that this rolling is not a simple grid transformation and that the atomic structure readjusts to attain the minimum system energy during the rolling process. In the numerical simulations, we need to consider the transformation from the reference configuration to the deformed configuration (the current configuration). For an arbitrary given deformation description (deformation gradients $\mathbf{F}$ and $\mathbf{G}$), we need to determine the constitutive response of the system (the stress tensors $\mathbf{P}$ and $\mathbf{Q}$ and the modulus $\mathbf{M}$), and these three geometrical parameters no longer have a clear physical meaning. We should find an appropriate planar sheet structure as the reference configuration so that the transformation from this sheet to the initial equilibrium SWCNT is a rigid map. From the previous analysis, it can be seen that Fig. 4.7C is such a sheet.

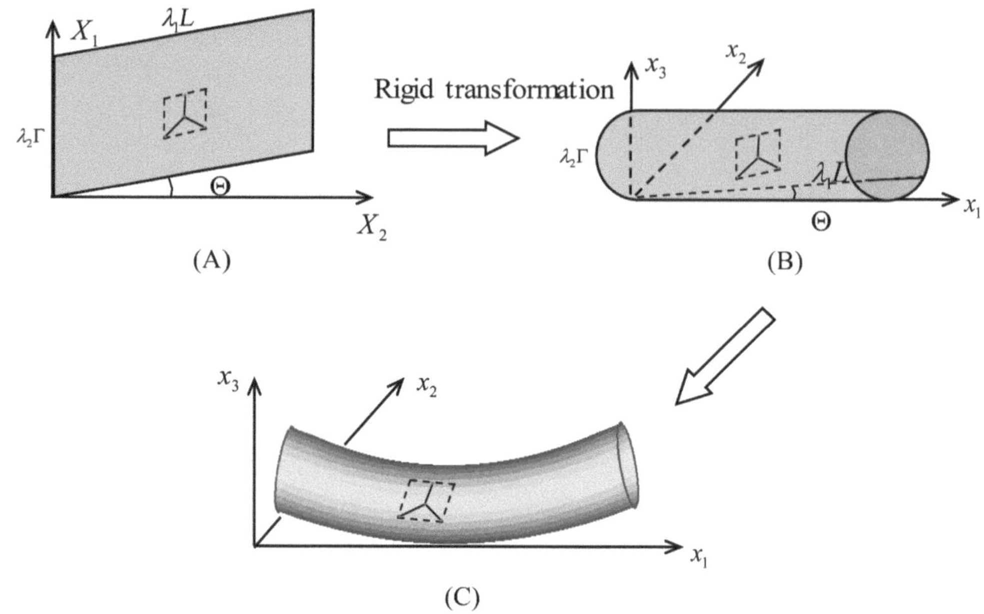

FIGURE 4.14

With an appropriate graphite sheet as the reference configuration, the transformation from the reference configuration to the initial equilibrium SWCNT is a rigid map. The deformation from the reference configuration to the deformed SWCNT is divided into two steps: one is from the reference configuration to the initial equilibrium SWCNT (transformed from (A) to (B)), and the other is from the initial equilibrium SWCNT to the current configuration (transformed from (B) to (C)).

In the future numerical simulations, we use Fig. 4.7C as the reference configuration. Thus, the deformation from the reference configuration to the deformed SWCNT can be divided into two steps (see Fig. 4.14): one is from the reference configuration to the initial equilibrium SWCNT, and the other is from the initial equilibrium SWCNT to the current configuration. The first step can be exactly expressed as

$$\begin{cases} x_1 = X_1 \\ x_2 = R\sin\dfrac{X_2}{R} \\ x_3 = R - R\cos\left(\dfrac{X_2}{R}\right) \end{cases} \tag{4.56}$$

Eq. (4.56) also maps the cell structure in the reference configuration to that in the initial SWCNT by minimizing the energy of the cell structure with only the inner shift being used as the optimization parameter. The cell structure in the reference configuration can be determined with the method described in Section 4.4.1. This cell structure is the original image of the representative cell.

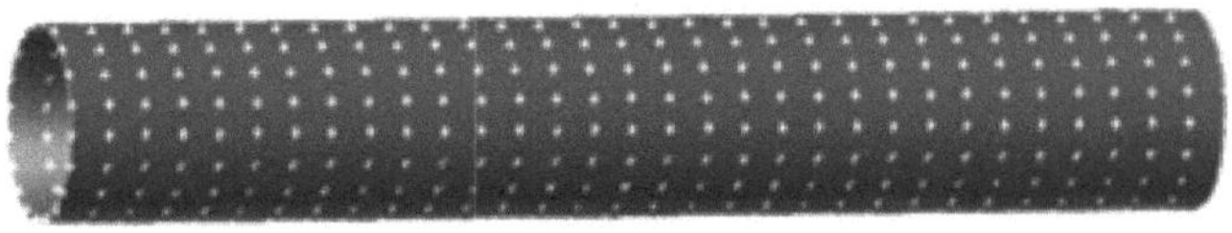

FIGURE 4.15

The mesh-free node collocation on the surface of an undeformed SWCNT.

The transformation map from the reference configuration to the current configuration is

$$\begin{cases} x_1 = X_1 + u_1 \\ x_2 = R \sin\dfrac{X_2}{R} + u_2 \\ x_3 = R - R\cos\left(\dfrac{X_2}{R}\right) + u_3 \end{cases} \tag{4.57}$$

where u_1, u_2, and u_3 are the displacements of an SWCNT relative to its undeformed counterpart (Fig. 4.14B), whereas Eq. (4.56) denotes the first step of the deformation of an SWCNT (the reference configuration to the initial SWCNT), u_1, u_2, and u_3 describe the remaining step (from the initial SWCNT to the current configuration), which is numerically computed in the later mesh-free numerical simulations. To determine the cell structure at any point of the current configuration, the energy of the cell structure is minimized only with respect to the inner shift. Such a minimizing algorithm also gives the results of the stresses and tangent modulus with the application of the formulas (4.35) and (4.36).

In mesh-free methods, the problem domain is discretized by a set of nodes, and each node has compact supports (shown in Fig. 4.1). For a 2-D domain, the two basic support types are discs and rectangles. The mesh-free nodes are collocated on the reference configuration, and the displacement u_1, u_2, and u_3 is interpolated with the mesh-free nodal parameters. As the transformation from the reference configuration to an undeformed SWCNT is a simple grid mapping, the mesh-free nodes can also be viewed as having been collocated on the surface of the undeformed SWCNT. Fig. 4.15 shows a node collocation on an undeformed SWCNT.

Mesh-free shape functions are always one of the key points in the development of mesh-free methods. The present research introduces two interpolation methods in detail, i.e., MLS approximation and moving Kriging interpolation, to construct the mesh-free shape function, and discusses their basic properties that are relative to the future application.

4.5.1 MOVING LEAST-SQUARES APPROXIMATION

In the MLS approximation, any variable $\psi(\mathbf{X})$ that is defined in a domain Ω can be approximated by $\psi^h(\mathbf{X})$ in the subdomain $\Omega_{\mathbf{X}}$, as follows.

$$\psi^h(\mathbf{X}) = \sum_{i=1}^{m} g_i(\mathbf{X})a_i(\mathbf{X}) = \mathbf{g}^T(\mathbf{X})\mathbf{a}(\mathbf{X}) \tag{4.58}$$

where $g_i(\mathbf{X})$ is the monomial basis function, m is the number of terms in the basis, and $a_i(\mathbf{X})$ are the coefficients of the basis function. The commonly used basis is the linear basis

$$\mathbf{g}^T = (1, X_1), \quad m = 2, \text{ in 1D} \tag{4.59}$$

$$\mathbf{g}^T = (1, X_1, X_2), \quad m = 3, \text{ in 2D} \tag{4.60}$$

and the quadratic basis is

$$\mathbf{g}^T = (1, X_1, X_1^2), \quad m = 3, \text{ in 1D} \tag{4.61}$$

$$\mathbf{g}^T = (1, X_1, X_2, X_1^2, X_1X_2, X_2^2), \quad m = 6, \text{ in 2D} \tag{4.62}$$

Certain basis functions can be used to satisfy the special properties of the variable. For example, the following basis function can capture stress singularity in linear-elastic fracture mechanics [116].

$$\mathbf{g}^T = \left(1, X_1, X_2, \sqrt{r}\cos\frac{\theta}{2}, \sqrt{r}\sin\frac{\theta}{2}, \sqrt{r}\sin\frac{\theta}{2}\cos\theta, \sqrt{r}\cos\frac{\theta}{2}\sin\theta\right), \quad m = 7 \tag{4.63}$$

where r and θ are polar coordinates with the crack tip as the origin.

The unknown coefficients $a_i(\mathbf{X})$ in Eq. (4.58) can be determined by the minimization of the weighted discrete L_2 norm as

$$J = \sum_{I=1}^{NP} w(\mathbf{X} - \mathbf{X}_I)[\mathbf{g}^T(\mathbf{X}_I)\mathbf{a}(\mathbf{X}_I) - \psi_I]^2 \tag{4.64}$$

where $w(\mathbf{X} - \mathbf{X}_I)$ is the weight function with compact support, and NP is the number of nodes with $w(\mathbf{X} - \mathbf{X}_I) > 0$. The minimum of J in Eq. (4.64) with respect to $\mathbf{a}(\mathbf{X})$ leads to a set of linear equations:

$$\mathbf{A}(\mathbf{X})\mathbf{a}(\mathbf{X}) = \mathbf{B}(\mathbf{X})\boldsymbol{\psi} \tag{4.65}$$

where $\boldsymbol{\psi} = (\psi_1, \psi_2, \ldots, \psi_N)^T$, and

$$\mathbf{A}(\mathbf{X}) = \sum_{I=1}^{NP} w(\mathbf{X} - \mathbf{X}_I)\mathbf{g}(\mathbf{X}_I)\mathbf{g}^T(\mathbf{X}_I) \tag{4.66}$$

and

$$\mathbf{B}(\mathbf{X}) = [w(\mathbf{X} - \mathbf{X}_1)\mathbf{g}(\mathbf{X}_1), w(\mathbf{X} - \mathbf{X}_2)\mathbf{g}(\mathbf{X}_2), \ldots, w(\mathbf{X} - \mathbf{X}_N)\mathbf{g}(\mathbf{X}_N)] \tag{4.67}$$

Thus, the unknown coefficients $\mathbf{a}(\mathbf{X})$ can be obtained from Eq. (4.64) as

$$\mathbf{a}(\mathbf{X}) = \mathbf{A}^{-1}(\mathbf{X})\mathbf{B}(\mathbf{X})\boldsymbol{\psi} \tag{4.68}$$

Substituting Eq. (4.68) into Eq. (4.58), the MLS approximation can be written in standard form as

$$\psi^h(\mathbf{X}) = \sum_{I=1}^{NP} \varphi_I(\mathbf{X})\psi_I(\mathbf{X}) \tag{4.69}$$

where the MLS shape function $\phi_I(\mathbf{X})$ is defined as

$$\varphi_I(\mathbf{X}) = \sum_{j=1}^{m} g_j(\mathbf{X})[\mathbf{A}^{-1}(\mathbf{X})\mathbf{B}(\mathbf{X})]_{jI} \tag{4.70}$$

The partial derivatives of $\phi_I(\mathbf{X})$ with respect to X_K can be obtained as

$$\varphi_{I,K} = \sum_{j=1}^{m}[g_{j,K}(\mathbf{A}^{-1}\mathbf{B})_{jI} + g_j(\mathbf{A}^{-1}\mathbf{B}_K + \mathbf{A}_K^{-1}\mathbf{B})_{jI}] \tag{4.71}$$

in which

$$\mathbf{B}_K(\mathbf{X}) = [w_{1,K}(\mathbf{X})\mathbf{g}(\mathbf{X}_1), w_{2,K}(\mathbf{X})\mathbf{g}(\mathbf{X}_2), \ldots, w_{N,K}(\mathbf{X})\mathbf{g}(\mathbf{X}_N)] \tag{4.72}$$

$$\mathbf{A}_K^{-1} = -\mathbf{A}^{-1}\mathbf{A}_K\mathbf{A}^{-1} \tag{4.73}$$

$$A_K(\mathbf{X}) = \sum_{I=1}^{N} w_{I,k}(\mathbf{X})\mathbf{g}(\mathbf{X}_I)\mathbf{g}^T(\mathbf{X}_I) \tag{4.74}$$

The second partial derivatives of $\phi_I(\mathbf{X})$ are obtained as

$$\varphi_{I,KL} = \sum_{j=1}^{m}\left[g_{j,KL}(\mathbf{A}^{-1}\mathbf{B})_{jI} + g_{j,K}(\mathbf{A}_L^{-1}\mathbf{B} + \mathbf{A}^{-1}\mathbf{B}_L)_{jI} + g_{j,L}(\mathbf{A}^{-1}\mathbf{B}_K + \mathbf{A}_K^{-1}\mathbf{B})_{jI} + g_j(\mathbf{A}_L^{-1}\mathbf{B}_K + \mathbf{A}^{-1}\mathbf{B}_{KL} + \mathbf{A}_{KL}^{-1}\mathbf{B} + \mathbf{A}_K^{-1}\mathbf{B}_L)_{jI}\right] \tag{4.75}$$

in which

$$\mathbf{B}_{KL}(\mathbf{X}) = [w_{1,KL}(\mathbf{X})\mathbf{g}(\mathbf{X}_1), w_{2,KL}(\mathbf{X})\mathbf{g}(\mathbf{X}_2), \ldots, w_{N,KL}(\mathbf{X})\mathbf{g}(\mathbf{X}_N)] \tag{4.76}$$

$$\mathbf{A}_{KL}^{-1} = \mathbf{A}^{-1}\mathbf{A}_L\mathbf{A}^{-1}\mathbf{A}_K\mathbf{A}^{-1} - \mathbf{A}^{-1}\mathbf{A}_{KL}\mathbf{A}^{-1} + \mathbf{A}^{-1}\mathbf{A}_K\mathbf{A}^{-1}\mathbf{A}_L\mathbf{A}^{-1} \tag{4.77}$$

In the present research, the quadratic basis (Eq. 4.62) is used, and the weight function is chosen as the cubic spline function (Eq. 4.80). Shown in Fig. 4.16 is a plot of the MLS shape function. Fig. 4.16A presents the 2-D view, and Fig. 4.16B the variation of the shape function along the X_1 direction ($X_2 = 0$), which corresponds to the 1-D MLS shape function. Fig. 4.17 shows the first-order derivative of the shape function ϕ_1, and plotted in Figs. 4.18 and 4.19 are the second-order derivatives ϕ_{11} and ϕ_{12}.

4.5.2 WEIGHT FUNCTIONS

The choice of weight functions is an important ingredient in mesh-free methods, as these functions affect the continuity and accuracy of the resulting mesh-free approximation. Generally, the weight functions can result in C^2 or higher-order continuity of the shape functions. To generate a set of sparse discrete equations, the weight functions are defined as having a compact support, which means that they should be nonzero only over a small neighborhood of node $\mathbf{X}_I$.

The commonly used weight functions are in the following forms.

$$\text{Exponential:} \quad w(d) = \begin{cases} e^{-(d/\alpha)^2}, & d \le 1 \\ 0, & d > 1 \end{cases} \tag{4.78}$$

$$\text{Quadratic spline:} \quad w(d) = \begin{cases} 1 - 2d^2, & d \le \frac{1}{2} \\ 2(1-d)^2, & \frac{1}{2} < d \le 1 \\ 0, & d > 1 \end{cases} \tag{4.79}$$

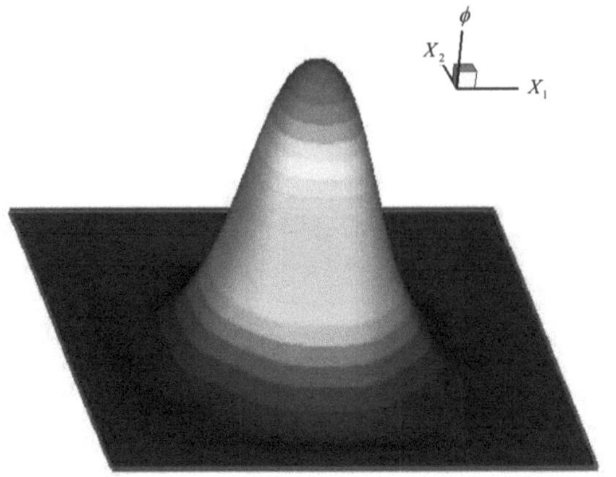

(A)

(B)

FIGURE 4.16

(A) The mesh-free shape function ϕ_1 constructed on the 2-D domain; (B) the variation of ϕ_1 along the X_1 direction ($X_2 = 0$).

$$\text{Cubic spline:} \quad w(d) = \begin{cases} \frac{2}{3} - 4d^2 + 4d^3, & d \leq \frac{1}{2} \\ \frac{4}{3} + 4d + 4d^2 - \frac{4}{3}d^3, & \frac{1}{2} < d \leq 1 \\ 0, & d > 1 \end{cases} \tag{4.80}$$

$$\text{Quartic spline:} \quad w(d) = \begin{cases} 1 - 6d^2 + 8d^3 - 3d^4, & d \leq 1 \\ 0, & d > 1 \end{cases} \tag{4.81}$$

where d is the normalized distance, and α is a dilation parameter of the weight function.

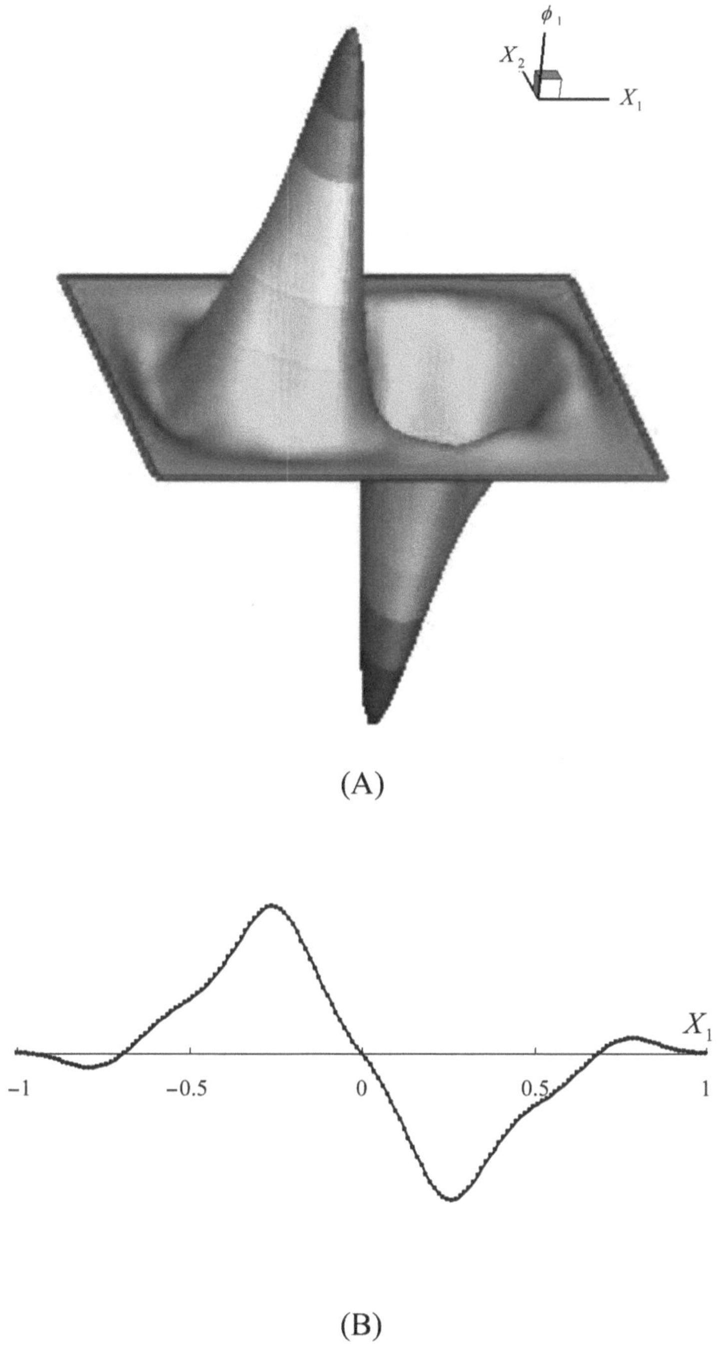

FIGURE 4.17

(A) The first-order derivative ϕ_1 of the mesh-free shape function constructed on a 2D domain; (B) the variation of ϕ_1 along the X_1 direction ($X_2 = 0$).

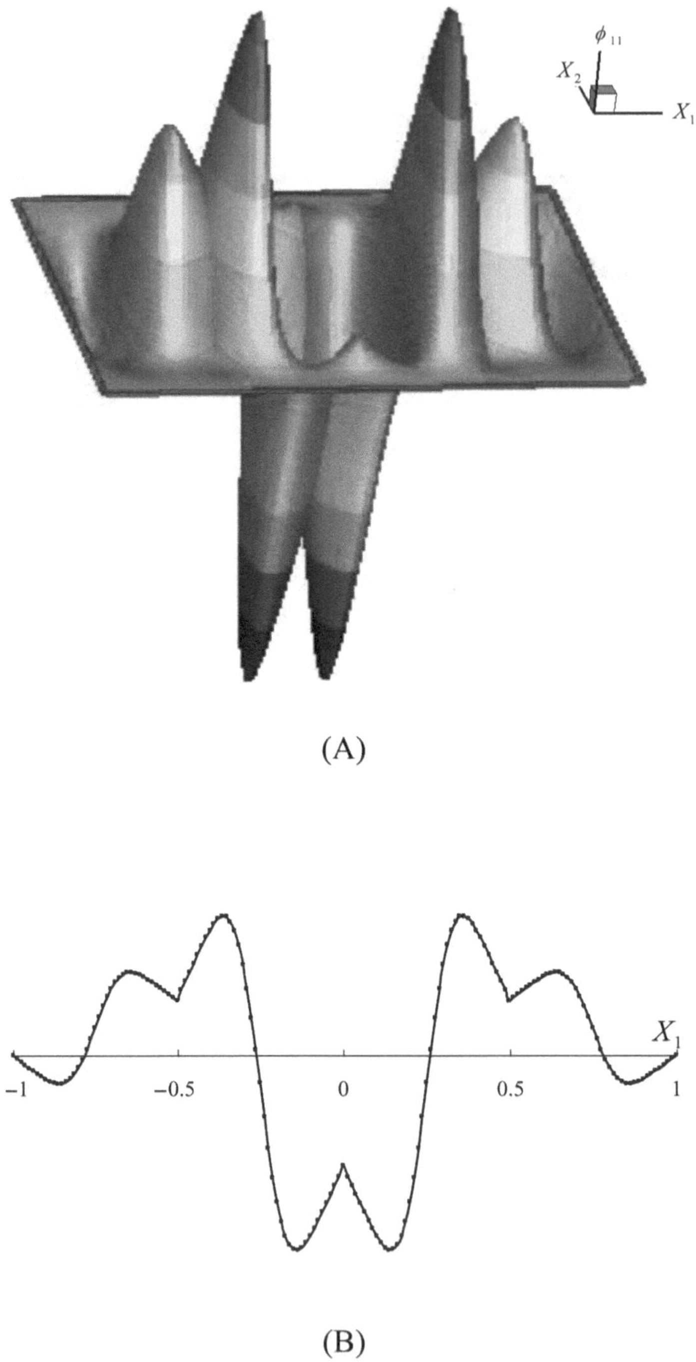

FIGURE 4.18

(A) The second-order derivative ϕ_{11} of the mesh-free shape function constructed on a 2D domain; (B) the variation of ϕ_{11} along the X_1 direction ($X_2 = 0$).

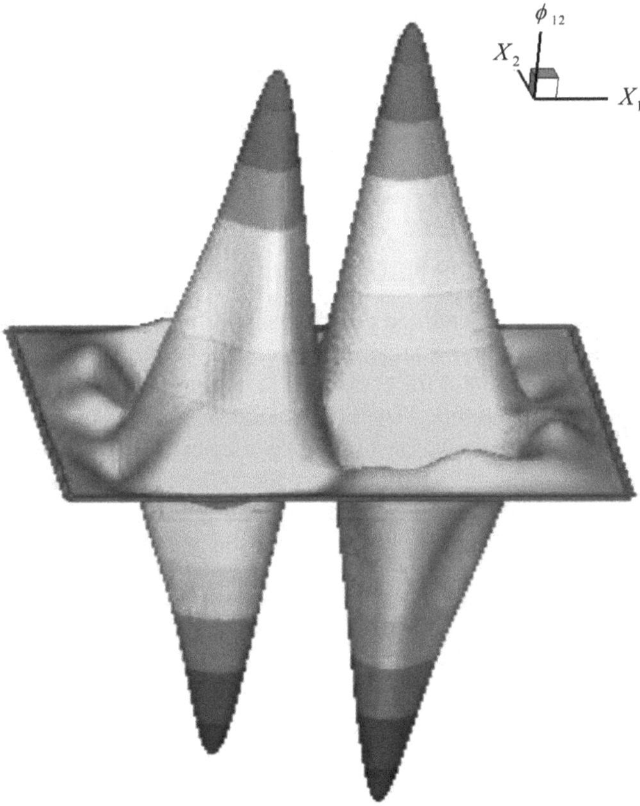

FIGURE 4.19

The second-order derivative ϕ_{12} of the mesh-free shape function.

In the 1-D case, the weight function is written as

$$w(X_1 - X_{1I}) = w\left(\frac{d_I}{c_I}\right) \tag{4.82}$$

where $d_I = |X_1 - X_{1I}|$, and c_I is the size of the support of node X_I.

In the 2-D case, there are two commonly used supports. For the circular compact support shown in Fig. 4.1A, the weight function can be written as

$$w(\mathbf{X} - \mathbf{X}_I) = w\left(\frac{d_I}{R_I}\right) \tag{4.83}$$

where $d_I = ||\mathbf{X} - \mathbf{X}_I|| = \sqrt{(X_1 - X_{1I})^2 + (X_2 - X_{2I})^2}$, and R_I is the radius of the support of node $\mathbf{X}_I$.

For the rectangular compact support shown in Fig. 4.1B, the product weight function is given by

$$w(\mathbf{X} - \mathbf{X}_I) = w\left(\frac{d_1}{a_{1I}}\right) w\left(\frac{d_2}{a_{2I}}\right) \tag{4.84}$$

where $d_1 = |X_1 - X_{1I}|$,$d_2 = |X_2 - X_{2I}|$, and (a_{1I}, a_{2I}) is the size of the rectangular support of node $\mathbf{X}_I$.

The MLS approximation is closely related to the weight functions and the basis functions. Dolbow and Belytschko [122] stated that if the continuity of the basis function is greater than the continuity of the weight function, then the resulting approximation would inherit the continuity of the weight function. The exponential weight can resemble a weight with C^1 or higher-order continuity. The cubic and quartic weights possess C^2 continuity, which is ideal for constructing the mesh-free shape function with C^2 continuity.

4.5.3 DOMAIN OF INFLUENCE OF NODES

The weight functions are defined as having a compact support, i.e., the subdomain over it is nonzero. Each subdomain Ω_I is associated with a node $\mathbf{X}_I$, which can be termed the node's domain of influence (DOI). The way in which the DOI for each node is determined has a tremendous impact on the total performance of a mesh-free method. The node's DOI is inherent in the weight function's compact support, which makes the interpolation limited on the several nodes whose supports cover the evaluated point. The most commonly used supports for 2-D problems are circular and rectangular supports, as shown in Fig. 4.1. The shape of the support has almost no effect on the results.

In the 1-D case, the size of the support can be defined as

$$c_I = D_{\max} c_{bI}, \quad I = 1, \ldots, n, \tag{4.85}$$

where $D_{\max}$ is a scaling factor, and c_{bI} is the basic support of node X_I. c_{bI} can be determined by the maximum distance to its nearest neighboring node.

The radius of the circular support can be defined as

$$R_I = D_{\max} r_I, \quad I = 1, \ldots, n, \tag{4.86}$$

where r_I is the radius of a basic circular support. The radius r_I is determined by searching for sufficient adjacent nodes to satisfy the basis. For example, we can designate the support to cover just the three nearest neighboring nodes.

The size of a rectangular support can be defined by

$$\begin{cases} a_{1I} = D_{\max} c_{1I} \\ a_{2I} = D_{\max} c_{2I} \end{cases}, \quad I = 1, \ldots, n, \tag{4.87}$$

where (c_{1I}, c_{2I}) defines a basic support that is determined by searching for sufficient neighboring nodes to satisfy the basis in both directions. For the regular collocation of nodes, c_{1I} and c_{2I} can be determined by the distance to the nearest neighboring nodes in two directions.

The size of the compact support (or DOI) determines the number of nodes N in the construction of matrix $\mathbf{A}(\mathbf{X})$. The number of nodes N should be large enough to ensure that $\mathbf{A}(\mathbf{X})$ is invertible. The size of each node's DOI can be different, and it depends on the spacing density of the nodes in the local context. Small support sizes may affect the convergence of the mesh-free solutions. Large support sizes ensure convergence, but increase the computational time. Furthermore, if the support of a node extends far outside the boundary of the problem domain, then the computation size of the domain that is decided by all of the supports will be very different from the actual, which will inevitably affect the computational results. Therefore, for a given $D_{\max}$, increasing the number of nodes will decrease the support size and make the computational size of the domain coincide with its actual size. According to the literature, $D_{\max}$ is typically in the range of 2.0–4.0.

In the later numerical simulations, the mesh-free nodes are uniformly collocated along the axial and circumferential directions (see Fig. 4.15). Two simple examples are tested to discuss the choice of the parameter D_{max} for the present problems.

4.5.4 MK INTERPOLATION

Kriging can be traced to the first application in geostatistics [123] for spatial interpolation and was used for construction of the shape function and its derivatives. Similar to the MLS approximation, Kriging interpolation can also be extended to any subdomain $\Omega(x_i)$—termed MK interpolation. We consider a distribution function $u(x_i)$ in a subdomain $\Omega(x_i)$. The subdomain is represented by a series of dispersive nodes x_i ($i = 1, 2, \ldots, n$), where n is the total number of nodes covered in the subdomain. With n given field values $u(x_1)$, $u(x_2)$, ... $u(x_n)$, the approximate field function $\hat{u}(x)$ can be interpolated by a weighted linear combination:

$$\hat{u}(x) = \sum_{i=1}^{n} \lambda_i(x) u(x_i) \tag{4.88}$$

where $\lambda_i(x)$ is the weight function of the ith node. There are two conditions which should be satisfied for the selection of coefficients $\lambda_i(x)$, i.e., unbiasedness and minimum variance.

The first term is unbiasedness, which implies that the expected value of approximate function $\hat{u}(x)$ must be in accordance with that of the given nodal values $u(x)$:

$$E[u(x)] = E[\hat{u}(x)] \tag{4.89}$$

Substitution of Eq. (4.88) into Eq. (4.89) yields:

$$E[u(x)] = E\left[\sum_{i=1}^{n} \lambda_i(x) u(x_i)\right] \tag{4.90}$$

A polynomial drift model $p_l(x)$ is summarized by:

$$\sum_{i=1}^{n} \lambda_i(x) p_l(x_i) = p_l(x), \quad 1 \le l \le m, \quad p_1(x) = 1 \tag{4.91}$$

A commonly used linear basis in one-, two-, and three-dimensional space is given as:

$$\mathbf{p}^T(\mathbf{x}) = (1, x_1), \quad m = 2 \tag{4.92}$$

$$\mathbf{p}^T(\mathbf{x}) = (1, x_1, x_2), \quad m = 3 \tag{4.93}$$

and a quadratic basis is:

$$\mathbf{p}^T(\mathbf{x}) = (1, x_1, x_1^2), \quad m = 3 \tag{4.94}$$

$$\mathbf{p}^T(\mathbf{x}) = (1, x_1, x_2, x_1^2, x_1 x_2, x_2^2), \quad m = 6 \tag{4.95}$$

Secondly, the condition of minimum variance requires a minimum estimation error in the mean square sense:

$$\prod = E[(u(x) - \hat{u}(x))^2] \tag{4.96}$$

Regarding Eq. (4.91) as a constraint by using a Lagrange multiplier μ_l in Eq. (4.11) yields:

$$
\begin{aligned}
\prod &= E[(u(x)-\hat{u}(x))^2] + 2\sum_{l=1}^{m}\mu_l\left[\sum_{i=1}^{n}\lambda_i p_l(x_i) - p_l(x)\right] \\
&= Var[\hat{u}_{(x)} - u_{(x)}] + 2\sum_{l=1}^{m}\mu_l\left[\sum_{i=1}^{n}\lambda_i p_l(x_i) - p_l(x)\right] \\
&= E\left[\sum_{i=1}^{n}\lambda_i u_{(x_i)} - u_{(x)}\right]^2 - E^2[\hat{u}_{(x)} - u_{(x)}] + 2\sum_{l=1}^{m}\mu_l\left[\sum_{i=1}^{n}\lambda_i p_l(x_i) - p_l(x)\right] \\
&= E\left[\sum_{i=1}^{n}\lambda_i u_{(x_i)} - u_{(x)}\right]^2 + 2\sum_{l=1}^{m}\mu_l\left[\sum_{i=1}^{n}\lambda_i p_l(x_i) - p_l(x)\right] \\
&= \sum_{i=1}^{n}\sum_{j=1}^{n}\lambda_i\lambda_j E[(u_{(x_i)} - u_{(x)})(u_{(x_j)} - u_{(x)})] - \sum_{i=1}^{n}\sum_{j=1}^{n}\lambda_i\lambda_j E[(u_{(x_i)} - u_{(x)})]E[(u_{(x_j)} - u_{(x)})] \\
&\quad + 2\sum_{l=1}^{m}\mu_l\left[\sum_{i=1}^{n}\lambda_i p_l(x_i) - p_l(x)\right] \\
&= \sum_{i=1}^{n}\sum_{j=1}^{n}\lambda_i\lambda_j Cov(u_{(x_i)} - u_{(x)}, u_{(x_j)} - u_{(x)}) + 2\sum_{l=1}^{m}\mu_l\left[\sum_{i=1}^{n}\lambda_i p_l(x_i) - p_l(x)\right]
\end{aligned}
\tag{4.97}
$$

By introducing a semivariogram model, covariance $E[u(x)u(x_i)]$ can be substituted by semivariogram $\gamma(h)$, defined as:

$$
\gamma(h) = \gamma(x, x_i) = E[u(x)u(x_i)] = \frac{1}{2}E\left[\{u(x) - u(x_i)\}^2\right] \tag{4.98}
$$

Many theoretical models are available for semivariograms. Due to the simplicity and wide application of the Gaussian function, we select this as the correlation function in the present numerical calculation:

$$
\gamma(x, x_i) = \gamma(h) = 2\left(1 - e^{-\theta\left(\frac{h}{a_0}\right)^2}\right) \quad (h \leq a_0) \tag{4.99}
$$

where $h = ||\mathbf{x} - \mathbf{x}_i||$ is the vectorial distance between x and x_i; θ is a correlation parameter and a_0 is the range of the compact support domain. It can be seen from the foregoing formulations that the selection of parameters for the correlation function plays an important role in the accuracy of MK interpolation and the numerical solution. Since trigonometric functions possess the higher-order continuity, the optimal value for correlation parameter θ is determined by fitting the trigonometric functions as well as their first- and second-order derivatives for both one- and two-dimensional problems. Finally, we concluded that a value of one for the correlation parameter in the numerical calculation would give a good prediction with the analytical solution. We select a circle with radius

$r = a_0$, as the compact support domain. In the same way, $E[u(x)u(x_j)]$ and $E[u(x_i)u(x_j)]$ can be replaced by $\gamma(x, x_j)$ and $\gamma(x_i, x_j)$, respectively. Thus, we obtain:

$$\begin{aligned} 2\gamma(x_i, x_j) &= E[u(x_i) - u(x_j)]^2 \\ &= E[(u(x_i) - u(x)) - (u(x_j) - u(x))]^2 \\ &= E[(u(x_i) - u(x))]^2 + E[(u(x_j) - u(x))]^2 \\ &\quad - 2E[(u(x_i) - u(x))(u(x_j) - u(x))] \\ &= 2\gamma(x, x_i) + 2\gamma(x, x_j) - 2\,\mathrm{cov}[u(x_i) - u(x),\ u(x_j) - u(x)] \end{aligned} \tag{4.100}$$

Eq. (4.100) can be rewritten as:

$$Cov(u_{(x)} - u_{(x_j)}, u_{(x_i)} - u_{(x_j)}) = \gamma_{(x,x_j)} + \gamma_{(x_i,x_j)} - \gamma_{(x,x_i)} \tag{4.101}$$

Substituting Eq. (4.101) into Eq. (4.97), we obtain the following equation:

$$\begin{aligned} \prod &= \sum_{i=1}^{n}\sum_{j=1}^{n} \lambda_i \lambda_j (\gamma_{(x_i,x)} + \gamma_{(x_j,x)} - \gamma_{(x_i,x_j)}) + 2\sum_{l=1}^{m} \mu_l \left[\sum_{i=1}^{n} \lambda_i p_l(x_i) - p_l(x)\right] \\ &= 2\sum_{i=1}^{n} \lambda_i \gamma_{(x_i,x)} - \sum_{i=1}^{n}\sum_{j=1}^{n} \lambda_i \lambda_j \gamma_{(x_i,x_j)} + 2\sum_{l=1}^{m} \mu_l \left[\sum_{i=1}^{n} \lambda_i p_l(x_i) - p_l(x)\right] \end{aligned} \tag{4.102}$$

The minimum mean square error is determined by equating all of the first derivatives with respect to $\lambda_i(x)$ and $\mu_i(x)$ to zero, which results in the following series of linear algebraic equations:

$$\begin{aligned} \frac{\partial \Pi}{\partial \lambda_i} &= 2\gamma_{(x,x_i)} - 2\sum_{j=1}^{n} \lambda_j \gamma_{(x_i,x_j)} + 2\sum_{l=1}^{m} \mu_l p_l(x_i) = 0 \quad 1 \le i \le n \\ \frac{\partial \Pi}{\partial \mu_l} &= \sum_{i=1}^{n} \lambda_i p_l(x_i) - p_l(x) = 0 \quad 1 \le l \le m \end{aligned} \tag{4.103}$$

For convenience, Eq. (4.103) is rewritten in matrix form as follows:

$$\mathbf{GW} = \mathbf{g} \tag{4.104}$$

where

$$\mathbf{G} = \begin{bmatrix} \mathbf{R} & \mathbf{P} \\ \mathbf{P^T} & \mathbf{0} \end{bmatrix} \tag{4.105}$$

$$\mathbf{R} = \begin{bmatrix} \gamma(x_1, x_1) & \cdots & \gamma(x_1, x_n) \\ \cdots & \cdots & \cdots \\ \gamma(x_n, x_1) & \cdots & \gamma(x_n, x_n) \end{bmatrix} \tag{4.106}$$

$$\mathbf{P} = \begin{bmatrix} p_1(x_1) & \cdots & p_m(x_1) \\ \cdots & \cdots & \cdots \\ p_1(x_n) & \cdots & p_m(x_n) \end{bmatrix} \tag{4.107}$$

$$\mathbf{W} = \begin{bmatrix} \lambda_1 & \lambda_2 & \cdots & \lambda_n & \mu_1 & \mu_2 & \cdots & \mu_m \end{bmatrix} \tag{4.108}$$

$$\mathbf{g}^T = \begin{bmatrix} \mathbf{r}^T(x) & \mathbf{p}^T(x) \end{bmatrix} \tag{4.109}$$

$$\mathbf{r}^T(x) = \begin{bmatrix} \gamma(x, x_1) & \cdots & \gamma(x, x_n) \end{bmatrix} \tag{4.110}$$

$$\mathbf{p}^T(x) = \begin{bmatrix} p_1(x) & \cdots & p_m(x) \end{bmatrix} \tag{4.111}$$

Based on Eqs. (4.88) and (4.104), the estimated value can be calculated numerically as:

$$\hat{u}(x) = \mathbf{p}^T(x)\hat{\beta} + \mathbf{r}^T(x)\mathbf{R}^{-1}(\mathbf{u} - \mathbf{P}\hat{\beta}) \tag{4.112}$$

where

$$\hat{\beta} = (\mathbf{P}^T\mathbf{R}^{-1}\mathbf{P})^{-1}\mathbf{P}^T\mathbf{R}^{-1}\mathbf{u} \tag{4.113}$$

The approximation function $\hat{u}(x)$ can be expressed in the form of $u(x)$:

$$\begin{aligned} \hat{u}(x) &= \mathbf{p}^T(x_i)\hat{\beta} + \mathbf{r}^T(x_i)\mathbf{R}^{-1}(\mathbf{u}(x_i) - \mathbf{P}\hat{\beta}) \\ &= [\mathbf{p}^T(x_i)\mathbf{A} + \mathbf{r}^T(x_i)\mathbf{B}]\mathbf{u}(x_i) \\ &= \sum_{i=1}^{n} \phi_i(x_i)u(x_i) = \mathbf{\Phi}(x)\mathbf{u}(x) \end{aligned} \tag{4.114}$$

where $\mathbf{\Phi}(x)$ is the interpolation function, which can be expressed by:

$$\mathbf{\Phi}(x) = \mathbf{p}^T(x)\mathbf{A} + \mathbf{r}^T(x)\mathbf{B} \tag{4.115}$$

$$\mathbf{u}(x) = \begin{bmatrix} u(x_1) & u(x_2) & \cdots & u(x_n) \end{bmatrix} \tag{4.116}$$

$$\mathbf{A} = (\mathbf{P}^T\mathbf{R}\mathbf{P})^{-1}\mathbf{P}^T\mathbf{R}^{-1} \tag{4.117}$$

$$\mathbf{B} = \mathbf{R}^{-1}(\mathbf{I} - \mathbf{P}\mathbf{A}) \tag{4.118}$$

where $\mathbf{I}$ is a $n \times n$ unit matrix. The first- and second-order derivatives of the shape function can be derived from Eq. (4.115):

$$\mathbf{\Phi}_i(x) = \mathbf{p}_i^T(x)\mathbf{A} + \mathbf{r}_i^T(x)\mathbf{B} \tag{4.119}$$

$$\mathbf{\Phi}_{ij}(x) = \mathbf{p}_{ij}^T(x)\mathbf{A} + \mathbf{r}_{ij}^T(x)\mathbf{B} \tag{4.120}$$

where $\mathbf{\Phi}_i(x)$ and $\mathbf{\Phi}_{ij}(x)$ denote the first- and second-order spatial derivatives, respectively. Compared with MLS approximation, the shape function $\mathbf{\Phi}(x)$ constructed using the MK interpolation in Eq. (4.115) has a distinct property:

$$\phi_i(x_j) = \delta_{ij} \tag{4.121}$$

Fig. 4.20A illustrates the one-dimensional shape functions which demonstrate that the MK interpolation has the delta function property; first- and second-order derivatives of the shape function are plotted in Fig. 4.20A and C, respectively.

4.5.5 THE MESH-FREE COMPUTATIONAL FRAMEWORK

The displacement of a deformed SWCNT is divided into two parts: the displacement of an undeformed SWCNT relative to the reference configuration and the displacement of the deformed SWCNT relative to the undeformed SWCNT. The first part can be exactly calculated (Fig. 4.21). The second part is approximated using the MLS approximation, i.e.,

$$\mathbf{u} = \begin{pmatrix} u_1 \\ u_2 \\ u_3 \end{pmatrix} = \sum_{I=1}^{NP} \phi_I \hat{\mathbf{u}}_I \tag{4.122}$$

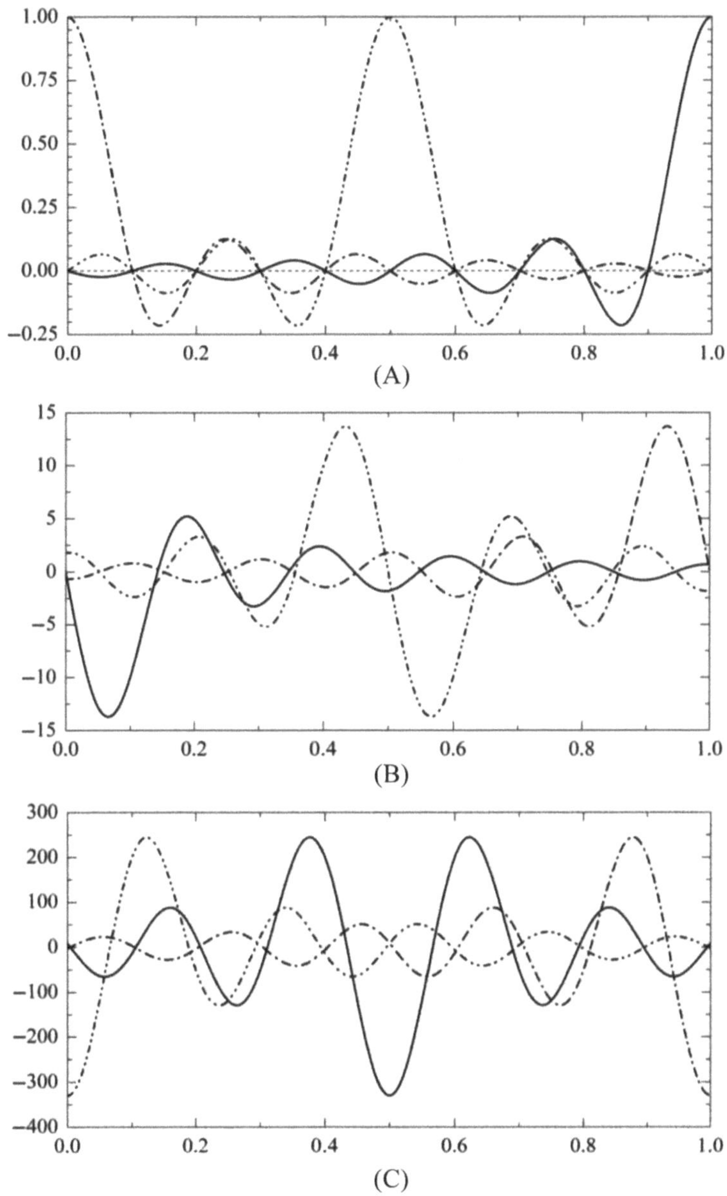

FIGURE 4.20

(A) Variation of MK shape functions $\phi(x)$ along the axial direction, constructed on a typical one-dimensional domain; (B) the corresponding first-order derivative $\phi_X(x)$ of the MK shape function; and (C) the second-order derivative $\phi_{XX}(x)$.

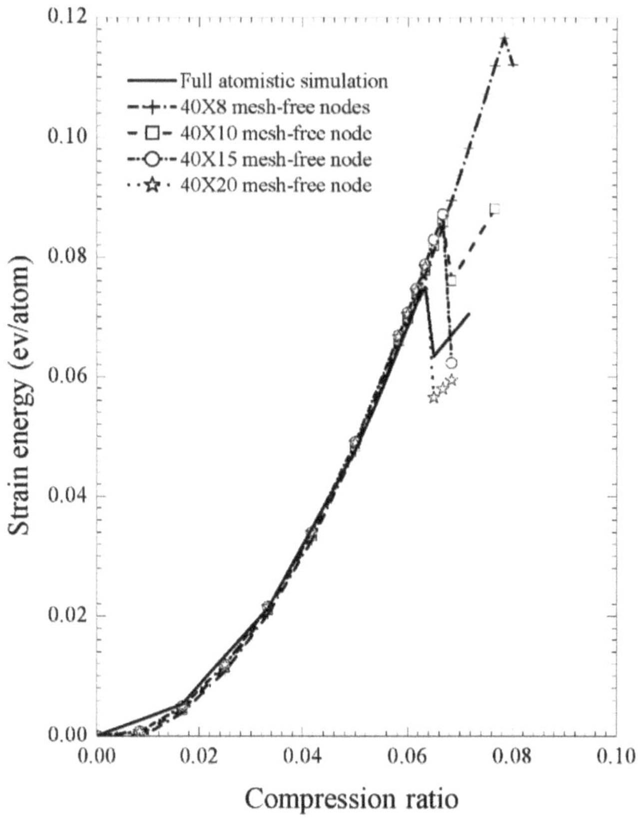

FIGURE 4.21

The average strain energy obtained from the continuum-based method with different nodes along with that obtained from the full atomistic simulation [124].

where $\hat{\mathbf{u}}_I = (\hat{u}_{I1}, \hat{u}_{I2}, \hat{u}_{I3})^T$ is the nodal parameter, ϕ_I is the mesh-free shape function, and NP is the number of nodes whose supports cover the evaluating point. It is noted that any node has 3 degrees of freedom, but the interpolation shape function is built on the 2-D domain (Fig. 4.22).

The first- and second-order gradients are computed using the following formulas.

$$\mathbf{F} = \mathbf{F}_{\text{ini}} + \mathbf{F}_{\text{apr}}, \quad \mathbf{G} = \mathbf{G}_{\text{ini}} + \mathbf{G}_{\text{apr}} \tag{4.123}$$

where $\mathbf{F}_{\text{ini}}$ and $\mathbf{G}_{\text{ini}}$ are the gradients of the evaluated point when no loading is applied, and can be analytically calculated using the mapping equation (4.56), and $\mathbf{F}_{\text{apr}}$ and $\mathbf{G}_{\text{apr}}$ are approximated as

$$\begin{cases} \mathbf{F}_{\text{apr}} = \sum_{I=1}^{NP} \nabla_X(\varphi_I \hat{\mathbf{u}}_I) = \sum_{I=1}^{NP} \begin{bmatrix} \varphi_{I,1} \\ \varphi_{I,2} \end{bmatrix} \hat{\mathbf{u}}_I \\ \mathbf{G}_{\text{apr}} = \sum_{I=1}^{NP} \nabla_{XX}(\varphi_I \hat{\mathbf{u}}_I) = \sum_{I=1}^{NP} \begin{bmatrix} \varphi_{I,11} & \varphi_{I,12} \\ \varphi_{I,21} & \varphi_{I,22} \end{bmatrix} \hat{\mathbf{u}}_I \end{cases} \tag{4.124}$$

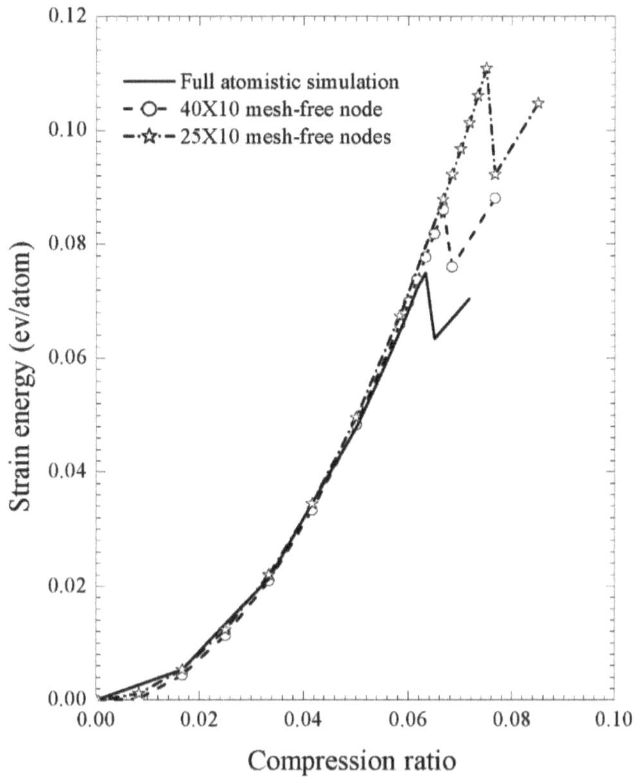

FIGURE 4.22

The average strain energy obtained from different longitudinal nodes (40 × 10 and 25 × 10) with fixed circumferential nodes along with that obtained from the full atomistic simulation [124].

with $\phi_{I,K}$ and $\phi_{I,KL}(K, L = 1, 2)$, respectively, being the first- and second-order derivatives of the mesh-free shape function ϕ_I.

Newton's method is used to solve the nonlinear governing equations in which the loading is decomposed into a series of time-steps, and at each iterative step the equilibrium state is determined by iteratively solving the incremental equation.

With the definition of the deformation gradient vector **grad** and the stress vector $\boldsymbol{\sigma}$, the weak form of the equilibrium equation can be written as

$$\int_{\Omega} \delta \mathbf{grad}^T \cdot \boldsymbol{\sigma} dV - \int_{\partial\Omega} \delta \mathbf{u}^T \cdot \mathbf{t}_0^P dS - \int_{\partial\Omega} (\nabla_N \delta \mathbf{u})^T \cdot \mathbf{t}_0^Q dS = 0 \tag{4.125}$$

Newton's method is adopted to solve the nonlinear governing equations. The stress at the $n + 1$th iterative step can be obtained by:

$$\boldsymbol{\sigma}_{n+1} = \boldsymbol{\sigma}_n + \Delta\boldsymbol{\sigma}_{n+1} \tag{4.126}$$

Thus, Eq. (4.125) can be rewritten as:

$$\int_{\Omega} \delta \mathbf{grad}_{n+1}^{T} \cdot \Delta \boldsymbol{\sigma}_{n+1} dV = -\int_{\Omega} \delta \mathbf{grad}_{n+1}^{T} \cdot \boldsymbol{\sigma}_{n} + \int_{\partial\Omega} \delta \mathbf{u}_{n+1}^{T} \cdot \mathbf{t}_{0}^{P} dS + \int_{\partial\Omega} (\nabla_{N} \delta \mathbf{u}_{n+1})^{T} \cdot \mathbf{t}_{0}^{Q} dS \tag{4.127}$$

At the $n + 1$th iterative step,

$$\delta \mathbf{u}_{n+1} = \delta(\mathbf{u}_{n} + \Delta \mathbf{u}_{n+1}) = \delta \Delta \mathbf{u}_{n+1} \tag{4.128}$$

$$\delta \mathbf{grad}_{n+1} = \delta(\mathbf{grad}_{n} + \Delta \mathbf{grad}_{n+1}) = \delta \Delta \mathbf{grad}_{n+1} \tag{4.129}$$

Based on Eqs. (4.128) and (4.129), Eq. (4.127) can be rewritten as:

$$\int_{\Omega} \delta \Delta \mathbf{grad}_{n+1}^{T} \cdot \Delta \boldsymbol{\sigma}_{n+1} dV = -\int_{\Omega} \delta \Delta \mathbf{grad}_{n+1}^{T} \cdot \boldsymbol{\sigma}_{n} + \int_{\partial\Omega} \delta \Delta \mathbf{u}_{n+1}^{T} \cdot \mathbf{t}_{0}^{P} dS + \int_{\partial\Omega} (\nabla_{N} \delta \Delta \mathbf{u}_{n+1})^{T} \cdot \mathbf{t}_{0}^{Q} dS \tag{4.130}$$

where $\Delta \boldsymbol{\sigma} = \mathbf{M} \cdot \Delta \mathbf{grad}$ and $\mathbf{M}$ is a combination of the tangential moduli which can be expressed as:

$$\mathbf{M} = \begin{bmatrix} \mathbf{M}_{\mathbf{FF}} & \mathbf{M}_{\mathbf{FG}} \\ \mathbf{M}_{\mathbf{GF}} & \mathbf{M}_{\mathbf{GG}} \end{bmatrix} \tag{4.131}$$

Consequently, Eq. (4.130) can be expressed as:

$$\int_{\Omega} \delta \Delta \mathbf{grad}_{n+1}^{T} \cdot \mathbf{M} \cdot \Delta \mathbf{grad}_{n+1} dV = -\int_{\Omega} \delta \Delta \mathbf{grad}_{n+1}^{T} \cdot \boldsymbol{\sigma}_{n} + \int_{\partial\Omega} \delta \Delta \mathbf{u}_{n+1}^{T} \cdot \mathbf{t}_{0}^{P} dS + \int_{\partial\Omega} (\nabla_{N} \delta \Delta \mathbf{u}_{n+1})^{T} \cdot \mathbf{t}_{0}^{Q} dS \tag{4.132}$$

The displacement increment of the evaluating point can be approximated as

$$\Delta \mathbf{u}_{n+1} = \begin{pmatrix} \Delta u_1 \\ \Delta u_2 \\ \Delta u_3 \end{pmatrix} = \boldsymbol{\Phi} \cdot \Delta \hat{\mathbf{U}}_{n+1}^{e} \tag{4.133}$$

$$\boldsymbol{\Phi} = \begin{bmatrix} \varphi_1 & 0 & 0 & \varphi_2 & 0 & 0 & \ddots & & & \varphi_{NP} & 0 & 0 \\ 0 & \varphi_1 & 0 & 0 & \varphi_2 & 0 & & \ddots & & 0 & \varphi_{NP} & 0 \\ 0 & 0 & \varphi_1 & 0 & 0 & \varphi_2 & & & \ddots & 0 & 0 & \varphi_{NP} \end{bmatrix} \tag{4.134}$$

$$\Delta \hat{\mathbf{U}}_{n+1}^{e} = \begin{bmatrix} \Delta \hat{u}_{11} \\ \Delta \hat{u}_{12} \\ \Delta \hat{u}_{13} \\ \Delta \hat{u}_{21} \\ \Delta \hat{u}_{22} \\ \Delta \hat{u}_{23} \\ \vdots \\ \vdots \\ \Delta \hat{u}_{NP1} \\ \Delta \hat{u}_{NP2} \\ \Delta \hat{u}_{NP3} \end{bmatrix} \tag{4.135}$$

The gradient increment can be calculated as

$$\Delta \mathbf{grad}_{n+1} = \mathbf{L} \cdot \Delta \mathbf{u} = \mathbf{L} \cdot \boldsymbol{\Phi} \cdot \Delta \hat{\mathbf{U}}_{n+1}^{e} \tag{4.136}$$

$$
\mathbf{L} = \begin{bmatrix}
\partial/\partial X_1 & 0 & 0 \\
\partial/\partial X_2 & 0 & 0 \\
0 & \partial/\partial X_1 & 0 \\
0 & \partial/\partial X_2 & 0 \\
0 & 0 & \partial/\partial X_1 \\
0 & 0 & \partial/\partial X_2 \\
\partial^2/\partial {X_1}^2 & 0 & 0 \\
\partial^2/\partial X_1 \partial X_2 & 0 & 0 \\
\partial^2/\partial {X_2}^2 & 0 & 0 \\
0 & \partial^2/\partial {X_1}^2 & 0 \\
0 & \partial^2/\partial X_1 \partial X_2 & 0 \\
0 & \partial^2/\partial {X_2}^2 & 0 \\
0 & 0 & \partial^2/\partial {X_1}^2 \\
0 & 0 & \partial^2/\partial X_1 \partial X_2 \\
0 & 0 & \partial^2/\partial {X_2}^2
\end{bmatrix} \tag{4.137}
$$

$$
\mathbf{L}\cdot\mathbf{\Phi} = \begin{bmatrix}
\varphi_{1,1} & 0 & 0 & \varphi_{2,1} & 0 & 0 & \ddots & \varphi_{NP,1} & 0 & 0 \\
\varphi_{1,2} & 0 & 0 & \varphi_{2,2} & 0 & 0 & \ddots & \varphi_{NP,2} & 0 & 0 \\
0 & \varphi_{1,1} & 0 & 0 & \varphi_{2,1} & 0 & \ddots & 0 & \varphi_{NP,1} & 0 \\
0 & \varphi_{1,2} & 0 & 0 & \varphi_{2,2} & 0 & \ddots & 0 & \varphi_{NP,2} & 0 \\
0 & 0 & \varphi_{1,1} & 0 & 0 & \varphi_{2,1} & \ddots & 0 & 0 & \varphi_{NP,1} \\
0 & 0 & \varphi_{1,2} & 0 & 0 & \varphi_{2,2} & \ddots & 0 & 0 & \varphi_{NP,2} \\
\varphi_{1,11} & 0 & 0 & \varphi_{2,11} & 0 & 0 & \ddots & \varphi_{NP,11} & 0 & 0 \\
\varphi_{1,12} & 0 & 0 & \varphi_{2,12} & 0 & 0 & \ddots & \varphi_{NP,12} & 0 & 0 \\
\varphi_{1,22} & 0 & 0 & \varphi_{2,22} & 0 & 0 & \ddots & \varphi_{NP,22} & 0 & 0 \\
0 & \varphi_{1,11} & 0 & 0 & \varphi_{2,11} & 0 & \ddots & 0 & \varphi_{NP,11} & 0 \\
0 & \varphi_{1,12} & 0 & 0 & \varphi_{2,12} & 0 & \ddots & 0 & \varphi_{NP,12} & 0 \\
0 & \varphi_{1,22} & 0 & 0 & \varphi_{2,22} & 0 & \ddots & 0 & \varphi_{NP,22} & 0 \\
0 & & \varphi_{1,11} & 0 & 0 & \varphi_{2,11} & \ddots & 0 & 0 & \varphi_{NP,11} \\
0 & & \varphi_{1,12} & 0 & 0 & \varphi_{2,12} & \ddots & 0 & 0 & \varphi_{NP,12} \\
0 & & \varphi_{1,22} & 0 & 0 & \varphi_{2,22} & \ddots & 0 & 0 & \varphi_{NP,22}
\end{bmatrix} \tag{4.138}
$$

The outward normal gradient is calculated as

$$
\nabla_N \Delta \mathbf{u}_{n+1} = N_1 \frac{\partial \Delta \mathbf{u}}{\partial X_1} + N_2 \frac{\partial \Delta \mathbf{u}}{\partial X_2} = (N_1 \mathbf{\Phi}_1 + N_2 \mathbf{\Phi}_2)\cdot \Delta \hat{\mathbf{U}}^e_{n+1} \tag{4.139}
$$

Substituting Eqs. (4.133), (4.136), and (4.139) into (4.132), we have

$$
\int_{\Omega} [(\mathbf{L}\cdot\mathbf{\Phi})^T \cdot \mathbf{M} \cdot (\mathbf{L}\cdot\mathbf{\Phi})] \Delta \hat{\mathbf{U}}^e_{n+1} dV = \int_{\partial\Omega} (\mathbf{\Phi})^T \cdot \mathbf{t}_0^P dS + \int_{\partial\Omega} ((N_1 \mathbf{\Phi}_1 + N_2 \mathbf{\Phi}_2))^T \cdot \mathbf{t}_0^Q dS - \int_{\Omega} (\mathbf{L}\cdot\mathbf{\Phi})^T \cdot \boldsymbol{\sigma}_n dV \tag{4.140}
$$

Thus, we can obtain the incremental system equation

$$
\mathbf{K}_{n+1} \Delta \hat{\mathbf{U}}_{n+1} = \mathbf{f}_{n+1} \tag{4.141}
$$

where $\Delta \hat{\mathbf{U}}_{n+1}$ contains the nodal parameters of all of the mesh-free nodes. The global stiffness matrix is

$$
\mathbf{K}_{n+1} = \int_{\Omega} [(\mathbf{L}\cdot\mathbf{\Phi})^T \cdot \mathbf{M} \cdot (\mathbf{L}\cdot\mathbf{\Phi})] dV \tag{4.142}
$$

The nonequilibrium force vector is

$$\mathbf{f}_{n+1} = \mathbf{f}_{n+1}^{\text{ext}} - \mathbf{f}_{n}^{\text{int}} \tag{4.143}$$

$$\mathbf{f}_{n+1}^{\text{ext}} = \int_{\partial\Omega} (\boldsymbol{\Phi})^T \cdot \mathbf{t}_0^P dS + \int_{\partial\Omega} ((N_1 \boldsymbol{\Phi}_1 + N_2 \boldsymbol{\Phi}_2))^T \cdot \mathbf{t}_0^Q dS \tag{4.144}$$

$$\mathbf{f}_{n}^{\text{int}} = \int_{\Omega} (\mathbf{L} \cdot \boldsymbol{\Phi})^T \cdot \boldsymbol{\sigma}_n dV \tag{4.145}$$

Eq. (4.144) expresses the external forces. The stable state is obtained by iteratively solving Eq. (4.141) until **f** reaches zero.

4.5.6 INTEGRATION SCHEME FOR A DISCRETE EQUATION

The evaluation of integrals in Eq. (4.142) requires an appropriate integral technique. In mesh-free methods, the evaluation of these integrations is a key point. Several integrations have been proposed in the literature.

One approach is to use the background mesh-free or background cell for quadrature integration, which is used in most mesh-free methods. In this approach, the domain is divided into a series of background cells, and the quadrature rule is used in each. Different from the FEM, the background cells are used only to implement the integration, and they need not be compatible with the mesh-free nodes. However, this approach is not very stable or accurate. The setup of the background cells and the distribution of the quadrature points at the global level do not ensure sufficient quadrature points in the compact support, which is crucial to the accuracy of the integration. It has been found that if quadrature integration is performed using background cells whose corners coincide with the nodes, then significant integration errors may occur in irregular cells. At the same time, if the background cells do not match the compact support of the mesh-free interpolations, then considerable integration errors may also arise. Nevertheless, this method is still widely used due to its simplicity both in theory and implementation. To alleviate integration error, one can use the integration cells that coincide with the nodal arrangements and apply a higher-order Gaussian quadrature, although this may result in significantly increased computational time. For example, 2×2 quadrature points can achieve a high level of accuracy, but the mesh-free method may require 4×4 quadrature points.

To avoid the "ghost" of the background cell, a nodal integration procedure was developed by Beissel and Belytschko [125]. This approach added a stabilization term consisting of the square of the residual of the equilibrium equation in the potential energy functional, and the resulting equations were integrated only at the nodes. However, introducing this additional term in the potential energy sacrifices variational consistency, and hence compromises the accuracy of the formulation.

Another development is the local support integration that was proposed by Li and Liu [126]. Due to the compact support of the mesh-free shape functions, effective integration should be performed for the common parts of the supports for the associated nodes. If each node takes on the same support shape, then we can fix the same quadrature rule for each support and integrate the weak form locally from one compact to another. By carrying out integration on the compact support, we are free from the background cell. One difficulty associated with local support integration is the integration of the compact that is cut by the boundary. A particular technique is needed for integration on these cut compacts because they are not regular. Moreover, in this approach, we still need to set up the quadrature points in the compact support.

To avoid the quadrature points completely, Chen et al. [84] put forward a stabilized conforming nodal integration method for the mesh-free Galerkin method. A smoothing stabilization technique was used to overcome the numerical instabilities occurring in the nodal integration.

From the above review, we can see that there are several integration strategies that can be used in the mesh-free method. However, in the present research, background quadrature meshes are used for integration due to their simplicity and easy implementation. The mesh-free nodes are collocated uniformly along the axial and circumferential directions, and the background cells are chosen to coincide with the nodal arrangements, i.e., each integral cell is a rectangle. The Gaussian quadrature rule is used for each cell. In the practical calculation, the Gaussian points can be globally marked. To avoid repeating the calculation for each Gaussian point, the information that includes the nodes whose supports cover the Gaussian points and the corresponding shape functions and their first- and second-order derivatives is calculated and deposited in advance. Assuming that the total number of Gaussian points for the entire problem domain is NQ, we can evaluate the stiffness matrix in Eq. (4.142) as follows.

$$\mathbf{K}_{n+1} = \int_{\Omega} [(\mathbf{L}\cdot\mathbf{\Phi})^T \cdot \mathbf{M}\cdot(\mathbf{L}\cdot\mathbf{\Phi})]dV = \sum_{q=1}^{NQ} w_q[(\mathbf{L}\cdot\mathbf{\Phi})^T \cdot \mathbf{M}\cdot(\mathbf{L}\cdot\mathbf{\Phi})]_q \tag{4.146}$$

where w_q is the Gaussian weight. For each Gaussian point, we can obtain the substiffness matrix

$$\mathbf{K}^q_{n+1} = w_q[(\mathbf{L}\cdot\mathbf{\Phi})^T \cdot \mathbf{M}\cdot(\mathbf{L}\cdot\mathbf{\Phi})]_q \tag{4.147}$$

Global stiffness can be obtained by assembling all of these substiffness matrices.

4.5.7 ENFORCEMENT OF ESSENTIAL BOUNDARY CONDITIONS

In comparison with the FEM, the mesh-free method has an obvious difficulty: the imposition of essential boundary conditions. Similar to most of the mesh-free shape functions, the MLS shape functions lack the Kronecker delta property, i.e., $\phi_I(X_J) \neq \delta_{IJ}$. Thus, the approximation that is generated by the MLS method does not pass through the nodal parameter values:

$$\psi^h(\mathbf{X}_I) = \sum_{iJ} \varphi_J(\mathbf{X}_I)\psi_J \neq \psi_I \tag{4.148}$$

where I and J denote the I-th and J-th mesh-free nodes, respectively. Therefore, the essential boundary conditions cannot be treated directly in mesh-free methods as they can in the FEM.

Some techniques have been proposed to impose the essential boundary conditions in mesh-free computations. Belytschko et al. [69] proposed the Lagrange multiplier method, in which not only the discrete field variables, but also the Lagrange multipliers, must be solved. Therefore, a separate set of interpolations for the Lagrange multipliers is required, which results in an increasing problem size. The introduction of the Lagrange multipliers to the method also leads to the singularity of the stiffness matrix, and thus a special solver is needed. Krongauz [127] introduced the EFG method coupled with the FEM method, in which finite elements are placed along the essential boundary so that the combined shape functions exactly satisfy the Kronecker delta property along the boundary. Ren and Liew [128] proposed the nodal interpolation method (NIM) and the direct imposition method (DIM). In the NIM, the polynomial interpolations are constructed near the essential boundary, and the combination of these interpolants with the mesh-free shape function allows the essential boundary conditions to be satisfied along the boundary. In the DIM, after a mathematical treatment is introduced to the resulting system matrix equations, the essential boundary conditions can be imposed directly as in the FEM. Chen et al. [85,91] developed a full transformation method, in which a global transformation matrix is introduced to transform the unknown variable vector and the stiffness matrix into those in the real deformation space. This method can achieve good efficiency and is very suitable to vibration problems. However, it is quite expensive. For nonlinear problems in particular, the repeated operations of the matrix multiplication requires a large amount of computational time. Zhu and Atluri [114] employ the penalty function method to impose the essential boundary conditions. The process in this method is similar to that in the FEM except that the mesh-free approximation at the boundary points is computed in advance. Other treatments of the essential boundary conditions include the Gram-Schmidt orthogonalization method [129], the singular weight function [130], and the development of the modified variational principle [115].

The full transformation method is found to be very accurate to impose the essential boundary conditions. However, the frequent matrix transformation often requires a large amount of computational resources for the present nonlinear problem. Therefore, the penalty function method was mainly used in preparing this thesis.

Assuming that the i-th displacement is prescribed at a boundary node Γ_K, this given displacement can be interpolated with the MLS approximation as follows.

$$\Delta\bar{u}_{iK} = \sum_{I=1,NP} \varphi_I(\mathbf{X}_{\Gamma_K})\Delta\hat{u}_{iI}, \quad at\ \Gamma_K \tag{4.149}$$

where Γ_K denotes a boundary node at which the i-th displacement is given, $\Delta\bar{u}_{iK}$ is the prescribed displacement increment, and $\Delta\hat{u}_{iI}$ is the nodal parameter.

Eq. (4.149) is performed at all of the boundary nodes at which the essential boundary conditions are prescribed, and all of these equations can be assembled as

$$\Delta\bar{\mathbf{u}} = \mathbf{\Psi}\Delta\hat{\mathbf{U}} \tag{4.150}$$

$$\mathbf{\Psi} = \begin{bmatrix} \varphi_1(\mathbf{X}_{\Gamma_1}) & \varphi_2(\mathbf{X}_{\Gamma_1}) & \cdots & \varphi_{NP}(\mathbf{X}_{\Gamma_1}) \\ \varphi_1(\mathbf{X}_{\Gamma_2}) & \varphi_2(\mathbf{X}_{\Gamma_2}) & \cdots & \varphi_{NP}(\mathbf{X}_{\Gamma_2}) \\ \vdots & \vdots & \ddots & \vdots \\ \varphi_1(\mathbf{X}_{\Gamma_K}) & \varphi_2(\mathbf{X}_{\Gamma_K}) & \cdots & \varphi_{NP}(\mathbf{X}_{\Gamma_K}) \\ \vdots & \vdots & \ddots & \vdots \end{bmatrix} \tag{4.151}$$

where $\varphi_I(\mathbf{X}_{\Gamma_K})$ is determined by the index i of the prescribed displacement $\Delta\bar{u}_{iK}$ at the boundary Γ_K, and it takes $(\phi_I(\mathbf{X}_{\Gamma_K}), 0,\ 0)$ (for $i = 1$), $(0, \phi_I(\mathbf{X}_{\Gamma_K}), 0)$ (for $i = 2$) or $(0,\ 0, \phi_I(\mathbf{X}_{\Gamma_K}))$ (for $i = 3$).

The weak equilibrium equation is

$$\begin{aligned}\int_\Omega (\delta\Delta\mathbf{grad}_{n+1})^T\cdot\mathbf{M}\cdot\Delta\mathbf{grad}_{n+1}dV &= \int_{\partial\Omega}(\delta\Delta\mathbf{u}_{n+1})^T\cdot\mathbf{t}_0^P dS + \int_{\partial\Omega}(\nabla_N\delta\Delta\mathbf{u}_{n+1})^T\cdot\mathbf{t}_0^Q dS \\ &\quad - \int_\Omega(\delta\Delta\mathbf{grad}_{n+1})^T\cdot\boldsymbol{\sigma}_n dV + \sum\nolimits_{\Gamma_K}\alpha(\delta\Delta\mathbf{u})^T(\Delta\mathbf{u}-\Delta\bar{\mathbf{u}})\end{aligned} \tag{4.152}$$

where α is the penalty number.

Using Eq. (4.150), the discretized equation can be obtained as

$$\{\mathbf{K}_{n+1} + \sum_{\Gamma_K}[\mathbf{\Psi}^T\alpha\mathbf{\Psi}]\}\Delta\hat{\mathbf{U}}_{n+1} = \mathbf{f}_{n+1} + \sum_{\Gamma_K}[\mathbf{\Psi}^T\alpha\Delta\bar{\mathbf{u}}] \tag{4.153}$$

The displacement in the first step can be precisely calculated, while that in the second step is accomplished by using the MK interpolation:

$$\hat{\mathbf{u}} = \hat{\mathbf{u}}(\hat{u}_1, \hat{u}_2, \hat{u}_3)^T = \sum_{i=1}^{n} \phi_i \mathbf{u}_i = \mathbf{\Phi U} \tag{4.154}$$

where the matrix $\mathbf{K}_{n+1}$ and the vectors $\Delta\hat{\mathbf{U}}_{n+1}$ and $\mathbf{f}$ are of the same form as Eq. (4.141).

4.5.8 STABILITY OF THE ALGORITHM

The stable state of the structures is determined by iteratively solving the incremental equation (4.141) with the nonequilibrium force and the stiffness matrix at each iterative step. The present algorithm is actually a standard Newton–Raphson method [131], which is faster than gradient-based methods, such as the conjugate gradient method [132] in which only the first-order derivative of the potential energy is required. However, the stiffness matrix $\mathbf{K}_{n+1}$ may not be positive definite in some cases, such as when buckling and material softening occur; thus, the iterative solution does not converge to the minimum point of the potential energy. A simple way to remedy this problem is to replace $\mathbf{K}_{n+1}$ with $\mathbf{K}_{n+1} + \beta\mathbf{I}$, where $\mathbf{I}$ is the identity matrix, and β is a positive number. The repeated replacement can ensure that the solution converges to the real one. The stiffness matrix $\mathbf{K}_{n+1}$ generally becomes positive definite after a few cycles of replacement, and the standard Newton–Raphson method can then be resumed. This modification has been used in the atomic-scale finite element [133,134].

The choice of β should ensure that $\mathbf{K}_{n+1} + \beta\mathbf{I}$ is positive definite. The ideal value of β is a positive number that is slightly larger than the magnitude of the most negative eigenvalue of $\mathbf{K}_{n+1}$ [131] because the larger the value of β, the slower the convergence of the solutions. Therefore, to achieve a good convergence rate, we can first calculate the eigenvalue of $\mathbf{K}_{n+1}$ at each iterative step. However, the calculation of this eigenvalue consumes additional computational time. Therefore, for small-sized structures, the value of β can be chosen by calculating the eigenvalue of $\mathbf{K}_{n+1}$. For large-scale structures, it can be determined by frequent attempts.

4.5.9 PROCEDURES FOR EQUILIBRIUM SOLUTION

Based on the above illustration, the solution procedure for a loading step can be summarized as follows (the MLS shape functions and their first- and second-order derivatives are calculated and deposited for each Gaussian point prior to the first loading step).

1. Compute the external force vector $\mathbf{f}_{n+1}^{\text{ext}}$ if the external force is prescribed on the boundary.
2. Compute the internal force vector $\mathbf{f}_n^{\text{int}}$ based on the results of the above iteration step.
3. Compute the nonequilibrium force $\mathbf{f}_{n+1} = \mathbf{f}_{n+1}^{\text{ext}} - \mathbf{f}_n^{\text{int}}$.
4. Compute the stiffness matrix.
   a. Compute $\mathbf{K}_{n+1}^q$ at each Gaussian point.
      i. Calculate $\mathbf{F}$ and $\mathbf{G}$ using the MLS approximation.
      ii. Optimize the energy of the representative cell.
      iii. Calculate $\mathbf{P}$, $\mathbf{Q}$, and $\mathbf{M}$.
   b. Assemble the global stiffness matrix $\mathbf{K}_{n+1}$.

5. Enforce the essential boundary conditions.
   a. The first iteration step?
      If yes, then displacement increment $\Delta\bar{\mathbf{u}}$ is applied.
      If no, then the displacement increment is set as $\Delta\bar{\mathbf{u}} = 0$.
   b. Compute $\boldsymbol{\Psi}$.
   c. Replace $\mathbf{K}_{n+1}$ with $\mathbf{K}_{n+1} + \sum_{\Gamma_K}[\boldsymbol{\Psi}^T \alpha \boldsymbol{\Psi}]$.
   d. Replace $\mathbf{f}_{n+1}$ with $\mathbf{f}_{n+1} + \sum_{\Gamma_K}[\boldsymbol{\Psi}^T \alpha \Delta\bar{\mathbf{u}}]$.
6. Solve the equation system $\mathbf{K}_{n+1}\Delta\hat{\mathbf{U}}_{n+1} = \mathbf{f}_{n+1}$.
   $\mathbf{K}_{n+1}$ is positive definite?
   a. If yes, then continue.
   b. If no, then modify the stiffness matrix as $\mathbf{K}_{n+1} \rightarrow \mathbf{K}_{n+1} + \beta\mathbf{I}$.
7. Check if the convergence criterion is satisfied.
   a. If yes, then go to the next loading step.
   b. If no, then go to (2).

4.5.10 VALIDATION STUDIES

The above computational strategy has been written in a standard code using the Fortran language. In the calculation, we use the internal routine DLSASF of Fortran software Powerstation 4.0 to solve the linear system of equations (i.e., Eqs. 4.141 and 4.153). In this section, two simple examples are provided to illustrate the validity of the developed mesh-free computational scheme. The mesh-free nodes are uniformly collocated along the circumferential and axial directions, and the background integral cells are in accordance with the nodal arrangements. In the future discussion, the nodal collation $n \times m$ means that the n nodes are uniform along the circumferential direction and the m nodes are uniform along the axial direction. Here, we mainly check the convergence of the method and discuss the dependence of the solutions on the DOI (the scaling factor $D_{\max}$).

4.5.11 UNIFORM TENSION

A tension test is first carried out. An (8, 8) SWCNT is considered, and it has 36 hexagonal cells along the axial direction. The total atom number is 1168, and the original length is 9.04 nm. In the numerical simulation, one end plane is fixed, and the tube is loaded by giving another end plane an axial movement at 0.1 nm per loading step. A total 10 loading times are applied during the test.

Different numbers of nodes and different values of $D_{\max}$ are used in the tests. First, let us consider the convergence of solutions. The cases of $D_{\max} = 2$ and $D_{\max} = 3$ are calculated. Shown in Tables 4.4 and 4.5 are the strain energy and axial stress when 10% axial strain is applied. Here, the axial stress is calculated as the sum of the nodal force on the end plane divided by the area of the cross-section that is evaluated as $2\pi Rt$ (R is the radius, and $t = 0.334$ nm is the thickness of the tube). With the increasing number of nodes, an obvious convergence can be seen from Tables 4.4 and 4.5. To reveal the variation of the system energy versus the axial strain, the energy–strain curve is first plotted for the case of 30×60 nodes with the use of $D_{\max} = 2$ along with the result that is obtained with the atomic simulation (see Fig. 4.23). The curves for the different numbers of nodes are so close that they cannot be clearly displayed in the same figure. A separate figure is provided to show the energy error in comparison with that obtained with the atomic simulation case

Table 4.4 Strain Energy and Axial Stress When 10% Axial Strain Is Applied. D_{max} Is Fixed as 2.0

Node Numbers	Energy (eV)	Stress (GPa)
6 × 6	213.5231	67.7021
10 × 10	211.7653	67.3328
20 × 20	210.7685	67.1753
20 × 40	210.2493	67.1385
30 × 30	210.4371	67.1474
30 × 60	210.0487	67.1273

Table 4.5 Strain Energy and Axial Stress When 10% Axial Strain Is Applied

Node Numbers	Energy (eV)	Stress (GPa)
6 × 6	211.7094	67.2891
10 × 10	210.9887	67.1708
20 × 20	210.3914	67.1273
20 × 40	209.9847	67.1149
30 × 30	210.1466	67.1231
30 × 60	209.8760	67.1115

D_{max} *is chosen as 3.0.*

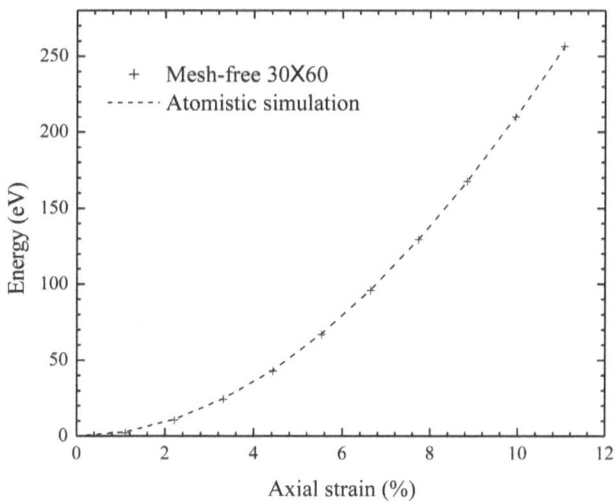

FIGURE 4.23

A comparison of the mesh-free results with those obtained by the atomic simulation. The mesh-free results are obtained with 30 × 60 nodes.

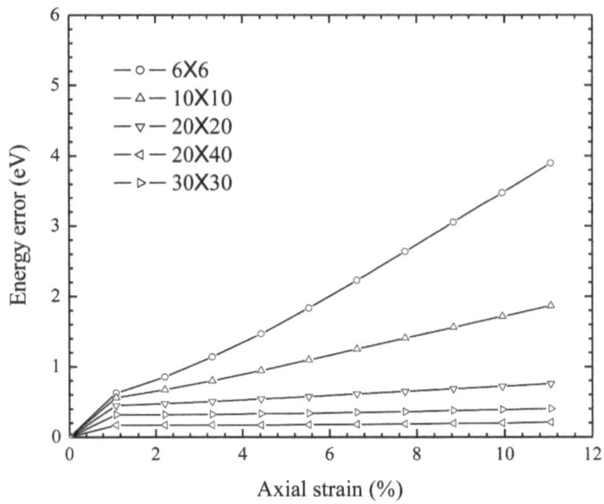

FIGURE 4.24

The energy error in comparing the mesh-free results with those of the atomic simulation for different numbers of nodes.

Table 4.6 Dependence of the Strain Energy and Axial Stress on the DOI

	10 × 10		20 × 40	
D_{max}	**Energy (eV)**	**Stress (GPa)**	**Energy (eV)**	**Stress (GPa)**
2.0	211.7653	67.3328	210.2493	67.1385
2.5	211.3532	67.2330	210.1631	67.1233
3.0	210.9887	67.1708	209.9847	67.1149
3.5	210.7740	67.1138	209.8499	67.0918

The results are obtained at the axial strain of 10%.

(see Fig. 4.24). Here, the energy error refers to the calculated energy with the present nodes minus that obtained with the atomic simulation. From Tables 4.4 and 4.5 and Figs. 4.23 and 4.24, it can be seen that the results are very close to those of the atomic simulation even when 10 × 10 nodes are used, which indicates that continuum simulation can efficiently reduce the degrees of freedom of the problem. This is an important advantage of the continuum method.

As illustrated before, the size of the compact support (DOI) generally has an effect on the computational results. The choice of the DOI depends on the properties of the practical problems. According to the literature, the efficient range of the scaling factor D_{max} is typically 2.0–4.0. A series of computations have been performed to test the effects of D_{max}. The results reveal that the different values of D_{max} in the range of 2.0–4.0 do not provide significant differences in the final results for the present tension test. This phenomenon can be seen from Table 4.6, which shows the dependence of the system energy and axial stress on D_{max}.

4.5.12 BENDING TEST

A bending test is carried out by incrementally rotating the two end planes of the SWCNTs in opposite directions (Fig. 4.25). During the rotation of the two end planes, the axial movement of one end is prohibited, but that of the other is unconstrained. A (15, 0) SWCNT is considered. It has 30 hexagonal cells along the axial direction, and the original length is 12.87 nm. The rotation angle is 1 degree per loading step, and a total of 21 steps are applied. The purpose of the present computations is to test the convergence of the solutions and the effect of the scaling factor D_{max}.

First, D_{max} is fixed as 2.0. Fig. 4.26 shows the variation of the system energy versus the bending angle for the cases of 6×6, 10×10, 20×20, and 30×30 nodes. Good convergence can be seen. To check the effect of the DOI, detailed computations are carried out for $D_{max} = 2.0$, 2.5, 3.0, and

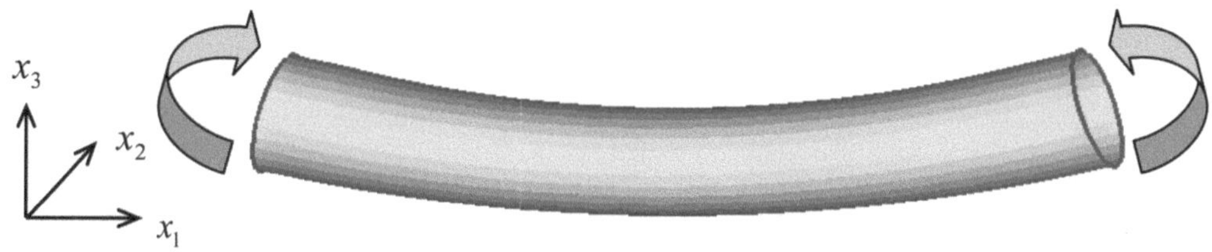

FIGURE 4.25

A SWCNT is bent by rotating the two end planes in opposite directions.

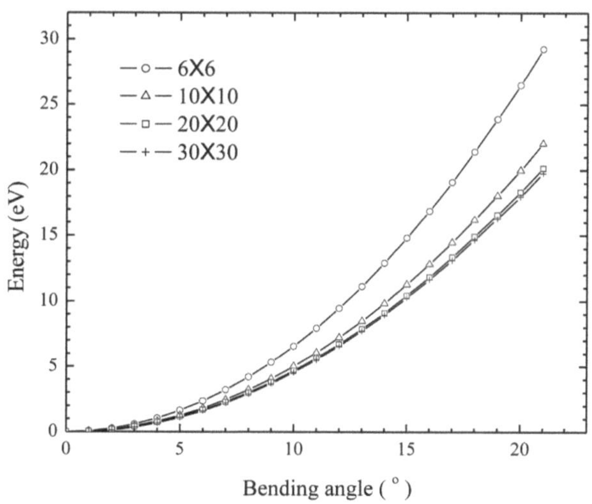

FIGURE 4.26

The strain energy versus the bending angle.

Table 4.7 The Dependence of the Strain Energy on the Scaling Factor and the Number of Nodes

Node Numbers	$D_{max} = 2.0$ (eV)	$D_{max} = 2.5$ (eV)	$D_{max} = 3.0$ (eV)	$D_{max} = 3.5$ (eV)
6×6	26.4863	19.2940	19.1104	19.0163
10×10	19.9848	18.1101	18.1191	17.9542
20×20	18.3417	17.9120	17.9631	17.8662
20×40	18.3100	17.8917	17.9441	17.8388
30×30	18.0450	17.8907	17.9241	17.8660
30×60	18.0212	17.8756	17.8947	17.8037

The results are obtained at a bending angle of 20 degrees.

3.5. Table 4.7 shows the dependence of the solution on D_{max} and the number of nodes. When fewer nodes are used, $D_{max} = 2.0$ has a bad result. However, when more nodes are used, D_{max} in the range of (2.0, 3.0) does not show very large differences.

4.6 BUCKLING AND POSTBUCKLING BEHAVIORS

4.6.1 HYDROSTATIC PRESSURE-INDUCED STRUCTURAL TRANSITIONS OF SWCNTs

In the previous study, we obtain a pressure–radial strain curve for SWCNTs through a semianalytical method. In fact, the structural transition appears when the pressure reaches the critical value. A number of high-pressure experiments have been carried out on bundles of SWCNTs [135–137], and they show pressure-induced structural transition in the range of 1–2 GPa. Recent experiments have also shown that pressure may induce transitions in the electrical and magneto-transport properties of SWCNT bundles [138], which correlates closely with pressure-induced structural shape transition. In parallel, extensive theoretical studies, ranging from first-principle calculations [139–141] and molecular dynamic simulation [142,143] to continuum mechanics modeling [144], have been performed by several groups to study the properties of both isolated single SWCNTs and bundles of SWCNTs under hydrostatic pressure. For an isolated single SWCNT, it has been shown that pressure induces a series of shape transitions, transforming its cross-sections from a circle to an elliptical and then a peanut shape [142,144]. In this section, we employ the mesh-free method to study the structural transition of an isolated single SWCNT.

As introduced before, the mesh-free formulation is based on the transformation from a planar sheet (2-D) to a spatial curved surface (3-D). When hydrostatic pressure is applied on an SWCNT, it uniformly deforms along the axial direction, and we need only consider the change of its cross-section. Thus, an SWCNT can be simplified as a circle, and its counterpart in the reference configuration is a straight line segment (see Fig. 4.27). In the simulation, this line segment (Fig. 4.27A) serves as the reference configuration, and the initial undeformed circle (Fig. 4.27B) can be formed by rigidly rolling up the line segment into a circle. The length of the line segment is $\lambda_2\Gamma$, and the cell structure in the reference configuration should be determined with the method described in

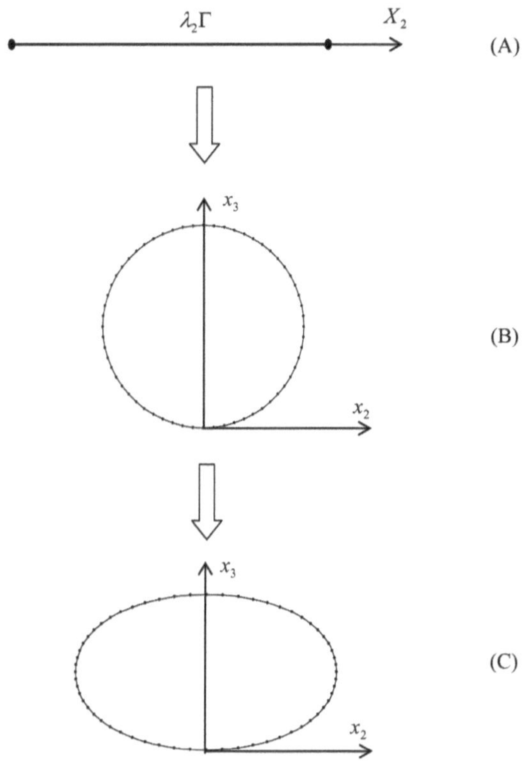

FIGURE 4.27

An SWCNT simplified as an elastic circle. It can be mapped from a line segment with the length $\lambda_2\Gamma$. The deformation from the reference to the current configuration is divided into two parts.

Section 4.1. The deformation from the reference configuration to the current configuration (Fig. 4.27C) can be divided into two parts: one is from the line segment to the circle, and the other is from the circle to the current configuration. The first part can be calculated exactly. The displacement of the current configuration relative to the initial circle can be completely expressed with u_2 and u_3, and we only need to consider the gradient components F_{22}, G_{222}, F_{32}, and G_{322} that can be calculated with u_2 and u_3. It should be noted that with the radial deformation of an SWCNT a slight axial strain takes place, and this axial strain can be described with λ_1. Moreover, a twisting angle may also occur, and this can be described with θ. Therefore, the energy of the representative cell is determined by F_{22}, G_{222}, F_{32}, G_{322}, λ_1, and θ, and the strain energy density is a function of F_{22}, G_{222}, F_{32}, G_{322}, λ_1, and θ:

$$\omega = \omega(F_{22}, G_{222}, F_{32}, G_{322}, \lambda_1, \theta) \tag{4.155}$$

The first- and second-order derivatives of ω with respect to F_{22}, G_{222}, F_{32}, G_{322}, λ_1, and θ can be calculated as described before.

The stable configurations of the system are identified with the minimization of the total energy, as follows.

$$E = \int_{B_0} \omega(F_{22}, G_{222}, F_{32}, G_{322}, \lambda_1, \theta)dS - \int_{B_0} u_N \cdot p dS \tag{4.156}$$

where p is the hydrostatic pressure applied on the outside surface of an SWCNT, and u_N is the normal displacement

$$u_N = u_2 \cos\varphi + u_3 \sin\varphi \tag{4.157}$$

where φ is the angle between the radius and the axis X_2.

The integral in Eq. (4.156) is performed along the circle, and E should be understood as the energy per unit length of SWCNTs.

In the mesh-free simulation, a series of nodes is collocated on the line segment (these can also be viewed as being on the undeformed circle). The displacement relative to the undeformed circle is approximated as

$$\begin{bmatrix} u_2 \\ u_3 \end{bmatrix} = \sum_{i=1}^{NP} \phi_i \begin{bmatrix} u_{2i} \\ u_{3i} \end{bmatrix} \tag{4.158}$$

where ϕ_i is the 1-D mesh-free shape function that is defined on the line segment, u_{2i} and u_{3i} are the nodal parameters, and NP is the number of nodes whose supporting domain covers the evaluated point. The deformation gradients are approximated as

$$\begin{aligned} F_{22} &= \overline{F}_{22} + \sum_{i=1}^{NP} \phi_{i,2} u_{2i}, \quad G_{222} = \overline{G}_{222} + \sum_{i=1}^{NP} \phi_{i,22} u_{2i} \\ F_{32} &= \overline{F}_{32} + \sum_{i=1}^{NP} \phi_{i,2} u_{3i}, \quad G_{322} = \overline{G}_{322} + \sum_{i=1}^{NP} \phi_{i,22} u_{3i} \end{aligned} \tag{4.159}$$

where $\overline{F}_{22}$, $\overline{G}_{222}$, $\overline{F}_{32}$, and $\overline{G}_{322}$ are the gradients of the undeformed circle and can be calculated exactly and $\phi_{i,2}$ and $\phi_{i,22}$ are the first- and second-order derivatives of ϕ_i with respect to X_2.

Assuming that the total N nodes are used, the total unknown vector can be written as

$$\mathbf{U} = (u_{21}, u_{31}, \ldots, u_{2i}, u_{3i}, \ldots, u_{2N}, u_{3N}, \lambda_1, \theta)^T \tag{4.160}$$

The nonequilibrium force and the stiffness matrix can be calculated as

$$\mathbf{f} = \frac{\partial E}{\partial \mathbf{U}}, \quad \mathbf{K} = \frac{\partial^2 E}{\partial \mathbf{U}^2} \tag{4.161}$$

The equilibrium configurations can be obtained by iteratively solving $\mathbf{K}\Delta\mathbf{U} = \mathbf{f}$ until $\mathbf{f}$ reaches zero. Moreover, around the structural transitions, the stiffness becomes nonpositive definite.

The numerical simulations are first carried out for a (10, 10) SWCNT. The computations reveal that the SWCNT displays the first structural transition when the hydrostatic pressure reaches a critical value. At this critical pressure, the cross-section transforms from a circle to an ellipse (see Fig. 4.28). The present computation reveals that 15 nodes can achieve good convergence prior to the structural transition. However, the critical pressure has a larger dependence on the node number. Given in Table 4.8 is the first critical pressure obtained for the different nodes when D_{max} is fixed as 3.0. This suggests that more than 50 nodes should be used along the line segment to obtain

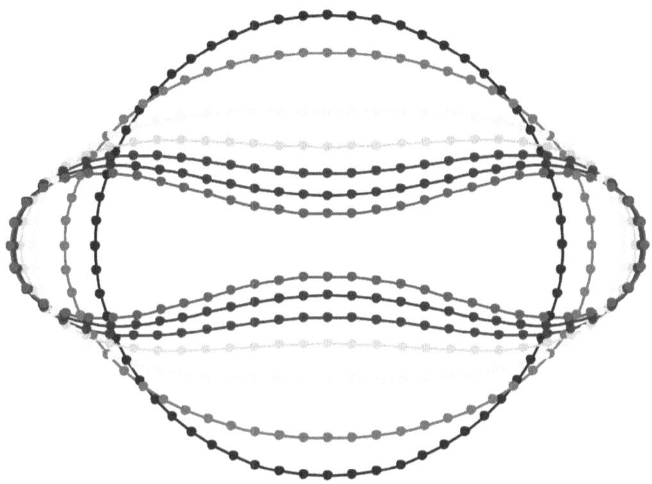

FIGURE 4.28

With increasing hydrostatic pressure, the shape of the cross-section transforms from a circle to an ellipse and then to a peanut shape [24].

Table 4.8 The First Critical Pressure Obtained With Different Nodes When D_{max} Is Fixed as 3.0

Number of Nodes	Critical Pressure (GPa)
20	1.968
30	1.424
40	1.248
50	1.174
60	1.168

an ideal critical pressure. Table 4.9 shows the dependence of the critical pressure on the scaling factor D_{max} when 60 nodes are used, which indicates that 2.5, 3.0, and 4.0 result in almost the same critical pressure, which confirms that a fewer number of nodes can achieve a good solution as discussed before.

Here, the long and short axes of the ellipse are used to describe the deformation. After the first structural transition, it is found that the length of the long axis increases with the increasing pressure, but the short axis reduces its length (see Figs. 4.28 and 4.29). At a pressure of 1.3 GPa, the second structural transition occurs, and the cross-section of the tube transforms from an elliptical to a peanut shape (see Fig. 4.28). When the pressure further increases, the short axis continues to reduce its length, but the length of the long axis remains almost unchanged. Fig. 4.28 shows the change of the cross-section at several pressure values. Fig. 4.29 shows the variation of the long and short axes with the pressure (prior to the first structural transition, the cross-section is a circle, and

Table 4.9 The Dependence of the First Critical Pressure on D_{max} When 60 Mesh-Free Nodes Are Used

D_{max}	Critical Pressure (GPa)
2.0	1.216
2.5	1.177
3.0	1.168
3.5	1.170

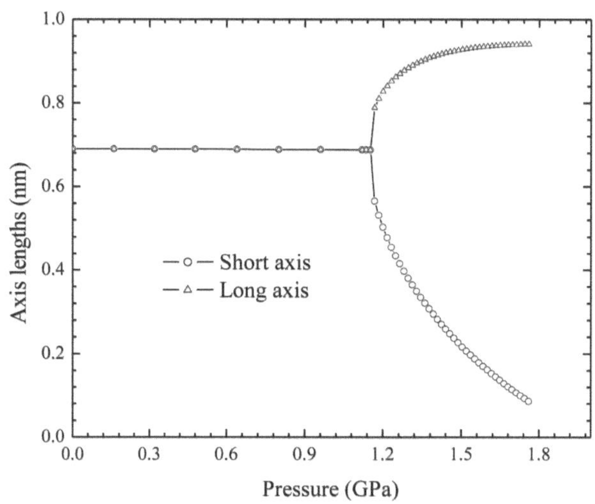

FIGURE 4.29

Variation of the lengths of the long and short axes with pressure for a (10, 10) SWCNT [24].

the two axes have the same lengths). To reveal the dependence of the critical pressure of the first structural transition on the chirality and radius of the CNTs, computations are carried out for (n, 0), (n, n), and ($3n$, $2n$) SWCNTs with different radii. Fig. 4.30 plots the variation of the first transition pressure versus the tube radius for (n, 0), (n, n), and ($3n$, $2n$) SWCNTs. It is found that the first transition pressure has almost no dependence on the chirality of the CNTs.

4.6.2 AXIAL BUCKLING OF SWCNTs

This section considers the mechanical response of SWCNTs under axial compression. The axial buckling of CNTs has been studied by many researchers. In continuum-based analysis, the attention is generally focused on a prediction of the critical strain. The present research presents a complete numerical simulation of buckling behavior.

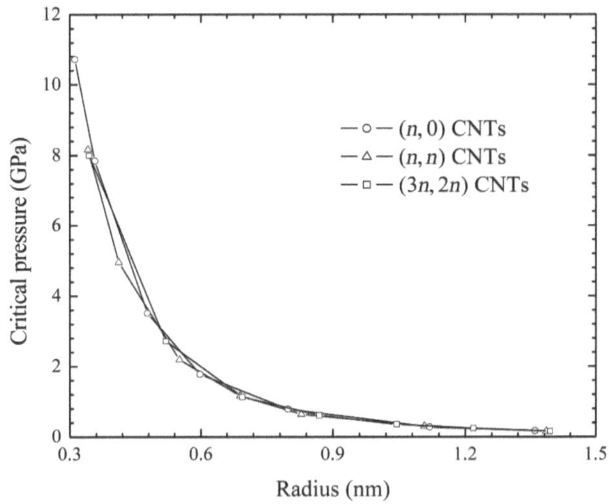

FIGURE 4.30

Critical pressure of the first structural transition versus the tube radius for $(n, 0)$, (n, n), and $(3n, 2n)$ SWCNTs [24].

In the compression simulation, one end of the SWCNT is completely fixed, and the in-plane displacement of the other end is prohibited. Axial compressive displacement is applied at the free end, the stable state is solved, and then the further compressive displacement is applied. In the present simulation, the penalty function method [114] is used to enforce the essential boundary condition.

Using the present mesh-free method, we have simulated the compression of an (18, 0) SWCNT. It has 20 hexagonal cells along the axis, and its initial length is 8.71 nm (this value is obtained by initializing the undeformed tube using the atomic simulation). The problem is treated as quasistatic. The length of the tube is reduced by 0.1 nm per loading step at the early stage and 0.01 nm per loading step near to the buckling. To test the effect of the number of mesh-free nodes on the simulation result, different numbers of nodes have been used in the computation. Similar to the numerical tests discussed before, even 10×10 nodes present a good simulation prior to buckling. However, the number of nodes has a large effect on the buckling strain. Fig. 4.31 plots the average energy per atom versus the compression ratio for the cases of 20×31 and 20×72 nodes, along with the results of the full atomic simulation. A distinct energy jump appears at all three curves in Fig. 4.31, which corresponds to the buckling of the CNT. Prior to buckling, the compression ratio–energy curve is a quadratic curve, but afterwards, it is almost a straight line. From Fig. 4.31, it can be seen that the calculated energy in the cases of 20×31 and 20×72 is very precise prior to buckling. However, the number of nodes has a larger effect on the compression ratio of the buckling. Buckling occurs at a smaller compression ratio when a fewer number of nodes are used, and the comparison ratio of buckling tends to that of the full atomic simulation with an increasing number of nodes. Moreover, our computation also reveals that the compression ratio of buckling hardly changes if we fix the number of nodes along the longitudinal direction but increase it along the circumferential direction. This phenomenon can be explained by the fact that the drastic deformation

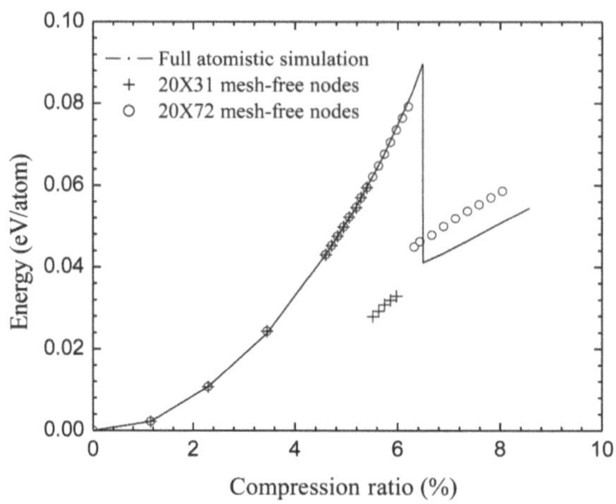

FIGURE 4.31

The average energy per atom versus the compression ratio obtained with the mesh-free method and atomic simulation [22].

is mainly along the axial direction for the compression test. Fig. 4.31 also shows that, even for the case of 20×72 nodes, the energy calculation becomes less accurate after buckling. This is unsurprising because the deformation becomes drastically inhomogeneous, and it is difficult for the continuum method to capture microscale energy changes.

The buckling deformation obtained with 20×72 nodes is shown in Fig. 4.32 along with that obtained with the full atomic simulation. The comparison shows that the present model can truly display the buckling pattern. Employing the mesh-free method, we also simulated the compression of the same SWCNT with the constitutive relations based on the standard Cauchy–Born rule, and the buckling deformation obtained is shown in Fig. 4.32C as a series of ridges scattered on the surface of the CNT. Obviously, this is a nonphysical buckling pattern, and it is the response of the structure in the case of no bending stiffness. The comparison shows that the model based on the standard Cauchy–Born rule does not describe the bending effect and thus cannot truly display the buckling behavior of CNTs. The present higher-order gradient theory considers the effect of the second deformation gradient, and thus the established model is more reasonable.

Based on the present atomistic-continuum simulation, the deformation of SWCNTs can be analyzed from the perspective of a continuum. Fig. 4.33 shows the contour plots of the axial and circumferential strains and the strain energy density (the energy per unit area) just after the buckling occurs. The axial and circumferential strains are calculated as

$$\begin{aligned} \varepsilon_a &= \sqrt{F^2(1,\ 1) + F^2(2, 1) + F^2(3,\ 1)} - 1 \\ \varepsilon_c &= \sqrt{F^2(1,\ 2) + F^2(2,\ 2) + F^2(3,\ 2)} - 1 \end{aligned} \tag{4.162}$$

From Fig. 4.33, it can be seen that the deformation and energy are highly concentrated on the local kinking region. This may explain the phenomenon: the compression ratio–energy curve is straight after buckling, but a quadratic curve prior to buckling. There is maximum axial tensile

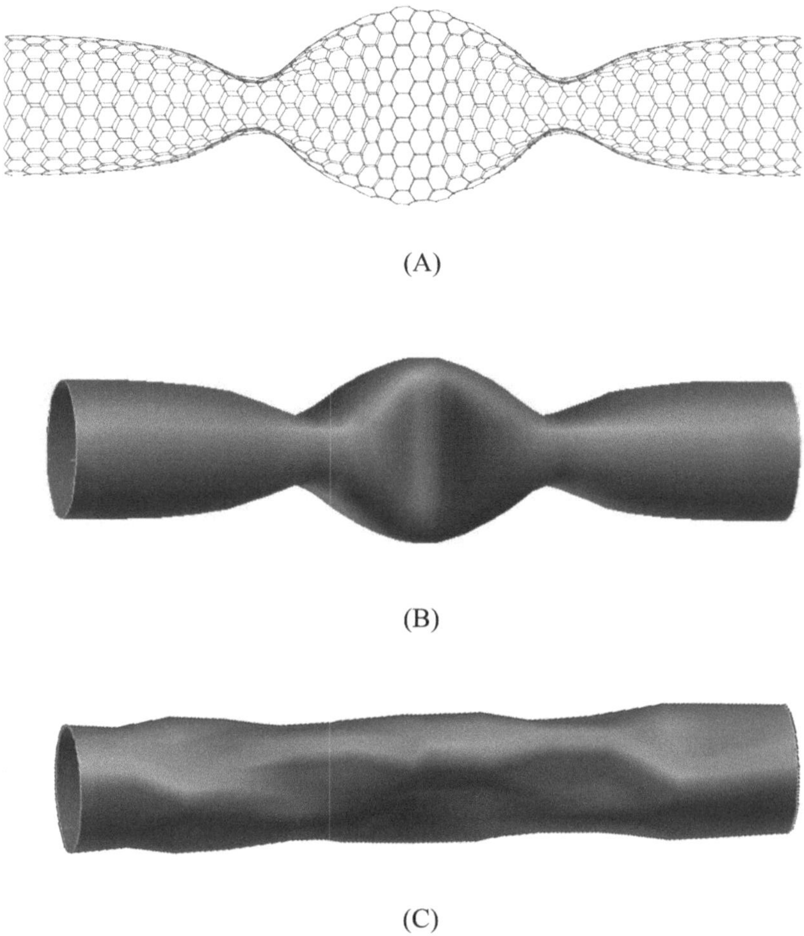

FIGURE 4.32

A comparison of the buckling deformations obtained with (A) atomic simulation; (B) the higher-order continuum model; and (C) the first-order continuum model [22].

strain at the convex regions, but the circumferential strain is negative. The distribution of the strain energy is similar to that of the axial strain, i.e., the maximum strain energy density is at the zone in which the axial tensile strain is largest.

4.6.3 TORSIONAL BUCKLING OF SWCNTs

The torsional collapse of CNTs has been found in experimental investigations [145,146] (see Fig. 4.34) and has been modeled by MD [147] and other simulation methods [17].

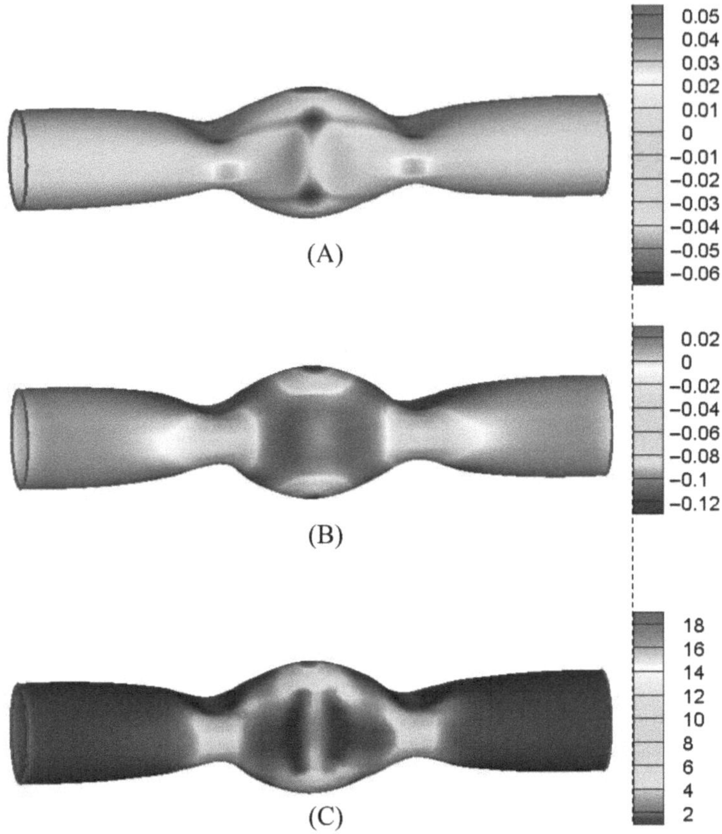

FIGURE 4.33

Contour figures on the surface of a compressed SWCNT: (A) axial strain; (B) circumferential strain; and (C) strain energy density.

The cross-section of a CNT remains a circle for the small twisting angle, but collapses when the twisting angle reaches a critical value.

In the modeling of twisting, the cross-sectional circular shape of the two ends of a CNT remains unchanged, and the twisting deformation is imposed by twisting the two ends in opposite directions without the prohibition of axial displacement.

A (10, 10) SWCNT is chosen for the twisting test. It has 47 hexagonal cells along the axis, and its initial length is 11.81 nm. The applied twisting angle is 1 degree per step. Plotted in Fig. 4.35 are the changes of the average energy per atom versus the twisting angle for the cases of 20×31 and 40×31 mesh-free nodes in a comparison with a full atomic simulation. When 40×31 nodes are used, the buckling angle is very close to that obtained by the full atomic simulation, whereas the buckling takes place at a larger twisting angle in the case of 20×31 nodes. Similar to the compression test, a fewer number of nodes can provide a good simulation for the homogeneous

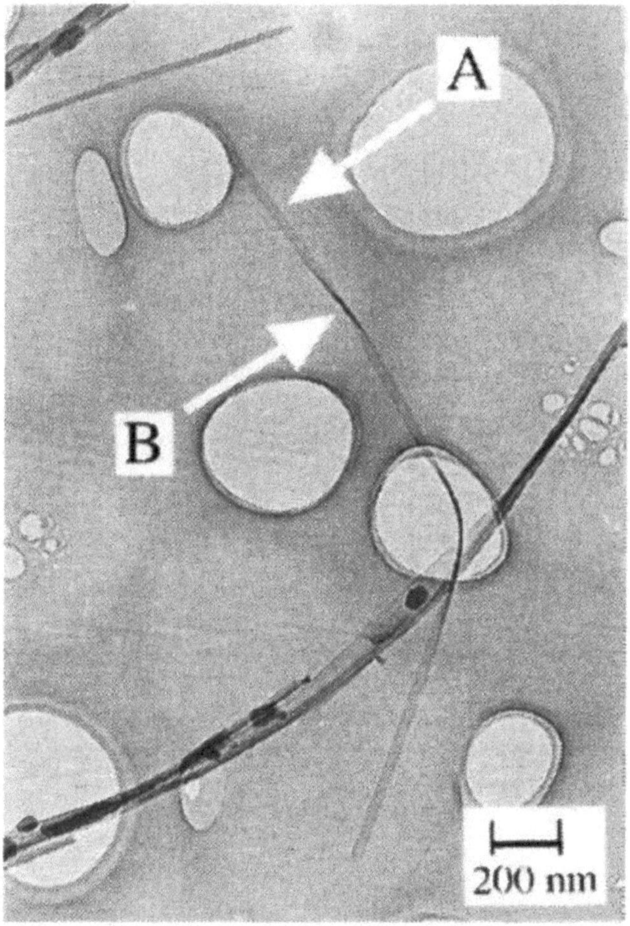

FIGURE 4.34

TEM image of a collapsed MWCNT with A and B marking the two twist regions [145].

deformation stage, but more nodes must be used to capture the critical twisting angle accurately. Moreover, the computations reveal that the critical twisting angle hardly changes if we fix the number of nodes along the circumferential direction but increase it along the longitudinal direction. This shows that the number of nodes along the circumferential direction has a large effect on the precision of the twisting simulation.

Fig. 4.36 shows a comparison of the twisting buckling patterns obtained with 40×31 nodes and the full atomic simulation, which demonstrates that such patterns can be truly displayed with the present model. Moreover, we also perform the torsion test with the constitutive model based on the standard Cauchy–Born rule, and the simulated buckling deformation is shown in Fig. 4.36C.

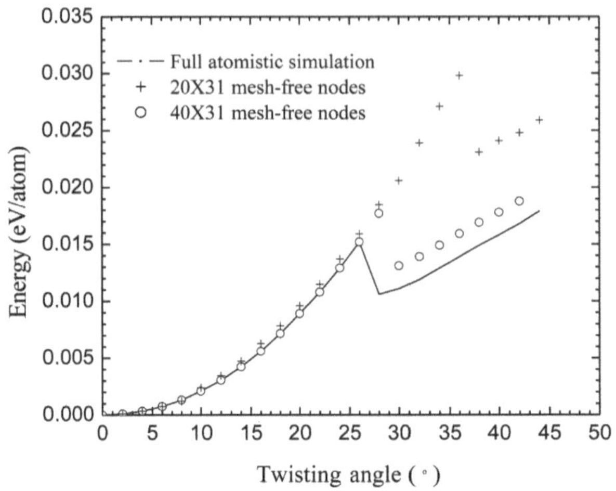

FIGURE 4.35

The average energy per atom versus the twisting angle obtained with the mesh-free method and the atomic simulation [22].

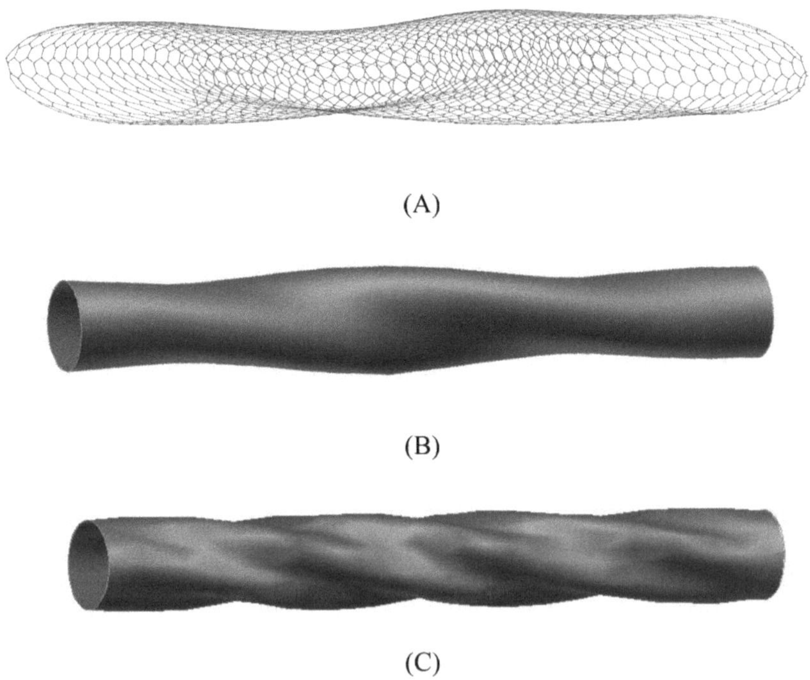

FIGURE 4.36

A comparison of twisting buckling deformations obtained with (A) atomic simulation; (B) the higher-order continuum model; and (C) the first-order continuum model [22].

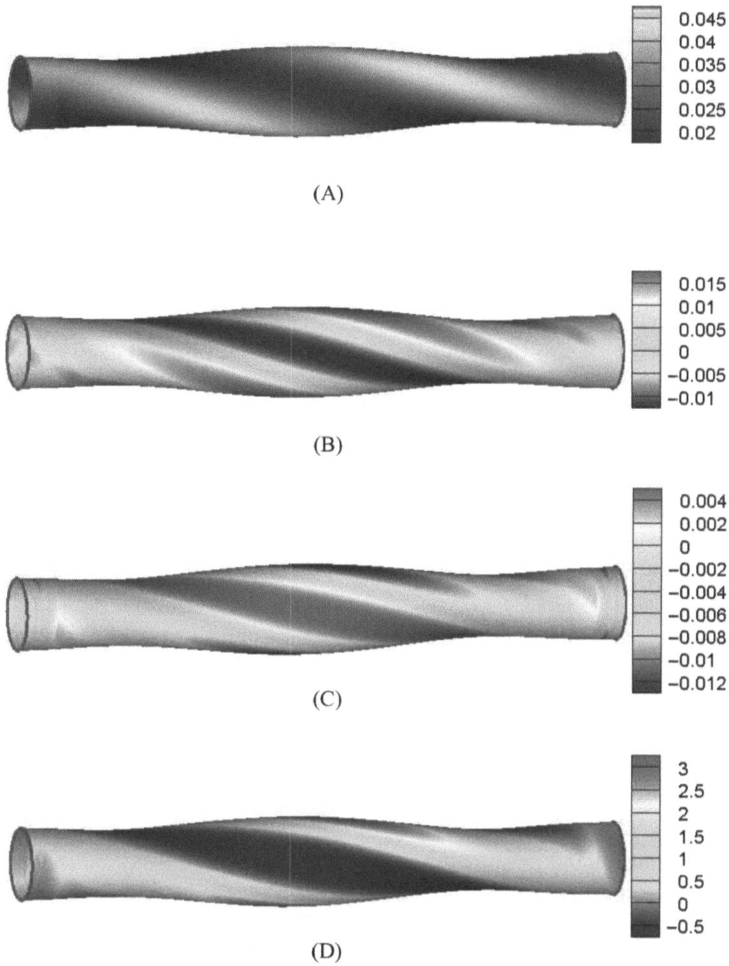

FIGURE 4.37

Contour figures of the surface of a twisted SWCNT: (A) the shear strain; (B) axial strain; (C) circumferential strain; and (D) strain energy density.

The deformation obtained is a nonphysical buckling pattern. This further demonstrates that the present model can accurately describe the constitutive response of CNTs, whereas the model based on the standard Cauchy–Born cannot.

The deformation of a twisted SWCNT is now analyzed from the perspective of a continuum. Fig. 4.37 shows the contour plots of the shear strain, axial and circumferential strains, and strain energy density at a twisting angle of 22 degrees. The axial and circumferential strains are calculated using Eqs. (4.162), and the shear strain is calculated as

$$\gamma = \frac{F(1,\ 1)F(1,\ 2) + F(2,\ 1)F(2,\ 2) + F(3,\ 1)F(3,\ 2)}{\sqrt{F^2(1,\ 1) + F^2(2,\ 1) + F^2(3,\ 1)}\sqrt{F^2(1,\ 1) + F^2(2,\ 1) + F^2(3,\ 1)}} \tag{4.163}$$

From Fig. 4.37, it can be seen that the distribution of the deformation and energy shows a series of twisted bands on the surface of the SWCNT. The large shear strain and axial tensile strain are distributed at the convex regions, and the large circumferential strain is distributed at the concave region. The distribution of the sreain energy is similar to that of the axial strain, i.e., the maximum strain energy density is at the zone in which the axial tensile strain is largest.

4.6.4 BUCKLING BEHAVIOR OF SWCNTs UPON BENDING

Due to the special structure of CNTs, bending buckling may be the type of buckling behavior that occurs most easily in CNTs in practice. In this section, the buckling behavior of SWCNTs upon bending is simulated and discussed in detail.

In most bending simulations of CNTs [147–149], the bending angle is imposed by incrementally rotating the two end planes in opposite directions. Both end sections remain circular and perpendicular to the deformed axis at each displacement increment. Because the axis is bent into a curve, the distance between the two end planes decreases with an increase in the rotating angle. The way to control this distance is not clearly specified in most of the literature. In [148], the researchers assume that an SWCNT takes on pure bending deformation, similarly to a beam under such loading conditions, and thus that the axis keeps a circular arc with its length being equal to the original length of the tube. The change in the distance between the two end sections is calculated exactly and imposed as the boundary condition. We can also treat this distance as an unknown and determine it by solving the equilibrium state of the tube. The detailed implementation is described as follows. At each loading step, the tube is bent by rotating the two end planes in opposite directions with respect to an axis that is perpendicular to the tube axis of the undeformed CNT and passing through its center. The rotating axis of one end plane is completely fixed, and the axial movement of the other end plane is not prohibited. Its axial position is treated as an unknown, and is determined with the minimization of the potential energy of the system along with the internal nodal unknowns. The above two methods of imposing bending deformation are respectively marked as the mode I and II loading methods. The latter method is mainly used in the present work, but the results of both methods are compared and discussed.

4.6.4.1 Bending buckling

A (15, 0) SWCNT is considered first. It has 18 hexagonal cells along the axis of the tube, and the original equilibrium length is 7.65 nm. The mode II loading method is used, and the two end planes of the tube are reversely rotated 1 degree per loading step at the early stage and 0.25 degree per loading step around the buckling. Fig. 4.38 plots the change in the total energy along with the result that is obtained with the atomic simulation. Fig. 4.38 indicates that the energy difference between the two methods becomes large with an increasing bending angle. Plotted in Fig. 4.39 is the change in the average energy per atom, which is equal to the total energy divided by the atom number. The atom number is 1080 for atomic simulation, whereas, in the mesh-free method, the total number of atoms should be understood from the continuum concept and is taken as S/V_0, where S is the surface area of the undeformed CNT, and $V_0 = 3\sqrt{3}r_0^2/4$ is the average area per atom (r_0 denotes the bond

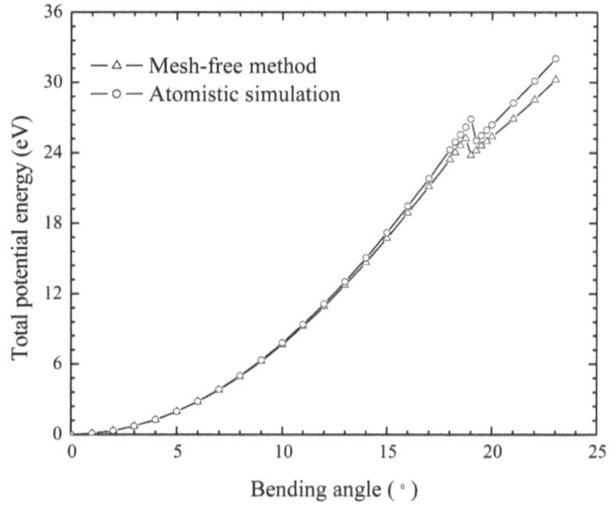

FIGURE 4.38

The variation of the total energy as a function of the bending angle for a (15, 0) SWCNT with a length of 7.65 nm [23].

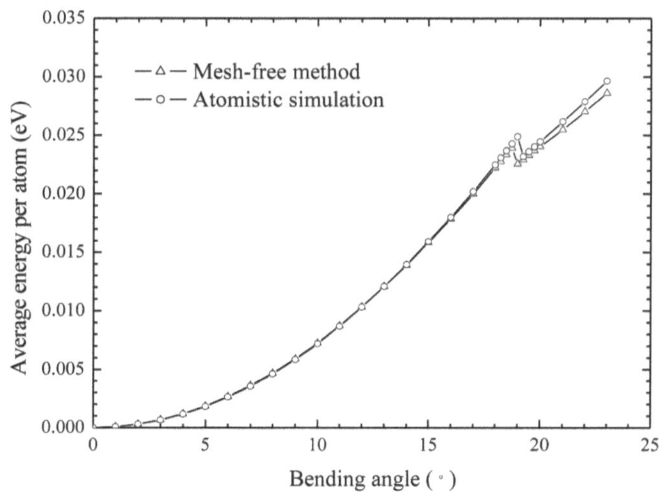

FIGURE 4.39

The variation of the average energy per atom as a function of the bending angle for a (15, 0) SWCNT with a length of 7.65 nm [23].

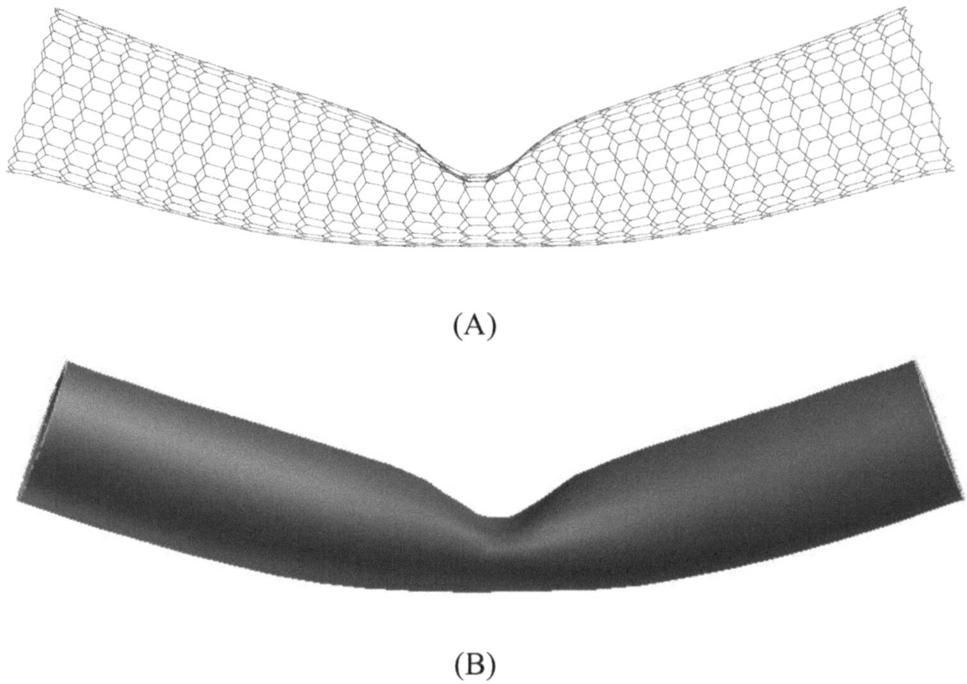

FIGURE 4.40

A comparison of the bending buckling deformations obtained with atomic simulation (A) and the mesh-free method (B) for a (15, 0) SWCNT with a length of 7.65 nm [23].

length). Fig. 4.39 reveals that the average energy per atom in the two methods agrees well. A distinct energy jump that corresponds to the SWCNT buckling appears at the energy curves of the two methods. The buckling bending angles in the atomic simulation and in the mesh-free method are 19.25 and 19 degrees, respectively, which are very close. Fig. 4.40 shows the buckling deformations for the two methods, and it can be seen that the buckling patterns are very similar.

We also perform the bending test on a longer (15, 0) CNT with a length of 30 hexagonal cells (12.87 nm). As plotted in Fig. 4.41, the change in the average bending energy shows good agreement with that in the full atomic simulation. The buckling angles in the atomic simulation and in the mesh-free method are 33.25 and 32.75 degrees, respectively. Fig. 4.42 provides a comparison of the buckling deformations of the two methods. Unlike in the short SWCNT, two symmetrical snap buckles are displayed in the long SWCNT.

The above examples show that the higher-order Cauchy–Born rule can precisely catch the constitutive response of CNTs and that the mesh-free method is an efficient numerical tool for modeling and predicting the mechanical behavior of CNTs. In [148], an SWCNT was viewed as a thin cylindrical shell with an effective thickness and an equivalent constant Young's modulus, and the FEM was used to model the bending buckling of SWCNTs. Their FEM simulations often overestimated the critical bending buckling curvature in comparison with their MD simulations. They

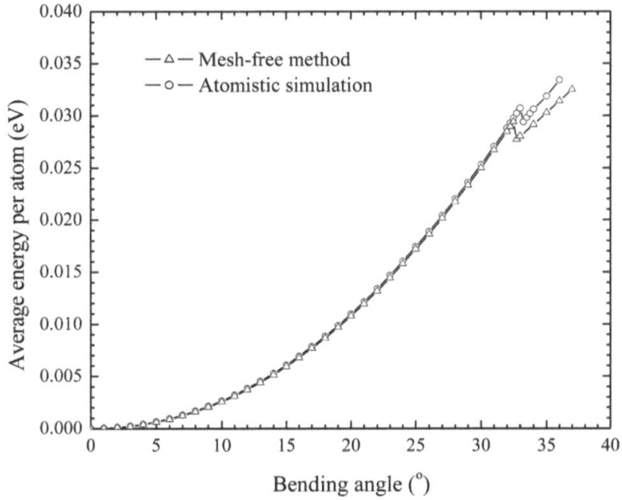

FIGURE 4.41

The variation of the average energy per atom as a function of the bending angle for a (15, 0) SWCNT with a length of 12.87 nm [23].

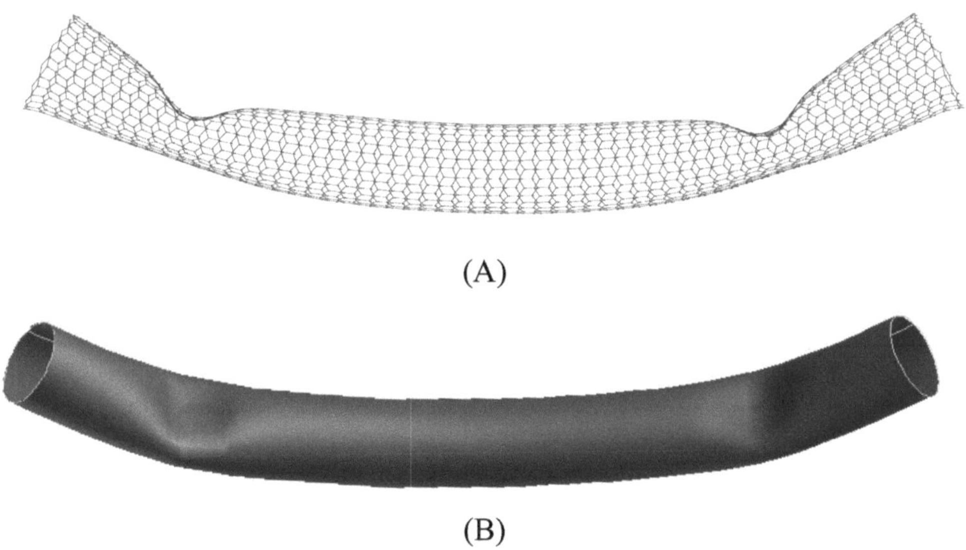

FIGURE 4.42

A comparison of the bending buckling deformations obtained with (A) atomic simulation and (B) the mesh-free method for a (15, 0) SWCNT with a length of 12.87 nm [23].

Table 4.10 Bending Buckling Angle for Different Cases

CNT	Length (nm)	Loading	Atomic Simulation (°)	Mesh-Free Method (°)
(15, 0)	7.65	Mode I	19.75	19.5
(15, 0)	7.65	Mode II	19.25	19.00
(15, 0)	12.87	Mode I	34.25	33.75
(15, 0)	12.87	Mode II	33.25	32.75
(10, 10)	6.28	Mode I	12.25	12.00
(10, 10)	6.28	Mode II	11.50	11.25
(10, 10)	12.56	Mode I	25.25	25.00
(10, 10)	12.56	Mode II	24.75	24.25

reasonably concluded that the reason is that their model does not catch the imperfection in the real atomic structure and does not reflect the local atomic rearrangements that may occur near a buckle site. The present computations show excellent agreement with the atomic simulation, which indicates that our model can precisely describe the constitutive response of the atomic structure. In particular, beyond and after the buckling, the deformation is large and drastic near the buckle site, and the structure behaves as a physical nonlinear material. The present work calculates the constitutive response of the atomic structure based on the current configuration at each iteration step, and thus the physical nonlinearity can be caught.

4.6.4.2 Effect of loading methods

In this section, zigzag (15, 0) and armchair (10, 10) SWCNTs are considered to study the difference of the two loading methods, and both short and long tubes are tested. Table 4.10 gives the bending buckling angle for different cases. It can be seen that the buckling angle for the mode I loading method is often slightly larger than that for the mode II method. This may be because the energy transmission from the ends to the middle is more difficult in mode I than it is in mode II. Fig. 4.43 shows the buckling deformation of the (15, 0) SWCNT with a length of 7.65 nm for the two loading methods. It can be seen that the snap in the case of mode I is slighter less than that in mode II, and that the local kinking in mode II is deeper than that in mode I. More energy is accordingly released in mode II. Fig. 4.44 gives the buckling deformations for the (15, 0) CNT with a length of 12.87 nm and shows that the snap buckle in mode I is closer to the ends than that in mode II. Moreover, our computations reveal that, prior to buckling, the optimized distance between the two end sections in mode II is always very close to the calculated value by assuming the tube is undergoing pure bending deformation, i.e., the SWCNT is undergoing approximately pure bending, as in a beam. After buckling, this optimized distance is largely smaller than the calculated value in the mode I loading method, which further shows that the energy release in mode II is easier than that in mode I.

4.6.4.3 Critical buckling curvature

In bending studies of CNTs, the exploration of the buckling mechanism is an interesting problem that remains a challenge because of its structural complexity. Theoretically, for a given type of

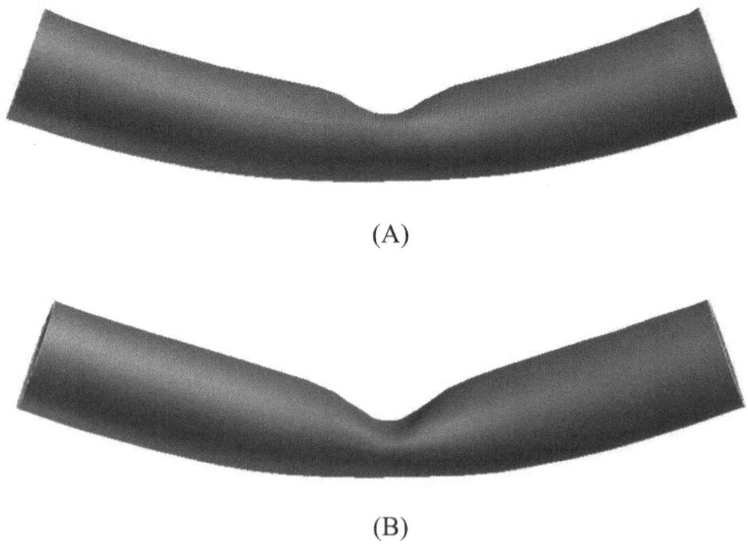

FIGURE 4.43

A comparison of the bending buckling deformations under the (A) mode I and (B) mode II loading methods for a (15, 0) SWCNT with a length of 7.65 nm [23].

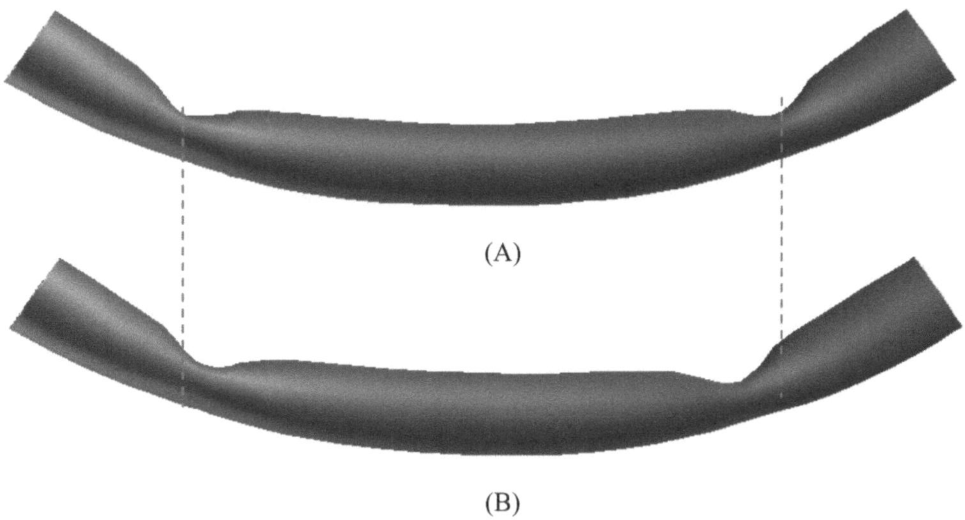

FIGURE 4.44

A comparison of the bending buckling deformations under the (A) mode I and (B) mode II loading methods for a (15, 0) SWCNT with a length of 12.87 nm [23].

SWCNT, it is believed that buckling will occur when the bending curvature reaches a critical value. Yakobson et al. [147] assumed that the local critical strain of snap buckling equals the critical buckling strain of axial compression. Under such an assumption, the critical bending buckling curvature can be estimated as

$$\kappa_{cr} = 2\varepsilon_{cr}^{com}/d = 2.352t/d^2 \tag{4.164}$$

where d is the tube diameter, and t is the thickness, being 0.066 nm [147]. Cao and Chen's MD simulation [148] revealed that the critical bending buckling curvature is dependent on the length, but is almost independent of the tube length for tubes with an aspect ratio of $L/d > 10$. Their numerical results also revealed that the critical buckling curvature varies as the inverse square of the CNT diameter. The relationship between the critical curvature and the diameter is fitted as

$$\kappa_{cr} = 0.0738/d^2 \tag{4.165}$$

Using the present mesh-free method, we also investigate the dependence of the critical buckling curvature on the tube diameter. Here, we consider (n, n) and $(n, 0)$ SWCNTs, and the mode II loading method is used. For each type of tube (i.e., a given n), a series of tubes of different lengths are tested, and the critical buckling curvature of each type is taken as the average of all of the buckling curvatures obtained. Plotted in Fig. 4.45 is the variation of the critical buckling curvature as a function of the tube diameter. It can be seen that the critical buckling curvature hardly depends on the chirality of the tube. The solid line is the fitting line using the formula $\kappa_{cr} = a/d^2$ line. The present fitted value of a is 0.78993 nm.

As we know, at a small bending angle, the tube deformation is smooth. The tube experiences simple compression on the inner side and tension on the outer side. When the bending angle reaches a critical value, the tube buckles, and kinking develops. In the present mesh-free method,

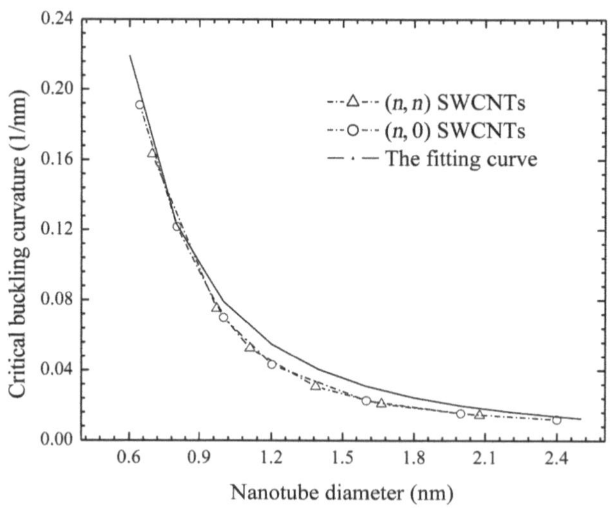

FIGURE 4.45

The variation of the critical bending buckling curvature as a function of diameter dependence [23].

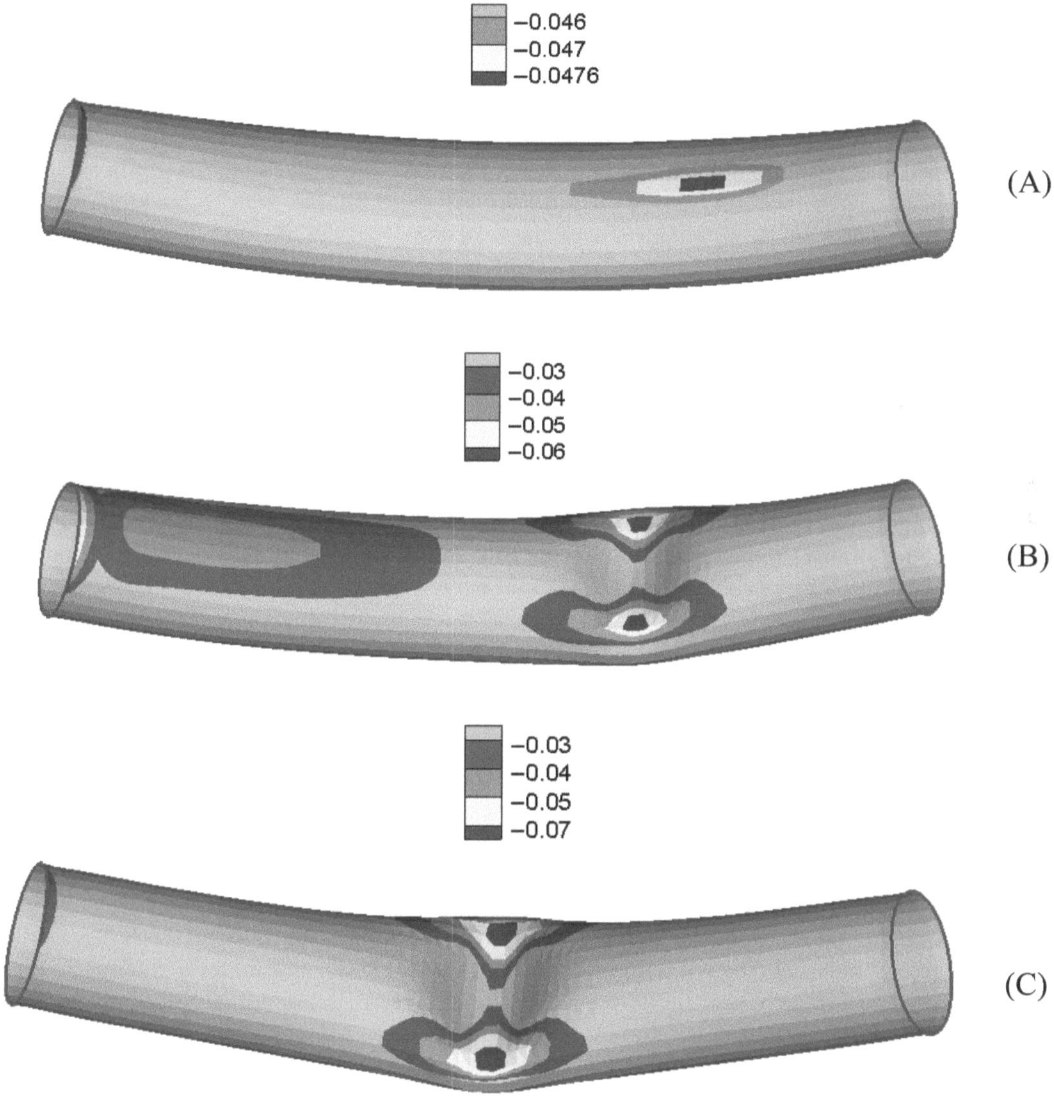

FIGURE 4.46

The contour plot of the axial compressive strain for a (15, 0) SWCNT with a length of 7.65 nm: (A) configuration just before buckling; (B) a slight snap originates in the compressive strain zone; (C) the final buckling pattern [23].

we can easily trace the change of the continuum variables, such as the strain and the curvature. Fig. 4.46 shows the deformation configuration with the contour of axial strain for the (15, 0) SWCNT of 7.65 nm under the mode II loading method. Fig. 4.46A shows the configuration prior to buckling. Although the compression strain is almost uniform along the innermost line, a slight compression zone can be seen on the inner side, in which the central compressive strain is slightly larger than that of the surrounding region. Fig. 4.46B shows that a slight snap first develops in this compressive zone. With the appearance of this snap, the strain distribution largely changes: the strains in the left and right regions of the snap become very small, but the largest compressive strain transits to the front and hind tips of the snap. With increasing iterative steps, the kinking moves to the middle and becomes large (Fig. 4.46C). Fig. 4.47 shows the contour of the axial compressive strain during the buckling of the (15, 0) SWCNT of 12.78 nm. Prior to buckling, there are two slight compression zones near the ends of the tube (see Fig. 4.47A). Subsequently, two snap buckles develop at the two compression zones (see Fig. 4.47B). The above analyses show that the maximum axial compressive strain is an important inducement for the development of kinking and thus can be used as the criteria for bending buckling.

4.6.5 AXIAL BUCKLING AND POSTBUCKLING BEHAVIORS OF SWCNTs

We first test the influence of the amount of nodes in the circumferential direction by using a fixed amount of nodes in the longitudinal direction. Fig. 4.21 depicts the average strain energy per atom versus the compression ratio in the case of 40×8, 40×10 40×15, and 40×20 nodes using the mesh-free computational framework, along with full atomistic simulations for comparison. The data demonstrate that a remarkable decrease in strain energy occurs at each curve, corresponding to the buckling behavior of the SWCNT in the physical meaning. It can also be observed that at the early stage, before buckling, the strain energy is a quadratic function of the compression ratio, while at the postbuckling stage it is almost a linear function. The results show that under the same number of longitudinal nodes, the simulation provides an accurate result at the prebuckling stage, even with only eight nodes along the circumferential direction. However, the number of circumferential nodes has an evident impact on the critical strain of the buckling. The buckling behavior occurs at a larger compression ratio when fewer nodes are used, whereas the compression ratio of buckling obtained by using an increasing number of nodes approaches that of the full atomistic simulation. When a minimum of 10 circumferential nodes are selected, simulation results are quite close to the full atomistic simulation. This phenomenon may be explained by that the tube constructed with only eight nodes along the circumferential direction is much more similar to a prism, while the shape of tubes constructed with more than 10 nodes closely resembles a column.

Accordingly, we further tested the influence of the number of longitudinal nodes on the buckling using an identical number of circumferential nodes. Fig. 4.22 depicts the strain energy versus compression ratio with 25×10 and 40×10 nodes, along with results of full atomistic simulations. It demonstrates that with a smaller number of nodes, the buckling compression ratio is far removed from that obtained from the full atomistic simulation. The reason for this phenomenon is that significant deformation in the axial direction occurs, caused by compressive buckling, and this requires a sufficient number of nodes to capture the nonhomogeneous deformation.

Due to the superior computational property of this mesh-free computational framework, we further adopt it to further study the postbuckling behavior of a (7, 7) SWCNT under an increasing compression. Observed from Figs. 4.21 and 4.22, we confirm a good computational performance of the numerical simulation for the analysis of axial buckling of the SWCNT by using 40 nodes in the longitudinal direction and 10 nodes in the circumferential direction. Therefore, we choose the 40×10 node scenario for postbuckling analysis. The strain energy versus compression ratio,

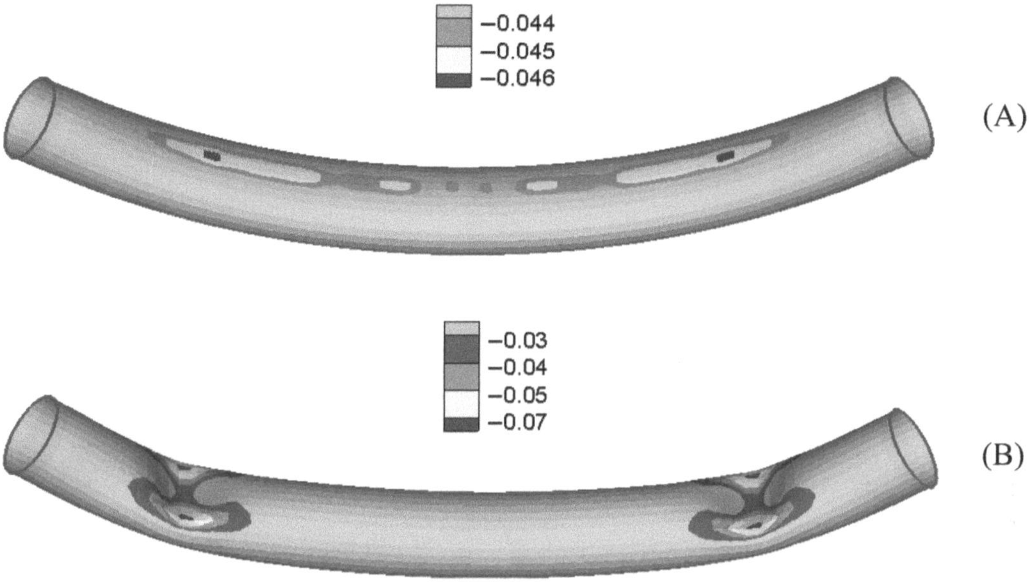

FIGURE 4.47

The contour plot of the axial compressive strain for a (15, 0) SWCNT with a length of 12.87 nm: (A) configuration just before buckling; (B) the final buckling pattern [23].

postbuckling, obtained by using the mesh-free computational framework, is shown in Fig. 4.48, along with data obtained from the full atomistic simulation. The buckling patterns obtained using this mesh-free computational framework, along with those from the full atomistic simulation, are plotted in Fig. 4.49. These are accurate physical buckling patterns for a CNT [17,133,134,147]. Fig. 4.49A shows cylindrical buckles resulting in bifurcation, with a drum in the middle position, at a compression ratio of 6.83%, in the continuum analysis. This result is slightly larger but quite close to that of 6.5% obtained from the full atomistic simulation. Given the increase in loading after buckling, at a compression ratio of 8.67%, the pattern changes into a sideways and centrosymmetrical configuration, and then at a compression ratio of 9.5%, it becomes entirely asymmetric, with the drum moving from the middle to the side, while in the full atomistic simulation, the two episodes of buckling occur at 8.17% and 9.5%, as shown in Fig. 4.49B and C. The comparison of the critical compression ratio and the corresponding buckling patterns confirms that the present quasicontinuum method can accurately capture mechanical behaviors of the nanostructure when only 400 mesh-free nodes are used to simulate the (7, 7) SWCNT. Each mesh-free node or atom has 3 degrees of freedom and thus there are 400×3 degrees of freedom in the quasicontinuum method. Meanwhile, the full atomistic simulation requires 686 atoms, that is to say, the number of degrees of freedom of the whole system is reduced by 286×3. Therefore, this method can save a large amount of computational resources and has great potential in the field of engineering.

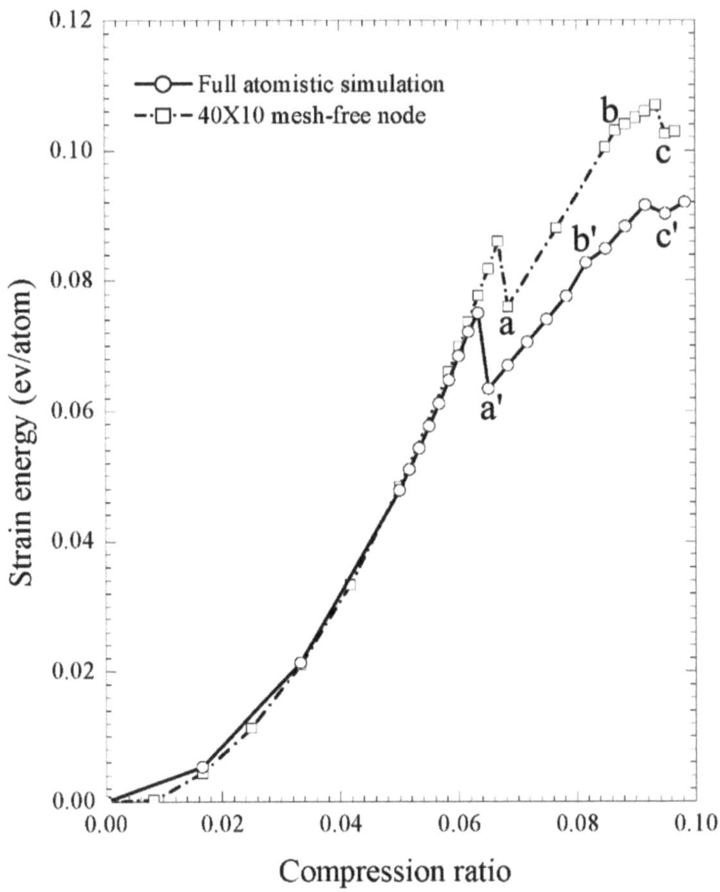

FIGURE 4.48

Average strain energy obtained from the mesh-free method and the full atomistic simulation. Prior to buckling, the curve of the strain energy–compression ratio is quadratic and beyond the buckling, the curve is almost a straight line which corresponds to the nonlinear response [124].

4.7 FRACTURE NUCLEATION

Continuum theory differs from MD simulation in that it does not keep track of the motion of every atom. The atomistic-based continuum theory represents the collective behavior of atoms via the constitutive model. The strain energy density on the continuum level is evaluated by the bond energy on the atomic level for all atomic bonds in a small representative cell via the Cauchy–Born rule [150], within which the deformation is uniform. Zhang et al. [14] developed an atomistic-based continuum theory for CNTs established from Brenner's empirical interatomic potential [120]. Here, a continuum theory established from a modified Morse potential is developed to analyze CNTs.

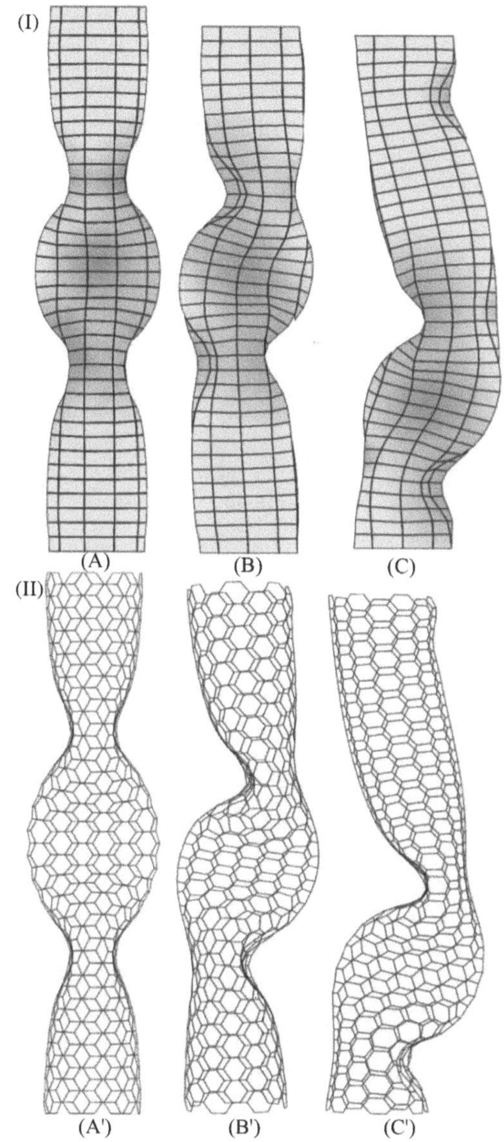

FIGURE 4.49

Buckling patterns of (7, 7) with a length of 6 nm obtained from the mesh-free method at compression ratios of 6.83%, 8.83%, and 9.5% (I) and the full atomistic simulation at compression ratios of 6.5%, 8.17%, and 9.5%, respectively (II). (A) The cylinder buckles to bifurcation with a drum in the middle position; (B) the pattern turns sideways and into a centrosymmetry configuration; and (C) transfers to an entirely asymmetric form with the drum moving from the middle to a side [124].

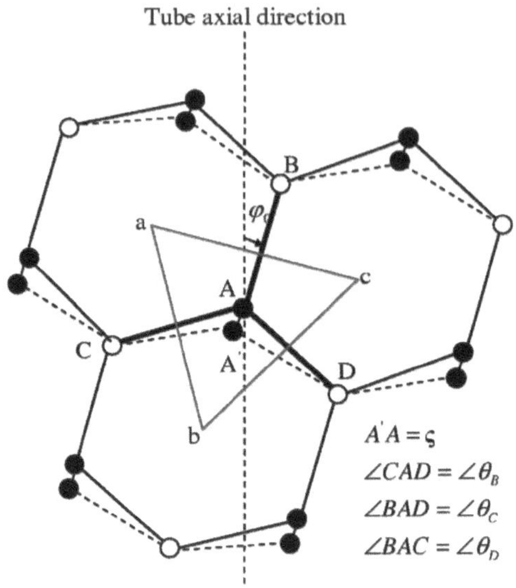

FIGURE 4.50

A schematic diagram of the atomic structure of a carbon nanotube within the tube surface [12].

Fig. 4.50 is a schematic diagram of the atomic structure of a CNT within the tube surface. The angle φ_0 (0 degree $\leq \varphi_0 \leq$ 30 degrees) shows the orientation of the tube axial direction. It is obvious that the angle $\varphi_0 = 0$ degree for zigzag nanotubes, $\varphi_0 = 30$ degrees for armchair nanotubes, and 0 degree $< \varphi_0 <$ 30 degrees for chiral nanotubes. For a simple Bravais lattice that has the centrosymmetric atomic structure, the Cauchy–Born rule ensures the atomic equilibrium. However, the CNT has a hexagonal lattice structure, which is a Bravais multilattice and does not possess centrosymmetry. To use the Cauchy–Born rule, the hexagonal lattice can be decomposed into two triangular sublattices marked by open and solid circles [14,16]. Each sublattice is a simple Bravais lattice that has centrosymmetry and therefore follows the Cauchy–Born rule. However, the two sublattices undergo a shift vector $\varsigma = A'A$ as shown in Fig. 4.50. The shift vector represents the relaxation of the atom positions in the hexagonal lattice to minimize the energy and reach equilibrium.

Let $\mathbf{r}_{ij}^{(0)} = r_{ij}^{(0)}\mathbf{n}_{ij}^{(0)}$ denote the vector between two atoms i and j on the CNT before deformation, where $r_{ij}^{(0)}$ and $\mathbf{n}_{ij}^{(0)}$ denote the bond length and unit vector, respectively. For atoms i and j within the same sublattice (e.g., $\mathbf{r}_{BC}$, $\mathbf{r}_{BD}$, and $\mathbf{r}_{CD}$ in Fig. 4.50), the vector $\mathbf{r}_{ij}$ after the deformation becomes

$$\mathbf{r}_{ij} = \mathbf{F}\cdot\mathbf{r}_{ij}^{(0)} \tag{4.166}$$

with its length

$$r_{ij}(\mathbf{E}) = r_{ij}^{(0)}\sqrt{1 + 2\mathbf{n}_{ij}^{(0)}\cdot\mathbf{E}\cdot\mathbf{n}_{ij}^{(0)}} \tag{4.167}$$

where $\mathbf{E} = 1/2(\mathbf{F}^T\cdot\mathbf{F} - \mathbf{I})$ is the Lagrangian strain. For atoms i and j from two different sublattices (e.g., $\mathbf{r}_{AC}$, $\mathbf{r}_{AB}$, and $\mathbf{r}_{AD}$ in Fig. 4.50), the vector $\mathbf{r}_{ij}$ after the deformation can be written as

$$\mathbf{r}_{ij} = \mathbf{F}\cdot\mathbf{r}_{ij}^{(0)} + \varsigma = r_{ij}^{(0)}\mathbf{F}\cdot(\mathbf{n}_{ij}^{(0)} + \boldsymbol{\xi}) \tag{4.168}$$

where $\boldsymbol{\xi} = (1/r_{ij}^{(0)})\mathbf{F}^{-1}$ is an internal degree of freedom. The bond length is

$$r_{ij}(\mathbf{E}, \boldsymbol{\xi}) = r_{ij}^{(0)}\sqrt{(\mathbf{n}_{ij}^{(0)} + \boldsymbol{\xi})\cdot(\mathbf{I} + 2\mathbf{E})\cdot(\mathbf{n}_{ij}^{(0)} + \boldsymbol{\xi})} \tag{4.169}$$

A modified Morse potential function [151] in which a bond angle-bending potential is added for the interatomic potential for carbon, is used,

$$\begin{aligned} V &= V(r) + V(\theta), \\ V(r) &= D_e\left[(1 - e^{-\beta(r-r_0)})^2 - 1\right], \\ V(\theta) &= \frac{1}{2}k_\theta(\theta - \theta_0)^2\left[1 + k_{\text{sextic}}(\theta - \theta_0)^4\right] \end{aligned} \tag{4.170}$$

where $V(r)$ is the bond energy due to bond stretch, $V(\theta)$ is the bond energy due to bond angle bending, r is the length of the bond, and θ is the current angle of the adjacent bond, which is a standard deformation measure in molecular mechanics. The parameters are

$$\begin{aligned} &r_0 = 0.139\ \text{nm}, \quad D_e = 3.7694\ \text{eV}, \qquad \beta = 26.25\ \text{nm}^{-1}, \\ &\theta_0 = 2.094\ \text{rad}, \quad k_\theta = 5.625\ \text{eV/rad}^2, \quad k_{\text{sextic}} = 0.754 \end{aligned} \tag{4.171}$$

This is the usual Morse potential except that the bond angle-bending energy has been added to stabilize the molecular structure in the tubular configuration.

The strain energy density on the continuum level is evaluated by the bond energy on the atomic level for all atomic bonds in a small representative cell (Δabc) as shown in Fig. 4.50) via the Cauchy–Born rule. The bond energy in the representative cell Δabc can be written as

$$V(\mathbf{E}, \boldsymbol{\xi}) = \frac{1}{2}[V(r_{AB}) + V(r_{AC}) + V(r_{AD})] + V(\theta_B) + V(\theta_C) + V(\theta_D) \tag{4.172}$$

where the factor 1/2 represents the equal split of the bond energy due to the bond stretch between two atoms. The strain energy density on the continuum level is thus obtained as

$$W(\mathbf{E}, \boldsymbol{\xi}) = \frac{V(\mathbf{E}, \boldsymbol{\xi})}{\Omega_{abc}} \tag{4.173}$$

where

$$\Omega_{abc} = \frac{3\sqrt{3}}{4}{r_0}^2 \tag{4.174}$$

The bond lengths and bond angles in the Eq. (4.172) are expressed in terms of $\mathbf{E}$ and $\boldsymbol{\xi}$ as

$$r_{AB}(\mathbf{E}, \boldsymbol{\xi}) = r_{ij}^{(0)}\sqrt{(\mathbf{n}_{AB}^{(0)} + \boldsymbol{\xi})\cdot(\mathbf{I} + 2\mathbf{E})}\sqrt{(\mathbf{n}_{AB}^{(0)} + \boldsymbol{\xi})\cdot(\mathbf{I} + 2\mathbf{E})} \tag{4.175}$$

$$r_{AC}(\mathbf{E}, \boldsymbol{\xi}) = r_{ij}^{(0)}\sqrt{(\mathbf{n}_{AC}^{(0)} + \boldsymbol{\xi})\cdot(\mathbf{I} + 2\mathbf{E})}\sqrt{(\mathbf{n}_{AC}^{(0)} + \boldsymbol{\xi})\cdot(\mathbf{I} + 2\mathbf{E})} \tag{4.176}$$

$$r_{AD}(\mathbf{E}, \boldsymbol{\xi}) = r_{ij}^{(0)}\sqrt{(\mathbf{n}_{AD}^{(0)} + \boldsymbol{\xi})\cdot(\mathbf{I} + 2\mathbf{E})}\sqrt{(\mathbf{n}_{AD}^{(0)} + \boldsymbol{\xi})\cdot(\mathbf{I} + 2\mathbf{E})} \tag{4.177}$$

$$\cos\theta_C = \frac{r_{AB}^2 + r_{AD}^2 - r_{BD}^2}{2r_{AB}r_{AD}} \tag{4.178}$$

$$\cos\theta_B = \frac{r_{AC}^2 + r_{AD}^2 - r_{CD}^2}{2r_{AC}r_{AD}} \tag{4.179}$$

$$\cos\theta_D = \frac{r_{AC}^2 + r_{AB}^2 - r_{BC}^2}{2r_{AC}r_{AB}} \tag{4.180}$$

$$r_{BC}(\mathbf{E},\boldsymbol{\xi}) = \sqrt{3}r_0\sqrt{(1 + 2\mathbf{n}_{BC}^{(0)}\cdot\mathbf{E}\cdot\mathbf{n}_{BC}^{(0)})} \tag{4.181}$$

$$r_{BD}(\mathbf{E},\boldsymbol{\xi}) = \sqrt{3}r_0\sqrt{(1 + 2\mathbf{n}_{BD}^{(0)}\cdot\mathbf{E}\cdot\mathbf{n}_{BD}^{(0)})} \tag{4.182}$$

$$r_{CD}(\mathbf{E},\boldsymbol{\xi}) = \sqrt{3}r_0\sqrt{(1 + 2\mathbf{n}_{CD}^{(0)}\cdot\mathbf{E}\cdot\mathbf{n}_{CD}^{(0)})} \tag{4.183}$$

For a given deformation gradient $\mathbf{F}$ (or equivalently, the Lagrangian $\mathbf{E}$), the internal degree of freedom $\boldsymbol{\xi}$ is determined by minimizing the strain energy density with respect to $\boldsymbol{\xi}$, i.e.,

$$\frac{\partial W(\mathbf{E},\boldsymbol{\xi})}{\partial\boldsymbol{\xi}} = \mathbf{0} \tag{4.184}$$

The above equation is nonlinear and needs to be solved numerically to determine $\boldsymbol{\xi}$ in terms of $\mathbf{E}$.

The second Piola-Kirchhoff stress $\mathbf{T}$ is then determined as the total derivative D of the strain energy density W,

$$\mathbf{T} = \frac{DW(\mathbf{E},\boldsymbol{\xi})}{D\mathbf{E}} = \frac{\partial W(\mathbf{E},\boldsymbol{\xi})}{\partial\mathbf{E}} \tag{4.185}$$

where Eq. (4.184) has been used to derive the equation(4.185).

The stress increment $\dot{\mathbf{T}}$ is related to the strain increment $\dot{\mathbf{E}}$ via the incremental modulus $\mathbf{C}$ as

$$\dot{\mathbf{T}} = \mathbf{C}:\dot{\mathbf{E}} \tag{4.186}$$

$$\mathbf{C} = \frac{D\mathbf{T}}{D\mathbf{E}} = \frac{D}{D\mathbf{E}}\left(\frac{\partial W}{\partial\mathbf{E}}\right) = \frac{\partial^2 W}{\partial\mathbf{E}\partial\mathbf{E}} + \frac{\partial^2 W}{\partial\mathbf{E}\partial\boldsymbol{\xi}}\frac{\partial\boldsymbol{\xi}}{\partial\mathbf{E}} \tag{4.187}$$

The term $\partial\boldsymbol{\xi}/\partial\mathbf{E}$ can be obtained from the total derivative of Eq. (4.184), which results in

$$\frac{\partial\boldsymbol{\xi}}{\partial\mathbf{E}} = -\left(\frac{\partial^2 W}{\partial\boldsymbol{\xi}\partial\boldsymbol{\xi}}\right)^{-1}\frac{\partial^2 W}{\partial\boldsymbol{\xi}\partial\mathbf{E}} \tag{4.188}$$

4.7.1 FRACTURE NUCLEATION IN SWCNTs

Atomistic study [152] shows that there is a breaking strain before which the CNT undergoes uniform deformation. Once the breaking strain is reached, one or a few atom bonds break, and this finally leads to the fracturing of the CNT. Unlike modeling the breaking of individual atomic bonds in MD simulation, Zhang et al. [14] modeled the fracture nucleation in CNTs under tension as a bifurcation. The Zhang et al. method is adopted here to study the fracture of CNTs via a continuum theory established from a modified Morse potential. The main procedures are as follows.

4.7.1.1 Equilibrium equation for the CNT

The equilibrium equation for the CNT is

$$(\mathbf{F}\cdot\mathbf{T})\cdot\nabla=\mathbf{0} \tag{4.189}$$

where ∇ is the gradient in the undeformed configuration, which can be written as

$$\nabla=e_R\frac{\partial}{\partial R}+e_\theta\frac{\partial}{\partial\theta}+e_Z\frac{\partial}{\partial Z} \tag{4.190}$$

in the cylindrical coordinates (R, θ, Z). The traction-free boundary condition can be written as

$$(\mathbf{F}\cdot\mathbf{T})\cdot\mathbf{N}=0 \tag{4.191}$$

where $\mathbf{N}$ is the unit normal to the surface in the undeformed configuration. On the inner and outer surfaces of the CNT, Eq. (4.191) becomes

$$(\mathbf{F}\cdot\mathbf{T})\cdot e_R=0 \tag{4.192}$$

For a single-wall CNT, Eq. (4.192) should hold for the entire nanotube, which leads to $\mathbf{T}\cdot e_R=0$. The equilibrium equation (4.189) can thus be written as

$$\begin{aligned}
&\frac{1}{R}\frac{\partial}{\partial\theta}(F_{R\alpha}T_{\alpha\theta})-\frac{1}{R}F_{\theta\alpha}T_{\alpha\theta}+\frac{\partial}{\partial Z}(F_{R\alpha}T_{\alpha Z})=0\\
&\frac{1}{R}F_{R\alpha}T_{\alpha\theta}+\frac{1}{R}\frac{\partial}{\partial\theta}(F_{\theta\alpha}T_{\alpha\theta})+\frac{\partial}{\partial Z}(F_{\theta\alpha}T_{\alpha Z})=0\\
&\frac{1}{R}\frac{\partial}{\partial\theta}(F_{Z\alpha}T_{\alpha\theta})+\frac{\partial}{\partial Z}(F_{Z\alpha}T_{\alpha Z})=0
\end{aligned} \tag{4.193}$$

where R is the nanotube radius, and the summation for α is over θ and Z.

4.7.2 PREBIFURCATION

The deformation before bifurcation is uniform. The nonzero components of the Lagrangian strain are $E_{\theta\theta}$ and E_{ZZ}. The nonzero second Piola-Kirchhoff stress components $T_{\theta\theta}$, T_{ZZ}, and $T_{\theta Z}=T_{Z\theta}$ are obtained from (4.185). The nonzero components of the incremental modulus $C_{\theta\theta\theta\theta}$, $C_{\theta\theta ZZ}=C_{ZZ\theta\theta}$, C_{ZZZZ}, $C_{\theta\theta\theta Z}=C_{\theta Z\theta\theta}$, $C_{\theta Z\theta Z}$, and $C_{\theta ZZZ}=C_{ZZZ\theta}$ are obtained from (4.185). The requirement that

$$T_{\theta\theta}=0 \tag{4.194}$$

gives an implicit equation to determine $E_{\theta\theta}$ in terms of E_{ZZ}.

4.7.3 ONSET OF BIFURCATION

At the onset of bifurcation, the increment of deformation is not uniform anymore. The nonuniform increment of deformation gradient $\dot{\mathbf{F}}$ on the tube surface is given in terms of the displacement increment $\dot{\mathbf{U}}$ by

$$\dot{F}_{R\theta} = \frac{1}{R}\frac{\partial \dot{U}_R}{\partial \theta} - \frac{\dot{U}_\theta}{R}, \quad \dot{F}_{\theta\theta} = \frac{1}{R}\frac{\partial \dot{U}_\theta}{\partial \theta} + \frac{\dot{U}_R}{R},$$
$$\dot{F}_{Z\theta} = \frac{1}{R}\frac{\partial \dot{U}_Z}{\partial \theta}, \quad \dot{F}_{RZ} = \frac{\partial \dot{U}_R}{\partial Z}, \quad \dot{F}_{\theta Z} = \frac{\partial \dot{U}_\theta}{\partial Z}, \quad \dot{F}_{ZZ} = \frac{\partial \dot{U}_Z}{\partial Z} \tag{4.195}$$

The corresponding nonzero components of the Lagrangian strain increment $\dot{\mathbf{E}}$ are

$$\begin{aligned} \dot{E}_{\theta\theta} &= F_{\theta\theta}\dot{F}_{\theta\theta} \\ \dot{E}_{ZZ} &= F_{ZZ}\dot{F}_{ZZ} \\ \dot{E}_{\theta Z} &= \dot{E}_{Z\theta} = \frac{1}{2}(F_{\theta\theta}\dot{F}_{\theta Z} + F_{ZZ}\dot{F}_{Z\theta}) \end{aligned} \tag{4.196}$$

The nonzero components of the stress increment are obtained from (4.186) as

$$\begin{aligned} \dot{T}_{\theta\theta} &= C_{\theta\theta\theta\theta}\dot{E}_{\theta\theta} + C_{\theta\theta ZZ}\dot{E}_{ZZ} + 2C_{\theta\theta\theta Z}\dot{E}_{\theta Z} \\ \dot{T}_{ZZ} &= C_{ZZ\theta\theta}\dot{E}_{\theta\theta} + C_{ZZZZ}\dot{E}_{ZZ} + 2C_{ZZZ\theta}\dot{E}_{\theta Z} \\ \dot{T}_{\theta Z} &= C_{Z\theta\theta\theta}\dot{E}_{\theta\theta} + C_{\theta ZZZ}\dot{E}_{ZZ} + 2C_{\theta Z\theta Z}\dot{E}_{\theta Z} \end{aligned} \tag{4.197}$$

The substitutions of (4.195)–(4.197) into the incremental form of (4.193) give three partial differential equations (with respect to θ and Z for $\dot{U}_R$, $\dot{U}_\theta$, and $\dot{U}_Z$).

However, the CNT is subjected to the axial displacement and vanishing shear stress tractions at the ends. At the onset of bifurcation, the increments of the axial displacement and shear stress tractions vanish at both ends, i.e.,

$$\dot{U}_Z = 0, \quad \dot{\mathbf{F}}\cdot\mathbf{T}\cdot e_R + \mathbf{F}\cdot\dot{\mathbf{T}}\cdot e_R = 0 \tag{4.198}$$

at $Z = 0, L$.

The homogeneous governing equations that are obtained from the substitutions of (4.195)–(4.197) into the incremental form of (4.193), and the boundary conditions (4.198) constitute an eigenvalue problem for the displacement increment $\dot{U}$. They have only the trivial solution $\dot{U} = 0$ until the axial strain E_{ZZ} reaches a critical value $(E_{ZZ})_{cr}$ for bifurcation.

4.7.4 YOUNG'S MODULUS OF AN SWCNT

The linear elastic modulus of a SWCNT can be obtained from (4.187) by taking $\mathbf{E} = 0$, $\boldsymbol{\xi} = 0$. For simple axial tension along the Z direction, Young's modulus can be calculated by

$$E = C_{ZZZZ} - \frac{C_{ZZ\theta\theta}^2}{C_{\theta\theta\theta\theta}} \tag{4.199}$$

This does not depend on the chirality of the CNT [153], which is consistent with the results from molecular dynamics simulations [154]. Assuming that the thickness of the single-walled nanotube $h = 0.335$ nm, then the calculated Young's modulus from the current theory is $E = 959$ GPa.

4.7.5 BIFURCATION STRAIN AND FRACTURE STRENGTH

For zigzag and armchair CNTs, the displacement increment takes the form [14]

$$\begin{bmatrix} \dot{U}_R \\ \dot{U}_\theta \\ \dot{U}_Z \end{bmatrix} = \begin{bmatrix} \dot{U}^{(n)}_{Rm} \cos \dfrac{m\pi Z}{L} \cos n\theta \\ \dot{U}^{(n)}_{\theta m} \cos \dfrac{m\pi Z}{L} \sin n\theta \\ \dot{U}^{(n)}_{Zm} \sin \dfrac{m\pi Z}{L} \cos n\theta \end{bmatrix} \tag{4.200}$$

where L is the length of the CNT, $n = 0, 1, 2, \ldots$ is the wave number in the circumferential direction, $m = 1, 2, 3, \ldots$ is the eigenmode number in the axial direction, $n = 0$ is for axisymmetric bifurcation, and $n = 1, 2, 3, \ldots$ is for nonaxisymmetric bifurcation. For chiral CNTs, the displacement increment can be written in the form

$$\begin{bmatrix} \dot{U}_R \\ \dot{U}_\theta \\ \dot{U}_Z \end{bmatrix} = \begin{bmatrix} \dot{U}^{(n)}_{Rs}(Z) \sin n\theta + \dot{U}^{(n)}_{Rc}(Z) \cos n\theta \\ \dot{U}^{(n)}_{\theta s}(Z) \sin n\theta + \dot{U}^{(n)}_{\theta c}(Z) \cos n\theta \\ \dot{U}^{(n)}_{Zs}(Z) \sin n\theta + \dot{U}^{(n)}_{Zc}(Z) \cos n\theta \end{bmatrix} \tag{4.201}$$

The fracture strength is defined as T_{ZZ} when the strain E_{ZZ} reaches the bifurcation strain $(E_{ZZ})_{\mathrm{cr}}$. Once $(E_{ZZ})_{\mathrm{cr}}$ is obtained, $(T_{ZZ})_{\mathrm{cr}}$ can be calculated without difficulty.

Figs. 4.51 and 4.52 illustrate how the bifurcation strain varies with the wave number n and the eigenmode number m for a zigzag single-wall CNT ($\varphi_0 = 0$ degree) and an armchair single-wall

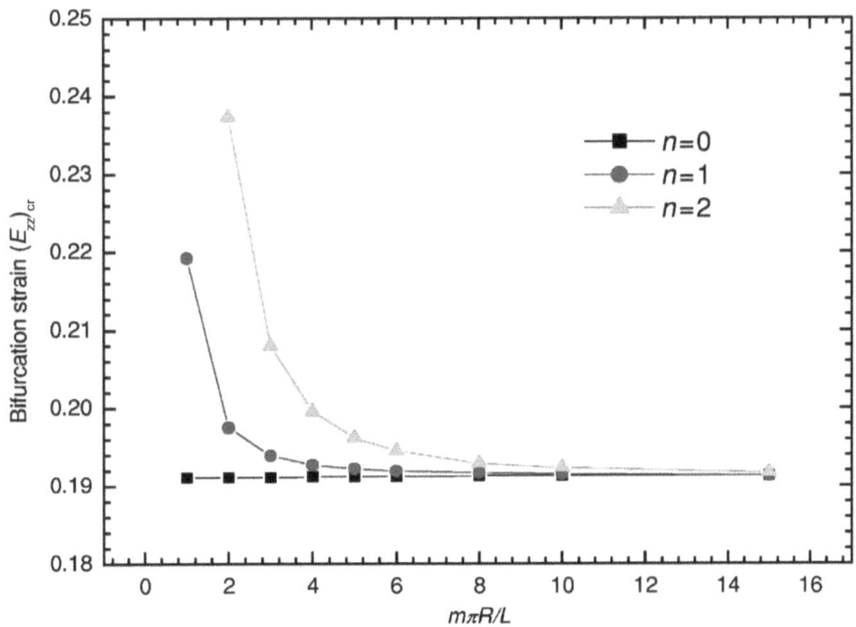

FIGURE 4.51

Bifurcation strain for zigzag single-walled carbon nanotubes [12].

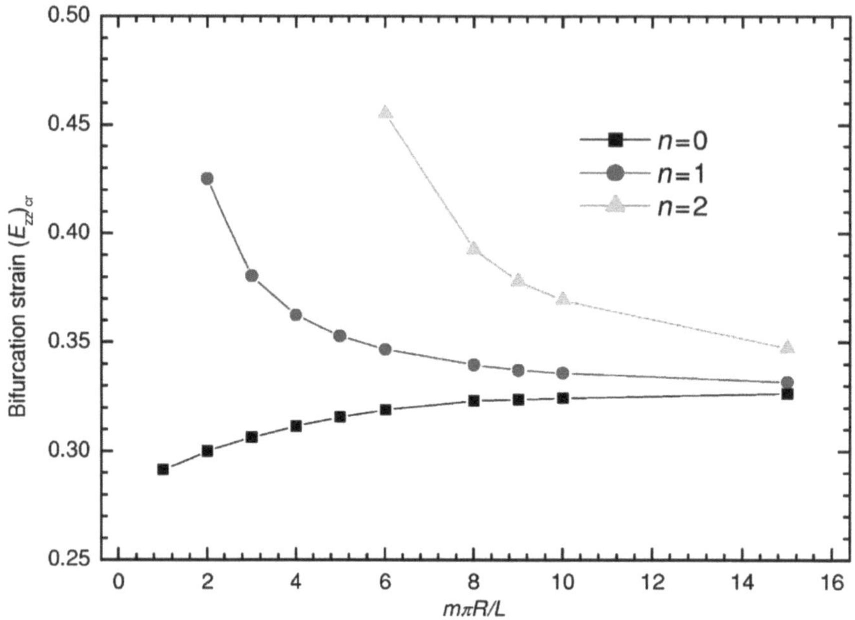

FIGURE 4.52

Bifurcation strain for armchair single-walled carbon nanotubes [12].

CNT ($\varphi_0 = 30$ degrees). It is easy to find that the bifurcation will take place at $n = 0$, $m = 1$, where the bifurcation strain $(E_{ZZ})_{cr}$ is the lowest.

The bifurcation strain for the chiral single-wall CNT is shown in Fig. 4.53 for the angle $\varphi_0 = 10$ degrees, and in Fig. 4.54 for $\varphi_0 = 20$ degrees. Unlike zigzag and armchair single-wall CNTs, the axisymetric bifurcation ($n = 0$) will take place with small tube radius-to-length ratio R/L. When the ratio R/L increases, the chiral SWCNT will bifurcate in the first nonaxisymmetric mode ($n = 1$).

Table 4.11 shows the bifurcation strain and fracture strength for a SWCNT with a diameter of 1 nm and length of 5 nm ($R/L = 0.1$). The smaller the angle φ_0, the easier it is for the SWCNT to bifurcate. The zigzag SWCNT has the lowest bifurcation strain and fracture strength, the armchair SWCNT has the highest bifurcation strain and fracture strength, and the bifurcation strain and fracture strength for the chiral SWCNT are in between.

4.8 BERNOULLI–EULER BEAM MODEL AND GLOBAL BUCKLING

4.8.1 BERNOULLI–EULER BEAM MODEL

The beam model is a convenient and efficient tool to study static and dynamic behaviors such as buckling and vibration of long SWCNTs. In the Bernoulli–Euler beam model, we only need to

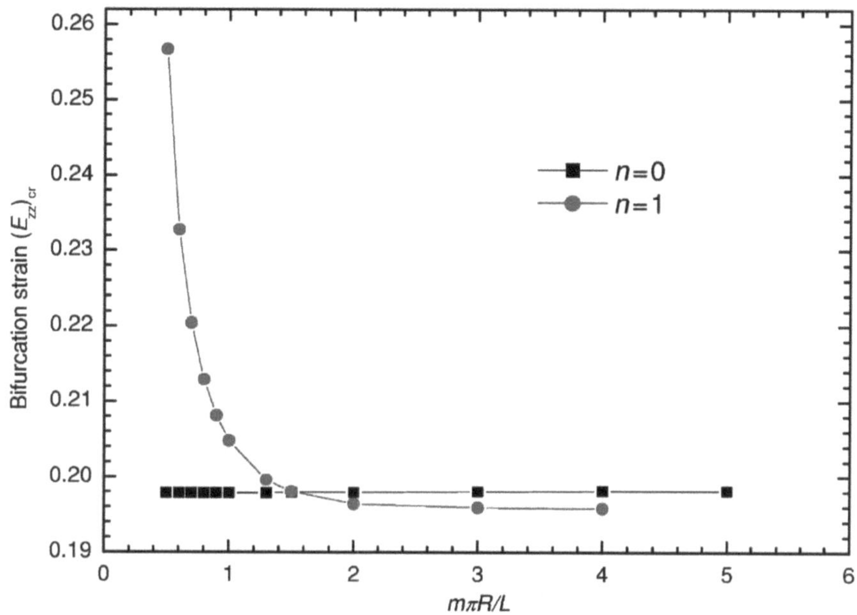

FIGURE 4.53

Bifurcation strain for chiral single-walled carbon nanotubes with $\varphi_0 = 10$ degrees [12].

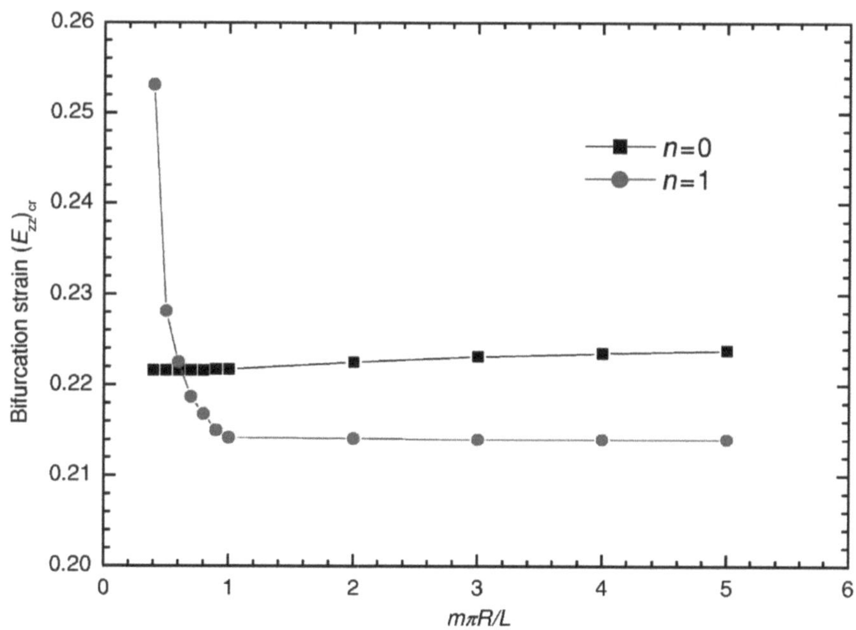

FIGURE 4.54

Bifurcation strain for chiral single-walled carbon nanotubes with $\varphi_0 = 20$ degrees [12].

Table 4.11 Bifurcation Strain and Fracture Strength for a SWCNT With $R/L = 0.1$ [12]

	Zigzag ($\varphi_0 = 0°$)	Armchair ($\varphi_0 = 30°$)	Chiral ($\varphi_0 = 10°$)	Chiral ($\varphi_0 = 20°$)
Bifurcation strain	0.19	0.28	0.20	0.22
Fracture strength (GPa)	83	99	87	94

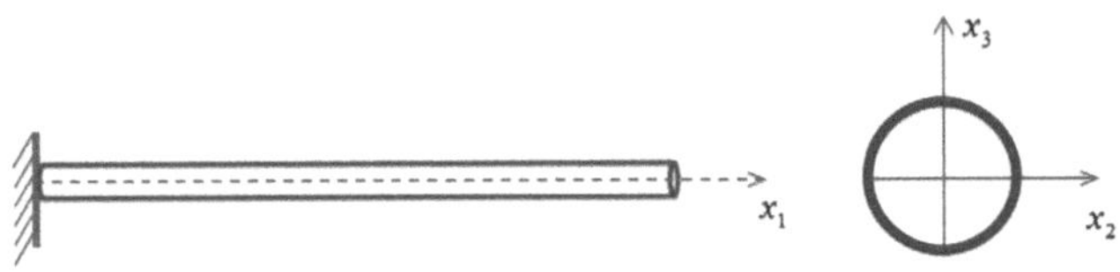

FIGURE 4.55

A cantilever beam model for the SWCNT.

consider the transverse displacement of the axis (Fig. 4.55), i.e., the deflection $u_3 = u_3(x_1)$; longitudinal displacement of location (x_2, x_3) on the cross-section is calculated as

$$u_1 = -x_3 \frac{\partial u_3}{\partial x_1} \tag{4.202}$$

In the classical Bernoulli–Euler beam theory, only stress component σ_{11} needs to be considered. Following [155,156], we treat the strain energy density as a function of first- and second-order deformation gradients F_{11}, G_{111}, and G_{311}. In [155,156], the involved higher-order term is the dilatation gradient ε_{113} which has the same expression as the second-order deformation gradient G_{311}, only a negative sign is added. But it is seen that the second-order deformation gradient G_{311} is endowed with a different meaning in later research.

Now, we return to calculation of the stiffness matrix relating the tress vector $\boldsymbol{\sigma} = (P_{11}, Q_{111}, Q_{311})$ and the deformation gradient vector **grad** $= (F_{11}, G_{111}, G_{311})$ at an evaluated point. To calculate the constitutive response, a representative cell is collocated at this point (Fig. 4.56). The relationship **σ-grad** can be written as

$$\boldsymbol{\sigma} = \mathbf{H} \cdot \mathbf{grad} \tag{4.203}$$

Using Eqs. (4.5), (4.14), and (4.38)–(4.40), deformation gradients F_{22}, G_{222}, F_{32}, and G_{322} can be exactly calculated and F_{11}, G_{111}, and G_{311} can be calculated as

$$F_{11} = 1 - x_3 \frac{\partial^2 u_3}{\partial X_1^2}, \quad G_{111} = -x_3 \frac{\partial^3 u_3}{\partial X_1^3}, \quad G_{311} = \frac{\partial^2 u_3}{\partial X_1^2} \tag{4.204}$$

All three bond vectors in the representative cell are approximated by using Eq. (4.4). It is emphasized that the initial bond vectors in the planar configuration should be precisely calculated, in advance, using an optimization method [23,157]. The strain energy density can be calculated using Eq. (4.14); it is a function of γ, η_1, η_2, F_{11}, G_{111}, and G_{311}

$$\omega = \omega(\gamma, \eta_1, \eta_2, F_{11}, G_{111}, G_{311}) \tag{4.205}$$

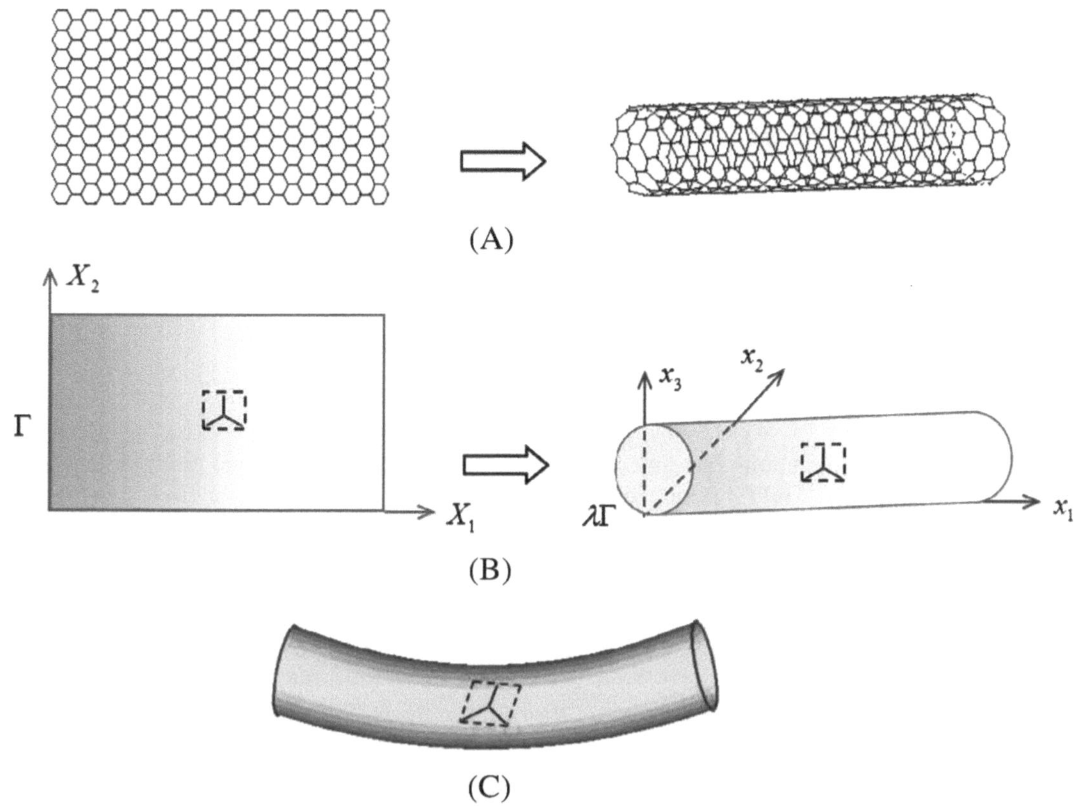

FIGURE 4.56

A graphene sheet is viewed as a continuum plane and an SWCNT is viewed as a hollow cylindrical tube with a thin smooth surface. Initially the SWCNT is formed by rolling up a planar sheet into a cylindrical shape. (A) Atomic structure; (B) continuum view; and (C) the deformed SWCNT.

The equilibrium state can be determined with the Newton's method. Let us introduce the unknown vector as

$$\boldsymbol{\lambda} = \boldsymbol{\lambda}(\gamma, \eta_1, \eta_2, F_{11}, G_{111}, G_{311})^T \tag{4.206}$$

The nonequilibrium force vector and the stiffness matrix are

$$\mathbf{f}_{\lambda} = -\frac{\partial \omega}{\partial \boldsymbol{\lambda}}, \quad \mathbf{K}_{\lambda} = \frac{\partial^2 \omega}{\partial \boldsymbol{\lambda}^2} \tag{4.207}$$

The unknown vector can be determined by iteratively solving the incremental equation

$$\mathbf{K}_{\lambda}\Delta\boldsymbol{\lambda} = \mathbf{f}_{\lambda} \tag{4.208}$$

where $\Delta\boldsymbol{\lambda}$ is the incremental solution.

If matrix $\mathbf{K}_\lambda$ is written as

$$\mathbf{K}_\lambda = \begin{bmatrix} \mathbf{K}_{11} & \mathbf{K}_{12} \\ \mathbf{K}_{12}^T & \mathbf{K}_{22} \end{bmatrix} \tag{4.209}$$

where $\mathbf{K}_{11}$, $\mathbf{K}_{12}$, and $\mathbf{K}_{22}$ are the 3×3 matrix. The stiffness matrix $\mathbf{H}$ can be calculated as

$$\mathbf{H} = \mathbf{K}_{22} - \mathbf{K}_{12}^T \mathbf{K}_{11}^{-1} \mathbf{K}_{12} \tag{4.210}$$

The total strain energy can be written as

$$U = \frac{1}{2}\int_\Omega (H_{11}F_{11}^2 + H_{22}G_{111}^2 + H_{33}G_{311}^2 + 2H_{13}F_{11}G_{311})dV \tag{4.211}$$

Inserting Eq. (4.204) into (4.211), the system energy can be rewritten as

$$U = \frac{1}{2}\int_\Omega (H_{11}x_3^2(u_3^{(2)})^2 + H_{22}x_3^2(u_3^{(3)})^2 + H_{33}(u_3^{(2)})^2 - 2H_{13}x_3(u_3^{(2)})^2)dV \tag{4.212}$$

Since a constant deformation gradient contributes to a lift of total energy, the unit gradient associated with F_{11} is ignored while it is substituted into Eq. (4.211).

Using $dV = dAdx_1$, and performing the integration on the cross-section, we have

$$U = \frac{1}{2}\int_L [(B_1 + B_2 + B_3)(u_3^{(2)})^2 + D(u_3^{(3)})^2]dx_1 = \frac{1}{2}\int_L [B(u_3^{(2)})^2 + D(u_3^{(3)})^2]dx_1 \tag{4.213}$$

where

$$B_1 = \int_A x_3^2 H_{11}dA, \quad B_2 = \int_A H_{33}dA, \quad B_3 = -2\int_A x_3 H_{13}dA, \quad D = \frac{1}{2}\int_A x_3^2 H_{22}dA \tag{4.214}$$

In the present study, the strain energy density is denoted as the energy per unit surface area, and no thickness concept is involved. The integral in Eq. (4.214) is with respect to circumference, and can be calculated using the Gauss integral method. Information of the microstructure of SWCNTs has been incorporated into the multiscale model and all parameters can be directly calculated. But in most of extant literature [156,158–161], scale parameters need to be assumed to study the effect of the higher-order strain gradient. Table 4.12 presents the calculated parameters for different (n, n) SWCNTs. Figs. 4.57 and 4.58 plot the variation of parameters B and D against the tube radius for three types of tubes, i.e., (n, n), $(n, 0)$, and $(3n, 2n)$ tubes. Subsequently, we shall discuss these parameters' meanings and their effect on mechanical properties of SWCNTs in detail.

Parameter B_1 corresponds to the classical bending stiffness, which should be equal to EI (I is the moment of inertia). Table 4.12 shows the results are in good agreement with Guo and Zhang [163] and Wang et al. [21]. B_2 can be understood as the contribution of the local bending stiffness. Actually, H_{33} is the local bending stiffness, which can be well understood by the fact that a single curved graphene sheet can bear the bending moment. This problem has been studied in detail recently by Guo and Zhang [163]. Of course, H_{33} has a dependence on the angle between the tangent plane and the bending orientation. Fig. 4.59 gives a contour for the variation of H_{33} around the circumference for an (8, 8) SWCNT. It can be seen that locations (0, 0) and (0, $2R$) have the maximum local bending stiffness but the minimum value appears at locations ($-R$, 0) and (R, 0). If the tube radius is chosen as a very large value, the obtained bending stiffness per atom (i.e., $M_{33}\Omega$) is 0.037465 eV nm/atom at locations (0, 0). This value falls well within the estimated range of

Table 4.12 The Obtained Parameters for the (*n*, *n*) SWCNTs [162]

(*n*, *n*)	*R* (nm)	B_1 (eV nm)	B_2 (eV nm)	B_3 (eV nm)	*D* (eV nm^3)
(4, 4)	0.2711	89.7440	0.6972	0.08341	0.0423
(6, 6)	0.4115	318.3249	1.0459	0.37024	0.1493
(8, 8)	0.5511	768.2305	1.3923	0.6171	0.3592
(10, 10)	0.6902	1513.1418	1.7386	0.8448	0.7065
(12, 12)	0.8291	2626.7739	2.0849	1.0624	1.2255
(14, 14)	0.9680	4182.8548	2.4313	1.2739	1.9505
(16, 16)	1.1067	6255.1180	2.7778	1.4815	2.9158
(18, 18)	1.2454	8917.2987	3.1244	1.6866	4.1559
(20, 20)	1.3841	12243.1360	3.4711	1.8898	5.7050

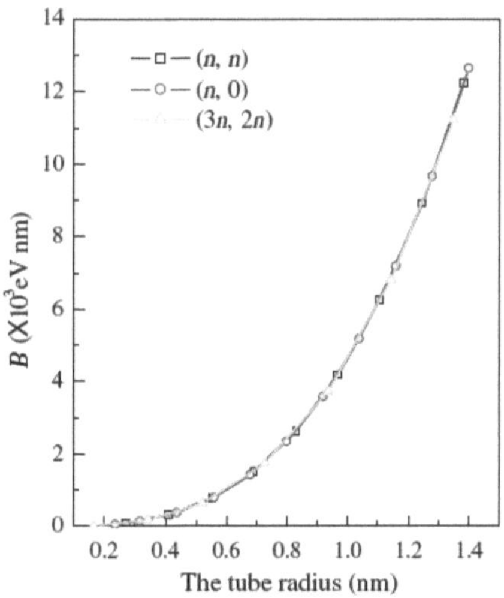

FIGURE 4.57

The calculated parameter *B* versus the tube radius for (*n*, *n*), (*n*, 0), and (3*n*, 2*n*) SWCNTs [162].

0.031–0.039 eV nm/atom [164]. Stress P_{11} is independent of deformation gradients G_{111} and G_{311} for the isotropic medium. Our calculation indicates that M_{12} is equal to zero but M_{13} is nonzero, which shows that P_{11} is coupled with G_{311}. This may be explained by the fact that the curvature along X_2-direction can lead to change of the Young's modulus along X_1-direction [164]. Moreover, an SWCNT is not a strict isotropic medium and there is a slight divergence between the longitudinal and circumferential Young's moduli [165]. Parameter *D* describes the contribution of the higher-order deformation gradient G_{111} and this may have a large effect on the SWCNT's mechanical behavior, which is discussed in the next section.

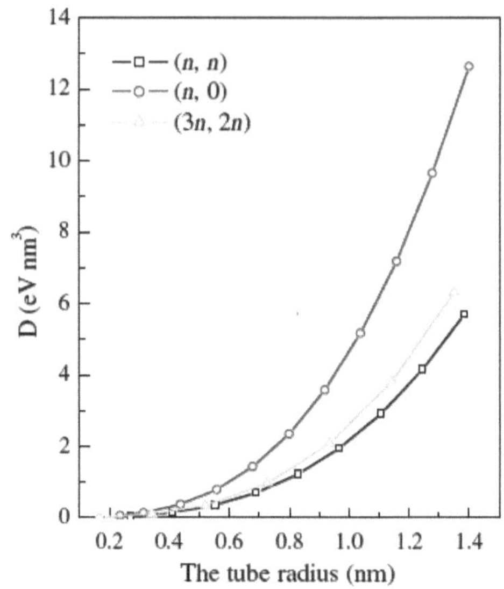

FIGURE 4.58

The calculated parameter D versus the tube radius for (n, n), $(n, 0)$, and $(3n, 2n)$ SWCNTs [162].

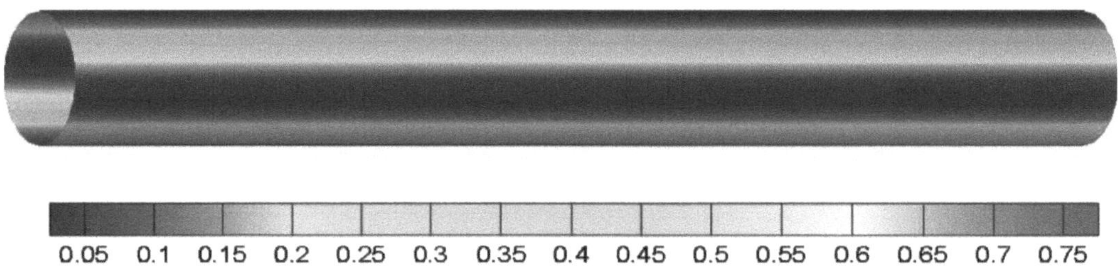

FIGURE 4.59

A contour plot for the variation of the local bending moment around the circumference (the unit is eV) [162].

4.8.2 GLOBAL BUCKLING OF SWCNTs UNDER AXIAL COMPRESSION

The above described multiscale model can be employed to study mechanical properties of SWCNTs, such as global buckling and vibration. This section presents the study about global buckling of SWCNTs under axial compression. Considering an SWCNT under axial compressive force N, one can obtain the total strain energy as [156]

$$U = \frac{1}{2}\int_L [B(u_3^{(2)})^2 + D(u_3^{(3)})^2 - N(u_3^{(1)})^2]dx_1 \tag{4.215}$$

The variation of Eq. (4.215) gives

$$\delta U = \int_L [Bu_3^{(4)} - Du_3^{(6)} + Nu_3^{(2)}]\delta u_3 dx_1 + [-Bu_3^{(3)} + Du_3^{(5)} - Nu_3^{(1)}]\delta u_3|_0^L + [Bu_3^{(2)} - Du_3^{(4)}]\delta u_3'|_0^L + [Du_3^{(3)}]\delta u''_3|_0^L \tag{4.216}$$

So, the governing equilibrium equation for global buckling of SWCNT is

$$Bu_3^{(4)} - Du_3^{(6)} + Nu_3^{(2)} = 0 \tag{4.217}$$

If the boundary condition is simply supported at the two ends of an SWCNT, the critical buckling force is [156]

$$N_{CR} = \frac{\pi^2}{L^2}\left(B + \frac{\pi^2 D}{L^2}\right) \tag{4.218}$$

If an SWCNT can be viewed as a cantilever beam, the critical buckling force is

$$N_{CR} = \frac{\pi^2}{4L^2}\left(B + \frac{\pi^2 D}{L^2}\right) \tag{4.219}$$

Since the critical buckling force of a cantilever beam is always a quarter of that of a simply supported beam with the same dimensions, we present the results only for the simply supported boundary condition. First, the tube length is fixed as 36 nm, to study the effect of its radius on the critical buckling force. Fig. 4.60 depicts the variation of critical buckling force versus the tube radius for (*n*, *n*), (*n*, 0), and (3*n*, 2*n*) SWCNTs; all three curves are almost overlapped, which shows that the effect

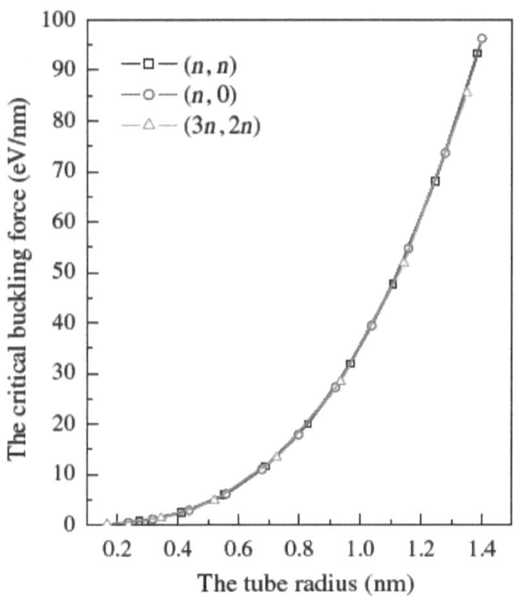

FIGURE 4.60

The critical buckling force versus the tube radius [162].

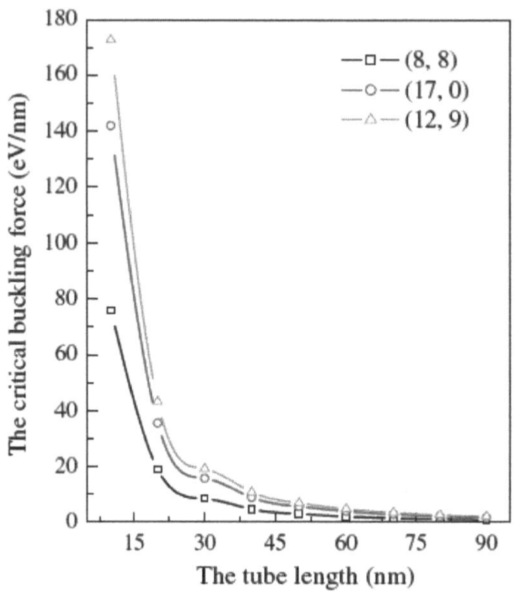

FIGURE 4.61

The critical buckling force versus the tube length [162].

of the higher-order term is generally small. Fig. 4.61 plots the variation of the critical buckling force versus lengths of (8, 8), (17, 0), and (12, 9) SWCNTs. In general, the critical buckling force decreases with increase in length in an approximately reciprocal of L2 that can be seen from Eq. (4.218).

Our main interest is how much the critical buckling force of SWCNT is affected by the higher-order terms. For this purpose, we denote the contribution ratio of the higher-order terms to the critical buckling force as

$$\text{Contribution ratio} = \frac{N_{CR} - N'_{CR}}{N_{CR}} \times 100\% \tag{4.220}$$

where N'_{CR} is the critical buckling force if the higher-order term is not involved, i.e., B_2, B_3, and D are set as zero.

Fig. 4.62 plots the ratio of contribution of the higher-order terms versus the tube radius for (n, n), $(n, 0)$, and $(3n, 2n)$ SWCNTs. It can be seen that the effect of the higher-order terms is very small if the tube radius is greater than 0.4 nm. But the contribution appears to be increasing rapidly when the tube radius is less than about 0.4 nm. Fig. 4.63 shows the ratio of contribution against the tube length for (6, 6), (11, 0), and (9, 6) SWCNTs. It is rather obvious that contribution of the higher-order terms has a larger dependence on tube length. When the tube length is less than about 5 nm, this effect increases very rapidly with decrease in tube length. But the effect can be ignored for a longer SWCNT. Although local buckling may happen in the case of an SWCNT with a larger radius if its length is less than 5 nm, we can draw an unswerving conclusion that the higher-order term has a noticeable effect on mechanical properties of SWCNTs, which shows a strong dependence on tube dimensions.

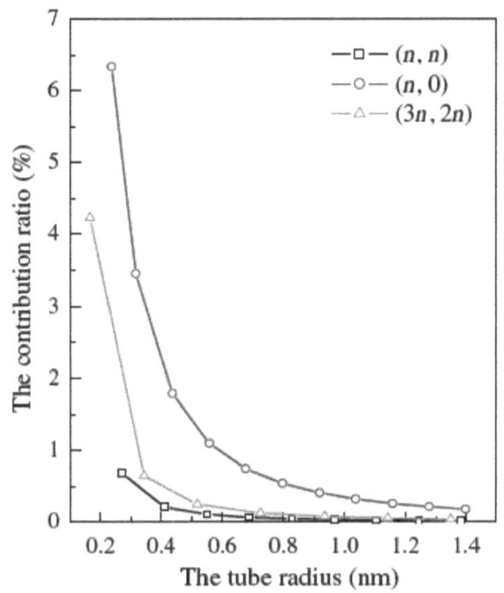

FIGURE 4.62

The contribution ratio of the higher-order terms to the critical buckling force versus the tube radius [162].

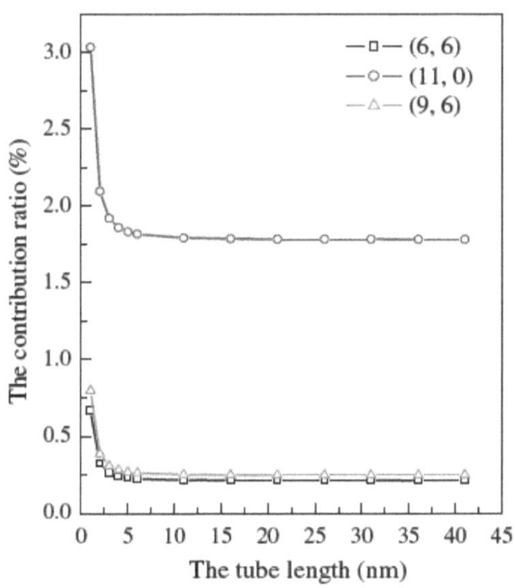

FIGURE 4.63

The contribution ratio of the higher-order terms to the critical buckling force versus the tube length [162].

4.9 VIBRATION CHARACTERISTICS

Similar to buckling behaviors, accurate estimation of vibration characteristics of carbon nanostructures is a key issue of great interest in material science and engineering applications such as resonators, oscillators, transducers, and detectors. Scholars have exerted themselves to establish many effective methods to unfold the dynamics properties of carbon nanostructures.

To show the possible application of carbon nanostructures in nanoresonators, Li and Chou [166,167] presented a molecular-structural-mechanics method to simulate vibration behaviors of CNTs. Their results demonstrated that SWCNTs have ultrahigh fundamental frequency, a level of 10 GHz–1.5 THz for cantilevered or bridged constraints. It was also observed that fundamental frequencies of DWCNTs are 10% lower than those of SWCNTs under the same outer diameter and the noncoaxial vibration started at the third vibration mode. He et al. [168] employed an elastic multiple shell model by considering vdW interaction between any two layers as the radius-dependent function for vibration analysis of MWCNTs. For an MWCNT with a small radius, their numerical results demonstrated that the interlayer vdW interaction exhibited a significant influence on vibration behaviors. Natsuki et al. [169] presented a wave propagation approach for vibration analysis of CNTs as well as embedded CNTs. Results showed that the order of vibration frequencies of CNTs reached a level of terahertz but in a wide range. Wang and Varadan [170] developed a nonlocal continuum mechanics model based on elastic beam theories to analyze the vibration characteristic of both SWCNTs and DWCNTs. Lu et al. [171] provided consistent equations of motion for the nonlocal Timoshenko beam model to study wave and vibration behaviors of CNTs. Hsu et al. [172] derived an analytical solution of fundamental frequency for chiral SWCNTs subjected to thermal vibration, using the Timoshenko beam model. On comparing results obtained by the Timoshenko beam model with the Euler beam model, it is found that the former's frequency is lower than the latter. Considering two nanotubes shells coupled by vdW interaction, Natsuki et al. [173] employed Euler–Bernoulli beam theory to analyze vibration characteristics of DWCNTs. Yang et al. [174] applied nonlocal Timoshenko beam theory to study nonlinear free vibration of SWCNTs and found the nonlocal parameters have a slight effect on the nonlinear mode shape but considerable effect on frequencies. Wang et al. [175] developed an elastic Bernoulli–Euler beam model based on the thermal elasticity mechanics theory to analyze the vibration and instability of fluid-conveying SWCNTs. It was found that the temperature change plays a significant role in buckling stability and natural frequency of a fluid-conveying SWCNT. Later, Yan et al. [176] treated TWCNTs as Euler–Bernoulli beam, considering the interlayer vdW interactions, to study natural frequencies and stability problems of the fluid-conveyed TWCNTs. It was found that interlayer vdW interactions have a large effect on natural frequencies and the internal moving fluid on instability of TWCNTs. Using the Timoshenko beam model, Chang and Lee [177] analyzed flexural vibration behaviors of SWCNTs which contain a fluid flow. A significant effect of rotary inertia and transverse shear deformation on vibration frequencies is observed. Chowdhury et al. [178] used an MM approach to investigate vibration properties of two types of SWCNTs and found natural frequencies of zigzag CNTs to be higher than those of the counterpart armchair cases. Georgantzinos and Anifantis [179] modeled MWCNTs using springs and lumped masses to study the vibration characteristic. Hashemnia et al. [180] determined fundamental frequencies of CNTs and graphene sheets by means of molecular structural mechanics approach. It is observed that fundamental frequencies of CNTs are far larger than those of graphene sheets. Reddy et al. [181] employed an atomistic simulation and a continuum shell modeling for the analysis of free vibrations of SWCNTs. They pointed out that a minimum potential configuration of CNT was important for an accurate computation of stiffness matrix.

4.9.1 ANALYSIS OF FREE VIBRATION CHARACTERISTIC OF CARBON NANOSTRUCTURES

Hamilton's principle is an important variational principle in elastodynamics and is adopted herein for the study of free vibration of carbon nanostructures. The real solution of histories of the motion in time period $t_1 \sim t_2$ is determined by minimization of the Lagrangian function, given by:

$$\delta \int_{t_1}^{t_2} (E(t) - W(t) + L(t))dt = 0 \tag{4.221}$$

where $E(t)$ and $W(t)$ denote kinetic energy and elastic energy stored in the carbon nanostructures, respectively, while $L(t)$ is total work done by the external force on the body and is not considered in analysis of free vibration since it is free of external force.

4.9.2 QUASICONTINUUM MODEL

Variations of kinetic energy of the system and elastic energy stored in carbon nanostructures are:

$$\delta \int_{t_1}^{t_2} E(t)dt = \delta \int_{t_1}^{t_2} \int_V \frac{1}{2} \rho \mathbf{U}'(t)^T \mathbf{U}'(t) dV dt = -\int_{t_1}^{t_2} \delta(\hat{\mathbf{U}}(t)^T) \int_V \rho \mathbf{\Phi}^T \mathbf{\Phi} \hat{\mathbf{U}}''(t) dV dt \tag{4.222}$$

$$\delta \int_{t_1}^{t_2} W(t)dt = \delta \int_{t_1}^{t_2} \int_V \frac{1}{2} \mathbf{grad}^T \frac{\partial^2 \omega_0}{\partial \mathbf{grad} \partial \mathbf{grad}} \mathbf{grad}\, dV dt = \int_{t_1}^{t_2} \delta(\hat{\mathbf{U}}(t)^T) \int_V (\mathbf{L\Phi})^T \mathbf{ML\Phi} \hat{\mathbf{U}}(t) dV dt \tag{4.223}$$

where ρ is the area mass density calculated by $\frac{m_c}{\Omega_I}$ with m_c being 1.9943×10^{-23} g.

Substituting Eqs. (4.222) and (4.223) into Eq. (4.221), we have:

$$\int_{t_1}^{t_2} \delta(\hat{\mathbf{U}}(t)^T) \int_V \rho \mathbf{\Phi}^T \mathbf{\Phi} \hat{\mathbf{U}}''(t) dV dt + \int_{t_1}^{t_2} \delta(\hat{\mathbf{U}}(t)^T) \int_V (\mathbf{L\Phi})^T \mathbf{ML\Phi} \hat{\mathbf{U}}(t) dV dt = 0 \tag{4.224}$$

Thus, the dynamic equation of the system is:

$$\int_V [\rho \mathbf{\Phi}^T \mathbf{\Phi} \hat{\mathbf{U}}''(t) + (\mathbf{L\Phi})^T \mathbf{ML\Phi} \hat{\mathbf{U}}(t)] dV = 0 \tag{4.225}$$

Considering the field function $\hat{\mathbf{U}}(t) = \mathbf{A} exp(i\omega t)$, Eq. (4.225) can be rewritten as:

$$\int_V [(\mathbf{L\Phi})^T \mathbf{ML\Phi} - \omega^2 \rho \mathbf{\Phi}^T \mathbf{\Phi}] \hat{\mathbf{U}} dV = 0 \tag{4.226}$$

where $\omega = 2\pi f$ is angular frequency. For simplifying, we define:

$$\mathbf{K} = \int_V (\mathbf{L\Phi})^T \mathbf{ML\Phi} dV \tag{4.227}$$

$$\mathbf{M}_A = \int_V \rho \mathbf{\Phi}^T \mathbf{\Phi} dV \tag{4.228}$$

The solutions of natural frequency of free vibration as well as the associated vibration mode become a matter of solving eigenvalue problems.

4.9.3 ATOMISTIC SIMULATION

To validate the accuracy of the application of the higher-order deformation gradient theory on analysis of free vibration of SWCNTs, we also derived the dynamic equation using atomistic simulation, which has the same solution framework as the mesh-free computational framework. The governing equation of free vibration is given as:

$$\sum_{i=1}^{NN}\left(m_i\mathbf{U}_i'' + \left.\frac{\partial^2 E_{A(i)}}{\partial\mathbf{x}_i\partial\mathbf{x}_i}\right|_{x_0}\mathbf{U}\right) = 0 \tag{4.229}$$

where NN is the total number of atoms, E_A is the potential stored in C–C bond, m_i is the mass per atom, $\mathbf{U}$ is the displacement field, and x_i is the atom's coordinate.

Considering the field function $\mathbf{U}(t) = \mathbf{A}exp(i\omega t)$, we obtain the dynamic equation:

$$\sum_{i=1}^{NN}\left(\left.\frac{\partial^2 E_{A(i)}}{\partial\mathbf{x}_i\partial\mathbf{x}_i}\right|_{\mathbf{x}_0} - \omega^2 m_i\right)\mathbf{U} = 0 \tag{4.230}$$

The vibration characteristic of SWCNTs can be obtained from the solution of Eq. (4.230). Free vibration frequencies of carbon nanostructures with various constraints, lengths, chiralities, and diameters are investigated. The aforementioned two initial equilibrium configurations of SWCNTs with and without considering edge effect are simulated using the present method in the next section. Atomistic simulation is performed for comparison.

4.9.4 FREE VIBRATION OF SWCNTs

In this section, performance of the numerical framework on the analysis of free vibration of SWCNTs, in terms of accuracy, convergence, and stability, is comprehensively evaluated. It is well known that accuracy of the mesh-free computational framework is greatly dependent on the arrangement of the mesh-free nodes and the value of the compact support factors. The number of mesh-free nodes required in circumferential and longitudinal directions is discussed, along with various compact support factors. For the purpose of validating and appraising the proposed approach, two types of CNTs are selected, a 6 nm-long armchair (7, 7) CNT and a 2 nm-long zigzag (8, 0) CNT. The vibration modes and free vibration frequencies are calculated by the higher-order deformation gradient continuum theory along with atomistic simulation for comparison. In most literature, the compact support factor is generally in the range of 2.0–4.0. In this research, three values of the compact support factor, i.e., 2.0, 3.0, and 4.0, are used to test its influence on the final numerical results. The number of mesh-free nodes arranged in circumferential direction for both types of CNTs is within a range of 10–14. The number of nodes arranged in longitudinal direction for the longer (7, 7) CNT is within a range of 23–45 and it is within 12–20 for the shorter (8, 0) CNT. In accordance with vibration modes obtained from atomistic simulation, selected values of the compact support factor affecting computational results can be listed as follows.

In the study of the (7, 7) CNT, the obtained vibration modes demonstrate that the number of mesh-free nodes should not be less than 13 for values of 2.0 as well as 4.0 and an appropriate ratio, between circumferential and longitudinal nodes, is required for the case of 4.0, otherwise some of the vibration modes would appear in an incorrect order. On the other hand, an exciting result is obtained for the case of 3.0, where a minimum number of 10 is sufficient to accurately display vibration modes which are in agreement with atomistic simulation. This suggests that a value of 3.0 for the compact support factor is most suitable, providing good accuracy and stability for the numerical solution. The same phenomenon occurs for the zigzag CNT, i.e., the value of the compact support factor plays an important role in the arrangement of the number of mesh-free nodes in the circumferential direction. For the case of a value of 2.0, the number of circumferential mesh-free nodes should be chosen as either 12 or 13. Either less than 12 or more than 13 would cause inaccurately higher modes of vibration. For the case of a value of 3.0 or 4.0, if the number of circumferential nodes is set at over 11, accurate vibration modes can be obtained in agreement with those obtained from atomistic simulation. However, the number of longitudinal nodes has a strict limit of less than 15 if the compact support factor is chosen as 4.0. In summary, the compact support factor with a value of 3.0 is the most suitable parameter for analysis of free vibration of SWCNTs. For the settled compact support factor, dependence of the vibration frequency on the number of mesh-free nodes is described in Figs. 4.64–4.66.

The fold lines show the obtained frequency differences between atomistic simulation and the continuum approach with different parameters. Vibration frequencies of (7, 7) SWCNT are depicted in Figs. 4.64 and 4.65. From several sets of data plotted in Fig. 4.64, it is found that most of the fold lines demonstrate the same trend as the number of vibration modes increases but the case of using 10 mesh-free nodes in the circumferential direction, which certifies the present method, is stable once more. Sufficient mesh-free nodes arranged in circumferential direction are required in the free vibration analysis of CNTs and 10 nodes are too insufficient and may cause an inaccurate frequency. Comparison of fundamental frequencies shows that in the use of 11 mesh-free nodes in the circumferential direction, the difference is only 0.011%. Therefore, in the following study, the number of mesh-free nodes in the circumferential direction is chosen as 11. With a fixed number of mesh-free nodes in the circumferential direction, investigation of the influence of longitudinal direction on the vibration frequency is carried out and the numerical results are depicted in Fig. 4.65. Data in both two figures show that the free vibration frequencies tend to be close to those obtained from atomistic simulation as the number of vibration modes increases. The fold lines reveal clearly that the computational results obtained from the case of 11×23 mesh-free nodes are the closest to those obtained from atomistic simulation, compared with the others. This phenomenon can be explained by the fact that the vibration deformation is uniform and thus the displacement between neighboring nodes in circumferential and longitudinal directions should be equivalent. The data in Fig. 4.65 also suggests that the present method can provide an accurate vibration frequency, i.e., the difference of the fundamental frequency is 0.42%, while the highest difference, corresponding to the third or fourth frequency, is 5.14%. The same phenomenon is also found in investigation of vibration frequency of the short 2 nm (8, 0) CNT. Fig. 4.66 shows that using an appropriate ratio of mesh-free nodes between circumferential and longitudinal directions is more helpful to improve the accuracy of numerical results rather than using an increasing number of mesh-free nodes.

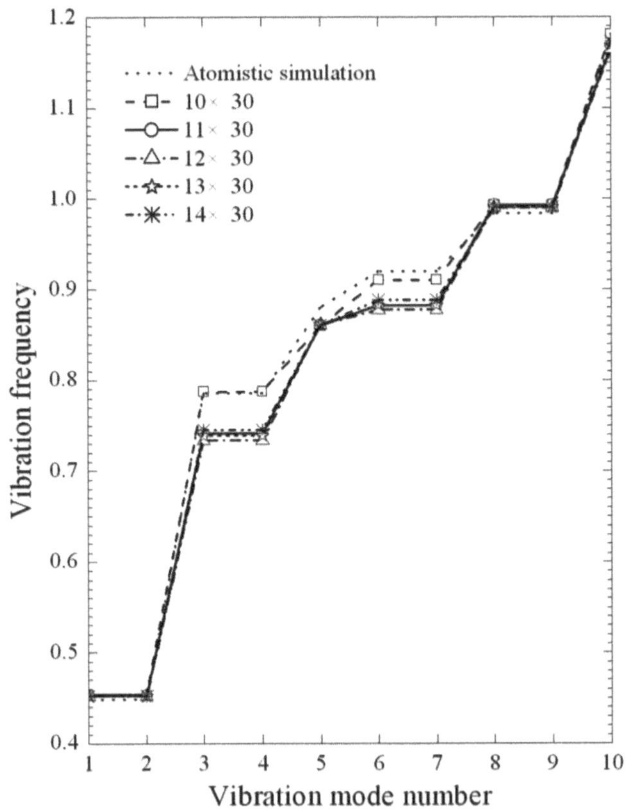

FIGURE 4.64

With a set value of 3.0 and 30 for the compact support factor and longitudinal nodes, the fold lines show the relationships of obtained free vibration frequencies differences, corresponding to the increased number of vibration mode of a fixed-fixed 6 nm (7, 7) CNT, between atomistic simulation and continuum approach using different number of circumferential nodes [182].

4.9.5 THE EDGE EFFECT ON THE FREE VIBRATION FREQUENCY

Due to the extremely small size of nanostructures, edge effect plays an important role in certain mechanical properties. In this section, the edge effect on the free vibration frequency is discussed in detail. The data listed in Tables 4.13 and 4.14 demonstrate that the edge effect has a distinct influence on the free vibration frequency.

Considering the edge effect on the tube, the predicted frequency approaches that of atomistic simulation much more. In the matter of fixed-fixed constraint, predicted free vibration frequencies of the two types of CNTs are quite close to those obtained from atomistic simulation. In order to clearly explain this point, we take the fundamental frequency as an example, where the frequency difference between a fixed-fixed 6 nm (7, 7) and a 2 nm (8, 0) CNT is 0.63% and 0.16%,

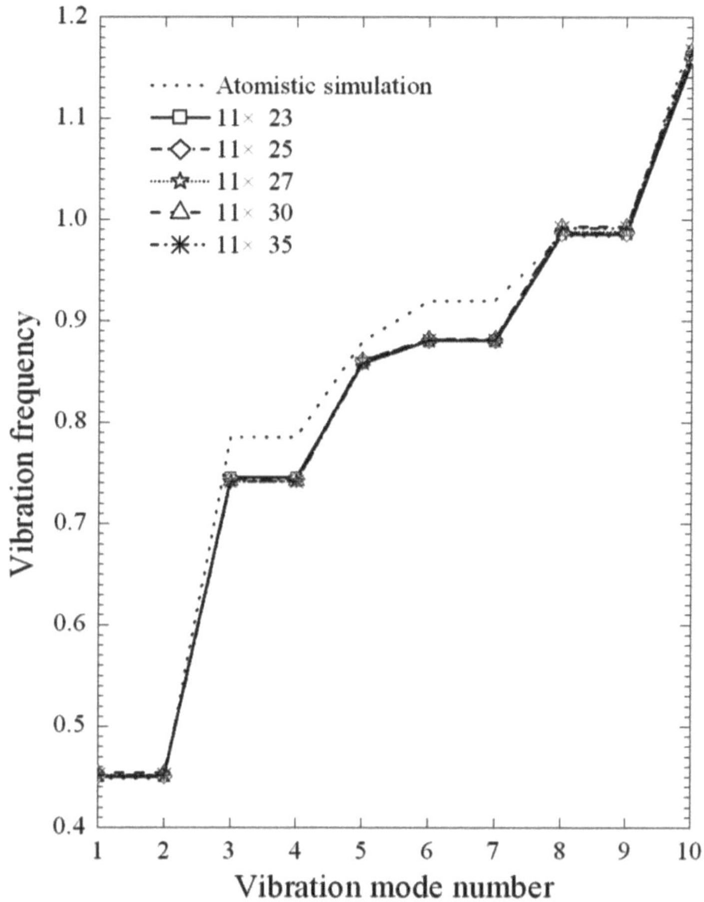

FIGURE 4.65

The relationships of free vibration frequencies differences for a 6 nm (7, 7) CNT between atomistic simulation and continuum approach using different combinations of mesh-free nodes in circumferential and longitudinal directions [182].

respectively. However, the predicted frequency differences increase to 2.07% and 4.6% if the edge effect is not considered, indicating that the edge effect is larger on a shorter tube. This conclusion can especially be extended to the free vibration study of fixed-free CNTs. For the longer CNT, the frequency difference is only 0.19%, while it is 13.2% for the shorter one.

A reasonable explanation of this phenomenon is that the representative cell actually possesses two neighboring atoms rather than three in the present model, thus largely affecting the free end. An increasing tube length leads to a decreasing proportion of edge atoms in the whole system and thus a decreasing influence on the free vibration frequency of CNTs. The 2 nm-long (8, 0) CNT has 160 atoms and 16 edge atoms, meaning a very large proportion of edge atoms from the total

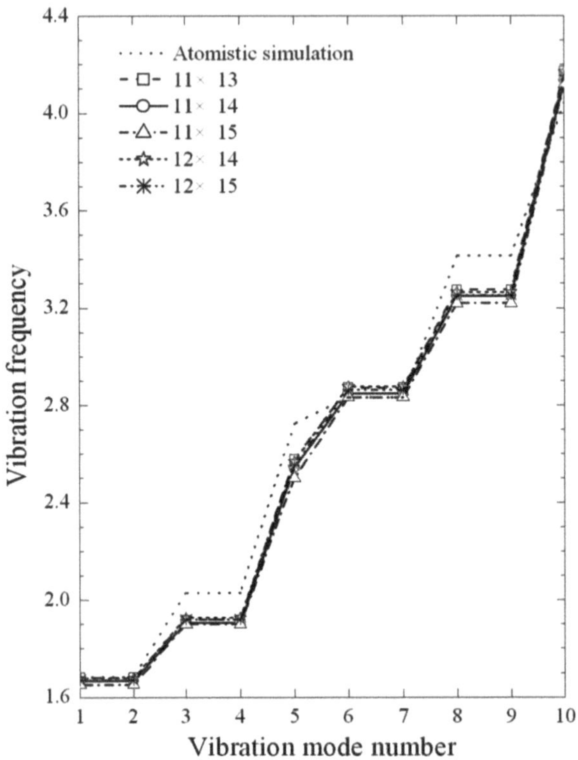

FIGURE 4.66

The free vibration frequencies of a 2 nm (8, 0) CNT obtained from atomistic simulation and continuum approach using different combinations of mesh-free nodes in circumferential and longitudinal directions [182].

atoms, and this causes a large difference. The accuracy of the present method increases as the tube length increases and thus it is undoubtedly promising for treatment for a larger system problem. The first ten modes of vibration plotted in Fig. 4.67A obtained using the present method are found in good agreement with atomistic simulation results (Fig. 4.67B). It is noteworthy that the edge effect results in a slightly radial warp on ends of the tube.

4.9.6 CHIRAL EFFECT AND CRITICAL DIAMETER

Although there are many research works focused on potential application of CNTs as highly sensitive resonators, most concentrate on the investigation of armchair or zigzag CNTs. To the best knowledge of the author, few research works have focused on the investigation of free vibration of chiral CNTs, because of structural complexities. It is clear that chirality of CNTs can be exhibited by that of the representative cell. Therefore, the present method provides an easy approach, using

Table 4.13 The Free Vibration Frequency (THz) of a 6 nm-long (7, 7) CNT [182]

	Fixed-Fixed			Fixed-Free		
Mode No.	Edge Effect	Ideal Tube	Atomistic Simulation	Edge Effect	Ideal Tube	Atomistic Simulation
1	0.451117	0.439001	0.448282	0.090060	0.088903	0.089887
2	0.451117	0.439001	0.448282	0.090060	0.088903	0.089887
3	0.743779	0.743296	0.785692	0.393696	0.393097	0.430134
4	0.743779	0.743296	0.785692	0.469199	0.462419	0.460079
5	0.860923	0.859385	0.88023	0.469199	0.462419	0.460079
6	0.881329	0.869775	0.920132	0.717427	0.679361	0.733827
7	0.881329	0.869775	0.920132	0.717427	0.679361	0.752279
8	0.986929	0.966139	0.983911	0.750227	0.7378	0.752279
9	0.986929	0.966139	0.983911	0.765586	0.7378	0.791594
10	1.161059	1.131084	1.181721	0.765586	0.746202	0.791594

Table 4.14 The Free Vibration Frequency (THz) of a 2 nm-long (8, 0) CNT [182]

	Fixed-Fixed			Fixed-Free		
Mode No.	Edge Effect	Ideal Tube	Atomistic Simulation	Edge Effect	Ideal Tube	Atomistic Simulation
1	1.682302	1.607504	1.685023 (1.68586)[a]	0.48393	0.476434	0.427508
2	1.682302	1.607504	1.685023 (1.68586)[a]	0.48393	0.476434	0.427508
3	1.918366	1.86787	2.031057 (2.03224)[a]	0.995104	1.012554	1.281989
4	1.918366	1.86787	2.031057 (2.03224)[a]	1.671384	1.458563	1.782553
5	2.579104	2.566002	2.725368 (2.72686)[a]	1.671384	1.458563	1.782553
6	2.872692	2.762155	2.833036 (2.83496)[a]	1.794172	1.797961	1.789452
7	2.872692	2.762155	2.833036 (2.83496)[a]	1.794172	1.797961	1.789452
8	3.274427	3.244976	3.413184 (3.41544)[a]	2.14179	1.969586	2.029901
9	3.274427	3.244976	3.413184 (3.41544)[a]	2.164217	1.969586	2.121707
10	4.173275	4.104967	4.046778 (4.04955)[a]	2.164217	2.13572	2.121707

[a]*The values in Ref. [181].*

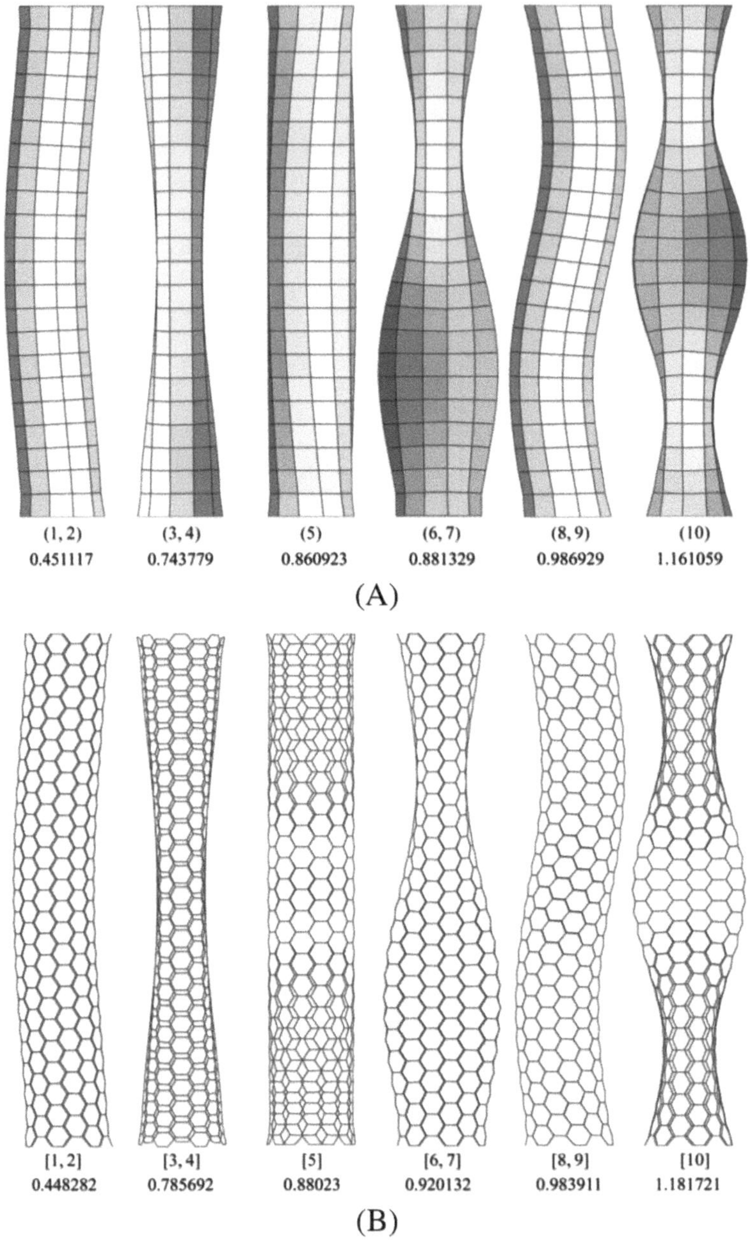

FIGURE 4.67

First 10 natural frequencies (THz) and associated vibration modes of a 6 nm (7, 7) SWCNT with fixed-fixed constraint obtained using the continuum approach (A) and atomistic simulation (B) [182].

different chiral representative cells, to simulate the vibration characteristic of CNTs of different types of chirality.

SWCNTs of various types of chirality are taken for investigation and computational results are given in Figs. 4.68 and 4.69. For a CNT with a fixed length, it is found that fundamental frequency increases initially as the tube diameter increases but then decreases. For the higher modes of vibrations, there is no regular pattern. However, compared with the two different lengths of tubes, it can be found that natural frequencies decrease generally as the tube length increases. Especially, as the length increases, the change of natural frequencies of CNTs with the former constraint is larger than the latter one.

We study the effect of chirality on the frequency by using a series of CNTs, i.e., (n, 0), (n, n), and (2n, n), with the same length of 12 nm. The relationships of fundamental frequencies and tube diameters are plotted in Figs. 4.70 and 4.71.

For the case of fixed-fixed constraint, fundamental frequencies of all CNTs of various types of chirality are quite close to each other, while for the case of fixed-free constraint, the chirality has a slightly increasing effect. Zigzag CNTs possess larger chirality and thus have larger fundamental frequency compared with the other counterpart chiral CNTs. As the tube diameter increases, the effect of chirality on fundamental frequency becomes inconspicuous The maximum difference is only 1.01% when the diameter reaches 2.5 nm. For fixed-fixed constraint, however, it is noteworthy that two distinct trends occur prior to and beyond the critical diameter. Prior to the critical diameter, fundamental frequency increases as the tube diameter increases, while beyond the critical point, it decreases as the diameter keeps increasing. This phenomenon can be explained by the fact that mechanical properties of a CNT with an increasing diameter approach to those of graphite sheet. Fundamental frequency of a 12 nm (15, 15) graphite sheet is approximately 0.08 THz [44], which is lower than the counterpart SWCNT. This indicates that fundamental frequency of CNTs should decrease as the diameter increases. It can be predicted that the fundamental frequency of SWCNTs will be close to that of graphite sheet as the diameter continues to increase. It should be noted that both the constraint and tube length have a significant influence on the critical diameter. As far as constraints are concerned, fixed-free style has a larger critical diameter than the fixed-fixed one. The increase of tube length gives rise to the increase of the critical diameter. In view of these issues, CNTs with the same length, 2 nm, are selected for the fundamental frequency analysis using the present method, along with atomistic simulation. Compared with SWCNTs of various types of chirality, it is found that the critical diameter is independent of chirality. Therefore, we only chose armchair CNTs as samples for the investigation of critical diameter associated with the length. The results are plotted in Fig. 4.72. It clearly shows that the latter constraint has a larger critical diameter than the former style and for such shorter CNTs, the critical diameter is generally smaller than the longer one. The critical diameter of a SWCNT versus the tube length is depicted in Fig. 4.73. It is found that the critical diameter is an approximately linear function of its corresponding tube length.

In conclusion, initial equilibrium configurations with consideration of the edge effect are chosen for free vibration analysis of SWCNTs with different lengths, diameters, chiralities, and constraints. Numerical results obtained from the present continuum method are found to be in agreement with those obtained from atomistic simulation. Especially, in the treatment of a large system, the present method exhibits promising application. It is found that there is a critical diameter. Fundamental frequency increases prior to it as the tube diameter increases but decreases beyond it. The value of

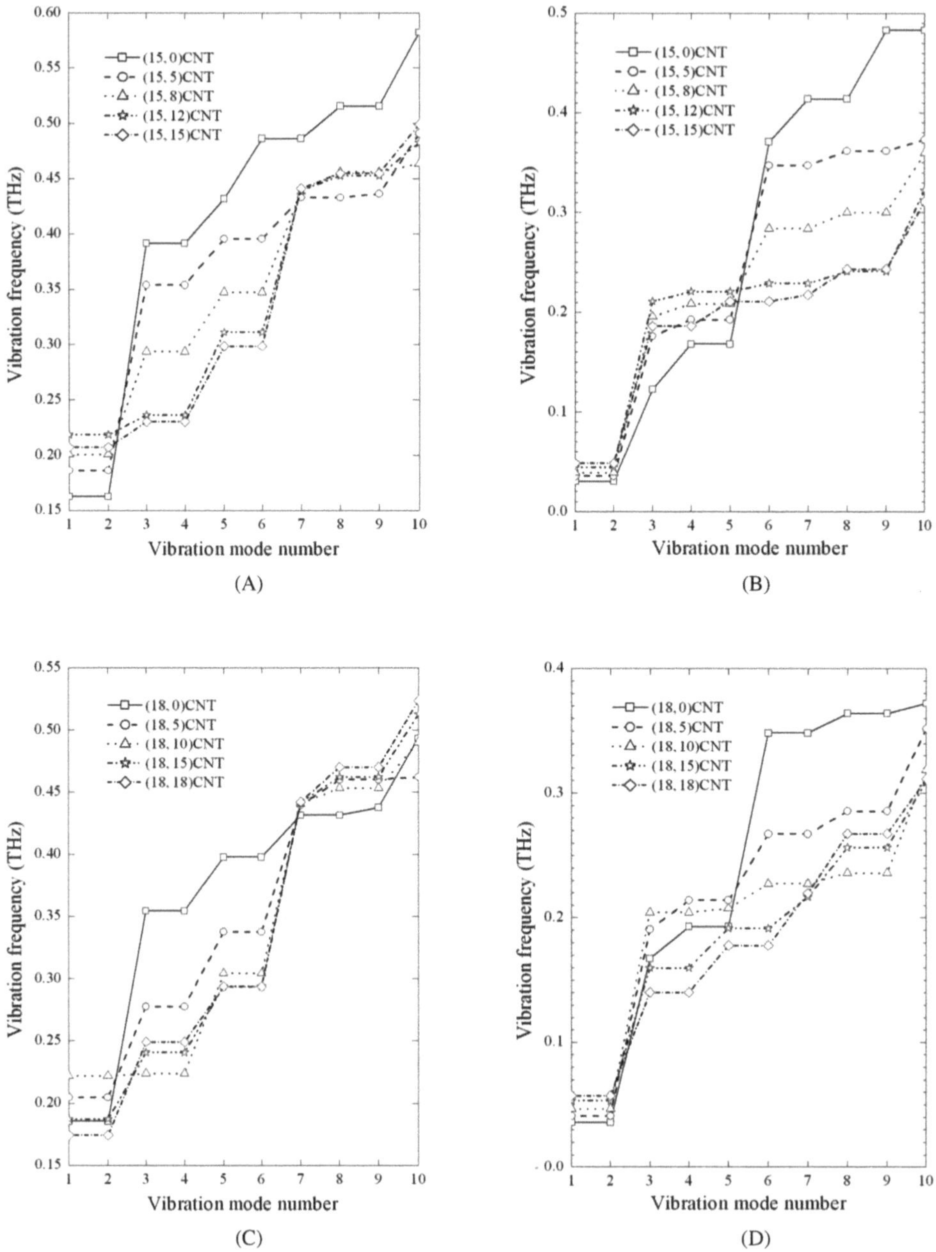

FIGURE 4.68

The vibration frequencies of various kinds of CNTs with the same length, 12 nm, and different constraints: (A) and (C) fixed-fixed, (B) and (D) fixed-free [182].

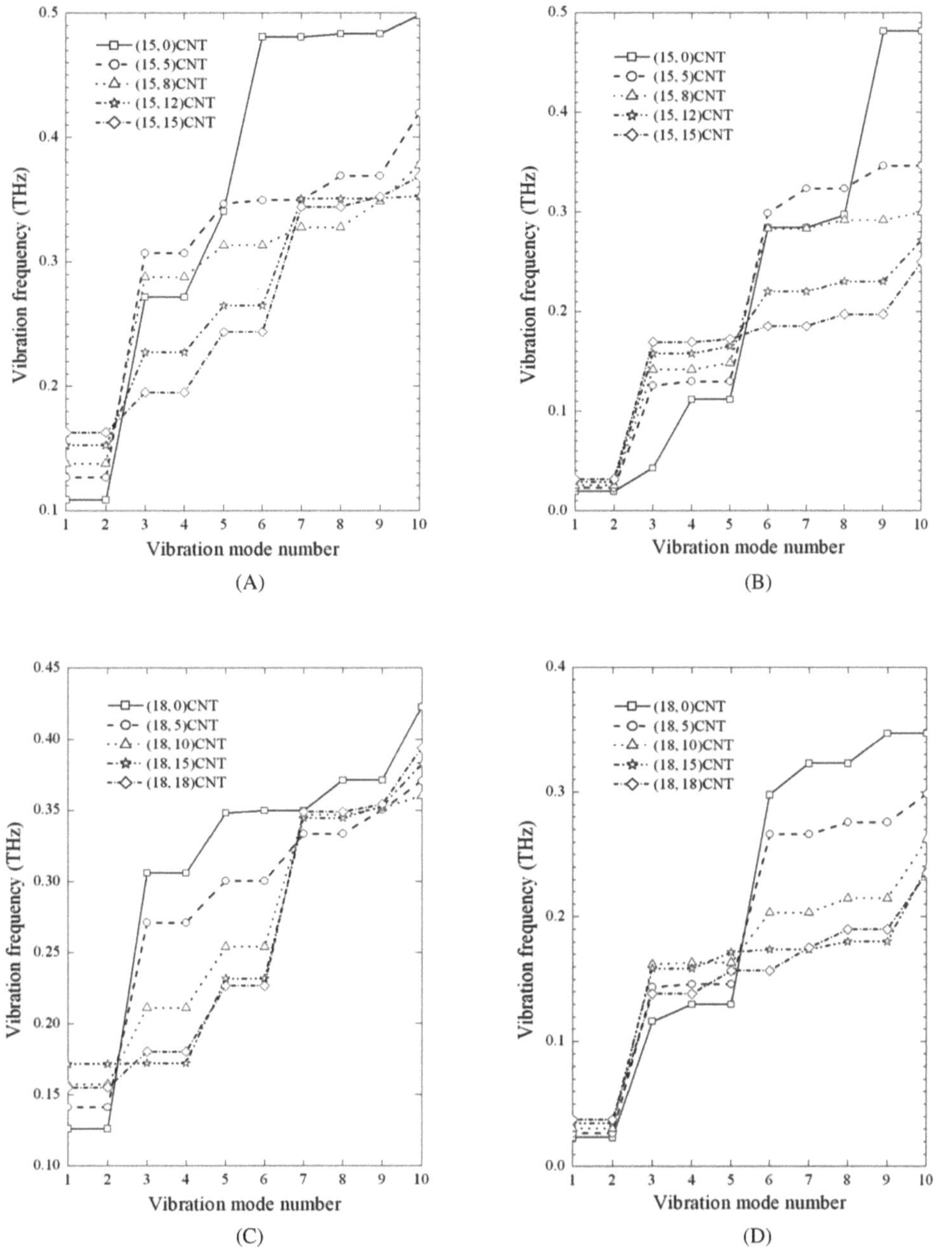

FIGURE 4.69

The vibration frequencies of various kinds of CNTs with the same length, 15 nm, and different constraints: (A) and (C) fixed-fixed, (B) and (D) fixed-free [182].

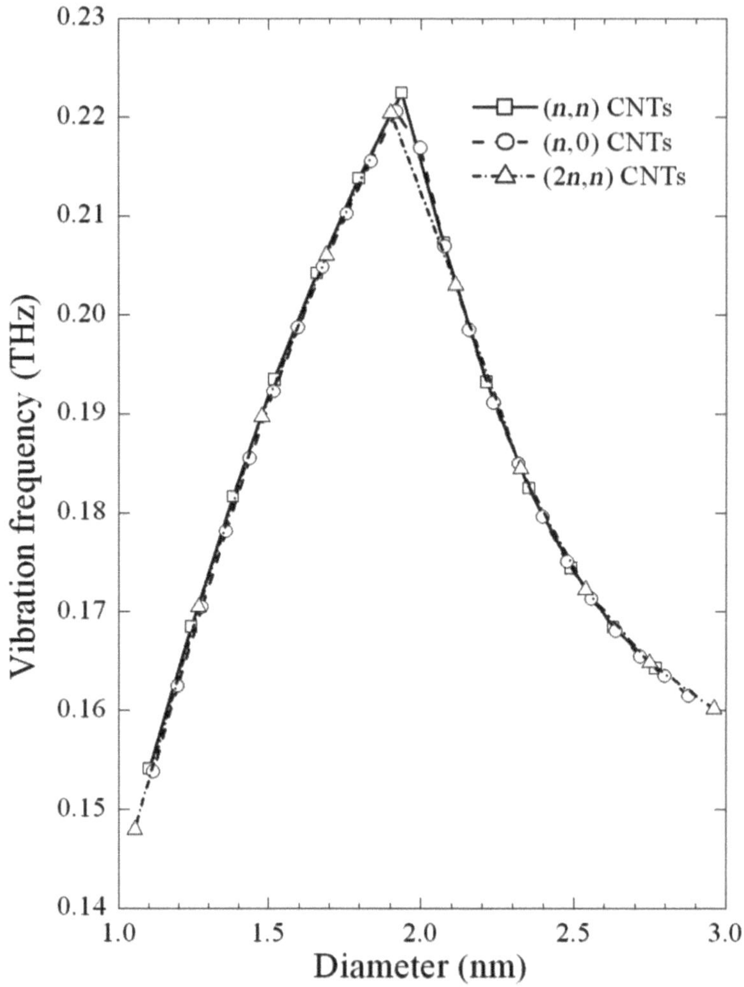

FIGURE 4.70

Fundamental frequencies of kinds of fixed-fixed CNTs, i.e., (n, n), $(n, 0)$, and $(2n, n)$, with a length of 12 nm versus tube diameter [182].

critical diameter is largely dependent on the tube length and constraint but independent of the chirality.

1. The increase of length results in an increase in the critical diameter, approximately a linear function relationship.
2. Fixed-free constraint has a larger critical diameter than the fixed-fixed style.

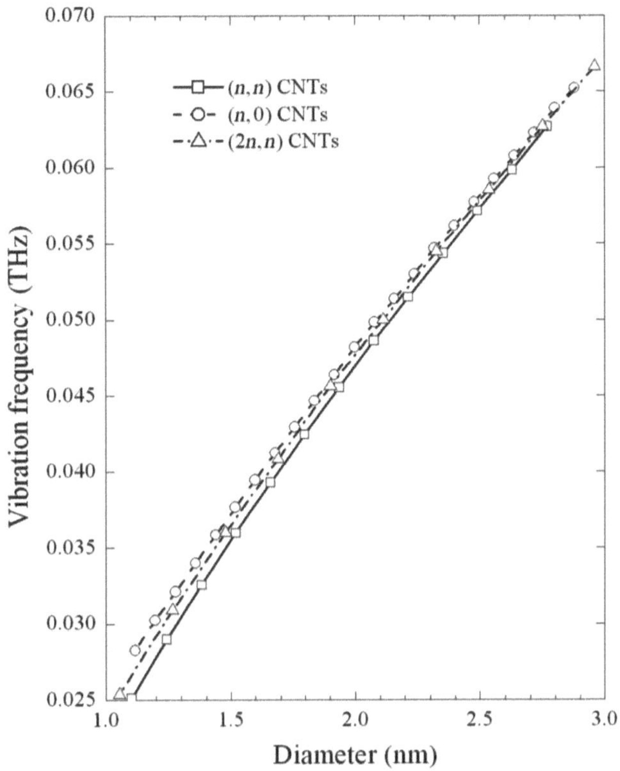

FIGURE 4.71

Fundamental frequencies of kinds of fixed-free CNTs, i.e., (*n*, *n*), (*n*, 0), and (2*n*, *n*), with a length of 12 nm versus tube diameter [182].

3. For the case of fixed-fixed constraint, the chirality has little effect on vibration frequency, while there is a slight effect for fixed-free style. However, the effect becomes negligible as tube diameter increases.

As the tube diameter continues to increase, mechanical properties of SWCNTs converge to graphite sheets and this principle is also applicable to free vibration frequencies: The change of fundamental frequency becomes smaller and smaller as the tube diameter increases and frequency converges to that of the counterpart graphite sheet.

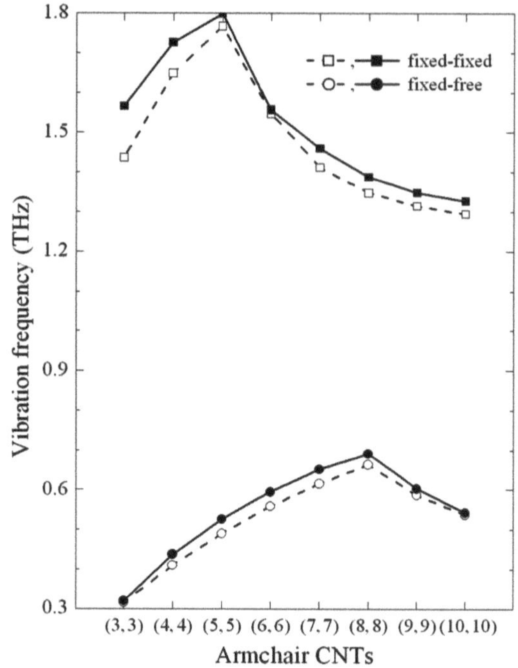

FIGURE 4.72

Fundamental frequencies of armchair CNTs with a length of 2 nm and different constraints. The solid and open lines are the results obtained from continuum approach and atomistic simulation [182].

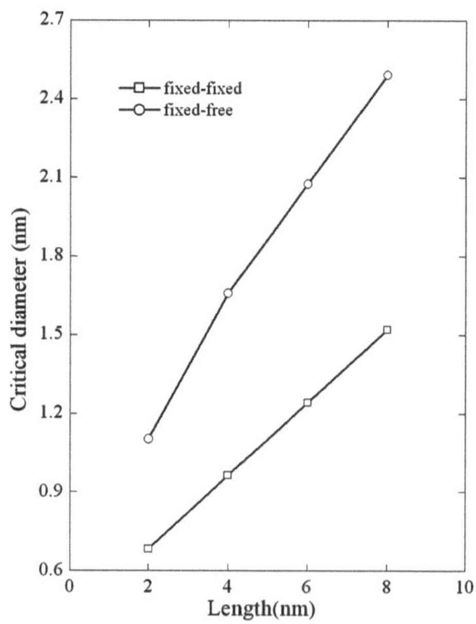

FIGURE 4.73

Critical diameters versus the tube lengths for armchair SWCNTs [182].

REFERENCES

[1] Govindjee S, Sackman JL. On the use of continuum mechanics to estimate the properties of nanotubes. Solid State Commun 1999;110:227–30.
[2] Ru CQ. Effective bending stiffness of carbon nanotubes. Phys Rev B 2000;62:9973–6.
[3] Ru CQ. Axially compressed buckling of a doublewalled carbon nanotube embedded in an elastic medium. J Mech Phys Solids 2001;49:1265–79.
[4] Li CY, Chou TW. Elastic properties of single-walled carbon nanotubes in transverse directions. Phys Rev B 2004;69:073401.
[5] He XQ, Kitipornchai S, Liew KM. Buckling analysis of multi-walled carbon nanotubes: a continuum model accounting for van der Waals interaction. J Mech Phys Solids 2005;53:303–26.
[6] Liew KM, He XQ, Kitipornchai S. Buckling characteristics of embedded multi-walled carbon nanotubes. Proc Royal Soc a-Math Phys Eng Sci 2005;461:3785–805.
[7] Wang LF, Hu HY. Flexural wave propagation in single-walled carbon nanotubes. Phys Rev B 2005;71:195412.
[8] Wang LF, Hu HY, Guo WL. Validation of the non-local elastic shell model for studying longitudinal waves in single-walled carbon nanotubes. Nanotechnology 2006;17:1408–15.
[9] Tadmor EB, Smith GS, Bernstein N, Kaxiras E. Mixed finite element and atomistic formulation for complex crystals. Phys Rev B 1999;59:235–45.
[10] Tadmor EB, Ortiz M, Phillips R. Quasicontinuum analysis of defects in solids. Philos Mag a-Phys Condens Matter Struct Defects Mechan Propert 1996;73:1529–63.
[11] Shenoy VB, Miller R, Tadmor EB, Rodney D, Phillips R, Ortiz M. An adaptive finite element approach to atomic-scale mechanics—the quasicontinuum method. J Mech Phys Solids 1999;47:611–42.
[12] Liew KM, Chen BJ, Xiao ZM. Analysis of fracture nucleation in carbon nanotubes through atomistic-based continuum theory. Phys Rev B 2005;71:235424.
[13] Zhang P, Huang Y, Geubelle PH, Klein PA, Hwang KC. The elastic modulus of single-wall carbon nanotubes: a continuum analysis incorporating interatomic potentials. Inter J Solids Struct 2002; 39:3893–906.
[14] Zhang P, Jiang H, Huang Y, Geubelle PH, Hwang KC. An atomistic-based continuum theory for carbon nanotubes: analysis of fracture nucleation. J Mech Phys Solids 2004;52:977–98.
[15] Zhang P, Huang Y, Gao H, Hwang KC. Fracture nucleation in single-wall carbon nanotubes under tension: a continuum analysis incorporating interatomic potentials. J Appl Mech-Trans Asme 2002; 69:454–8.
[16] Arroyo M, Belytschko T. An atomistic-based finite deformation membrane for single layer crystalline films. J Mech Phys Solids 2002;50:1941–77.
[17] Arroyo M, Belytschko T. Finite element methods for the non-linear mechanics of crystalline sheets and nanotubes. Int J Numer Methods Eng 2004;59:419–56.
[18] Arroyo M, Belytschko T. Finite crystal elasticity of carbon nanotubes based on the exponential Cauchy-Born rule. Phys Rev B 2004;69:115415.
[19] Sunyk R, Steinmann P. On higher gradients in continuum-atomistic modelling. Inter J Solids Struct 2003;40:6877–96.
[20] Guo X, Wang JB, Zhang HW. Mechanical properties of single-walled carbon nanotubes based on higher order Cauchy-Born rule. Inter J Solids Struct 2006;43:1276–90.
[21] Wang JB, Guo X, Zhang HW, Wang L, Liao JB. Energy and mechanical properties of single-walled carbon nanotubes predicted using the higher order Cauchy-Born rule. Phys Rev B 2006;73:115428.
[22] Sun YZ, Liew KM. Mesh-free simulation of single-walled carbon nanotubes using higher order Cauchy-Born rule. Comput Mater Sci 2008;42:444–52.

[23] Sun YZ, Liew KM. The buckling of single-walled carbon nanotubes upon bending: the higher order gradient continuum and mesh-free method. Computer Methods Appl Mechan Eng 2008;197:3001–13.

[24] Liew KM, Sun YZ. Elastic properties and pressure-induced structural transitions of single-walled carbon nanotubes. Phys Rev B 2008;77:205437.

[25] Xia ZC, Hutchinson JW. Crack tip fields in strain gradient plasticity. J Mech Phys Solids 1996; 44:1621–48.

[26] Tang Z, Shen S, Atluri SN. Analysis of materials with strain-gradient effects: a Meshless Local Petrov-Galerkin(MLPG) approach, with nodal displacements only. Cmes-Computer Model Eng Sci 2003; 4:177–96.

[27] Ferreira AJM, Roque CMC, Martins PALS. Radial basis functions and higher-order shear deformation theories in the analysis of laminated composite beams and plates. Composite Struct 2004;66:287–93.

[28] Ferreira AJM, Roque CMC, Jorge RMN. Analysis of composite plates by trigonometric shear deformation theory and multiquadrics. Comput Struct 2005;83:2225–37.

[29] Ferreira AJM, Roque CMC, Jorge RMN. Free vibration analysis of symmetric laminated composite plates by FSDT and radial basis functions. Computer Methods Appl Mech Eng 2005;194:4265–78.

[30] Ferreira AJM, Fasshauer GE, Batra RC. Natural frequencies of thick plates made of orthotropic, monoclinic, and hexagonal materials by a meshless method. J Sound Vibration 2009;319:984–92.

[31] Gu L. Moving kriging interpolation and element-free Galerkin method. Int J Numer Methods Eng 2003;56:1–11.

[32] Bui TQ, Nguyen TN, Nguyen-Dang H. A moving Kriging interpolation-based meshless method for numerical simulation of Kirchhoff plate problems. Int J Numer Methods Eng 2009;77:1371–95.

[33] Chen L, Liew KM. A local Petrov-Galerkin approach with moving Kriging interpolation for solving transient heat conduction problems. Comput Mech 2011;47:455–67.

[34] Zhu P, Liew KM. Free vibration analysis of moderately thick functionally graded plates by local Kriging meshless method. Composite Struct 2011;93:2925–44.

[35] Lucy LB. A numerical approach to the testing of the fission hypothesis. Astronom J 1977;82:1013–24.

[36] Gingold RA, Monaghan JJ. Smoothed particle hydrodynamics—theory and application to non-spherical stars. Mon Not R Astron Soc 1977;181:375–89.

[37] Johnson GR, Beissel SR. Normalized smoothing functions for SPH impact computations. Int J Numer Methods Eng 1996;39:2725–41.

[38] Swegle JW, Hicks DL, Attaway SW. Smoothed particle hydrodynamics stability analysis. J Comput Phys 1995;116:123–34.

[39] Dyka CT, Randles PW, Ingel RP. Stress points for tension instability in SPH. Int J Numer Methods Eng 1997;40:2325–41.

[40] Chen JK, Beraun JE, Jih CJ. An improvement for tensile instability in smoothed particle hydrodynamics. Computat Mech 1999;23:279–87.

[41] Vignjevic R, Campbell J, Libersky L. A treatment of zero-energy modes in the smoothed particle hydrodynamics method. Computer Methods Appl Mech Eng 2000;184:67–85.

[42] Monaghan JJ. Why particle methods work. Siam J Sci Statis Comput 1982;3:422–33.

[43] Monaghan JJ. An introduction to sph. Comput Phys Commun 1988;48:89–96.

[44] Monaghan JJ. Smoothed particle hydrodynamics. Annu Rev Astron Astrophys 1992;30:543–74.

[45] Monaghan JJ, Gingold RA. Shock simulation by the particle method sph. J Comput Phys 1983; 52:374–89.

[46] Gomez-Gesteira M, Dalrymple RA. Using a three-dimensional smoothed particle hydrodynamics method for wave impact on a tall structure. J Waterway Port Coastal Ocean Eng-Asce 2004;130:63–9.

[47] Jiang T, Ouyang J, Li Q, Ren JL, Yang BX. A corrected smoothed particle hydrodynamics method for solving transient viscoelastic fluid flows. Appl Math Model 2011;35:3833–53.

[48] Lahooti M, Pishevar A, Saidi MS. A novel 2D algorithm for fluid solid interaction based on the smoothed particle hydrodynamics (SPH) method. Scientia Iranica 2011;18:358–67.
[49] Rodriguez-Paz MX, Bonet J. A corrected smooth particle hydrodynamics method for the simulation of debris flows. Numer Methods Partial Differ Equ 2004;20:140–63.
[50] Gingold RA, Monaghan JJ. Kernel estimates as a basis for general particle methods in hydrodynamics. J Comput Phys 1982;46:429–53.
[51] Swegle JW, Attaway SW. On the feasibility of using smoothed particle Hydrodynamics for underwater explosion calculations. Computat Mech 1995;17:151–68.
[52] Johnson GR, Stryk RA, Beissel SR. SPH for high velocity impact computations. Computer Methods Appl Mech Eng 1996;139:347–73.
[53] Johnson GR, Beissel SR, Stryk RA. A Generalized Particle Algorithm for high velocity impact computations. Computat Mech 2000;25:245–56.
[54] Libersky LD, Petschek AG, Carney TC, Hipp JR, Allahdadi FA. High-strain lagrangian hydrodynamics—a 3-dimensional sph code for dynamic material response. J Comput Phys 1993;109:67–75.
[55] Randles PW, Libersky LD. Smoothed particle hydrodynamics: some recent improvements and applications. Computer Methods Appl Mech Eng 1996;139:375–408.
[56] Nayroles B, Touzot G, Villon P. Generalizing the finite element method: diffuse approximation and diffuse elements. Comput Mech 1992;10:307–18.
[57] Nayroles B, Touzot G, Villon P. The diffuse approximation. Comptes Rendus De L Academie Des Sciences Serie Ii 1991;313:293–6.
[58] Nayroles B, Touzot G, Villon P. The diffuse elements method. Comptes Rendus De L Academie Des Sciences Serie Ii 1991;313:133–8.
[59] Lancaster P, Salkauskas K. Surfaces generated by moving least-squares methods. Math Computat 1981;37:141–58.
[60] Belytschko T, Lu YY, Gu L. Element-free Galerkin methods. Int J Numer Methods Eng 1994;37:229–56.
[61] Chung HJ, Belytschko T. An error estimate in the EFG method. Computat Mech 1998;21:91–100.
[62] Dolbow J, Belytschko T. Numerical integration of the Galerkin weak form in meshfree methods. Computat Mech 1999;23:219–30.
[63] Belytschko T, Krongauz Y, Fleming M, Organ D, Liu WKS. Smoothing and accelerated computations in the element free Galerkin method. J Computat Appl Math 1996;74:111–26.
[64] Krongauz Y, Belytschko T. EFG approximation with discontinuous derivatives. Int J Numer Methods Eng 1998;41:1215–33.
[65] Krysl P, Belytschko T. Analysis of thin plates by the element-free Galerkin method. Computat Mech 1995;17:26–35.
[66] Krysl P, Belytschko T. Analysis of thin shells by the element-free Galerkin method. Inter J Solids Struct 1996;33:3057–78.
[67] Belytschko T, Gu L, Lu YY. Fracture and crack-growth by Element Free Galerkin Methods. Model Simulat Mater Sci Eng 1994;2:519–34.
[68] Belytschko T, Lu YY, Gu L. Crack-propagation by Element-Free Galerkin Methods. Eng Fracture Mechan 1995;51:295–315.
[69] Belytschko T, Lu YY, Gu L, Tabbara M. Element-free Galerkin methods for static and dynamic fracture. Inter J Solids Struct 1995;32:2547–70.
[70] Lu YY, Belytschko T, Tabbara M. Element-free Galerkin method for wave-propagation and dynamic fracture. Computer Methods Appl Mech Eng 1995;126:131–53.
[71] Belytschko T, Tabbara M. Dynamic fracture using element-free Galerkin methods. Int J Numer Methods Eng 1996;39:923–38.

[72] Fleming M, Chu YA, Moran B, Belytschko T. Enriched element-free Galerkin methods for crack tip fields. Int J Numer Methods Eng 1997;40:1483–504.
[73] Sukumar N, Moran B, Black T, Belytschko T. An element-free Galerkin method for three-dimensional fracture mechanics. Computat Mech 1997;20:170–5.
[74] Krysl P, Belytschko T. The Element Free Galerkin method for dynamic propagation of arbitrary 3-D cracks. Int J Numer Methods Eng 1999;44:767–800.
[75] Ponthot JP, Belytschko T. A combined Element-Free Galerkin method/arbitrary Lagrangian-Eulerian formulation for dynamic crack propagation. Iutam Symposium on Non-Linear Singularities in Deformation and Flow, Proceedings. 1999:205-16.
[76] Belytschko T, Organ D, Gerlach C. Element-free Galerkin methods for dynamic fracture in concrete. Computer Methods Appl Mech Eng 2000;187:385–99.
[77] Ventura G, Xu JX, Belytschko T. A vector level set method and new discontinuity approximations for crack growth by EFG. Int J Numer Methods Eng 2002;54:923–44.
[78] Belytschko T, Krysl P, Krongauz Y. A three-dimensional explicit element-free Galerkin method. Inter J Numer Methods Fluids 1997;24:1253–70.
[79] Ponthot JP, Belytschko T. Arbitrary Lagrangian-Eulerian formulation for element-free Galerkin method. Computer Methods Appl Mech Eng 1998;152:19–46.
[80] Du CJ. An element-free Galerkin method for simulation of stationary two-dimensional shallow water flows in rivers. Computer Methods Appl Mech Eng 2000;182:89–107.
[81] Cordes LW, Moran B. Treatment of material discontinuity in the Element-Free Galerkin method. Computer Methods Appl Mech Eng 1996;139:75–89.
[82] Liu WK, Jun S, Zhang YF. Reproducing kernel particle methods. Inter J Numer Methods Fluids 1995;20:1081–106.
[83] Liu WK, Jun S, Li SF, Adee J, Belytschko T. Reproducing kernel particle methods for structural dynamics. Int J Numer Methods Eng 1995;38:1655–79.
[84] Chen JS, Wu CT, Yoon S, You Y. A stabilized conforming nodal integration for Galerkin mesh-free methods. Int J Numer Methods Eng 2001;50:435–66.
[85] Chen JS, Pan C, Wu CT. Large deformation analysis of rubber based on a reproducing kernel particle method. Computat Mech 1997;19:211–27.
[86] Chen JS, Pan CH, Wu CT. Application of reproducing kernel particle method to large deformation contact analysis of elastomers. Rubber Chem Technol 1998;71:191–213.
[87] Uras RA, Chang CT, Chen Y, Liu WK. Multiresolution Reproducing Kernel Particle Methods in acoustics. J Computat Acoust 1997;5:71–94.
[88] Liu WK, Jun S, Sihling DT, Chen YJ, Hao W. Multiresolution reproducing kernel particle method for computational fluid dynamics. Inter J Numer Methods Fluids 1997;24:1391–415.
[89] Gunther F, Liu WK, Diachin D, Christon MA. Multi-scale meshfree parallel computations for viscous, compressible flows. Computer Methods Appl Mech Eng 2000;190:279–303.
[90] Lee SH, Kim HJ, Jun S. Two scale meshfree method for the adaptivity of 3-D stress concentration problems. Computat Mech 2000;26:376–87.
[91] Chen JS, Pan CH, Wu CT, Liu WK. Reproducing kernel particle methods for large deformation analysis of non-linear structures. Computer Methods Appl Mech Eng 1996;139:195–227.
[92] Jun S, Liu WK, Belytschko T. Explicit reproducing kernel particle methods for large deformation problems. Int J Numer Methods Eng 1998;41:137–66.
[93] Chen JS, Yoon S, Wang HP, Liu WK. An improved reproducing kernel particle method for nearly incompressible finite elasticity. Computer Methods Appl Mech Eng 2000;181:117–45.
[94] Li S, Hao W, Liu WK. Numerical simulations of large deformation of thin shell structures using meshfree methods. Computat Mech 2000;25:102–16.

[95] Duarte CA, Oden JT. An h-p adaptive method using clouds. Computer Methods Appl Mech Eng 1996;139:237–62.
[96] Garcia O, Fancello EA, de Barcellos CS, Duarte CA. hp-Clouds in Mindlin's thick plate model. Int J Numer Methods Eng 2000;47:1381–400.
[97] Mendonca PDR, de Barcellos CS, Duarte A. Investigations on the hp-Cloud Method by solving Timoshenko beam problems. Computat Mech 2000;25:286–95.
[98] Zhu T, Zhang J, Atluri SN. A meshless local boundary integral equation (LBIE) method for solving nonlinear problems. Computat Mech 1998;22:174–86.
[99] Zhu T, Zhang JD, Atluri SN. A local boundary integral equation (LBIE) method in computational mechanics, and a meshless discretization approach. Computat Mech 1998;21:223–35.
[100] Zhu TL, Zhang JD, Atluri SN. A meshless numerical method based on the local boundary integral equation (LBIE) to solve linear and non-linear boundary value problems. Eng Anal Boundary Elements 1999;23:375–89.
[101] Atluri SN, Sladek J, Sladek V, Zhu T. The local boundary integral equation (LBIE) and it's meshless implementation for linear elasticity. Computat Mech 2000;25:180–98.
[102] Atluri SN, Zhu T. A new meshless local Petrov-Galerkin (MLPG) approach in computational mechanics. Computat Mech 1998;22:117–27.
[103] Atluri SN, Zhu TL. The meshless local Petrov-Galerkin (MLPG) approach for solving problems in elasto-statics. Computat Mech 2000;25:169–79.
[104] Onate E, Idelsohn S, Zienkiewicz OC, Taylor RL, Sacco C. A stabilized finite point method for analysis of fluid mechanics problems. Computer Methods Appl Mech Eng 1996;139:315–46.
[105] De S, Bathe KJ. The method of finite spheres. Computat Mech 2000;25:329–45.
[106] De S, Bathe KJ. The method of finite spheres with improved numerical integration. Comput Struct 2001;79:2183–96.
[107] De S, Bathe KJ. Displacement/pressure mixed interpolation in the method of finite spheres. Int J Numer Methods Eng 2001;51:275–92.
[108] Yagawa G, Yamada T. Free mesh method: a new meshless finite element method. Computat Mech 1996;18:383–6.
[109] Sulsky D, Schreyer HL. Axisymmetric form of the material point method with applications to upsetting and Taylor impact problems. Computer Methods Appl Mech Eng 1996;139:409–29.
[110] Sulsky D, Chen Z, Schreyer HL. A particle method for history-dependent materials. Computer Methods Appl Mech Eng 1994;118:179–96.
[111] Sulsky D, Zhou SJ, Schreyer HL. Application of a particle-in-cell method to solid mechanics. Comput Phys Commun 1995;87:236–52.
[112] Liu GR, Gu YT. A point interpolation method for two-dimensional solids. Int J Numer Methods Eng 2001;50:937–51.
[113] Wang JG, Liu GR. A point interpolation meshless method based on radial basis functions. Int J Numer Methods Eng 2002;54:1623–48.
[114] Zhu T, Atluri SN. A modified collocation method and a penalty formulation for enforcing the essential boundary conditions in the element free Galerkin method. Computat Mech 1998;21:211–22.
[115] Lu YY, Belytschko T, Gu L. A new implementation of the element free Galerkin method. Computer Methods Appl Mech Eng 1994;113:397–414.
[116] Belytschko T, Krongauz Y, Organ D, Fleming M, Krysl P. Meshless methods: an overview and recent developments. Computer Methods Appl Mech Eng 1996;139:3–47.
[117] Chen JS, Pan C, Roque CMOL, Wang HP. A Lagrangian reproducing kernel particle method for metal forming analysis. Computat Mech 1998;22:289–307.

[118] Geng Y, Wang SJ, Kim JK. Preparation of graphite nanoplatelets and graphene sheets. J Colloid Interface Sci 2009;336:592–8.
[119] Feng SQ, Yu DP, Hu G, Zhang XF, Zhang Z. The HREM observation of cross-sectional structure of carbon nanotubes. J Phys Chem Solids 1997;58:1887–92.
[120] Brenner DW. Empirical Potential for Hydrocarbons for Use in Simulating the Chemical Vapor-Deposition of Diamond Films. Phys Rev B 1990;42:9458–71.
[121] Chandraseker K, Mukherjee S. Coupling of extension and twist in single-walled carbon nanotubes. J Appl Mech-Trans Asme 2006;73:315–26.
[122] Dolbow J, Belytschko T. An introduction to programming the meshless Element Free Galerkin method. Archiv Computat Methods Eng 1998;5:207–41.
[123] Matheron G. Principles of geostatistics. Econom Geol 1963;58:1246–66.
[124] Yan JW, Liew KM, He LH. Analysis of single-walled carbon nanotubes using the moving Kriging interpolation. Computer Methods Appl Mech Eng 2012;229:56–67.
[125] Beissel S, Belytschko T. Nodal integration of the element-free Galerkin method. Computer Methods Appl Mech Eng 1996;139:49–74.
[126] Li S, Liu WK. Meshfree and particle methods and their applications. Appl Mech Rev 2002;55:1–34.
[127] Krongauz Y, Belytschko T. Enforcement of essential boundary conditions in meshless approximations using finite elements. Computer Methods Appl Mech Eng 1996;131:133–45.
[128] Ren J, Liew K. Mesh-free method revisited: two new approaches for the treatment of essential boundary conditions. Inter J Computat Eng Sci 2002;3:219–33.
[129] Gunther FC, Liu WK. Implementation of boundary conditions for meshless methods. Computer Methods Appl Mech Eng 1998;163:205–30.
[130] Kaljevic I, Saigal S. An improved element free Galerkin formulation. Int J Numer Methods Eng 1997;40:2953–74.
[131] Dennis Jr JE, Schnabel RB. Numerical methods for unconstrained optimization and nonlinear equations. Philadelphia: Siam; 1996.
[132] Qian D, Wagner GJ, Liu WK, Yu M-F, Ruoff RS. Mechanics of carbon nanotubes. Appl Mech Rev 2002;55:495–533.
[133] Liu B, Huang Y, Jiang H, Qu S, Hwang KC. The atomic-scale finite element method. Computer Methods Appl Mech Eng 2004;193:1849–64.
[134] Liu B, Jiang H, Huang Y, Qu S, Yu MF, Hwang KC. Atomic-scale finite element method in multiscale computation with applications to carbon nanotubes. Phys Rev B 2005;72:035435.
[135] Peters MJ, McNeil LE, Lu JP, Kahn D. Structural phase transition in carbon nanotube bundles under pressure. Phys Rev B 2000;61:5939–44.
[136] Tang J, Qin LC, Sasaki T, Yudasaka M, Matsushita A, Iijima S. Compressibility and polygonization of single-walled carbon nanotubes under hydrostatic pressure. Phys Rev Lett 2000;85:1887–9.
[137] Venkateswaran UD, Rao AM, Richter E, Menon M, Rinzler A, Smalley RE, et al. Probing the single-wall carbon nanotube bundle: Raman scattering under high pressure. Phys Rev B 1999;59:10928–34.
[138] Cai JZ, Lu L, Kong WJ, Zhu HW, Zhang C, Wei BQ, et al. Pressure-induced transition in magnetoresistance of single-walled carbon nanotubes. Phys Rev Lett 2006;97:026402.
[139] Sluiter MHF, Kumar V, Kawazoe Y. Symmetry-driven phase transformations in single-wall carbon-nanotube bundles under hydrostatic pressure. Phys Rev B 2002;65:161402.
[140] Chan SP, Yim WL, Gong XG, Liu ZF. Carbon nanotube bundles under high pressure: transformation to low-symmetry structures. Phys Rev B 2003;68:075404.
[141] Sun DY, Shu DJ, Ji M, Liu F, Wang M, Gong XG. Pressure-induced hard-to-soft transition of a single carbon nanotube. Phys Rev B 2004;70:165417.

[142] Zang J, Aldas-Palacios O, Liu F. MD simulation of structural and mechanical transformation of single-walled carbon nanotubes under pressure. Communicat Computat Phys 2007;2:451–65.

[143] Zhang XH, Sun DY, Liu ZF, Gong XG. Structure and phase transitions of single-wall carbon nanotube bundles under hydrostatic pressure. Phys Rev B 2004;70:165417.

[144] Zang J, Treibergs A, Han Y, Liu F. Geometric constant defining shape transitions of carbon nanotubes under pressure. Phys Rev Lett 2004;92:105501.

[145] Chopra NG, Benedict LX, Crespi VH, Cohen ML, Louie SG, Zettl A. Fully collapsed carbon nanotubes. Nature 1995;377:135–8.

[146] Yu MF, Kowalewski T, Ruoff RS. Structural analysis of collapsed, and twisted and collapsed, multiwalled carbon nanotubes by atomic force microscopy. Phys Rev Lett 2001;86:87–90.

[147] Yakobson BI, Brabec CJ, Bernholc J. Nanomechanics of carbon tubes: instabilities beyond linear response. Phys Rev Lett 1996;76:2511–14.

[148] Cao GX, Chen X. Buckling of single-walled carbon nanotubes upon bending: molecular dynamics simulations and finite element method. Phys Rev B 2006;73.

[149] Mylvaganam K, Vodenitcharova T, Zhang LC. The bending-kinking analysis of a single-walled carbon nanotube—a combined molecular dynamics and continuum mechanics technique. J Mater Sci 2006;41:3341–7.

[150] Milstein F. Theoretical elastic behavior of crystals at large strains. J Mater Sci 1980;15:1071–84.

[151] Belytschko T, Xiao SP, Schatz GC, Ruoff RS. Atomistic simulations of nanotube fracture. Phys Rev B 2002;65.

[152] Yakobson BI, Campbell MP, Brabec CJ, Bernholc J. High strain rate fracture and C-chain unraveling in carbon nanotubes. Computat Mater Sci 1997;8:341–8.

[153] Saito R, Dresselhaus G, Dresselhaus MS. Physical properties of carbon nanotubes. Singapore: World Scientific; 1998.

[154] Lu JP. Elastic properties of carbon nanotubes and nanoropes. Phys Rev Lett 1997;79:1297.

[155] Civalek Ö, Demir C. Buckling and bending analyses of cantilever carbon nanotubes using the euler-bernoulli beam theory based on non-local continuum model. Asian J Civil Eng (Building and Housing) 2011;12:651–61.

[156] Akgoz B, Civalek O. Strain gradient elasticity and modified couple stress models for buckling analysis of axially loaded micro-scaled beams. Inter J Eng Sci 2011;49:1268–80.

[157] Sun Y, Liew KM. Application of the higher-order Cauchy-Born rule in mesh-free continuum and multiscale simulation of carbon nanotubes. Int J Numer Methods Eng 2008;75:1238–58.

[158] Liew KM, Wong CH, He XQ, Tan MJ, Meguid SA. Nanomechanics of single and multiwalled carbon nanotubes. Phys Rev B 2004;69:115429.

[159] Lim CW. Equilibrium and Static Deflection for Bending of a Nonlocal Nanobeam. Adv Vibrat Eng 2009;8:277–300.

[160] Lim CW. On the truth of nanoscale for nanobeams based on nonlocal elastic stress field theory: equilibrium, governing equation and static deflection. Appl Math Mech-English Edition 2010;31:37–54.

[161] Gao YW, Lei FM. Small scale effects on the mechanical behaviors of protein microtubules based on the nonlocal elasticity theory. Biochem Biophys Res Commun 2009;387:467–71.

[162] Sun YZ, Liew KM. Effect of higher-order deformation gradients on buckling of single-walled carbon nanotubes. Composite Struct 2014;109:279–85.

[163] Guo X, Zhang T. A study on the bending stiffness of single-walled carbon nanotubes and related issues. J Mech Phys Solids 2010;58:428–43.

[164] Ma TH, Li B, Chang TC. Chirality- and curvature-dependent bending stiffness of single layer graphene. Appl Phys Lett 2011;99.

[165] Sun YZ, Liew KM. A Precise Model to Predict the Structural and Elastic Properties of Single-Walled Carbon Nanotubes. J Computat Theoret Nanosci 2010;7:583–93.
[166] Li CY, Chou TW. Single-walled carbon nanotubes as ultrahigh frequency nanomechanical resonators. Phys Rev B 2003;68:073405.
[167] Li CY, Chou TW. Vibrational behaviors of multiwalled-carbon-nanotube-based nanomechanical resonators. Appl Phys Lett 2004;84:121–3.
[168] He XQ, Eisenberger M, Liew KM. The effect of van der Waals interaction modeling on the vibration characteristics of multiwalled carbon nanotubes. J Appl Phys 2006;100:124317-1-12.
[169] Natsuki T, Endo M, Tsuda H. Vibration analysis of embedded carbon nanotubes using wave propagation approach. J Appl Phys 2006;99.
[170] Wang Q, Varadan VK. Vibration of carbon nanotubes studied using nonlocal continuum mechanics. Smart Materials Struct 2006;15:659–66.
[171] Lu P, Lee HP, Lu C, Zhang PQ. Application of nonlocal beam models for carbon nanotubes. Inter J Solids Struct 2007;44:5289–300.
[172] Hsu JC, Chang RP, Chang WJ. Resonance frequency of chiral single-walled carbon nanotubes using Timoshenko beam theory. Phys Lett A 2008;372:2757–9.
[173] Natsuki T, Ni QQ, Endo M. Analysis of the vibration characteristics of double-walled carbon nanotubes. Carbon NY 2008;46:1570–3.
[174] Yang J, Ke LL, Kitipornchai S. Nonlinear free vibration of single-walled carbon nanotubes using nonlocal Timoshenko beam theory. Phys E-Low-Dimens Sys Nanostruct 2010;42:1727–35.
[175] Wang L, Ni Q, Li M, Qian Q. The thermal effect on vibration and instability of carbon nanotubes conveying fluid. Phys E-Low-Dimens Sys Nanostruct 2008;40:3179–82.
[176] Yan Y, He XQ, Zhang LX, Wang CM. Dynamic behavior of triple-walled carbon nanotubes conveying fluid. J Sound Vibration 2009;319:1003–18.
[177] Chang WJ, Lee HL. Free vibration of a single-walled carbon nanotube containing a fluid flow using the Timoshenko beam model. Phys Lett A 2009;373:982–5.
[178] Chowdhury R, Adhikari S, Wang CY, Scarpa F. A molecular mechanics approach for the vibration of single-walled carbon nanotubes. Computat Mater Sci 2010;48:730–5.
[179] Georgantzinos SK, Anifantis NK. Vibration analysis of multi-walled carbon nanotubes using a spring-mass based finite element model. Computat Mater Sci 2009;47:168–77.
[180] Hashemnia K, Farid M, Vatankhah R. Vibrational analysis of carbon nanotubes and graphene sheets using molecular structural mechanics approach. Computat Mater Sci 2009;47:79–85.
[181] Reddy CD, Rajendran S, Liew KM. Vibration properties of single walled carbon nanotubes-a comparison between the atomistic simulations and continuum shell modelling. J Computat Theoret Nanosci 2010;7:1400–6.
[182] Yan JW, Liew KM, He LH. Free vibration analysis of single-walled carbon nanotubes using a higher-order gradient theory. J Sound Vibration 2013;332:3740–55.

CHAPTER

ATOMIC FINITE ELEMENT METHOD AND COUPLING WITH ATOMISTIC-CONTINUUM METHOD

5

5.1 INTRODUCTION

In this chapter, the multiscale coupling of the developed mesh-free computational scheme with an atomic simulation technique is investigated. The multiscale method takes advantage of both continuum and atomic simulations and is currently a popular technique in the modeling of large-scale nanostructures. In most multiscale studies, research is focused on the approach to bridging two different scales, and less attention is paid to the efficiency of continuum simulations. For example, continuum simulation based on the standard Cauchy–Born rule has been applied in multiscale analysis. The present study mainly investigates the application of the developed mesh-free method in multiscale analysis. A previous coupling technique, the bridging domain method, is employed to bridge two scales. Multiscale computational examples are tested, and good computational efficiency is obtained.

As illustrated in previous chapters, atom-based methods consume a huge amount of computational resources and are restricted to very small systems. Continuum modeling can significantly reduce the degrees of freedom of problems, and the theoretical analysis and numerical simulation of large-scale systems become possible. However, continuum simulations cannot completely substitute for the atom-based method since it they employ a homogenization technique. In particular, the continuum methods are unable to capture the microscale physical laws of nanostructures. Multiscale analysis is thus emerging as a feasible and efficient approach to large-size problems. The multiscale method couples continuum simulation and atomic simulation and makes use of both approaches. The basic idea of the multiscale method is to use atomic simulation for the localized region in which the discrete motion of atoms is important and to use the continuum method for the remaining regions in which the deformation is homogeneous and smooth. A key issue associated with the multiscale method is the way to bridge the two different scales smoothly. Currently, researchers have developed several techniques to couple the continuum and atomic methods. Here, three popular approaches, the quasicontinuum method [1–9], the bridging domain method [10–12], and the bridging scale method [13–17], are introduced. The bridging domain method is used in the present research.

5.2 ATOMIC FINITE ELEMENT METHOD

The multiscale method requires an appropriate atomic simulation approach. An atomic simulation method that has the same solution framework as the present mesh-free method is introduced in this

Mechanical Behaviors of Carbon Nanotubes. DOI: http://dx.doi.org/10.1016/B978-0-323-43137-8.00005-6

section. This atomic simulation method will be employed in the present multiscale computation. In the previous chapters, this method has been used in testing the efficiency and accuracy of the developed mesh-free method.

Let us consider a system that contains NA carbon atoms. The energy that is stored in the atomic bonds is a function of the positions of all atoms, as follows:

$$U_{nb} = U_{nb}(\mathbf{q}_1, \mathbf{q}_2, \ldots, \mathbf{q}_N) = \sum_{I<J}^{NA} V_{IJ}(\mathbf{q}_J - \mathbf{q}_I) \tag{5.1}$$

where $\mathbf{q}_I$ is the position of atom I, and $V_{IJ}(\mathbf{q}_J - \mathbf{q}_I)$ is the Brenner potential.

The total potential energy of the atomic system is

$$E_A(\mathbf{q}) = U_{nb}(\mathbf{q}) - \sum_{I=1}^{NA} \overline{\mathbf{F}}_I \cdot \mathbf{q}_I \tag{5.2}$$

where $\mathbf{q} = (\mathbf{q}_1, \mathbf{q}_2, \ldots, \mathbf{q}_{NA})^T$, and $\overline{\mathbf{F}}_I$ is the external force (if any) that is exerted on atom I. The state of the stable configuration corresponds to

$$\frac{\partial E_A(\mathbf{q})}{\partial \mathbf{q}} = 0 \tag{5.3}$$

The Taylor expansion of $E_A(\mathbf{q})$ around an initial guess of $\mathbf{q}^0 = (\mathbf{q}_1^0, \mathbf{q}_2^0, \ldots, \mathbf{q}_{NA}^0)^T$ for the equilibrium state gives

$$E_A(\mathbf{q}) \approx E_A(\mathbf{q}^0) + \left.\frac{\partial E_A(\mathbf{q})}{\partial \mathbf{q}}\right|_{\mathbf{q}=\mathbf{q}^0} \cdot (\mathbf{q} - \mathbf{q}^0) + \frac{1}{2}(\mathbf{q} - \mathbf{q}^0)^T \cdot \left.\frac{\partial E_A(\mathbf{q})}{\partial \mathbf{q}\partial \mathbf{q}}\right|_{\mathbf{q}=\mathbf{q}^0} \cdot (\mathbf{q} - \mathbf{q}^0) \tag{5.4}$$

Its substitution into Eq. (5.3) yields the following governing equation for the displacement $\Delta\mathbf{q} = (\mathbf{q} - \mathbf{q}^0)$,

$$\mathbf{K}_A \Delta\mathbf{q} = \mathbf{P} \tag{5.5}$$

where the stiffness matrix is

$$\mathbf{K}_A = \frac{\partial^2 E_A}{\partial \mathbf{q}\partial \mathbf{q}} = \frac{\partial^2 U_{nb}}{\partial \mathbf{q}\partial \mathbf{q}} \tag{5.6}$$

and the nonequilibrium force vector is

$$\mathbf{P} = -\frac{\partial E_A}{\partial \mathbf{q}} = \overline{\mathbf{F}} - \frac{\partial U_{nb}}{\partial \mathbf{q}} \tag{5.7}$$

where $\overline{\mathbf{F}} = (\overline{\mathbf{F}}_1, \overline{\mathbf{F}}_2, \ldots, \overline{\mathbf{F}}_{NP})^T$ is the external force vector. The equilibrium state is obtained by iteratively solving Eq. (5.5) until $\mathbf{P}$ reaches zero.

In [18,19], Liu et al. introduced the concept of the atomic-scale finite element to assemble the global stiffness matrix and force vector, and they employed finite element software to solve large-scale problems. In the present computation, all work, including assembling the stiffness matrix and force vector and solving the equation system, is performed with our Fortran codes. This atomic simulation approach has been used in previous chapters to test the efficiency of the continuum method. It is noted that the stiffness matrix $\mathbf{K}_A$ lacks the positive definiteness beyond the buckling, and the method that is described in Chapter 4, Atomistic-Continuum Theory, should be used to ensure the convergence of the algorithm.

5.3 COUPLING OF ATOMIC FINITE ELEMENT METHOD WITH ATOMISTIC-CONTINUUM METHOD

5.3.1 QUASICONTINUUM METHOD

The quasicontinuum method was proposed by Tadmor et al. [2,3] for modeling 2-D crystallites, and it has been used to study a variety of fundamental aspects of deformation in crystalline solids, including fracture, grain boundary structure and deformation, nano-indentation, and 3-D dislocation junctions [1–9].

The quasicontinuum method introduced the concept of representative atoms. Shown in Fig. 5.1A is a crystalline solid that involves a Lomer dislocation [9]: the filled circles denote the chosen representative atoms. Around the core of the dislocation, the deformation is drastic, the microcosmic change of the atomic structure needs to be described and traced, and all atoms are chosen as the representative atoms. Far away from the core of the dislocation, the field is slowly varying and the deformation is homogeneous, and certain sparse atoms are chosen as the representative atoms. In the former region, the atoms are called nonlocal atoms, and their interactions with the neighboring atoms are reflected. The atoms in the latter region are called local atoms, and a continuum discretization scheme can be established based on them.

The total potential energy of the quasicontinuum model can be obtained as

$$E_{\mathrm{QC}} = \sum_{\text{nonlocal}} E_i + \sum_{\text{local}} w_\mu E_\mu \tag{5.8}$$

The first and second terms in Eq. (5.8) denote the energies of the nonlocal and local atoms, respectively. w_μ is the weight function and refers to the number of atoms that this representative

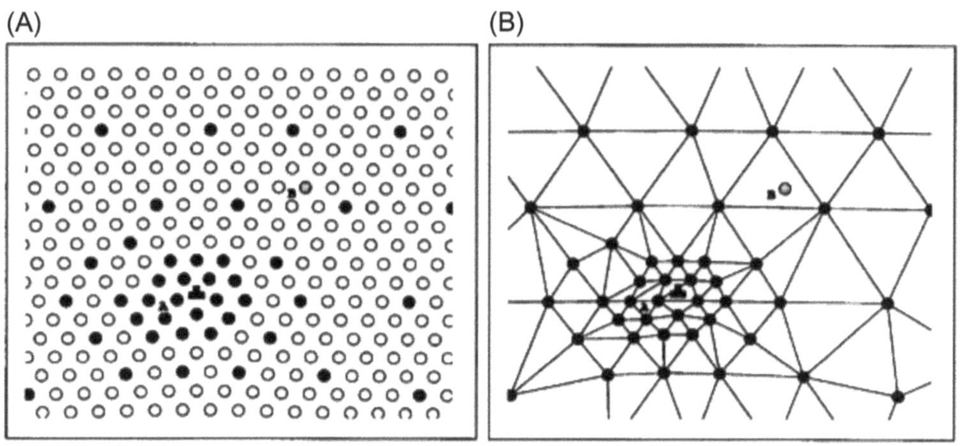

FIGURE 5.1

(A) Around the core of a Lomer dislocation, all atoms are chosen as the representative atoms (filled circles), whereas sparse atoms are chosen in the far field. (B) The constructed FEM is based on the chosen representative atoms [9].

atom can represent. In practice, the local region can be discretized as a series of elements by using the representative atoms as nodes (see Fig. 5.1B).

In the quasicontinuum method, a ghost force often arises in the transition zone between the local and nonlocal regions, even when the crystal is undeformed. To avoid the effect of this ghost force, the energy is augmented by a term associated with the work done by the ghost force:

$$\hat{E}_{QC} = E_{QC} - \sum_{g} F^g u_g \tag{5.9}$$

The ghost force F^g is calculated each time the status of the representative atoms is updated and is held constant until an update is required due to the evolving state of the deformation [9].

5.3.2 BRIDGING DOMAIN METHOD

The bridging domain method was proposed by Belytschko and Xiao [10,11], was used to simulate the fracture of graphite sheets [10], and was recently extended to simulate the fracture of large-diameter carbon nanotubes (CNTs) [11,12]. In this technique, the entire domain is decomposed into three types of regions (see Fig. 5.2): an atomic region Ω^A in which the atomic movement is described, a continuum region $\Omega^C = \Omega^{CL} + \Omega^{CR}$ in which the lattice undergoes a smooth deformation and the continuum simulation is used, and an overlapping region $\Omega^O = \Omega^A \cap \Omega^C$ in which the atomic and continuum models overlap. In the atomic and continuum regions, the energy is computed using the atomic and continuum methods, respectively. In the overlapping region, the potential energy is expressed as a linear combination of the continuum and atomic energies. This ensures smooth bridging between the continuum and atomic deformation fields.

5.3.3 BRIDGING SCALE METHOD

The bridging scale method was developed by Liu and his coworkers [13–17]. In this method, the total displacement field is decomposed into two different components, as follows:

$$\mathbf{u}(\mathbf{x}, t) = \bar{\mathbf{u}}(\mathbf{x}, t) + \mathbf{u}'(\mathbf{x}, t), \tag{5.10}$$

where $\bar{\mathbf{u}}(\mathbf{x}, t)$ is the coarse scale component and can be established in a continuum discretization frame such as the finite element or mesh-free method. $\mathbf{u}'(\mathbf{x}, t)$ is the fine scale component that must

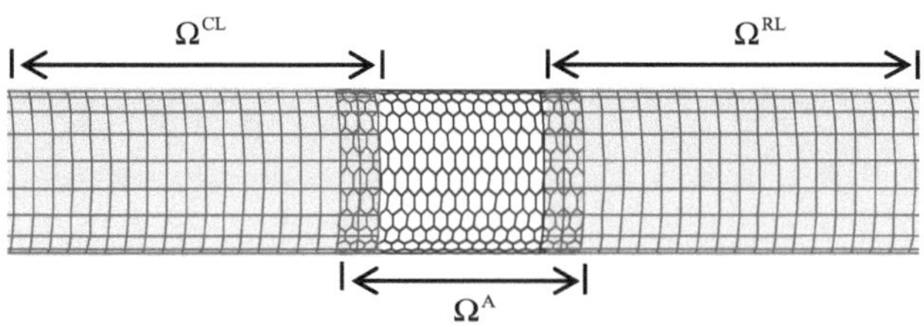

FIGURE 5.2

The entire domain is decomposed into a continuum simulation region $\Omega^C = \Omega^{CL} + \Omega^{CR}$ and a full atomic region Ω^A with an overlapping region $\Omega^O = \Omega^A \cap \Omega^C$ [20].

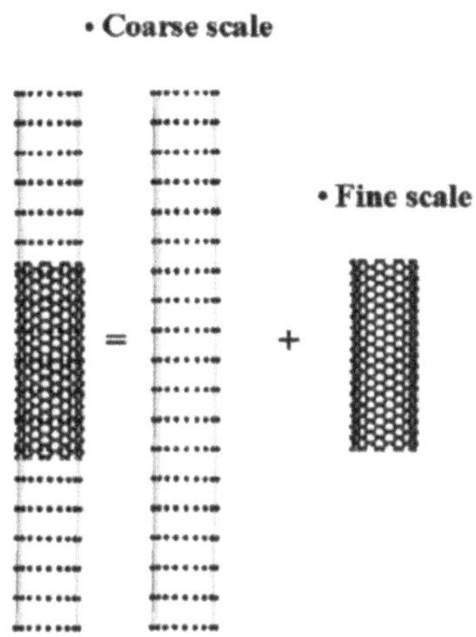

FIGURE 5.3

Schematic illustration of the multiscale discretization scheme in the bridging scale method [13].

be determined by the atom-based method. Far away from the localized domain, $\mathbf{u}'(\mathbf{x}, t)$ becomes insignificant, or, mathematically, $\mathbf{u}'(\mathbf{x}, t)$ has a vanishing projection onto the coarse scale basis function. Therefore, atomic simulation is only needed in the localized regions. Fig. 5.3 shows the computational scheme for the bridging scale method. There are two different computational domains. The first is the coupled domain in which the atomic structure coexists with the continuum discretization. The second domain is the coarse scale domain in which only the continuum discretization exists. The coarse and fine scale solutions are coupled. In the computation, the coarse scale solution is solved in the entire domain, whereas the fine scale solution is only solved in the localized region. Bridging between the coarse and fine scales is realized by transparently exchanging information between the coarse and fine scale regions.

The present multiscale computational scheme couples the developed mesh-free method and the atomic simulation method presented in Section 5.2. The coupling approach used is the bridging domain method. The main purpose is to investigate the validity of the developed mesh-free computational framework in the multiscale simulation.

Referring to Fig. 5.2, the entire domain is decomposed into three regions: Ω^{A}, $\Omega^{\mathrm{C}} = \Omega^{\mathrm{CL}} + \Omega^{\mathrm{CR}}$, and $\Omega^{\mathrm{O}} = \Omega^{\mathrm{A}} \cap \Omega^{\mathrm{C}}$. The mesh-free method developed in Chapter 4, Atomistic-Continuum Theory, is used in the continuum region. As specified in Chapter 4, Atomistic-Continuum Theory, the mesh-free nodes are uniformly collocated along the axial and circumferential directions, and the background integral cells are in accordance with the nodal arrangements. The crossing points of the axial and circumferential lines denote the mesh-free nodes in Fig. 5.2.

The total energy of the coupled system can be written as a weighted sum of energies for the continuum and atomic regions:

$$\mathrm{U}_{\mathrm{tot}} = \sum_{I \in \Omega^{\mathrm{A}}} \left[\beta^{\mathrm{A}} \left(\frac{1}{2} (\mathbf{q}_I^{\mathrm{A}} + \mathbf{q}_J^{\mathrm{A}}) \right) \sum_{I<J} V_{IJ}(r_{IJ}) \right] + \int_{\Omega^c} \beta^c(\mathbf{X}) W(\mathbf{F},\mathbf{G}) dV \tag{5.11}$$

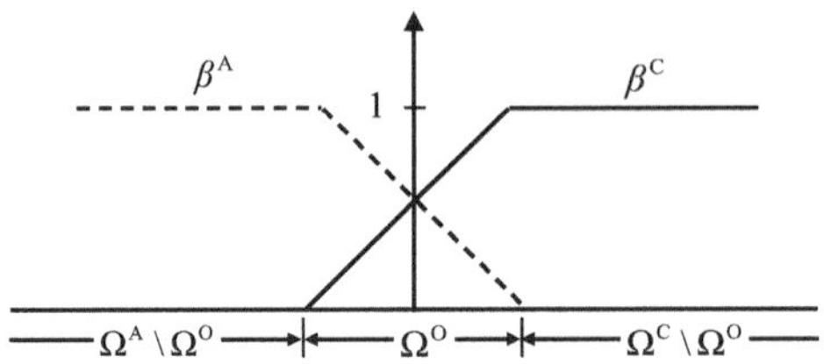

FIGURE 5.4

The definition of weight functions β^{A} and β^{C} in the three regions.

where $\mathbf{q}_I^A$ is the position of atom I in the undeformed single-walled CNT (SWCNT), and the weight functions β^{A} and β^{c} take the forms [10–12]:

$$\beta^{c}(\mathbf{X}) = 1 - \beta^{A}(\mathbf{X}) = \begin{cases} 1, & x \in \Omega^{c}\backslash\Omega^{A}, \\ \alpha, & x \in \Omega^{A}, \\ 0, & x \in \Omega^{A}\backslash\Omega^{c}, \end{cases} \tag{5.12}$$

where the symbol "\" denotes the set-minus operation, and the parameter α varies linearly from 0 to 1 across the overlapping region, as shown in Fig. 5.4. This method thus allows the minimization of the continuum and atomic configurations concurrently, and the two scale solutions can simultaneously be solved.

The total potential energy of the coupled system is

$$E_{\text{tot}} = U_{\text{tot}} - \sum_{i \in \Omega^{A}} \beta^{A}(\mathbf{q}_I^A)\overline{\mathbf{F}}_I \cdot \mathbf{q}_I - \int_{\partial\Omega^{c}} \beta^{c}(\mathbf{X})\mathbf{u} \cdot \mathbf{t}_0^P dS - \int_{\partial\Omega^{c}} \beta^{c}(\mathbf{X})\nabla_N \mathbf{u} \cdot \mathbf{t}_0^Q dS. \tag{5.13}$$

The nonequilibrium force vector and stiffness matrix can be obtained by

$$\mathbf{P}_{\text{tot}} = \frac{\partial E_{\text{tot}}}{\partial \mathbf{X}_{\text{tot}}} \tag{5.14}$$

$$\mathbf{K}_{\text{tot}} = \frac{\partial^2 E_{\text{tot}}}{\partial \mathbf{X}_{\text{tot}} \partial \mathbf{X}_{\text{tot}}}. \tag{5.15}$$

In [10–12], the augmented Lagrange multiplier method was used to ensure that the continuum displacements conformed to the atomic displacements at the discrete positions of the atoms in the overlapping region. In the present research, the atomic displacement in the overlapping region is directly interpolated with the mesh-free nodal parameters at each iterative step, which means that the vector $\mathbf{X}_{\text{tot}}$ contains all of the mesh-free modal parameters, but only the current positions of the atoms outside the overlapping region. The stable configuration can be obtained by iteratively solving $\mathbf{K}_{\text{tot}}\Delta\mathbf{X}_{\text{tot}} = \mathbf{P}_{\text{tot}}$ until $\mathbf{P}_{\text{tot}}$ reaches zero. Moreover, the treatment described in Section 4.5.8 is also effective for the present problem.

5.4 TENSILE FAILURE

Two examples are chosen to test the application of the mesh-free method in multiscale analysis and illustrate the validity of the present multiscale computational scheme. The first example is the bending test, for which the Brenner potential with the second set of parameters is used. The second example considers the fracture of SWCNTs with a single-atom vacancy. As described before, the Brenner potential always results in overestimated fracture stress. Therefore, the modified second-generation Brenner potential is used for the second example.

5.4.1 BENDING TEST

A (12, 12) SWCNT is chosen upon which to perform the bending test. The SWCNT has 42 hexagonal cells (a total of 2040 atoms) in its length. Atomic simulation is used for the middle portion, whereas the left and right parts are modeled using a mesh-free method in which 396 mesh-free nodes are used. In the overlapping region, the atomic position is interpolated using the mesh-free nodes, and the pure atomic region contains 552 atoms. The degrees of freedom of the system are reduced from 3×2024 to 3×948. The bending deformation is imposed using the mode II loading method described in Chapter 4, Atomistic-Continuum Theory. For each loading step, the tube is bent by 1 degree per loading step. Fig. 5.5 shows a plot of the changes in the average energy per atom against the bending angle for the multiscale simulation and the full atomic simulation. Buckling occurs at a bending angle of 13 degrees in both simulations, and the energy change from the two methods also agrees quite well both before and after the buckling occurs. Fig. 5.6 shows the deformations at a bending angle of 20 degrees, and a comparison of the two methods reveals that a precise simulation can be achieved with the multiscale method. In Chapter 4, Atomistic-Continuum Theory, the mesh-free simulation presents very precise results prior to buckling, but

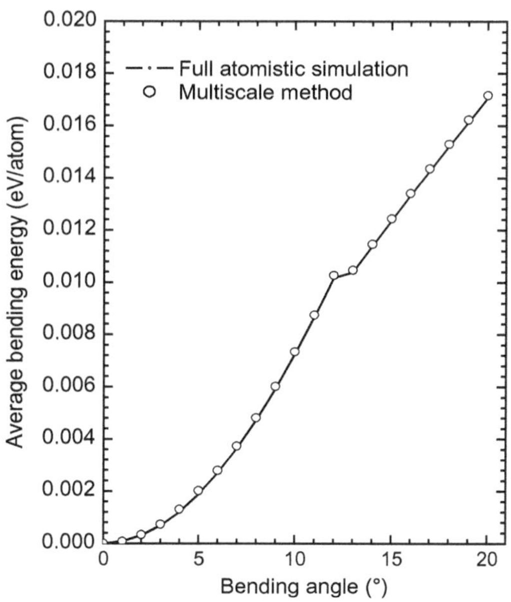

FIGURE 5.5

Comparison of the average bending energy derived from the multiscale method and the full atomic simulation [20].

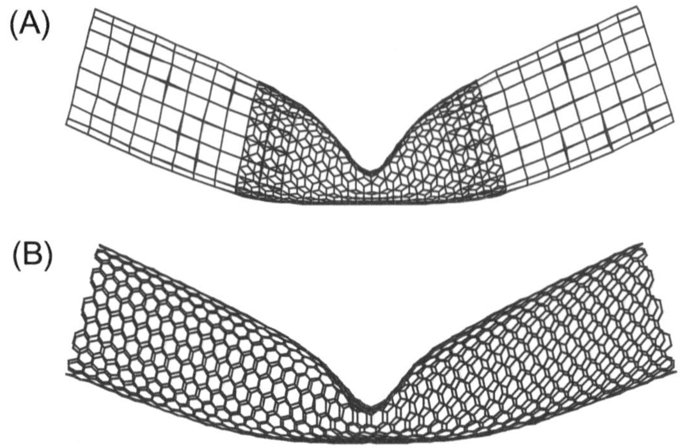

FIGURE 5.6

Deformation of the SWCNT at a bending angle of 20 degrees, as obtained by (A) the multiscale method and (B) the full atomic simulation [20].

after buckling, the results are less precise than those of the full atomic simulation. The present multiscale simulation presents good agreement with the atomic simulation before and after buckling, which shows that the multiscale method is very efficient for localized problems.

5.4.2 TENSILE FAILURE OF SWCNTs WITH A SINGLE-ATOM VACANCY DEFECT

In this example, the multiscale method is used to study the tensile failure of an SWCNT with a single-atom vacancy defect. This kind of defect has a significant effect on the strength of a CNT and thus has attracted considerable research attention. In the computation, one atom is removed from the hexagonal network in the atomic region, which leads to the reconstruction of the atomic structure near the vacancy in the initial equilibrium configuration [12]. Here, this initial equilibrium configuration is determined with the full atomic simulation before the multiscale computation is performed. The SWCNT is stretched until the fracture occurs. Fig. 5.7 shows the stress–strain curves for (18, 0) and (12, 12) SWCNTs. The tensile force can be computed by summing the nodal force along the end of the CNT, and the stress is calculated by dividing the tensile force by the cross-sectional area $2\pi Rt$. Fig. 5.8 shows the cracking procedure for an (18, 0) SWCNT, with Fig. 5.8A showing the initial equilibrium configuration of the undeformed SWCNT, Fig. 5.8B highlighting the breakage of the bonds around the defect, and Fig. 5.8C displaying the completely fractured SWCNT in which all of the bonds around the circumference have failed. Fig. 5.9 shows the fracture progression for a (12, 12) SWCNT, from which it can be seen that the failure path displays a slight difference from that of the (18, 0) SWCNT. In the present simulation, the failure is actually a brittle fracture. To show the cracking progression, a larger positive number β is added to the diagonal elements of the stiffness matrix after the occurrence of the fracture.

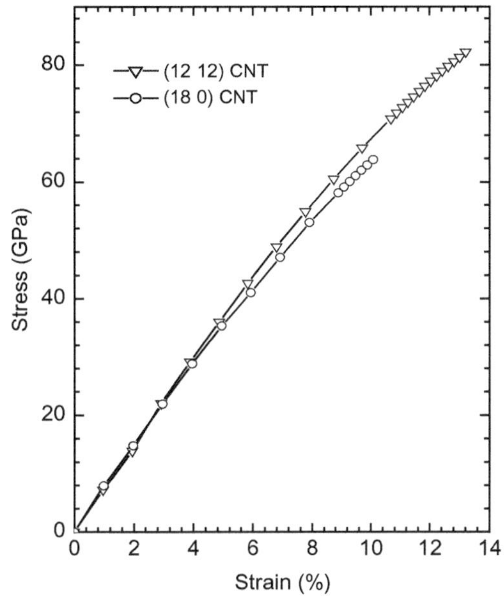

FIGURE 5.7

Stress–strain curves for the tension of (18, 0) and (12, 12) SWCNTs with a single-atom vacancy defect [20].

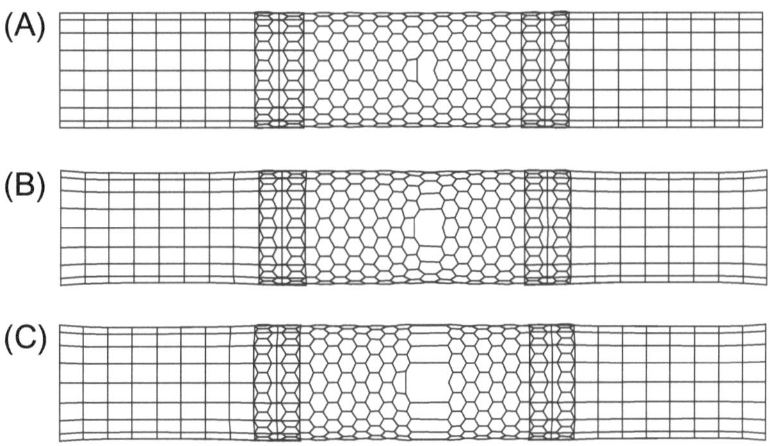

FIGURE 5.8

Fracture evolution of an (18, 0) SWCNT with a single-atom vacancy defect: (A) the initial equilibrium configuration of the undeformed SWCNT; (B) bond failure spreading from the defect; and (C) the fractured SWCNT, in which all of the bonds around the circumference have failed [20].

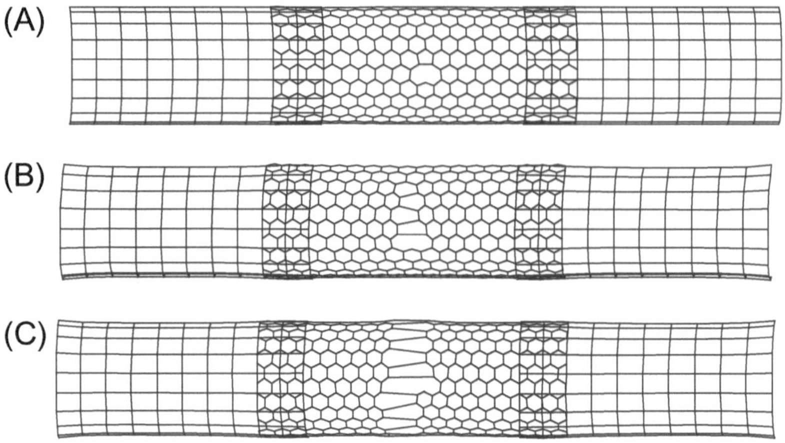

FIGURE 5.9

Fracture evolution of a (12, 12) SWCNT with a single-atom vacancy defect: (A) the initial equilibrium configuration of the undeformed SWCNT; (B) bond failure spreading from the defect; and (C) the fractured CNT in which all of the bonds around the circumference have failed [20].

REFERENCES

[1] Tadmor EB, Smith GS, Bernstein N, Kaxiras E. Mixed finite element and atomistic formulation for complex crystals. Phys Rev B 1999;59:235–45.

[2] Tadmor EB, Ortiz M, Phillips R. Quasicontinuum analysis of defects in solids. Philosl Mag A 1996;73:1529–63.

[3] Tadmor EB, Phillips R, Ortiz M. Mixed atomistic and continuum models of deformation in solids. Langmuir 1996;12:4529–34.

[4] Shenoy VB, Miller R, Tadmor EB, Phillips R, Ortiz M. Quasicontinuum models of interfacial structure and deformation. Phys Rev Lett 1998;80:742–5.

[5] Miller R, Tadmor EB, Phillips R, Ortiz M. Quasicontinuum simulation of fracture at the atomic scale. Modelling Simul Mater Sci Eng 1998;6:607–38.

[6] Ortiz M, Cuitino AM, Knap J, Koslowski M. Mixed atomistic continuum models of material behavior: The art of transcending atomistics and informing continua. Mrs Bull 2001;26:216–21.

[7] Knap J, Ortiz M. An analysis of the quasicontinuum method. J Mech Phys Solids 2001;49:1899–923.

[8] Shenoy V, Shenoy V, Phillips R. Finite temperature quasicontinuum methods. Multiscale Model Mater 1999;538:465–71.

[9] Shenoy VB, Miller R, Tadmor EB, Rodney D, Phillips R, Ortiz M. An adaptive finite element approach to atomic-scale mechanics—the quasicontinuum method. J Mech Phys Solids 1999;47:611–42.

[10] Belytschko T, Xiao SP. Coupling methods for continuum model with molecular model. Inter J Multiscale Comput Eng 2003;1:115–26.

[11] Xiao SP, Belytschko T. A bridging domain method for coupling continua with molecular dynamics. Computer Methods Appl Mechan Eng 2004;193:1645–69.

[12] Zhang SL, Mielke SL, Khare R, Troya D, Ruoff RS, Schatz GC, et al. Mechanics of defects in carbon nanotubes: Atomistic and multiscale simulations. Phys Rev B 2005;71:115403.

[13] Qian D, Wagner GJ, Liu WK. A multiscale projection method for the analysis of carbon nanotubes. Computer Methods Appl Mechan Eng 2004;193:1603–32.

[14] Wagner GJ, Liu WK. Coupling of atomistic and continuum simulations using a bridging scale decomposition. J Comput Phys 2003;190:249–74.

[15] Park HS, Karpov EG, Liu WK, Klein PA. The bridging scale for two-dimensional atomistic/continuum coupling. Philos Mag 2005;85:79–113.

[16] Park HS, Karpov EG, Klein PA, Liu WK. Three-dimensional bridging scale analysis of dynamic fracture. J Comput Phys 2005;207:588–609.

[17] Liu WK, Park HS, Qian D, Karpov EG, Kadowaki H, Wagner GJ. Bridging scale methods for nanomechanics and materials. Computer Methods Appl Mechan Eng 2006;195:1407–21.

[18] Liu B, Huang Y, Jiang H, Qu S, Hwang KC. The atomic-scale finite element method. Computer Methods Appl Mechan Eng 2004;193:1849–64.

[19] Liu B, Jiang H, Huang Y, Qu S, Yu MF, Hwang KC. Atomic-scale finite element method in multiscale computation with applications to carbon nanotubes. Phys Rev B 2005;72:035435.

[20] Sun Y, Liew KM. Application of the higher-order Cauchy-Born rule in mesh-free continuum and multiscale simulation of carbon nanotubes. Int J Numer Methods Eng 2008;75:1238–58.

CHAPTER

CONTINUUM MODELS

6

6.1 INTRODUCTION

Carbon nanotubes (CNTs) exhibit superior mechanical properties and hold substantial promise as a class of materials by providing strong, light, and high toughness characteristics. Due to their superior properties, CNTs are particularly attractive for structural elements in nanoscale devices. Much research has been carried out on CNTs since their first discovery in 1991 by Iijima of the NEC Laboratory in Tsukuba, Japan [1]. In this research activity, atomistic-based methods such as classical molecular dynamics (MD) [2–4], tight-binding MD [5], and density functional theory [6] have been used to study single-walled CNTs (SWCNTs) or multiwalled CNTs (MWCNTs) under axial compression, axial tension, bending, and torsion deformation. However, the computational time and effort needed for these methods limited the size of MWCNTs studied. For example, in our MD simulation [7] of buckling behavior, the calculation for a (10, 10) SWCNT with 2000 atoms would have required 36 h computational time on the SGI Origin 2000 system. The computational time increases as the number of atoms increases, and thus the computational time explodes exponentially for MWCNTs. In another example, for an MWCNT under tension [8], a four-walled (5, 5), (10, 10), (15, 15), and (20, 20) CNT with length-to-diameter ratio $L/D = 9.1$ contains 15097 atoms, and the calculation for the elasto-plastic deformation up to failure of CNT took up over 2 months on the SGI Origin 3000 system. Therefore, researchers have been seeking more efficient computational methods with which to analyze MWCNTs.

It was reported in the literature that continuum structural models might be applicable for the analysis of CNTs. In recent years, many continuum structural models have been proposed. Govindjee and Sackman [9] applied the Bernoulli–Euler beam-bending theory to examine Young's modulus of an MWCNT. A space truss/frame model was developed by Li and Chou [10,11] for modeling an SWCNT, and Young's modulus and shear modulus of an SWCNT were examined using their proposed model. Odegard et al. [12] proposed a truss model and an equivalent-plate model as a link between computational chemistry and solid mechanics. The proposed model was directly applied to predict the mechanical behavior of an SWCNT. Yakobson et al. [3] used a traditional continuum shell model to predict the buckling of an SWCNT, and compared it with the MD simulation. Their results show that the continuum shell model can obtain the buckling pattern. Pantano et al. [13] proposed a finite element approach to model the structure and deformation of SWCNTs and MWCNTs. Large bending deformation of MWCNTs was carried out in their work. However, to ensure that the shell theory is applicable to the analysis of CNTs, different estimates for the combination of Young's modulus, effective wall thickness, and Poisson ratio have to be tested by comparing MD simulation with the finite element approach.

Mechanical Behaviors of Carbon Nanotubes. DOI: http://dx.doi.org/10.1016/B978-0-323-43137-8.00006-8

The classical cylindrical shell model was also used in the buckling and postbuckling analysis of MWCNTs. For the study of MWCNT, a proper representation of the vdW forces is critical to establish a continuum model. Liu et al. [14] developed a nanoscale finite element method for the multiscale simulation of CNT. The vdW force between layers is modeled as nonlinear spring elements. By modeling the vdW interaction as truss rods, Li and Chou [10] simulated the elastic properties of MWCNT by using an equivalent space framelike structure model and the finite element method. Ru [15–19] proposed a continuum shell model considering vdW interaction to study the buckling of double-walled CNTs (DWCNTs). He suggested a linear proportional relationship between the variation of the vdW force and the normal deflection to model the vdW forces, and obtained a simple relationship for the vdW interaction coefficient, $c = (200\ \text{ergs/cm}^2)/(0.16d^2)$, where $d = 0.142$ nm. This relationship cannot reflect the effect of the radius of the CNT on the vdW interaction coefficient. In addition, he did not discuss the influence of the ignorance of the terms related to the ratio of $(R_O - R_I)/R_I$ on the buckling load, and concluded that the vdW forces do not increase the critical axial strain for infinitesimal buckling of DWCNTs. Wang et al. [20] extended Ru's vdW interaction model to the buckling analysis of MWCNTs, but Ru's model can only be applied to the interaction of two adjacent layers, as is the case of a DWCNT. Thus, only the effect of vdW interaction between two adjacent layers was taken into consideration. Shen [21] recently presented a traditional cylindrical shell model for the postbuckling of a DWCNT using Ru's vdW interaction model.

So far, a rigorous vdW interaction model is lacking. In this chapter, we aim to develop a new vdW interaction model for not only the postbuckling pressure but also the prebuckling pressure between any two layers of an MWCNT. In contrast to the models of Ru [16] and Wang [20], where the vdW interaction coefficient is only for the adjacent layers and independent of the radii of the tubes, the formulas derived herein account for the dependence of the vdW interaction (before and after buckling) not only on the change of interlayer spacing but also on the tube radius. Hence, it is a better model and can be applied to the MWCNTs. Based on the derived formulas of vdW interaction, a continuum cylindrical shell model is presented to study the buckling of an MWCNT. An accurate formula and a simple approach formula are also derived for the buckling of a DWCNT. The validity of the newly derived formulas for the buckling analysis of MWCNTs is demonstrated through comparison studies with existing models. The effect of the tube size is also examined on the critical buckling loads of MWCNTs. In addition, based on this refined model, the vibration characteristics of MWCNTs [22–24] have been studied and compared with MD simulations [25,26], for DWCNTs, triple-walled CNTs (TWCNTs), and four-walled CNTs with excellent agreement.

In recent years, many researchers have also focused their attention on the vibration of CNT by using elastic beam, truss, and shell theory. Based on a multiple-elastic beam model, Yoon et al. [27] investigated resonant frequencies and the associated vibrational modes of an individual MWCNT embedded in an elastic medium. Mahan [28] applied the equations of thin hollow cylinder to the low frequency vibration modes of an SWCNT. The prediction for the breathing mode was in good agreement with the experimental result [29]. A multiple-elastic beam model [27] was applied to study resonant frequencies and the associated vibration modes of an individual MWCNT. The work showed the phenomenon of coaxial and noncoaxial vibrations. Based on the Bernoulli–Euler beam theory and nonlocal elasticity, a nonlocal double-elastic beam model [30] is presented for the free transverse vibrations of DWCNT, and their results show that the classic double-elastic beam model could overestimate the natural frequencies and that the amplitude ratio is dependent upon the lower natural frequency. The vibrational behavior of a DWCNT has been investigated using a framelike structure model [31]. The simulations indicated that the non-coaxial vibration begins at the third resonant frequency. Based on Ru's vdW model [16], Wang et al. [32,33] investigated the free vibration of SWCNT and MWCNT using Donnell's and Flugge's shell theories. Radial breathing modes for DWCNT and TWCNT have been simulated and compared with MD simulations [34]. Their results show excellent agreement with the MD simulations. Also, their numerical calculations indicate that the vdW interaction has a crucial influence on the radial modes of vibration, but negligible influence on the torsional and longitudinal modes of vibration. Based on the He et al.'s vdW model [22], Natsuki et al. [35] investigated the vibrational properties of DWCNT embedded in an elastic matrix using wave propagation approach. Flugge's shell equations have been used in their work to study the natural frequency and vibration modes of CNT.

Here Donnell's shell model is applied to study the vibration of MWCNT based on He et al.'s vdW model [22]. Explicit expressions are obtained for the natural frequencies and mode shapes of DWCNT and TWCNT based on the simplified Donnell shell equation, in which the role of vdW interaction is clearly indicated. In contrast to Ru's vdW model [16] which is independent of the size of the MWCNT, and only reflects the effect of vdW interaction between one adjacent layer in the two directions, the main objective of the present study is to predict the influence of the vdW interaction on the vibration of MWCNTs for various radii and number of layers. The numerical simulations show that the vdW interaction has very small influence on the fundamental natural frequency, but plays a significant role in all higher radial-dominated natural frequencies. Especially, due to the effect of vdW interaction, there exists a change over from radial mode to longitudinal mode or from longitudinal mode to torsional mode, for MWCNTs in which the innermost radius of the CNT is very small.

6.2 EXPLICIT FORMULAS FOR VAN DER WAALS INTERACTION

To model vdW interaction that exists between any two layers of MWCNTs, the Lennard–Jones pair potential V_{LJ} [36–38] is adopted as

$$V_{LJ}(\bar{d}) = 4\varepsilon\left[\left(\frac{\sigma}{\bar{d}}\right)^{12} - \left(\frac{\sigma}{\bar{d}}\right)^{6}\right] \tag{6.1}$$

where $\bar{d}$ is the distance between the interacting atoms, ε the depth of the potential, and σ a parameter that is determined by the equilibrium distance. The vdW force F is obtained from taking the derivative of the Lennard–Jones pair potential, i.e.,

$$F(\bar{d}) = -\frac{dV_{LJ}(\bar{d})}{d\bar{d}} = \frac{24\varepsilon}{\sigma}\left[2\left(\frac{\sigma}{\bar{d}}\right)^{13} - \left(\frac{\sigma}{\bar{d}}\right)^{7}\right] \tag{6.2}$$

It should be noted that the negative value of the vdW force represents the attractive force of an approaching pair of atoms from a certain distance, whereas the positive value represents the repulsive force between a pair of atoms.

As only the infinitesimal buckling of the CNT is of interest, the vdW force can be estimated by the Taylor expansion to the first order around the equilibrium position prior to buckling, i.e.,

$$\begin{aligned} F(\overline{d}_0) &= F(\overline{d}_0) + \frac{dF(\overline{d}_0)}{d\overline{d}}(\overline{d}-\overline{d}_0) \\ &= \frac{24\varepsilon}{\sigma}\left[2\left(\frac{\sigma}{\overline{d}_0}\right)^{13} - \left(\frac{\sigma}{\overline{d}_0}\right)^{7}\right] - \frac{24\varepsilon}{\sigma^2}\left[26\left(\frac{\sigma}{\overline{d}_0}\right)^{14} - 7\left(\frac{\sigma}{\overline{d}_0}\right)^{8}\right](\overline{d}-\overline{d}_0) \end{aligned} \tag{6.3}$$

where $\overline{d}_0 = \sqrt{(R_j \cos\theta - R_i)^2 + R_j^2 \sin^2\theta + x^2}$ is the initial distance between atoms of different tubes prior to buckling.

The vdW force exerted on any atom of a tube can be estimated by summing all forces between the atom and all atoms on the other tube. To simplify the calculations, we consider the nanotube as a continuum cylindrical shell and note that each carbon atom corresponds to the area of $9a^2/4\sqrt{3}$ [39]. Thus, with assuming inward pressure positive, the integration of Eq. (6.3) over the entire nanotube leads to an analytical representation for the initial pressure contribution p_{ij} caused by vdW interaction

$$\begin{aligned} p_{ij} &= \left(\frac{4\sqrt{3}}{9a^2}\right)^2 \frac{24\varepsilon}{\sigma}\int_{-\pi}^{\pi}\int_{-L/2}^{L/2}\left[2\left(\frac{\sigma}{\overline{d}_0}\right)^{13} - \left(\frac{\sigma}{\overline{d}_0}\right)^{7}\right] R_j dx d\theta \\ &= \left[\frac{2048\varepsilon\sigma^{12}}{9a^4}\sum_{k=0}^{5}\frac{(-1)^k}{2k+1}\binom{5}{k}E_{ij}^{12} - \frac{1024\varepsilon\sigma^6}{9a^4}\sum_{k=0}^{2}\frac{(-1)^k}{2k+1}\binom{2}{k}E_{ij}^{6}\right]R_j \end{aligned} \tag{6.4}$$

and the pressure increment due to the buckling deflection

$$\begin{aligned} \Delta p_{ij} &= -\left(\frac{4\sqrt{3}}{9a^2}\right)^2 \frac{24\varepsilon}{\sigma^2}\int_{-\pi}^{\pi}\int_{-L/2}^{L/2}\left[26\left(\frac{\sigma}{\overline{d}_0}\right)^{14} - 7\left(\frac{\sigma}{\overline{d}_0}\right)^{8}\right](\overline{d}-\overline{d}_0) R_j dx d\theta \\ &= -\left[\frac{1001\pi\varepsilon\sigma^{12}}{3a^4}E_{ij}^{13} - \frac{1120\pi\varepsilon\sigma^6}{9a^4}E_{ij}^{7}\right]R_j(w_i - w_j) \end{aligned} \tag{6.5}$$

where $\alpha = 1.42$ Å is the C–C bond length, R_j the radius of jth layer, the subscripts i and j denote ith and jth layers, respectively, and E_{ij}^6, E_{ij}^7, E_{ij}^{12}, and E_{ij}^{13} are the elliptic integrals defined by

$$E_{ij}^m = (R_j + R_i)^{-m}\int_0^{\pi/2}\frac{d\theta}{[1 - K_{ij}\cos^2\theta]^{m/2}} \tag{6.6}$$

where m is an integer and

$$K_{ij} = \frac{4R_jR_i}{(R_j + R_i)^2}. \tag{6.7}$$

6.3 CONTINUUM SHELL MODEL

The elastic cylindrical shell theory has been adopted to study the buckling MWCNTs. An MWCNT consists of two or more single CNTs of radius R_i, thickness h, and modulus of elasticity E. The considered MWCNT is empty inside. No internal or external lateral pressures are applied to the tube except for the pressure caused by vdW interaction. We also assume that no sliding occurs between walls. Based on the classical thin-shell theory [40], the governing equations for the elastic buckling of an MWCNT can be derived as the N coupled equations, i.e.,

$$\begin{aligned} L_1 w_1 &= \nabla_1^4 p_1 \\ &\vdots \\ L_i w_i &= \nabla_i^4 p_i \\ &\vdots \\ L_N w_N &= \nabla_N^4 p_N, \end{aligned} \tag{6.8}$$

where w_i ($i = 1, 2, \ldots, N$) is the deflection of the ith nanotube, p_i is the pressure exerted on the nanotube i due to vdW interaction between walls, and L_i is the differential operator that is given by

$$L_i = D\nabla_i^8 - N_x \frac{\partial^2}{\partial x^2} \nabla_i^4 - \frac{N_\theta}{R_i^2} \frac{\partial^2}{\partial \theta^2} \nabla_i^4 + \frac{Eh}{R_i^2} \frac{\partial^4}{\partial x^4} \tag{6.9}$$

in which x and θ are the axial coordinate and the circumferential coordinate, $N_x = \sigma_x h$ and $N_\theta = \sigma_\theta h$ are the uniform forces per unit length in the axial and circumferential directions of the nanotube prior to buckling, σ_x and σ_θ are the axial and circumferential stresses, D is the bending stiffness of the nanotube, and

$$\nabla_i^2 = \frac{\partial^2}{\partial x^2} + \frac{1}{R_i^2} \frac{\partial^2}{\partial \theta^2} \tag{6.10}$$

Note that the attractive vdW force is negative and the repulsive vdW force is positive, and the inward pressure is assumed to be positive in Eq. (6.8). Thus, with the infinitesimal buckling deflection between two layers, the net pressure due to vdW interaction can be expressed as

$$p_i(x, \theta) = -\sum_{j=1}^{i-1} \overline{p}_{ij} + \sum_{j=i+1}^{N} \overline{p}_{ij} + \Delta p_i(x, \theta), \tag{6.11}$$

where $\overline{p}_{ij}$ is the initial uniform vdW pressure contribution to the ith layer from the jth layer prior to buckling, N is the total number of layers of the MWCNT, and $\Delta p_i(x, \theta)$ is the pressure increment after buckling. As only infinitesimal buckling is considered, $\Delta p_i(x, \theta)$ is assumed to be linearly proportional to the buckling deflection between two layers, i.e.,

$$\Delta p_i = \sum_{j=1}^{N} \Delta p_{ij} = \sum_{j=1}^{N} c_{ij}(w_i - w_j) = w_i \sum_{j=1}^{N} c_{ij} - \sum_{j=1}^{N} c_{ij} w_j, \tag{6.12}$$

where Δp_{ij} is the contribution to the pressure increment Δp_i exerted on the ith layer from the jth layer, and c_{ij} is the vdW interaction coefficient.

Comparing Eqs. (6.12) and (6.5), we have

$$c_{ij} = -\left[\frac{1001\pi\varepsilon\sigma^{12}}{3a^4} E_{ij}^{13} - \frac{1120\pi\varepsilon\sigma^6}{9a^4} E_{ij}^{7} \right] R_j \tag{6.13}$$

Eq. (6.13) furnishes the mathematical expression for the interaction coefficient that models vdW interaction after buckling in an MWCNT, where each tube has been treated as an individual continuum cylindrical shell. This more refined vdW model captures the effects of all layers, not just the effect between two adjacent layers. Moreover, the expression takes into account the size effect.

Finally, we obtain the following governing equations for the buckling of an MWCNT:

$$
\begin{aligned}
L_1 w_1 &= \nabla_1^4 w_1 \sum_{j=1}^{N} c_{1j} - \sum_{j=1}^{N} c_{1j} \nabla_1^4 w_j \\
&\vdots \\
L_i w_i &= \nabla_i^4 w_i \sum_{j=1}^{N} c_{ij} - \sum_{j=1}^{N} c_{ij} \nabla_i^4 w_j \\
&\vdots \\
L_N w_N &= \nabla_N^4 w_N \sum_{j=1}^{N} c_{Nj} - \sum_{j=1}^{N} c_{Nj} \nabla_N^4 w_j .
\end{aligned}
\tag{6.14}
$$

The equations are coupled with each other due to vdW interaction.

6.4 BUCKLING OF CNTs

The present section describes the solutions to the buckling of MWCNTs using the proposed vdW force model. We note that the buckling analysis is not sensitive to the boundary conditions [40]. Thus, throughout this section, we assume that the ends of all nanotubes are simply supported.

6.4.1 GENERAL MWCNTs

First, let us examine the buckling solutions for an MWCNT with any number of layers. The deflection of all tubes can be approximated by

$$w_k = A_k \sin\frac{m\pi x}{L} \cos n\theta, \tag{6.15}$$

where $A_k (k = 1, 2, \ldots, N)$ are N unknown coefficients, L is the length of the MWCNT, m is the axial half wavenumber, and n the circumferential wavenumber. It is obvious that these approximate solutions meet the following simply supported boundary conditions exactly

$$w_k = \frac{\partial^2 w_k}{\partial x^2}, \quad \text{at } x = 0 \quad \text{and} \quad x = L. \tag{6.16}$$

Substituting Eq. (6.15) into Eq. (6.14) yields N equations

$$
\left\{ D\left[\left(\frac{m\pi}{L}\right)^2 + \left(\frac{n}{R_k}\right)^2\right]^2 - \sum_{j=1}^{N} c_{kj} + N_x\left(\frac{m\pi}{L}\right)^2 - p_k R_k\left(\frac{n}{R_k}\right)^2 + \frac{Eh}{R_k^2}\left[\frac{1}{1+\left(\frac{Ln}{m\pi R_k}\right)^2}\right]^2 \right\} A_k
$$
$$
+ \sum_{j=1}^{N} c_{kj} A_j = 0 \quad (k = 1, 2, \ldots, N), \tag{6.17}
$$

where $p_k = -N_\theta/R_k$ and the net pressure exerted on the kth tube is assumed to be inward. Eq. (6.17) can be rewritten in the matrix form

$$-N_x\left(\frac{m\pi}{L}\right)^2\begin{pmatrix}1 & 0 & 0 & \cdots & 0\\ 0 & 1 & 0 & \cdots & 0\\ 0 & 0 & 1 & \cdots & 0\\ \vdots & \vdots & \vdots & \vdots & \vdots\\ 0 & 0 & 0 & \cdots & 1\end{pmatrix}\begin{Bmatrix}A_1\\ A_2\\ A_3\\ \vdots\\ A_n\end{Bmatrix} = \begin{bmatrix}b_{11} & c_{12} & c_{13} & \cdots & c_{1n}\\ c_{21} & b_{22} & c_{23} & \cdots & c_{2n}\\ c_{31} & c_{32} & b_{33} & \cdots & c_{3n}\\ \vdots & \vdots & \vdots & \vdots & \vdots\\ c_{n1} & c_{n2} & c_{n3} & \cdots & b_{nn}\end{bmatrix}\begin{Bmatrix}A_1\\ A_2\\ A_3\\ \vdots\\ A_n\end{Bmatrix} \tag{6.18}$$

or, equivalently,

$$\left[-N_x\left(\frac{m\pi}{L}\right)^2\mathbf{I}_{N\times N} - \mathbf{C}_{N\times N}\right]\begin{Bmatrix}A_1\\ A_2\\ \vdots\\ A_N\end{Bmatrix} = 0 \tag{6.19}$$

where

$$b_{kk} = \left[D\left(\frac{m\pi}{L}\right)^2 + \left(\frac{n}{R_k}\right)^2\right]^2 - \sum_{\substack{j=1\\ j\neq k}}^{N} c_{kj} - p_k R_k\left(\frac{n}{R_k}\right)^2\frac{Eh}{R_k^2}\left[\frac{1}{1+\left(\frac{Ln}{m\pi R_k}\right)^2}\right]^2 \tag{6.20}$$

To determine nonzero solutions of A_k, it is necessary to equate its determinant to zero. Thus, we have the characteristic equation of [**C**], i.e.,

$$\det\left(-[\mathbf{C}] - N_x\left(\frac{m\pi}{L}\right)[\mathbf{I}]\right) = 0 \tag{6.21}$$

Solving Eq. (6.21) yields the buckling load of the MWCNT relative to wavenumber of m and n.

6.4.2 AN EXPLICIT SOLUTION FOR DWCNTs

As a particular case of MWCNT, a DWCNT with an inner radius R_I and an outer radius R_O is considered for the buckling analysis. Then, Eq. (6.17) reduces to

$$\left\{D\left[\left(\frac{m\pi}{L}\right)^2 + \left(\frac{n}{R_\mathrm{I}}\right)^2\right]^2 - c_{12} + N_x\left(\frac{m\pi}{L}\right)^2 - p_1 R_1\left(\frac{n}{R_\mathrm{I}}\right)^2 + \frac{Eh}{R_\mathrm{I}^2}\left[\frac{1}{1+\left(\frac{Ln}{m\pi R_\mathrm{I}}\right)^2}\right]^2\right\}A_1 + c_{12}A_2 = 0 \tag{6.22}$$

and

$$\left\{D\left[\left(\frac{m\pi}{L}\right)^2 + \left(\frac{n}{R_\mathrm{O}}\right)^2\right]^2 - c_{21} + N_x\left(\frac{m\pi}{L}\right)^2 - p_2 R_\mathrm{O}\left(\frac{n}{R_\mathrm{O}}\right)^2 + \frac{Eh}{R_\mathrm{O}^2}\left[\frac{1}{1+\left(\frac{Ln}{m\pi R_\mathrm{O}}\right)^2}\right]^2\right\}A_2 + c_{21}A_1 = 0 \tag{6.23}$$

Here the inner layer is the 1st layer and the outer layer is the 2nd layer. With the existence condition for a nonzero solution of A_1 and A_2, the buckling load of the DWCNTs can be obtained by solving

$$\left[N_x\left(\frac{m\pi}{L}\right)^2\right]^2 + (B_1 + B_2)\left[N_x\left(\frac{m\pi}{L}\right)^2\right] + B_1B_2 - c_{12}c_{21} = 0, \tag{6.24}$$

where

$$B_1 = D\left[\left(\frac{m\pi}{L}\right)^2 + \left(\frac{n}{R_I}\right)^2\right]^2 - c_{12} - p_1R_I\left(\frac{n}{R_I}\right)^2 + \frac{Eh}{R_I^2}\left[\frac{\left(\frac{m\pi}{L}\right)^2}{\left(\frac{m\pi}{L}\right)^2 + \left(\frac{n}{R_I}\right)^2}\right]^2 \tag{6.25}$$

and

$$B_2 = D\left[\left(\frac{m\pi}{L}\right)^2 + \left(\frac{n}{R_O}\right)^2\right]^2 - c_{21} - p_2R_O\left(\frac{n}{R_O}\right)^2 + \frac{Eh}{R_O^2}\left[\frac{\left(\frac{m\pi}{L}\right)^2}{\left(\frac{m\pi}{L}\right)^2 + \left(\frac{n}{R_O}\right)^2}\right]^2 \tag{6.26}$$

Because the equilibrium distance between a carbon atom and a flat monolayer is around 0.34 nm [37], any increase ($w_i - w_j > 0$, $i, j = 1, 2$) or decrease ($w_i - w_j < 0$) in space between the inner and outer tubes would cause an attractive or repulsive vdW force, respectively. Note that we assume the inward pressure to be positive. Thus, from Eq. (6.5) we have $c_{12} < 0$ and $c_{21} < 0$. In contrast,

$$(B_1 + B_2)^2 - 4(B_1B_2 - c_{12}c_{21}) = (B_1 - B_2)^2 + 4c_{12}c_{21} > 0. \tag{6.27}$$

Thus, the solution to Eq. (6.24) is given by

$$-N_x\left(\frac{m\pi}{L}\right)^2 = \frac{1}{2}\left(B_1 + B_2 \mp \sqrt{(B_1 - B_2)^2 + 4c_{12}c_{21}}\right), \tag{6.28}$$

where the sign before the square root depends on the combination of integers n and m, as well as the constants c_{12} and c_{21}.

6.4.3 THE PARTICULAR CASE OF DWCNTs WITHOUT vdW INTERACTION

Considering a DWCNT without vdW interaction, we have $c_{12} = c_{21} = p_1 = p_2 = 0$. Note that $R_O > R_I$, and comparing Eqs. (6.25) and (6.26), we have $B_1 > B_2$. Then, when we take the positive sign before the square root, Eq. (6.28) reduces to the classical equation for the buckling load of cylindrical shells [40]

$$-N_x = D\left(\frac{L}{m\pi}\right)^2\left[\left(\frac{m\pi}{L}\right)^2 + \left(\frac{n}{R_I}\right)^2\right]^2 + \frac{Eh}{R_I^2}\left(\frac{L}{m\pi}\right)^2\left[\frac{\left(\frac{m\pi}{L}\right)^2}{\left(\frac{m\pi}{L}\right)^2 + \left(\frac{n}{R_I}\right)^2}\right]^2. \tag{6.29}$$

The axial load determined by Eq. (6.29) corresponds to the axial load exerted on the inner tube, which is modeled as an individual cylindrical shell. Similarly, when we take the negative sign before the square root, Eq. (6.28) reduces to the same classical solution as that of the buckling load of the outer tube

$$-N_x = D\left(\frac{L}{m\pi}\right)^2\left[\left(\frac{m\pi}{L}\right)^2+\left(\frac{n}{R_O}\right)^2\right]^2+\frac{Eh}{R_O^2}\left(\frac{L}{m\pi}\right)^2\left[\frac{\left(\frac{m\pi}{L}\right)^2}{\left(\frac{m\pi}{L}\right)^2+\left(\frac{n}{R_O}\right)^2}\right]^2. \tag{6.30}$$

The axial load determined by Eq. (6.30) is lower than that determined by Eq. (6.29). Thus, if we take the DWCNT as a whole system, the critical buckling load should be determined by minimizing the axial load solution of Eq. (6.30).

6.4.4 DWCNTs WITH vdW INTERACTION

Now, let us consider the buckling load for DWCNTs with vdW interaction. Because B_1 and B_2 depend on the wavenumber of m and n, the sign before the square root of Eq. (6.28) depends on the following cases.

In the case of $B_1B_2 \leq c_{12}c_{21}$

Based on the assumption that $B_1B_2 \leq c_{12}c_{21}$, one can easily arrive at

$$(B_1+B_2)^2-(B_1-B_2)^2 \leq 4c_{12}c_{21}. \tag{6.31}$$

In addition, the vdW forces between the two tubes are equal and opposite as shown in Fig. 6.1, and the pressures exerted on the two tubes should satisfy the equilibrium condition

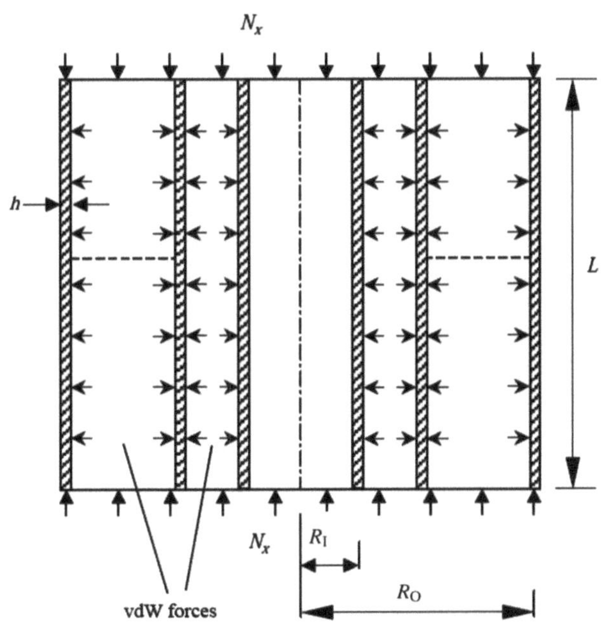

FIGURE 6.1

A continuum cylindrical shell model of an MWCNT under axial compression and vdW interaction [22].

$p_1R_I = -p_2R_O$. Hence, it is obvious from Eqs. (6.25) and (6.26) that $B_1 + B_2 > 0$, and Eq. (6.31) becomes

$$B_1 + B_2 \le \sqrt{(B_1 - B_2)^2 + 4c_{12}c_{21}} \tag{6.32}$$

To ensure that the right-hand side of Eq. (6.28) is positive, we take the positive sign before the square root in Eq. (6.28) and the solution to the axial load is given by

$$-N_x\left(\frac{m\pi}{L}\right)^2 = \frac{1}{2}\left(B_1 + B_2 + \sqrt{(B_1 - B_2)^2 + 4c_{12}c_{21}}\right). \tag{6.33}$$

For DWCNTs with large radii, the ratio $(R_O - R_I)/R_I$ is very small and can be neglected. Thus, we have

$$\begin{aligned}\left[\left(\frac{m\pi}{L}\right)^2 + \left(\frac{n}{R_I}\right)^2\right]^2 &= \left[\left(\frac{m\pi}{L}\right)^2 + \left(\frac{n}{R_O}\right)^2 + \left(\frac{n}{R_O}\right)^2\left(\frac{R_O + R_I}{R_I}\right)\left(\frac{R_O - R_I}{R_I}\right)\right]^2 \\ &\approx \left[\left(\frac{m\pi}{L}\right)^2 + \left(\frac{n}{R_O}\right)^2\right]^2\end{aligned} \tag{6.34}$$

and

$$\frac{n}{R_I} = \frac{n}{R_O}\frac{R_O}{R_I} = \frac{n}{R_O} + \frac{n}{R_O}\left(\frac{R_O - R_I}{R_I}\right) \approx \frac{n}{R_O}. \tag{6.35}$$

Substitution of Eqs. (6.34) and (6.35) into Eqs. (6.25) and (6.26), and then Eqs. (6.25) and (6.26) into Eq. (6.33) leads to an approximate formula for the buckling load of DWCNTs with large radii as

$$\begin{aligned}-N_x = &\left(\frac{L}{m\pi}\right)^2 D\left[\left(\frac{m\pi}{L}\right)^2 + \left(\frac{n}{R_O}\right)^2\right]^2 + \frac{Eh}{R_O^2}\left(\frac{L}{m\pi}\right)^2\left[\frac{\left(\frac{m\pi}{L}\right)^2}{\left(\frac{m\pi}{L}\right)^2 + \left(\frac{n}{R_O}\right)^2}\right]^2 \\ &+ \frac{1}{2}\left(\frac{L}{m\pi}\right)^2\left[-(c_{12} + c_{21}) + \sqrt{(c_{12} + c_{21})^2 - 4(c_{12} - c_{21})p_2\frac{n^2}{R_O} + 4p_2^2n^4R_O^2}\right].\end{aligned} \tag{6.36}$$

Let us examine the effect of vdW interaction on the axial load. Note that, as stated above, the vdW constants are negative, i.e., $c_{12} < 0$ and $c_{21} < 0$. Thus, we can verify that the last term of the right-hand side of Eq. (6.36) is positive. Comparing Eqs. (6.30) and (6.36), it is obvious that the presence of vdW interaction will significantly increase the buckling load in the case of $B_1B_2 \le c_{12}c_{21}$.

The case of $B_1B_2 > c_{12}c_{21}$

Because the equilibrium distance between a carbon atom and a flat monolayer is around 0.34 nm [37], the initial pressure $p_1 (p_1 < 0)$ is very small when the interlayer separation is taken as 0.34 nm. Note that $p_2 > 0$ ($p_2R_O = -p_1R_I$). It is clear from Eqs. (6.25) and (6.26) that we have $B_1 > -c_{12}$ and $B_2 > -c_{21}$ or $B_1B_2 > c_{12}c_{21}$, which gives the following relation:

$$B_1 + B_2 > \sqrt{(B_1 - B_2)^2 + 4c_{12}c_{21}} \tag{6.37}$$

To obtain the lowest positive solution to Eq. (6.28), we should take the negative before the square root in Eq. (6.28) so that it becomes

$$-N_x\left(\frac{m\pi}{L}\right)^2=\frac{1}{2}\left(B_1+B_2-\sqrt{(B_1-B_2)^2+4c_{12}c_{21}}\right). \tag{6.38}$$

The substitution of Eqs. (6.25) and (6.26) into Eq. (6.38) for B_1 and B_2 and noted Eqs. (6.34) and (6.35) yields an approximate solution for the buckling load of DWCNTs

$$-N_x=\left(\frac{L}{m\pi}\right)^2 D\left[\left(\frac{m\pi}{L}\right)^2+\left(\frac{n}{R_O}\right)^2\right]^2+\frac{Eh}{R_O^2}\left(\frac{L}{m\pi}\right)^2\left[\frac{\left(\frac{m\pi}{L}\right)^2}{\left(\frac{m\pi}{L}\right)^2+\left(\frac{n}{R_O}\right)^2}\right]^2$$
$$+\frac{1}{2}\left(\frac{L}{m\pi}\right)^2\left[-(c_{12}+c_{21})-\sqrt{(c_{12}+c_{21})^2-4(c_{12}-c_{21})p_2\frac{n^2}{R_O}+4p_2^2\frac{n^4}{R_O^2}}\right]. \tag{6.39}$$

Our numerical results demonstrate that the prebuckling vdW forces between any two layers are attractive, as shown in Table 6.1, and the inward pressure is assumed to be positive. Hence, we

Table 6.1 Prebuckling Pressure p_{ij} (MPa) Due to vdW Interaction for a Nine-walled CNT With Various Innermost Radii R_I [22]

R_I (nm)	p_{51}	p_{52}	p_{53}	p_{54}	p_{56}	p_{57}	p_{58}	p_{59}
0.34	−3.4978	−19.537	−173.80	−522.01	−635.30	−263.53	−37.790	−9.5832
0.68	−3.4978	−21.417	−182.31	−529.42	−624.19	−256.82	−36.488	−9.1808
1.36	−5.0877	−23.619	−192.61	−539.37	−610.69	−248.29	−34.830	−8.6677
2.04	−5.5085	−24.881	−198.63	−545.66	−602.81	−243.10	−33.816	−8.3533
2.72	−5.7785	−25.702	−202.58	−549.98	−597.64	−239.60	−33.132	−8.1406
3.40	−5.9667	−26.277	−205.37	−553.12	−593.99	−237.08	−32.638	−7.9868
4.08	−6.1054	−26.704	−207.45	−555.50	−591.27	−235.18	−32.264	−7.8704
4.76	−6.2120	−27.032	−209.05	−557.37	−589.17	−233.70	−31.972	−7.7792
5.44	−6.2963	−27.293	−210.33	−558.88	−587.51	−232.51	−31.737	−7.7058
6.12	−6.3648	−27.505	−211.37	−560.12	−586.15	−231.53	−31.544	−7.6455
6.80	−6.4215	−27.681	−212.23	−561.16	−585.02	−230.71	−31.383	−7.5949
9.52	−6.5758	−28.162	−214.60	−564.04	−581.93	−228.46	−30.936	−7.4551
16.32	−6.7507	−28.709	−217.31	−567.39	−578.41	−225.84	−30.416	−7.2913
26.52	−6.8517	−29.026	−218.89	−569.38	−576.36	−224.29	−30.108	−7.1945
42.84	−6.9159	−29.228	−219.89	−570.66	−575.06	−223.30	−29.910	−7.1320
68.00	−6.9554	−29.353	−220.51	−571.45	−574.26	−222.69	−29.787	−7.0931
95.20	−6.9749	−29.414	−220.82	−571.82	−573.84	−222.38	−29.726	−7.0738
122.40	−6.9858	−29.448	−220.99	−573.04	−574.62	−222.21	−29.692	−7.0630
152.32	−6.9934	−29.472	−221.11	−570.26	−571.44	−222.09	−29.669	−7.0555

have $p_2 > 0$. In contrast, as stated above, the vdW constant $c_{12} < 0$ and $c_{21} < 0$. In addition, we have $|c_{12}| > |c_{21}|$ because the surface area of the outer tube is larger than that of the inner tube, which means that the sum of vdW forces from all atoms on the surface of the outer tube is larger than that of the inner tube. Thus, it is verified that the second term in the square root is positive and then the last term on the right-hand side of Eq. (6.39) is negative. Hence, comparing Eq. (6.30) with Eq. (6.39) indicates that Eq. (6.39) will result in a lower buckling load than that obtained without vdW interaction. However, as stated before, when the outer (inner) tube deflects outward (inward), the vdW force caused would be attractive, whereas when the outer (inner) tube deflects inward (outward), the vdW force would be repulsive. Hence, vdW interaction always has an effect against buckling. This verifies that vdW interaction will raise the buckling load. It should be noted that Eq. (6.39) is only an approximate expression for the buckling load of a DWCNT, and cannot accurately describe the influence of vdW interaction on buckling behavior. In our numerical examples, the exact solution given by Eq. (6.38) shows that vdW force results in a higher buckling load than is obtained without vdW force, while the approximate Eq. (6.39) leads to a lower buckling load. This difference is caused by ignorance of the terms related to the ratio of $(R_O - R_I)/R_I$ in Eq. (6.39). If we take $c_{12} = c_{21}$, then Eq. (6.39) reduces to the Ru's result [16].

We first examine vdW interaction for MWCNTs with various innermost radii and make a comparison with Wang et al.'s result [20]. The buckling loads for DWCNTs are also calculated and compared with the results of Ru [16]. Having validated the derived formula, the buckling loads are investigated for various MWCNTs. The effect of vdW interaction on the buckling load is also highlighted. Now let us consider an MWCNT with an innermost radius R_I and an outermost radius R_O as shown in Fig. 6.1. Suppose that each tube has the same length L and thickness h, and is modeled as an individual cylindrical shell. The MWCNT is subjected to the combined action of axial compression and vdW interaction. No internal or external lateral pressures are considered for the buckling analysis of the MWCNT. Following most published papers on CNTs, the initial interlayer separation between the two adjacent layers is assumed to be 0.34 nm. The parameters used in the vdW potential (given by Eq. (6.1)) are taken as $\varepsilon = 2.968$ meV and $\sigma = 3.407$ Å [41].

6.4.5 vdW INTERACTION BEFORE BUCKLING

To illustrate vdW interaction prior to buckling, the prebuckling vdW pressures between two adjacent layers are calculated using the derived Eq. (6.4) for a nine-walled CNT with innermost radius range of 0.34–152.32 nm and are presented in Table 6.1. The value of p_{ij} in Table 6.1 represents the pressure contribution to layer i from layer j due to vdW interaction before buckling. Note that the negative sign in Table 6.1 represents an attraction between two layers. It can be seen that the greatest influence of vdW interaction before buckling occurs between two adjacent layers, and the pressure decreases monotonically with increasing distance between two layers until it is very small, or is negligible when the distance is large enough. The curvature effect on the vdW pressure is also reflected in Table 6.1. With a small tube radius (or large curvature), the prebuckling vdW pressure varies quickly and the curvature effect becomes significant. However, with a large radius (or small curvature), the pressure varies only slightly, and the curvature effect can be neglected. The prebuckling pressures due to vdW interaction approach constant values as the radius increases. These constants are around -570 MPa for p_{54} and p_{56}, -221 MPa for p_{53} and p_{57}, -29 MPa for p_{52} and p_{58}, and -7 MPa for p_{51} and p_{59}.

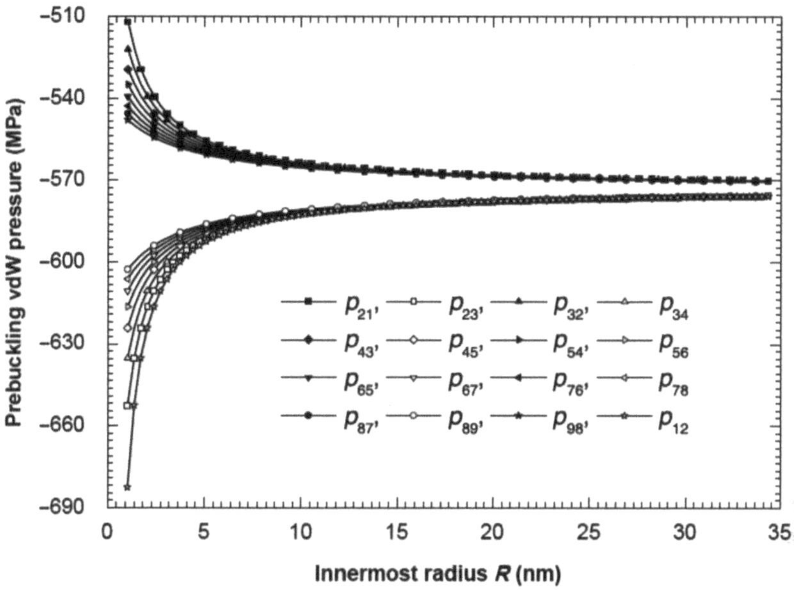

FIGURE 6.2

Prebuckling vdW pressure between two adjacent layers versus the innermost radius of a nine-walled CNT [22].

To show the dependence of all vdW pressures between two adjacent layers of a nine-walled CNT on the radii of tubes, the prebuckling pressure contributions to the 1st and 9th layers from one side layer, the 2nd, 3rd, 4th, 5th, 6th, 7th, and 8th layers from two side layers are presented in Fig. 6.2 for various innermost radii of the nine-walled CNT. The innermost layer is the 1st layer and the outermost layer is the 9th layer. For any two adjacent layers, the pressure contribution to the outer tube from inner tube is plotted in the curve with solid symbol, as shown in Fig. 6.2. The largest absolute value of the pressure contribution (p_{98}) occurs on the outermost tube for any innermost radius of the nine-walled CNT, and then the pressure contribution decreases from the outermost tube to the innermost tube until the pressure contribution to the tube next to the innermost tube is smallest (p_{21}). For the pressure contribution to the inner tube from outer tube (the curves with empty symbols in Fig. 6.2), the largest pressure contribution (p_{12}) occurs on the innermost tube and the smallest contribution (p_{89}) occurs on the tube next to the outermost tube. For the small radius, the pressure contribution to the outer tube from inner tube between any two adjacent layers drops sharply, whereas the pressure contribution to the inner tube from the outer tube rises sharply with the increase of radius, which means that the size of the radius significantly affects the prebuckling vdW pressure when the radius is small. As the radius increases, the pressures decrease (or increase) slowly and stably until the pressures between all of the adjacent layers approach a constant of around -570 MPa.

6.4.6 vdW INTERACTION AFTER BUCKLING

To illustrate the derived Eq. (6.13), the vdW interaction coefficients c_{ij} are calculated and presented in Table 6.2 for a nine-walled CNT with an innermost radii range of 0.34–152.32 nm. The values

Table 6.2 vdW Interaction Coefficients C_{ij} (GPa/nm) for a Nine-Walled CNT With Various Innermost Radii R_I [22]

R_I (nm)	c_{51}	c_{52}	c_{53}	c_{54}	c_{56}	c_{57}	c_{58}	c_{59}
0.34	0.0149	0.1105	1.463	97.14	118.98 −99.19[a]	2.2234	0.2152	0.0409
0.68	0.0183	0.1216	1.5371	99.15	117.33 −99.19[a]	2.1676	0.2080	0.0392
1.36	0.0217	0.1345	1.6257	101.62	115.23 −99.19[a]	2.0964	0.1987	0.0371
2.04	0.0235	0.1419	1.6771	103.07	113.94 −99.19[a]	2.0529	0.1930	0.0358
2.72	0.0247	0.1467	1.7108	104.02	113.08 −99.19[a]	2.0236	0.1891	0.0349
3.40	0.0256	0.1500	1.7345	104.69	112.46 −99.19[a]	2.0025	0.1863	0.0342
4.08	0.0262	0.1524	1.7522	105.19	111.99 −99.19[a]	1.9865	0.1842	0.0337
4.76	0.0266	0.1543	1.7658	105.59	111.63 −99.19[a]	1.9740	0.1826	0.0333
5.44	0.0270	0.1558	1.7766	105.90	111.33 −99.19[a]	1.9640	0.1812	0.0330
6.12	0.0273	0.1571	1.7855	106.15	111.09 −99.19[a]	1.9558	0.1801	0.0328
6.80	0.0275	0.1581	1.7928	106.36	110.89 −99.19[a]	1.9489	0.1792	0.0326
9.52	0.0282	0.1608	1.8129	106.94	110.33 −99.19[a]	1.9299	0.1767	0.0320
16.32	0.0289	0.1640	1.8358	107.60	109.69 −99.19[a]	1.9078	0.1737	0.0313
26.52	0.0294	0.1658	1.8491	107.99	109.31 −99.19[a]	1.8948	0.1720	0.0309
42.84	0.0297	0.1669	1.8576	108.23	109.07 −99.19[a]	1.8864	0.1708	0.0306
68.00	0.0298	0.1676	1.8629	108.28	108.91 −99.19[a]	1.8812	0.1701	0.0304
95.20	0.0299	0.1680	1.8655	108.32	108.84 −99.19[a]	1.8786	0.1698	0.0303
122.40	0.0300	0.1682	1.8669	108.35	108.80 −99.19[a]	1.8772	0.1696	0.0303
152.32	0.0300	0.1683	1.8679	108.40	108.75 −99.19[a]	1.8762	0.1694	0.0303

[a] *Results obtained from [20].*

of the coefficients c_{ij} in Table 6.2 represent the pressure increment contribution (due to vdW interaction after buckling) to layer i from layer j. It can be seen that vdW interaction between two adjacent layers (c_{54} and c_{56}) is much larger than that between two nonadjacent layers (such as c_{53} and c_{59}), and the coefficient decreases rapidly from the adjacent layer to the innermost or outermost layer. Thus, when the distance between the two layers is large enough, the value of the coefficient is very small and vdW interaction can be neglected. It should be noted that the negative sign in Table 6.2 represents an attraction between two layers, whereas the positive sign represents repulsion. The coefficients between two adjacent layers are compared with the results obtained by Wang et al. [20] for various innermost radii, as shown in Table 6.2. It can be seen that the present results (based on the newly proposed vdW interaction model) are about 10% higher than Wang's results. However, the value of the percentage differences in the coefficient values depends on the radius of the tube. Note that Wang assumed the interaction coefficient c to be curvature independent, and only vdW interaction between two adjacent layers was taken into consideration. However, as shown in Table 6.2, the coefficients c_{ij} are curvature dependent. This is because the surface area of the tube varies with respect to the radius, and the force exerted on each carbon atom of the surface of tube i is the sum of vdW forces from all atoms on the surface of tube j. Thus, the coefficients c_{ij} vary with the different radii of tubes. It is interesting to note that with the increase of the radius, when $j<i$, the absolute values of coefficient c_{ij} increase, whereas when $j>i$, the absolute values of coefficient c_{ij} decrease. However, when the radius is large enough, the coefficient c_{ij} is curvature independent [42,43] and approaches a constant. For example, as shown in Table 6.2, the coefficients almost approach the same constant values of -108 GPa/nm for c_{54} and c_{56}, 1.87 GPa/nm for c_{53} and c_{57}, 0.169 GPa/nm for c_{52} and c_{58}, and 0.03 GPa/nm for c_{51} and c_{59}.

The relationships of the coefficients and the innermost radius of the nine-walled CNT are plotted in Fig. 6.3. Similar to the prebuckling pressure, the coefficients corresponding to the pressure increment contribution to outer (or inner) tube from inner (or outer) tube between any two adjacent layers drop (rise) sharply with the increase of radius when the radius of the tube is small. This implies that the vdW interaction coefficients are highly dependent on the size of the radius for MWCNTs with small radii. Large radii have little influence on the coefficients, as shown in Fig. 6.3. Similarly, the coefficients approach the same constant of -108 GPa/nm, which is about 10% higher than the Wang et al. result [20].

6.4.7 BUCKLING OF A DWCNT

The buckling behavior of MWCNTs is investigated using the vdW interaction model we have developed. Comparison between the present model and Ru's model is carried out for the buckling of DWCNTs. For all the considered MWCNTs, the bending stiffness $D=0.85$ eV, $Eh=360$ J/m^2 [3], and the length to the outermost radius ratio $L/R_O=10$.

To examine the validity of the approximate formula (Eq. 6.39), we simulated the buckling loads with the circumferential wavenumber $n=6$ for a DWCNT with inner-tube radius $R_I=11.9$ nm, and compared them with results calculated using the proposed exact solution (Eq. 6.38), the classical shell model (Eq. 6.30) and Ru's model [16], as shown in Fig. 6.4. The relative errors for the lowest buckling loads from Eqs. (6.38) and (6.39), and from Eq. (6.39) and the Ru'model [16] are 2.46% and 0.49% respectively. It can be seen from Fig. 6.4 that the buckling loads obtained from

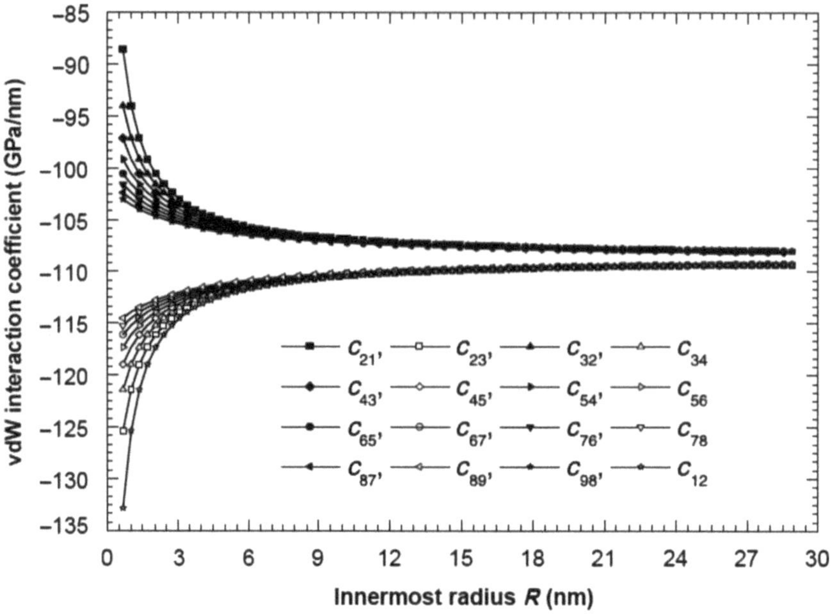

FIGURE 6.3

vdW interaction coefficients between two adjacent layers versus the innermost radius of a nine-walled CNT [22].

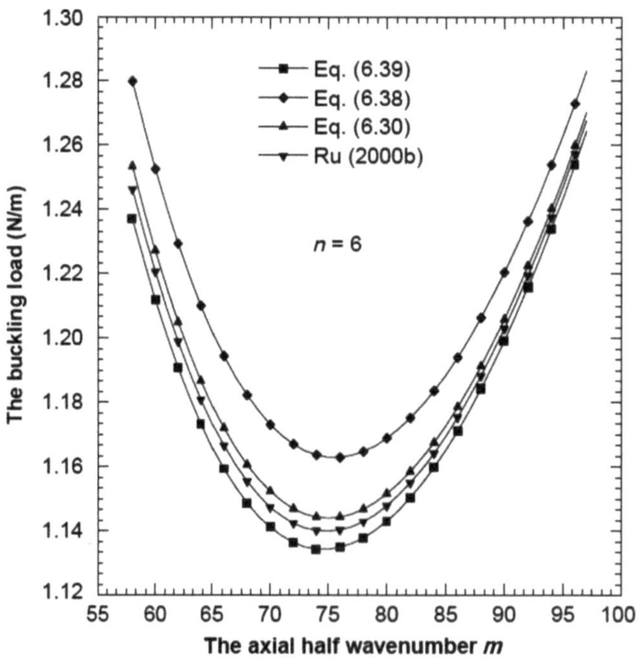

FIGURE 6.4

Comparison study of buckling loads for DWCNT with inner radices $R_{\mathrm{I}} = 11.9$ nm [22].

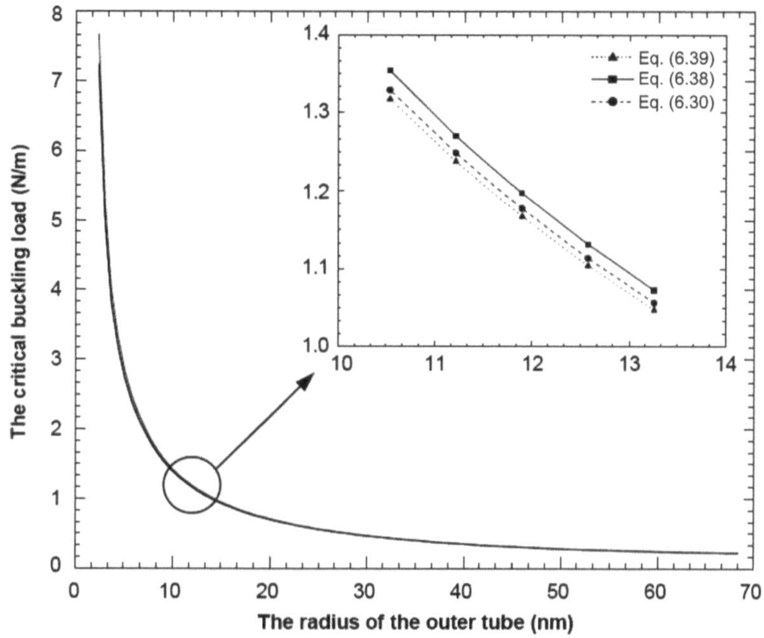

FIGURE 6.5

Critical buckling loads of the DWCNT with various radii of the outer tube [22].

Eq. (6.30) (without vdW interaction) are larger than those from Eq. (6.39) and Ru's model (taking vdW interaction into account), but smaller than those from Eq. (6.38) (taking vdW interaction into consideration). As mentioned earlier, vdW interaction always has an effect against buckling deflection. Thus, the presence of vdW interaction will lead to a higher buckling load. With the vdW forces taken into account, the numerical simulation from Eq. (6.38) illustrates that the buckling load is indeed bigger than when the vdW forces are not considered. As for the lower buckling loads from Eq. (6.39), as shown in Fig. 6.4, they occur because the ignorance of the terms related to the ratio of $(R_O - R_I)/R_I$ in Eq. (6.39) leads to the results smaller than those in Eq. (6.30). The buckling loads in Ru's model [16] are also smaller than those in the classical shell model without vdW forces. Ru reported that the vdW forces do not increase the critical axial strain for infinitesimal buckling of DWCNTs. It is clear that the influence of the terms related to the ratio of $(R_O - R_I)/R_I$ on the buckling load is not considered in his paper.

Fig. 6.5 shows the curvature effects on the critical buckling loads of DWCNTs. The critical buckling loads are obtained by minimizing the results from Eqs. (6.38) and (6.39) with respect to the wavenumbers n and m. It can be seen that the critical buckling load is a function of the tube radius. With a very small radius, say less than 5 nm, the critical load drops rapidly, indicating that the curvature effect plays an important role in the critical buckling load for a small radius. For DWCNTs with a large radius, say bigger than 40 nm, the critical load varies very slowly, showing that the dependence of the critical load on the radius can be neglected. The relative errors between the critical buckling loads obtained by Eqs. (6.38) and (6.39) are examined and

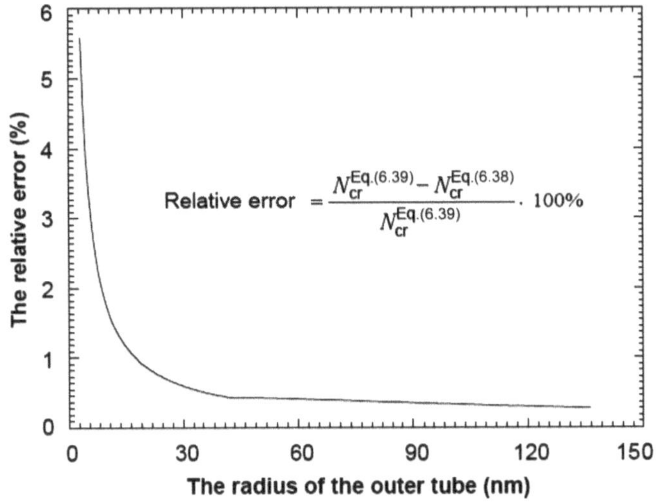

FIGURE 6.6

Relative errors between the critical axial loads obtained from Eqs. (6.38) and (6.39) for various outer tube radii of the DWCNT [22].

shown in Fig. 6.6. As can be seen, the relative error is significant with very small radii. However, when the radius is big, say bigger than 40 nm, the relative error is very small and is negligible.

6.4.8 BUCKLING OF MWCNTs

Based on the proposed vdW interaction model, the buckling loads for MWCNTs with 2, 5, 10, and 17 layers have been calculated and are presented in Fig. 6.7(A–D). For all of the considered MWCNTs in Fig. 6.7(A–D), the innermost radius $R_I = 11.9$ nm. Fig. 6.7(A–D) show the relationship between the buckling loads and the wavenumbers (m, n). The effect of vdW interaction on the buckling loads is examined by comparing the buckling loads with (curves with solid line) and without (curves with dash line) the vdW force taken into consideration. As can be observed in Fig. 6.7A–D, the influence of vdW interaction is more obvious as the number of layers increases. This phenomenon is expected because the critical buckling load always occurs on the outermost layer of an MWCNT, and for a CNT with a fixed innermost radius the more layers there are, the bigger the surface area of the outermost tube. As mentioned above, the force exerted on each carbon atom of the surface of a tube is the sum of the vdW forces from all atoms on the surface of all the other tubes. Thus, the influence of vdW interaction is greater for an MWCNT with a larger radius. With minimization of the buckling loads with respect to the wavenumber m and n, the critical buckling loads are obtained as $N_{cr} = 1.1605$ N/m for the DWCNT; $N_{cr} = 1.1138$ N/m for the five-walled CNT; $N_{cr} = 1.0432$ N/m for the 10-walled CNT; and $N_{cr} = 0.9520$ N/m for the 17-walled CNT.

Fig. 6.8 shows the dependency of critical buckling loads on the number of layers of an MWCNT. The critical buckling loads decrease as the number of layers increases. This is because the more layers that there are, the larger the outermost radius of an MWCNT, and a larger radius

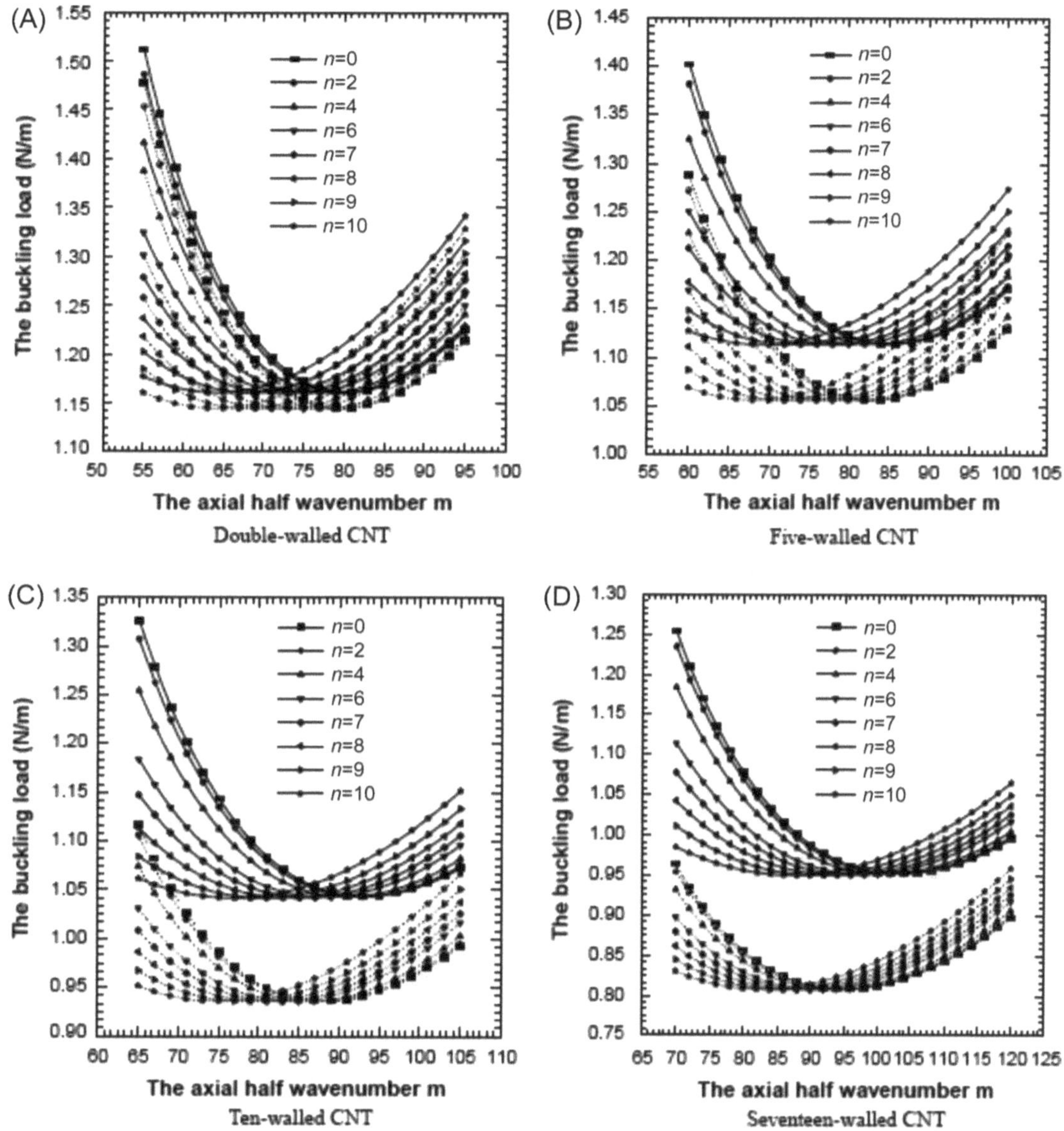

FIGURE 6.7

Axial loads versus the half wavenumbers (m, n) for (A) DWCNT, (B) five-walled CNT, (C) 10-walled CNT, and (D) 17-walled CNT with innermost radius $R_{\mathrm{I}} = 11.9$ nm. The upper solid line curves are obtained from the present model and the lower dash line curves are obtained from the classical shell model with the absence of vdW interaction [22].

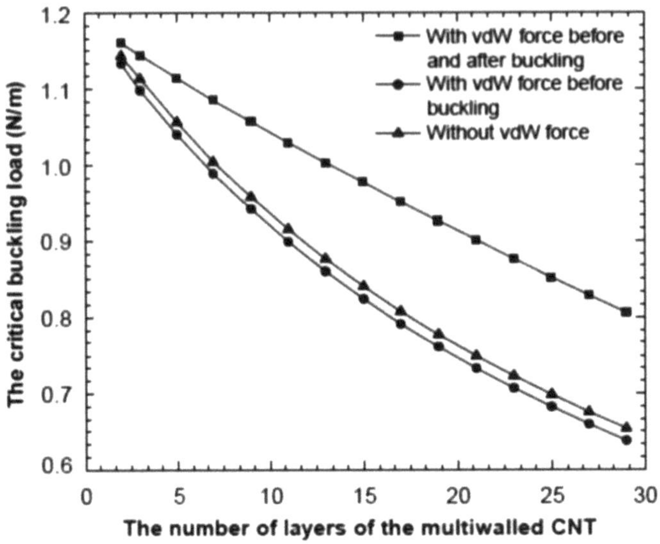

FIGURE 6.8

Influence of vdW interaction on the critical buckling loads of the MWCNT with various layers [22].

leads to smaller critical axial stress [44]. The effects of vdW interaction before and after buckling are also clearly illustrated in Fig. 6.8. Taking only the initial pressure p_{ij} before buckling into account, vdW interaction will be slightly lower for the critical load compared to the critical load without vdW force. This result is coincident with the fact that the presence of the external pressure will reduce the critical axial load of a cylindrical shell [40]. Furthermore, taking into account the prebuckling pressure p_{ij} as well as the interaction coefficient c_{ij}, vdW interaction results in higher critical loads than those without considering vdW force. For DWCNTs, vdW interaction is insignificant. As the number of layers of an MWCNT increases, the influence of vdW interaction becomes more significant. As explained above, this is because the vdW force is greater for larger CNT surfaces.

In summary, original explicit formulas have been derived for predicting the vdW interaction before and after buckling for MWCNTs under compression. The derived formulas capture the vdW interaction between any two layers of an MWCNT. This refined vdW model indicates that the prebuckling pressures and the vdW interaction coefficients between different two layers are different and strongly dependent on the radius of the tube when the radius is small enough. Moreover, the prebuckling pressures and the vdW interaction coefficients between any two layers approach constant values with increasing radii of the MWCNT. However, the vdW interaction coefficients (or prebuckling pressures) between any two adjacent layers approach the same constant as the radius increases. The derived formulas of the vdW interaction are not only for the buckling problem of MWCNTs, but also for other related topics, such as vibration and bending problems.

Based on the proposed vdW interaction formulas, a continuum cylindrical shell model is presented to investigate the buckling of an MWCNT, in which each tube is treated as an individual cylindrical shell. An exact formula Eq. (6.38) and a simple approximate formula Eq. (6.39) have been derived for the buckling analysis in the special case of a DWCNT. Comparison between the present approximate formula, Eq. (6.39), and the approximate formula of Ru [16] is carried out, and the relative error is 0.49. Hence, the difference between these two approximate formulas is very small. The results from these two approximate formulas show that vdW interaction does not increase the critical buckling load of a DWCNT. However, the results from the exact Eq. (6.38) indicate that vdW interaction does increase the critical buckling load. It is obvious that these two approximate formulas cannot correctly describe the influence of vdW interaction on the buckling load because they ignore the terms related to the ratio of $(R_O - R_I)/R_I$.

Analysis of the buckling of MWCNTs was carried out using the proposed model. In contrast to the work of Wang et al. [20], where only vdW interaction between two adjacent layers was considered, the buckling behavior was simulated taking into consideration vdW interaction not just between two adjacent layers but between all layers. The effect of tube size on the critical buckling loads of MWCNTs was also examined.

6.5 VIBRATION CHARACTERISTICS OF CNTs

6.5.1 DONNELL SHELL MODEL FOR THE VIBRATION OF MWCNT

The Donnell shell model has been adopted to analyze the free vibration of MWCNT. An MWCNT consists of two or more single CNTs of radius R_i, thickness h, and modulus of elasticity E. The considered MWCNT is empty inside. No internal or external lateral pressures are applied to the tubes except for the pressure caused by vdW interaction [24].

Based on the Donnell shell theory [40], the governing equations for the harmonic vibrations of an MWCNT can be derived as the $3N$ coupled equations, i.e.,

$$\begin{aligned}
L_{11}u_i + L_{12}v_i - L_{13}w_i &= \frac{(1-\nu^2)R_i^2\rho h}{Eh}\frac{\partial^2 u_i}{\partial t^2},\\
L_{21}u_i + L_{22}v_i - L_{23}w_i &= \frac{(1-\nu^2)R_i^2\rho h}{Eh}\frac{\partial^2 v_i}{\partial t^2},\\
L_{31}u_i + L_{32}v_i - L_{33}w_i &= \frac{(1-\nu^2)R_i^2}{Eh}\left(\rho h\frac{\partial^2 w_i}{\partial t^2} - p_i\right),\\
i &= 1,\ 2, \ldots N,
\end{aligned} \tag{6.40}$$

where u_i, v_i, and w_i ($i = 1, 2, \ldots, N$ with first layer being the innermost layer and the Nth layer being the outermost layer) are the longitudinal, circumferential, and radial displacement of the ith tube, t is the time, R_i is the radius of tube i, Eh is the in-plane stiffness, ρh is the mass density per unit area, ν is Poisson's ratio, p_i is the pressure exerted on the tube i due to vdW interaction between walls, and L_{ij} are the differential operators that are given by

$$\begin{aligned}
&L_{11} = R_i^2\frac{\partial^2}{\partial x^2} + \frac{1-\nu}{2}\frac{\partial^2}{\partial \theta^2}, \quad L_{12} = L_{21} = \frac{1+\nu}{2}R_i\frac{\partial^2}{\partial x \partial \theta},\\
&L_{13} = L_{31} = \nu R_i\frac{\partial}{\partial x}, \quad L_{22} = \frac{1-\nu}{2}R_i^2\frac{\partial^2}{\partial x^2} + \frac{\partial^2}{\partial \theta^2},\\
&L_{23} = L_{32} = \frac{\partial}{\partial \theta}, \quad L_{33} = 1 + \frac{(1-\nu^2)D}{Eh}R_i^2\nabla^4,
\end{aligned} \tag{6.41}$$

where x and θ are the axial coordinate and the circumferential coordinate, respectively, and D is the effective bending stiffness of a tube. Note that the attractive vdW force is negative and the repulsive vdW force is positive, and the inward pressure is assumed to be positive in Eq. (6.40). Thus, the net pressure due to vdW interaction can be expressed as

$$p_i(x, \theta) = \sum_{j=1}^{N} c_{ij}(w_i - w_{ij}) \tag{6.42}$$

where c_{ij} is the vdW interaction coefficient [22] and is dependent on R_i.

Let us examine the vibration solutions for MWCNT with any number of layers. The deflection of all tubes is approximated by

$$\begin{aligned} u_i &= U_i \cos\frac{m\pi x}{L} \cos n\theta e^{\bar{j}\omega t}, \\ v_i &= V_i \sin\frac{m\pi x}{L} \sin n\theta e^{\bar{j}\omega t}, \\ w_i &= W_i \sin\frac{m\pi x}{L} \cos n\theta e^{\bar{j}\omega t}, \end{aligned} \tag{6.43}$$

where U_i, V_i, W_i ($i = 1, 2, \ldots, N$) are $3N$ unknown coefficients, L is the length of the MWCNT, $\bar{j} = \sqrt{-1}$, m is the axial half wave number, and n is the circumferential wave number.

Substituting Eq. (6.43) into Eq. (6.40) yields $3N$ equations:

$$\left[\rho h \omega^2 \mathbf{I}_{3N\times 3N} - \frac{Eh}{1-\nu^2}\mathbf{H}_{3N\times 3N}\right]\begin{bmatrix} U_1 \\ V_1 \\ W_1 \\ \vdots \\ U_N \\ V_N \\ W_N \end{bmatrix} = 0 \tag{6.44}$$

where I is an identity matrix and the elements in matrix H are

$$\begin{aligned} h_{(3i-2)(3i-2)} &= \left(\frac{m\pi}{L}\right)^2 + \frac{1-\nu}{2R_i^2}n^2 \\ h_{(3i-2)(3i-1)} = h_{(3i-1)(3i-2)} &= -\frac{1+\nu}{2R_i}\frac{m\pi}{L}n \\ h_{(3i-2)(3i)} = h_{(3i)(3i-2)} &= \frac{\nu}{R_i}\frac{m\pi}{L} \\ h_{(3i-1)(3i-1)} &= \frac{1-\nu}{2}\left(\frac{m\pi}{L}\right)^2 + \frac{n^2}{R_i^2} \\ h_{(3i-1)(3i)} = h_{(3i)(3i-1)} &= -\frac{n}{R_i^2} \\ h_{(3i)(3i)} &= \frac{1}{R_i^2} + \frac{(1-\nu^2)D}{Eh}\left[\left(\frac{m\pi}{L}\right)^2 + \left(\frac{n}{R_i}\right)^2\right]^2 - \frac{(1-\nu^2)}{Eh}\sum_{j=1}^{N} c_{ij} \end{aligned} \tag{6.45}$$

To determine nonzero solutions of U_i, V_i, and W_i, it is necessary to equate its determinant to zero. Thus, we have the characteristic equation, i.e.,

$$\det\left\{\frac{1-\nu^2}{Eh}\rho h\omega^2[\mathbf{I}]-[\mathbf{H}]\right\}=0 \tag{6.46}$$

Solving Eq. (6.46) yields the natural frequencies of the MWCNT for a given set of wave numbers m and n.

6.5.2 RADIAL VIBRATION ANALYSIS OF MWCNT

It is noted that the interlayer friction is very small [45] and can be neglected. There is no tangential external loading along the axial or the circumferential directions of the CNT. In this case, Donnell's equation for the radial deflection w_i of the ith layer of the MWCNT can be simplified as

$$\frac{w_i}{R_i^2}+\frac{(1-\nu^2)D}{Eh}\left(\frac{\partial^4 w_i}{\partial x^4}+\frac{2}{R_i^2}\frac{\partial^4 w_i}{\partial x^2\partial\theta^2}+\frac{1}{R_i^4}\frac{\partial^4 w_i}{\partial\theta^4}\right)=-\frac{(1-\nu^2)}{Eh}\left(\rho h\frac{\partial^2 w_i}{\partial t^2}-p_i\right) \tag{6.47}$$

$i=1,\ 2,\ldots N$

Substitution of the proposed solution in Eq. (6.43) into Eq. (6.47) yields a set of N equations

$$(B_i-\rho h\omega^2)W_i+\sum_{j=1}^{N}c_{ij}W_j=0,\quad i=1,\ 2,\ \ldots N \tag{6.48}$$

where

$$B_i=\frac{Eh}{(1-\nu^2)}\frac{1}{R_i^2}+D\left[\left(\frac{m\pi}{L}\right)^2+\left(\frac{n}{R_i}\right)^2\right]^2-\sum_{j=1}^{N}c_{ij} \tag{6.49}$$

6.5.2.1 Explicit solution for DWCNT

Considering a DWCNT, the system of Eq. (6.48) is reduced to two equations:

$$\begin{aligned}(B_1-\rho h\omega^2)W_1+c_{12}W_2&=0,\\ c_{21}W_1+(B_2-\rho h\omega^2)W_2&=0,\end{aligned} \tag{6.50}$$

and the natural frequencies of the DWCNT are obtained by solving

$$(\rho h\omega^2)^2-(B_1+B_2)\rho h\omega^2+B_1B_2-c_{12}c_{21}=0 \tag{6.51}$$

Due to the fact that from the equilibrium distance, any increase ($w_i-w_j\geq 0$, i, $j=1,\ 2$) or decrease ($w_i-w_j\leq 0$) in the distance between the inner and outer tubes will cause an attractive or repulsive vdW force, respectively. Note that we assume the inward pressure to be positive. Thus, from Eq. (6.42) we have $c_{12} \leq 0$ and $c_{21} \geq 0$, and it is obvious that

$$(B_1 - B_2)^2 + 4c_{12}c_{21} \geq 0 \tag{6.52}$$

and note that

$$\begin{aligned} B_1B_2 - c_{12}c_{21} = &\left\{\frac{Eh}{(1-\nu^2)}\frac{1}{R_1^2} + D\left[\left(\frac{m\pi}{L}\right)^2 + \left(\frac{n}{R_1}\right)^2\right]^2\right\} \times \left\{\frac{Eh}{(1-\nu^2)}\frac{1}{R_2^2} + D\left[\left(\frac{m\pi}{L}\right)^2 + \left(\frac{n}{R_2}\right)^2\right]^2\right\} \\ &- c_{21}\left\{\frac{Eh}{(1-\nu^2)}\frac{1}{R_1^2} + D\left[\left(\frac{m\pi}{L}\right)^2 + \left(\frac{n}{R_1}\right)^2\right]^2\right\} \\ &- c_{12}\left\{\frac{Eh}{(1-\nu^2)}\frac{1}{R_2^2} + D\left[\left(\frac{m\pi}{L}\right)^2 + \left(\frac{n}{R_2}\right)^2\right]^2\right\} \geq 0 \end{aligned} \tag{6.53}$$

Hence, one can easily arrive at

$$(B_1 - B_2)^2 + 4c_{12}c_{21} = (B_1 + B_2)^2 - 4(B_1B_2 - c_{12}c_{21}) < (B_1 + B_2)^2 \tag{6.54}$$

Thus, the solution to Eq. (6.51) is given by

$$\rho h\omega_1^2 = \frac{1}{2}\left[B_1 + B_2 - \sqrt{(B_1 - B_2)^2 + 4c_{12}c_{21}}\right] \tag{6.55}$$

and

$$\rho h\omega_2^2 = \frac{1}{2}\left[B_1 + B_2 + \sqrt{(B_1 - B_2)^2 + 4c_{12}c_{21}}\right] \tag{6.56}$$

It is worth mentioning that when the innermost radius is very large, $c_{12} \approx c_{21}$ and $R_1 \approx R_2$, and thus $B_1 \approx B_2$. Hence, we have

$$\rho h\omega_1^2 \approx \frac{1}{2}\left\{\frac{Eh}{(1-\nu^2)}\frac{1}{R_1^2} + D\left[\left(\frac{m\pi}{L}\right)^2 + \left(\frac{n}{R_1}\right)^2\right]^2 + \frac{Eh}{(1-\nu^2)}\frac{1}{R_2^2} + D\left[\left(\frac{m\pi}{L}\right)^2 + \left(\frac{n}{R_2}\right)^2\right]^2\right\} \tag{6.57}$$

Eq. (6.57) indicates that the effect of the vdW interaction on the lower frequency is very small and then can be neglected when the innermost radius is very large.

The ratios of the amplitudes of the vibration modes associated with ω_1 and ω_2 can be obtained by substituting Eqs. (6.55) and (6.56) into one of Eq. (6.50) and are expressed, respectively, as

$$R_{2,1} = \left.\frac{W_2}{W_1}\right|_{\omega=\omega_1} = \frac{1}{2c_{12}}\left[B_2 - B_1 - \sqrt{(B_2 - B_1)^2 + 4c_{12}c_{21}}\right] \tag{6.58}$$

$$R_{2,2} = \left.\frac{W_2}{W_1}\right|_{\omega=\omega_2} = \frac{1}{2c_{12}}\left[B_2 - B_1 + \sqrt{(B_2 - B_1)^2 + 4c_{12}c_{21}}\right] \tag{6.59}$$

for the second mode. Note that $\sqrt{(B_2 - B_1)^2 + 4c_{12}c_{21}} > B_2 - B_1$ and $c_{12} < 0$, that means the amplitude ratio associated with ω_1 is $R_{2,1} = W_2/W_1 > 0$ and the amplitude ratio associated with ω_2 is

$R_{2,2} = W_2/W_1 < 0$. Thus, for the vibration associated with ω_1, both layers are always vibrating in the same direction. In contrast, for the vibration associated with ω_2, both layers are vibrating in the opposite direction always.

6.5.2.2 Explicit solution for TWCNT

Considering a TWCNT, the system of Eq. (6.48) is reduced to the following three equations:

$$\begin{aligned}
&(B_1 - \rho h\omega^2)W_1 + c_{12}W_2 + c_{13}W_3 = 0,\\
&c_{21}W_1 + (B_2 - \rho h\omega^2)W_2 + c_{23}W_3 = 0,\\
&c_{31}W_1 + c_{32}W_2 + (B_3 - \rho h\omega^2)W_3 = 0,
\end{aligned} \tag{6.60}$$

Using the condition for the nonzero solutions of $W_1 - W_3$, one obtains the following determinant equation for the natural frequencies of a TWCNT:

$$(\rho h\omega^2)^3 + a_1(\rho h\omega^2)^2 + a_2(\rho h\omega^2) + a_3 = 0, \tag{6.61}$$

where

$$\begin{aligned}
&a_1 = -B_1 - B_2 - B_3,\\
&a_2 = B_1B_2 + B_1B_3 + B_2B_3 - c_{31}c_{13} - c_{21}c_{12} - c_{23}c_{32},\\
&a_3 = -B_1B_2B_3 - c_{21}c_{32}c_{13} - c_{31}c_{12}c_{23} + c_{13}c_{31}B_2 + c_{21}c_{12}B_3 + c_{23}c_{32}B_1
\end{aligned} \tag{6.62}$$

The solutions to Eq. (6.61) are

$$\begin{aligned}
&\rho h\omega_1^2 = -\frac{2}{3}\sqrt{a_1^2 - 3a_2}\cos\frac{\alpha}{3} - \frac{a_1}{3},\\
&\rho h\omega_2^2 = -\frac{2}{3}\sqrt{a_1^2 - 3a_2}\cos\frac{\alpha + 4\pi}{3} - \frac{a_1}{3},\\
&\rho h\omega_3^2 = -\frac{2}{3}\sqrt{a_1^2 - 3a_2}\cos\frac{\alpha + 2\pi}{3} - \frac{a_1}{3}
\end{aligned} \tag{6.63}$$

where

$$\alpha = \cos^{-1}\frac{27a_3 + 2a_1^3 - 9a_1a_2}{2\sqrt{(a_1^2 - 3a_2)^3}} \tag{6.64}$$

The associated amplitude ratios $R_{2,i}$ and $R_{3,i}$ of the TWCNT can be determined as

$$R_{2,i} = \left.\frac{W_2}{W_1}\right|_{\omega=\omega_i} = \frac{c_{23}(B_1 - \rho h\omega_i^2) - c_{21}c_{13}}{c_{13}(B_2 - \rho h\omega_i^2) - c_{12}c_{23}} \tag{6.65}$$

and

$$R_{3,i} = \left.\frac{W_3}{W_1}\right|_{\omega=\omega_i} = -\frac{(B_1 - \rho h\omega_i^2)(B_2 - \rho h\omega_i^2) - c_{12}c_{21}}{c_{13}(B_2 - \rho h\omega_i^2) - c_{12}c_{23}} \tag{6.66}$$

In the examples MWCNTs with an innermost radius R_I and an outermost radius R_O are analyzed. Each tube has the same length L and is modeled as an individual cylindrical shell. No internal or external lateral pressures are considered for the vibration analysis of the MWCNTs. Following most published papers on CNT, the initial interlayer separation between two adjacent layers is taken as 0.34 nm. The parameters used to calculate the vdW interaction coefficient c_{ij}

(Ref. [22]) are taken as $\varepsilon = 2.968$ meV and $\sigma = 3.407$ Å [41]. For all examples, the bending and axial stiffnesses of MWCNTs are $D = 0.85$ eV and $Eh = 360$ J/m^2 [3], respectively, and the length to the outermost radius ratio is $L/R_O = 10$.

Let us first consider a 10-walled MWCNT with innermost radius $R_I = 5$ nm. Based on the proposed vdW interaction model, the natural frequencies $f_i = \omega_i/2\pi$ which are associated with the radial mode for a 10-walled MWCNT have been calculated by using Eq. (6.48) and are presented in Table 6.3 for

Table 6.3 Radial-mode Associated Natural Frequencies (THz) for a 10-Layered MWCNT With Innermost Radius $R_I = 5$ nm [24]

m	n	f_1	f_2	f_3	f_4	f_5	f_6	f_7	f_8	f_9	f_{10}
1	0	0.5236	0.8189	1.2606	1.7583	2.2443	2.6907	3.0787	3.3939	3.6254	3.7657
	1	0.5236	0.8189	1.2606	1.7583	2.2443	2.6907	3.0787	3.3939	3.6254	3.7657
	6	0.5262	0.8221	1.2623	1.7595	2.2452	2.6914	3.0793	3.3945	3.6259	3.7662
	11	0.5515	0.8547	1.2803	1.7717	2.2545	2.6991	3.086	3.4005	3.6314	3.7712
	16	0.6312	0.9685	1.3485	1.8177	2.2899	2.7283	3.1113	3.4231	3.652	3.7908
	21	0.778	1.1915	1.5195	1.9329	2.3781	2.8012	3.1744	3.4794	3.7029	3.8439
6	0	0.5236	0.8189	1.2606	1.7584	2.2443	2.6907	3.0787	3.3939	3.6254	3.7657
	1	0.5237	0.8189	1.2606	1.7584	2.2443	2.6907	3.0787	3.3939	3.6254	3.7657
	6	0.5266	0.8225	1.2625	1.7597	2.2453	2.6915	3.0794	3.3946	3.626	3.7662
	11	0.5526	0.8557	1.2809	1.7721	2.2549	2.6994	3.0862	3.4007	3.6316	3.7714
	16	0.6331	0.9703	1.3497	1.8186	2.2906	2.7289	3.1118	3.4236	3.6524	3.7912
	21	0.7804	1.1939	1.5215	1.9343	2.3792	2.8021	3.1752	3.4802	3.7036	3.8446
11	0	0.5238	0.819	1.2607	1.7584	2.2443	2.6907	3.0787	3.3939	3.6254	3.7657
	1	0.5238	0.819	1.2607	1.7584	2.2443	2.6907	3.0787	3.3939	3.6254	3.7657
	6	0.5275	0.8233	1.263	1.76	2.2456	2.6917	3.0796	3.3947	3.6261	3.7664
	11	0.5553	0.8581	1.2824	1.7731	2.2557	2.7001	3.0868	3.4012	3.6321	3.7719
	16	0.6376	0.9747	1.3527	1.8207	2.2922	2.7303	3.113	3.4247	3.6534	3.7922
	21	0.7864	1.1996	1.5263	1.9378	2.382	2.8044	3.1772	3.482	3.7053	3.8464
16	0	0.5243	0.8193	1.2609	1.7585	2.2445	2.6908	3.0788	3.394	3.6255	3.7658
	1	0.5243	0.8194	1.2609	1.7585	2.2445	2.6908	3.0788	3.394	3.6255	3.7658
	6	0.5293	0.8247	1.2639	1.7606	2.2461	2.6921	3.08	3.395	3.6264	3.7667
	11	0.5598	0.8622	1.2849	1.7749	2.257	2.7012	3.0878	3.4021	3.6329	3.7727
	16	0.6451	0.9819	1.3576	1.8241	2.2949	2.7325	3.115	3.4265	3.6551	3.7939
	21	0.7959	1.2088	1.534	1.9435	2.3865	2.8082	3.1805	3.485	3.708	3.8493
21	0	0.5255	0.8201	1.2614	1.7589	2.2448	2.691	3.079	3.3942	3.6256	3.7659
	1	0.5257	0.8203	1.2615	1.759	2.2448	2.6911	3.079	3.3942	3.6257	3.766
	6	0.5324	0.8271	1.2654	1.7617	2.2469	2.6928	3.0805	3.3956	3.6269	3.7671
	11	0.5665	0.868	1.2885	1.7774	2.259	2.7029	3.0893	3.4034	3.6341	3.7738
	16	0.6556	0.992	1.3645	1.829	2.2988	2.7357	3.1178	3.429	3.6574	3.7962
	21	0.8093	1.2215	1.5446	1.9513	2.3927	2.8134	3.1851	3.4891	3.7118	3.8533

Table 6.4 Radial Natural Frequencies (THz) of a Five-Walled CNT With $L/R_O = 10$: Present vdW Interaction Model Compared to the Model in Ref. [32] for $m = 1$, $n = 0$ [24]

	$R_I = 0.5$ nm			$R_I = 1$ nm			$R_I = 5$ nm		
	Present	**Ref. [32]**	**%**	**Present**	**Ref. [32]**	**%**	**Present**	**Ref. [32]**	**%**
ω_{R1}	2.1850	2.1620	1.06	1.7510	1.7370	0.81	0.6183	0.6181	0.03
ω_{R2}	3.3690	3.2500	3.66	2.7210	2.6490	2.72	1.2960	1.2710	1.97
ω_{R3}	4.1580	3.9520	5.21	3.4040	3.2770	3.88	2.2710	2.1820	4.08
ω_{R4}	5.0370	4.8320	4.25	3.9190	3.7260	5.18	3.1010	2.9430	5.37
ω_{R5}	7.3600	7.2600	1.38	4.4320	4.2020	5.47	3.6410	3.4370	5.94

various combinations of the wave numbers (m, n). It can be seen from Eq. (6.48) that there are 10 equations that give 10 different natural frequencies $f_1 \leq f_2 \leq \ldots f_i \ldots \leq f_9 \leq f_{10}$ for every combination of m and n. Table 6.3 shows that the natural frequencies of the 10-layered MWCNT are very high, in the order of terahertz. As the mode order increases, the lower natural frequencies such as f_1, f_2, and f_3 increase significantly. However, all the other higher natural frequencies, from f_4 to f_{10}, are insensitive to the mode order, and thus the effect of the mode order on the higher natural frequencies can be neglected. It should be noted that the natural frequencies for the breathing mode with $n = 0$ are almost the same for various m. This is because the first and third terms in Eq. (6.49) are around 10^7 times larger than the second term which is related to the wave number m, and thus the influence of the axial half wave number on the natural frequencies is negligible.

In Table 6.4 a comparison is made between the results for the radial modes of a five-walled CNT with three inner radii using the present multilayer interaction of vdW forces and the results which are obtained using the single neighboring layer on each side, as was used in Ref. [32]. It can be seen that the additional coupling that exists in the present model between remote layers will lead to higher frequencies for all the CNTs in this table. The differences are in the range of 0–5% and one cannot predict for which case the influence will be larger, as there are several coupling effects involved. In general the second, third, and fourth radial frequencies have larger differences as the radii are reduced.

To show the effect of vdW interaction on natural frequencies of an MWCNT of small radius, natural frequencies that are associated with the radial mode for a TWCNT with innermost radius $R_I = 0.65$ nm are calculated by using Eq. (6.63), and the comparison of the results by considering and ignoring the vdW interaction is presented in Table 6.5(A) for various combinations of the wave numbers (m, n). It is observed that the vdW interaction has a significant effect on the frequencies, especially on the lower frequencies when the innermost radius is very small. Also, it can be seen that the wave number n has a dominant influence on the frequencies. This is because all the three terms in Eq. (6.49) are in the same order when the innermost radius R_I is very small, e.g., $R_I \leq 1$ nm, and thus the effects of the vdW interaction and the mode order must be taken into consideration.

Table 6.5(B) shows the comparison of radial-mode associated frequencies of a TWCNT with large radius for various combinations of (m, n). In contrast to the TWCNT of small radius, the effect of vdW interaction on the lowest frequency is very small and is negligible due to the fact that the effect of the vdW interaction on the lowest frequency that is associated with the radial

Table 6.5 Comparison of the Influence of the vdW Interaction on the Radial-Mode Associated Natural Frequencies (THz) for a TWCNT With Innermost Radius [24]

		vdW Interaction Is Considered			Without vdW Interaction		
m	n	f_1	f_2	f_3	f_1	f_2	f_3
(A) $R_I = 0.65$ nm							
1	1	2.9603	4.462	5.8611	2.6381	3.5444	5.3996
	6	3.3122	5.117	8.1387	2.9697	4.3134	7.8499
	11	5.5419	9.3774	20.0049	5.2825	8.9855	19.896
	16	10.179	18.0186	40.9221	10.0319	17.8216	40.8694
	21	16.9671	30.4086	70.0237	16.8782	30.2927	69.993
6	1	2.9651	4.466	5.866	2.6433	3.5494	5.4051
	6	3.3705	5.18	8.2303	3.0319	4.3892	7.9452
	11	5.6505	9.4924	20.1299	5.396	9.1056	20.0217
	16	10.3033	18.1452	41.0514	10.158	17.9496	40.9988
	21	17.0954	30.5377	70.1539	17.0072	30.4224	70.1232
11	1	3.0006	4.4917	5.8901	2.6827	3.5815	5.4315
	6	3.5279	5.3435	8.457	3.2001	4.5834	8.1807
	11	5.9179	9.7733	20.4337	5.6747	9.3985	20.3272
	16	10.6056	18.4528	41.3653	10.4645	18.2605	41.3131
	21	17.4072	30.8515	70.4698	17.3205	30.7373	70.4393
16	1	3.1231	4.5776	5.9628	2.8184	3.6884	5.5108
	6	3.8181	5.6304	8.8287	3.5099	4.9182	8.5656
	11	6.352	10.2238	20.9171	6.125	9.8667	20.8131
	16	11.0875	18.942	41.8639	10.9525	18.7547	41.8124
	21	17.9026	31.3499	70.9717	17.8183	31.2376	70.9414
21	1	3.4065	4.7797	6.1286	3.1288	3.936	5.6905
	6	4.2793	6.07	9.3583	3.9989	5.4196	9.1118
	11	6.9619	10.8489	21.5811	6.7544	10.5137	21.4803
	16	11.7509	19.6135	42.5475	11.6235	19.4328	42.4968
	21	18.5823	32.0332	71.6595	18.5011	31.9233	71.6295
(B) $R_I = 5$ nm							
1	1	0.6574	1.9721	3.3364	0.6176	0.657	0.7016
	6	0.6629	1.974	3.3374	0.6222	0.6624	0.7082
	11	0.7166	1.9935	3.3486	0.6667	0.7157	0.7727
	16	0.8916	2.0664	3.3909	0.8143	0.89	0.9804
	21	1.2242	2.242	3.4962	1.1032	1.2251	1.3724
6	1	0.6575	1.9721	3.3364	0.6177	0.657	0.7017
	6	0.6638	1.9743	3.3376	0.6231	0.6633	0.7093
	11	0.7195	1.9945	3.3492	0.6694	0.7185	0.7758
	16	0.8964	2.0686	3.3922	0.819	0.8948	0.9854
	21	1.2302	2.2454	3.4983	1.1091	1.2312	1.3786

Table 6.5 Comparison of the Influence of the vdW Interaction on the Radial-Mode Associated Natural Frequencies (THz) for a TWCNT With Innermost Radius [24] *Continued*

		vdW Interaction Is Considered			Without vdW Interaction		
m	n	f_1	f_2	f_3	f_1	f_2	f_3
11	1	0.658	1.9723	3.3365	0.6182	0.6575	0.7022
	6	0.6665	1.9752	3.3382	0.6256	0.666	0.712
	11	0.7267	1.9972	3.3508	0.6763	0.7258	0.7833
	16	0.9082	2.0738	3.3954	0.8305	0.9067	0.9977
	21	1.2448	2.2537	3.5036	1.1236	1.2459	1.3936
16	1	0.6597	1.9728	3.3368	0.62	0.6592	0.7038
	6	0.6715	1.9769	3.3392	0.6306	0.671	0.7172
	11	0.7388	2.0017	3.3535	0.688	0.7379	0.796
	16	0.9273	2.0825	3.4006	0.8492	0.926	1.0175
	21	1.268	2.2672	3.5121	1.1467	1.2695	1.4175
21	1	0.6639	1.9742	3.3376	0.6245	0.6634	0.7078
	6	0.6803	1.9799	3.341	0.6392	0.6798	0.726
	11	0.7568	2.0085	3.3575	0.7055	0.7559	0.8146
	16	0.9542	2.0949	3.4081	0.8756	0.953	1.0453
	21	1.3001	2.2861	3.5241	1.1788	1.302	1.4505

mode is very small when the innermost radius is large, as indicated in Eq. (6.57) for a DWCNT of big radius. However, the effect of vdW interaction is significant on the higher frequencies for various combinations of (m, n), which is obvious from Eq. (6.49) that when the innermost radius is big, the term related to the vdW interaction coefficient dominates.

Natural frequencies of a DWCNT, a TWCNT, and a five-walled MWCNT are calculated from Eq. (6.44) for various m and n as well as the innermost radii $R_I = 5$, 1, and 0.5 nm, and shown in Table 6.6(A)–(C) for the DWCNT, Table 6.7(A)–(C) for the TWCNT, and Table 6.8(A)–(C) for the five-walled MWCNT. It should be noted that $n = 0$ represents the breathing vibration of MWCNTs. In Table 6.6(A)–(C) the modes are also identified by the letters L, T, and R and their combinations to indicate the dominance of the longitudinal, torsional, and radial modes in the various modes. In most cases one can see that the modes are not purely of type L, T, or R, but rather a combination of these where the more dominant one is put first. As can be seen from these tables the frequencies increase as n is increased except for the lowest frequency which decreases first and then increases with the increase of n. Similarly, all the frequencies increase as m is increased. It is clear from the tables that there are changes of mode shapes for the same order of frequency when m or n are changed. For example, in Table 6.6(A), the mode shape of the DWCNT associated with ω_1 is mainly torsional mode for the combination of $m = 1$ and $n = 1$, but the mode shape becomes mostly radial mode for the combination of $m = 1$ and $n = 2$. In addition, the frequencies increase significantly with the increase of the radius of MWCNTs, especially the frequencies associated with higher number of n. Also, it can be seen from Tables 6.6–6.8 that for MWCNTs with the same size but different numbers of layers, the highest frequency for the

Table 6.6 Natural Frequencies (THz) of a DWCNT With Innermost Radius [24]

Natural frequencies (THz)	$m=1$					$m=2$					$m=3$				
	$n=0$	$n=1$	$n=2$	$n=3$	$n=4$	$n=0$	$n=1$	$n=2$	$n=3$	$n=4$	$n=0$	$n=1$	$n=2$	$n=3$	$n=4$
(A) $R_I=5$ nm and $L/R_O=10$															
ω_1	...	0.0391	0.0163	0.0226	0.0393	...	0.1248	0.0509	0.0334	0.0424	...	0.2210	0.1021	0.0581	0.0519
		T&R	R&T	R&T	R&T		T&R&L	R&T	R&T	R&T		R&T&L	R&T	R&T	R&T
ω_2	...	0.4423	0.8440	1.2540	1.6670	...	0.4961	0.8782	1.2760	1.6830	...	0.5729	0.9266	1.3110	1.7080
		L&T	L	L	L		L&T	L&T	L&T	L&T		L&T	L&T&R	L&T&R	L&T
ω_3	0.2018	0.4793	0.8998	1.3380	1.7800	0.3999	0.5769	0.9372	1.3590	1.7940	0.5741	0.6912	0.9972	1.3940	1.8180
	L	L&R	L	L	L	L	L&R	L&T&R	L&T	L&T	L	L&R&T	L&T&R	L&T&R	L&T
ω_4	0.2061	0.6883	1.3130	1.9130	2.3750	0.4123	0.7773	1.3620	1.9440	2.3930	0.6184	0.9048	1.4380	1.9960	2.4210
	L	T&L	T&R&L	T&R	T&R	L	T&L	T&L&R	T&R&L	T&R&L	L	L&T	T&L&R	T&R&L	R&T&L
ω_5	0.6802	0.9708	1.5360	2.1670	2.7910	0.6865	1.0010	1.5670	2.1920	2.8100	0.7174	1.0590	1.6190	2.2330	2.8420
	R	T&R	T&R	T&R	T&R	R	R&T&L	T&R&L	T&R&L	T&R	R	T&R&L	T&R&L	T&R&L	T&R&L
ω_6	2.7570	2.7620	2.7830	2.8470	3.0810	2.7570	2.7620	2.7840	2.8490	3.0920	2.7570	2.7620	2.7840	2.8530	3.1090
	R	R	R	R&T	T&R	R	R	R&T	R&T	T&R	R	R	R&T	R&T	T&R&L
(B) $R_I=1$ nm and $L/R_O=10$															
ω_1	...	0.1508	0.1950	0.4455	0.7981	...	0.4749	0.2714	0.4635	0.8095	...	0.8515	0.4369	0.5106	0.8326
		T&R&L	R&T	R&T	R&T		T&R&L	R&T	R&T	R&T		R&T&L	R&T	R&T	R&T
ω_2	...	1.5790	2.3540	2.5780	2.7450	...	1.6230	2.3380	2.5760	2.7470	...	1.8150	2.3490	2.5790	2.7530
		T&R&L	R&T&L	R&T	R&T		T&R&L	R&T&L	R&T	R&T		T&R&L	R&T&L	R&T	R&T
ω_3	0.8046	1.9360	3.3890	5.0060	6.6490	1.5980	2.3510	3.5950	5.1170	6.7230	2.3300	2.8170	3.8760	5.2910	6.8440
	L	L&T&R	L&R	L	L	L	L&R&T	L&R	L&T&R	L&T	L	L&R&T	L&R&T	L&T&R	L&T&R
ω_4	0.8128	2.3990	4.4930	6.6850	8.8940	1.6230	2.7990	4.6490	6.7690	8.9500	2.4240	3.2800	4.8830	6.9030	9.0420
	L	L&T&R	L	L	L	L	L&R&T	L&R	L&T	L	L	L&R&T	L&R&T	L&T&R	L&T
ω_5	2.8960	3.9380	5.9540	8.3340	10.830	2.9160	4.0380	6.0740	8.4370	10.920	3.0010	0.2380	6.2780	8.6080	11.060
	R	R&T	T&R&L	T&R	T&R	R	R&T&L	T&R&L	T&R&L	T&R&L	R	T&R&L	T&R&L	T&R&L	T&R&L
ω_6	4.2450	0.2500	7.9340	11.140	14.500	4.2530	5.3150	8.0240	11.220	14.560	4.2700	5.4390	8.1760	11.350	14.670
	R	R&T	T&R	T&R	T&R	R	R&T&L	T&R&L	T&R&L	T&R&L	R	R&T&L	T&R&L	T&R&L	T&R&L
(C) $R_I=0.5$ nm and $L/R_O=10$															
ω_1	...	0.2602	0.6339	1.3380	2.0310	...	0.7415	0.7112	1.3640	2.0530	...	1.3150	0.9085	1.4270	2.0950
		T&R&L	R&T	R&T	R&T		T&R&L	R&T	R&T	R&T		R&T&L	R&T&L	R&T	R&T
ω_2	...	1.8250	2.5370	3.2720	4.7150	...	1.9080	2.5440	3.2870	4.7380	...	2.2110	2.5890	3.3180	4.7760
		T&R&L	R&T	R&T	R&T		T&R&L	R&T	R&T	R&T		T&R&L	R&T&L	R&T	R&T
ω_3	1.2840	2.9650	5.3950	7.9840	10.610	2.5500	3.6420	5.7020	8.1580	10.720	3.6990	4.3850	6.1370	8.4310	10.920
	L	L&R&T	L&R	L	L	L	L&R	L&R&T	L&T&R	L&T	L	L&R&T	L&R&T	L&T&R	L&T&R
ω_4	1.2900	4.6330	8.9420	13.280	17.270	2.5720	5.1380	9.1360	13.440	17.410	3.8390	5.8110	9.4400	13.630	17.630
	L	L&R	L	T&R	T&R	L	L&R	L&R	T&R&L	T&R&L	L	L&R	L&R&T	L&T&R	T&R&L
ω_5	4.4590	6.0840	9.4370	13.350	17.770	4.4930	6.2630	9.6350	13.460	17.850	4.6490	0.6220	9.9710	13.720	17.960
	R	R&T&L	T&R&L	L	L	R	T&R&L	T&R&L	L&T	L	R	T&R&L	T&R&L	T&R&L	L&T
ω_6	7.3670	10.080	15.770	22.240	28.980	7.3820	10.180	15.880	22.340	29.060	7.4150	10.360	16.080	22.510	29.200
	R	R&T	T&R	T&R	T&R	R	R&T&L	T&R&L	T&R&L	T&R	R	T&R&L	T&R&L	T&R&L	T&R&L

Table 6.7 Natural Frequencies (THz) of a TWCNT With Innermost Radius [24]

Natural frequencies (THz)	$m=1$					$m=2$					$m=3$				
	$n=0$	$n=1$	$n=2$	$n=3$	$n=4$	$n=0$	$n=1$	$n=2$	$n=3$	$n=4$	$n=0$	$n=1$	$n=2$	$n=3$	$n=4$
(A) $R_I = 5$ nm and $L/R_O = 10$															
ω_1	...	0.0359	0.0150	0.0212	0.0370	...	0.1155	0.0469	0.0310	0.0398	...	0.2059	0.0942	0.0536	0.0483
ω_2	...	0.4129	0.7918	1.1780	1.5660	...	0.4623	0.8205	1.1960	1.5750	...	0.5338	0.8630	1.2240	1.5900
ω_3	...	0.4395	0.8417	1.2530	1.6670	...	0.4887	0.8720	1.2720	1.6800	...	0.5575	0.9165	1.3030	1.7030
ω_4	0.1898	0.4692	0.8963	1.3360	1.7500	0.3764	0.5544	0.9239	1.3500	1.7440	0.5440	0.6609	0.9721	1.3740	1.7480
ω_5	0.1936	0.6355	1.1900	1.6100	1.7930	0.3872	0.7214	1.2370	1.6360	1.8240	0.5807	0.8428	1.3090	1.6760	1.8610
ω_6	0.1939	0.6843	1.3120	1.9200	2.4890	0.3878	0.7640	1.3540	1.9490	2.5100	0.5816	0.8795	1.4230	1.9950	2.5440
ω_7	0.6587	0.9413	1.4870	2.0570	2.6300	0.6642	0.9686	1.5150	2.0780	2.6480	0.6893	1.0220	1.5620	2.1130	2.6790
ω_8	1.9720	1.9860	2.0550	2.3330	2.9070	1.9720	1.9870	2.0580	2.3470	2.9230	1.9720	1.9880	2.0630	2.3700	2.9500
ω_9	3.3360	3.3390	3.3480	3.3710	3.4330	3.3360	3.3390	3.3490	3.3710	3.4350	3.3360	3.3390	3.3490	3.3720	3.4380
(B) $R_I = 1$ nm and $L/R_O = 10$															
ω_1	...	0.1145	0.1578	0.3601	0.6334	...	0.3628	0.2121	0.3719	0.6403	...	0.6562	0.3351	0.4041	0.6550
ω_2	...	1.1780	1.6720	1.8390	2.0070	...	1.2100	1.6660	1.8390	2.0080	...	1.3510	1.6770	1.8420	2.0130
ω_3	...	1.5020	2.6710	3.0920	3.2340	...	1.6350	2.6900	3.0840	3.2330	...	1.7760	2.6970	3.0780	3.2330
ω_4	0.6419	1.6580	2.8230	3.9940	5.3030	1.2770	1.9020	2.9320	4.0860	5.3630	1.8730	2.2730	3.1380	4.2270	5.4610
ω_5	0.6466	2.0030	3.3780	4.9940	6.6400	1.2900	2.3090	3.5460	5.0700	6.6890	1.9280	2.6960	3.7690	5.1900	6.7680
ω_6	0.6509	2.3390	4.4730	6.6580	8.6450	1.3010	2.6230	4.5750	6.7280	8.7120	1.9480	2.9850	4.7330	6.8160	8.8240
ω_7	2.3920	3.2030	4.7820	6.6750	8.8870	2.4070	3.2800	4.8740	6.7390	8.9230	2.4620	3.4330	5.0320	6.8740	8.9820
ω_8	3.5710	4.3270	6.0350	8.3510	10.830	3.5770	4.3630	6.1040	8.4150	10.890	3.5900	4.4330	6.2210	8.5210	10.980
ω_9	4.3960	5.2580	7.9230	11.130	14.490	4.4000	5.2960	7.9800	11.180	14.530	4.4070	5.3670	8.0750	11.260	14.600
(C) $R_I = 0.5$ nm and $L/R_O = 10$															
ω_1	...	0.1800	0.4563	0.8863	1.2800	...	0.5055	0.5025	0.8999	1.2910	...	0.8972	0.6289	0.9356	1.3120
ω_2	...	1.3130	1.8750	2.5080	3.0910	...	1.3660	1.8780	2.5130	3.0960	...	1.5540	1.9040	2.5260	3.1070
ω_3	...	1.9960	2.9260	3.4470	4.7250	...	2.0320	2.9130	3.4500	4.7360	...	2.1790	2.9180	3.4590	4.7550
ω_4	0.9142	2.1970	3.8460	5.6840	7.5500	1.8180	2.6440	4.0720	5.8100	7.6350	2.6560	3.1740	4.3880	6.0050	7.7710
ω_5	0.9170	2.8520	5.3440	7.9550	10.590	1.8310	3.3000	5.5170	8.0470	10.650	2.7380	3.8370	5.7760	8.1940	10.750
ω_6	0.9223	4.4220	6.7440	9.4590	12.300	1.8420	4.5390	6.8820	9.5760	12.400	2.7530	4.7750	7.1170	9.7700	12.560
ω_7	3.2990	4.5390	8.9090	13.270	17.260	3.3200	4.8180	9.0100	13.350	17.330	3.4110	5.2210	9.1710	13.470	17.440
ω_8	4.9960	6.2720	9.4550	13.330	17.760	5.0040	6.3420	9.5520	13.390	17.800	5.0240	6.4730	9.7150	13.490	17.860
ω_9	7.3620	10.060	15.750	22.230	28.970	7.3690	10.110	15.810	22.280	29.010	7.3820	10.200	15.900	22.360	29.080

Table 6.8 Natural Frequencies (THz) of a Five-walled CNT With Innermost Radius [24]

Natural frequencies (THz)	$m=1$					$m=2$					$m=3$				
	$n=0$	$n=1$	$n=2$	$n=3$	$n=4$	$n=0$	$n=1$	$n=2$	$n=3$	$n=4$	$n=0$	$n=1$	$n=2$	$n=3$	$n=4$
(A) $R_I = 5$ nm and $L/R_O = 10$															
ω_1	...	0.0309	0.0130	0.0189	0.0332	...	0.1005	0.0403	0.0270	0.0354	...	0.1807	0.0815	0.0463	0.0424
ω_2	...	0.3660	0.7049	1.0310	1.0890	...	0.4075	0.7248	1.0190	1.0880	...	0.4682	0.7545	1.0130	1.0850
ω_3	...	0.3861	0.7450	1.0680	1.4010	...	0.4274	0.7668	1.0920	1.4150	...	0.4889	0.8005	1.1210	1.4390
ω_4	...	0.4086	0.7893	1.1140	1.4790	...	0.4497	0.8127	1.1360	1.4910	...	0.5087	0.8478	1.1710	1.5130
ω_5	...	0.4325	0.8359	1.1770	1.5660	...	0.4767	0.8544	1.1920	1.5760	...	0.5334	0.8918	1.2160	1.5930
ω_6	0.1695	0.4555	0.8815	1.2520	1.6660	0.3366	0.5152	0.8911	1.2690	1.6770	0.4909	0.6078	0.9181	1.2940	1.6950
ω_7	0.1725	0.5470	0.9314	1.3370	1.7780	0.3448	0.6292	0.9836	1.3540	1.7880	0.5170	0.7395	1.0490	1.3830	1.8050
ω_8	0.1730	0.5925	1.1260	1.6330	1.9950	0.3460	0.6660	1.1650	1.6590	2.0090	0.5190	0.7713	1.2280	1.7010	2.0300
ω_9	0.1732	0.6308	1.2060	1.7600	2.2630	0.3463	0.7003	1.2430	1.7840	2.2800	0.5194	0.8014	1.3020	1.8240	2.3090
ω_{10}	0.1732	0.6785	1.3080	1.8450	2.3540	0.3464	0.7434	1.3420	1.8640	2.3710	0.5196	0.8394	1.3950	1.8970	2.3980
ω_{11}	0.6183	0.8813	1.3470	1.9230	2.4870	0.6227	0.9041	1.3690	1.9460	2.5040	0.6404	0.9475	1.4080	1.9830	2.5310
ω_{12}	1.2960	1.3440	1.6170	2.1590	2.6620	1.2970	1.3470	1.6320	2.1740	2.6750	1.2970	1.3520	1.6570	2.1990	2.6950
ω_{13}	2.2710	2.2780	2.3090	2.4310	2.8970	2.2710	2.2790	2.3100	2.4360	2.9100	2.2710	2.2790	2.3120	2.4460	2.9300
ω_{14}	3.1010	3.1040	3.1140	3.1390	3.2160	3.1020	3.1040	3.1140	3.1390	3.2190	3.1020	3.1040	3.1140	3.1400	3.2230
ω_{15}	3.6410	3.6430	3.6480	3.6600	3.6880	3.6410	3.6430	3.6480	3.6600	3.6880	3.6410	3.6430	3.6480	3.6600	3.6890
(B) $R_I = 1$ nm and $L/R_O = 10$															
ω_1	...	0.0773	0.1140	0.2561	0.4299	...	0.2456	0.1478	0.2626	0.4332	...	0.4466	0.2286	0.2819	0.4410
ω_2	...	0.7687	1.0500	1.1740	1.3360	...	0.7890	1.0480	1.1740	1.3380	...	0.8782	1.0570	1.1770	1.3400
ω_3	...	1.0160	1.8550	2.0970	2.2210	...	1.1030	1.8390	2.0930	2.2210	...	1.2300	1.8440	2.0900	2.2210
ω_4	...	1.1490	1.9690	2.8170	2.9890	...	1.2790	2.0730	2.8080	2.9870	...	1.3800	2.1810	2.8010	2.9840
ω_5	...	1.2740	2.2080	2.8790	3.4850	...	1.3830	2.2420	2.9420	3.4810	...	1.6380	2.3240	3.0380	3.4770
ω_6	0.4571	1.4240	2.5490	3.2860	3.7760	0.9099	1.5910	2.6250	3.2760	3.8210	1.3390	1.6520	2.6780	3.2690	3.8930
ω_7	0.4591	1.5950	2.6580	3.3690	4.4070	0.9165	1.6770	2.6920	3.4410	4.4480	1.3710	1.9490	2.7930	3.5420	4.5120
ω_8	0.4619	1.6750	2.9490	3.9810	5.2940	0.9231	1.8530	3.0140	4.0360	5.3260	1.3820	2.1190	3.1380	4.1220	5.3780

ω_9	0.4637	1.9440	3.3470	4.7650	6.1640	0.9271	2.1190	3.4420	4.8210	6.2110	1.3900	2.3590	3.5770	4.9140	6.2900
ω_{10}	0.4650	2.2820	3.4690	4.9810	6.6320	0.9298	2.3930	3.5310	5.0200	6.6560	1.3940	2.4980	3.6360	5.0840	6.6980
ω_{11}	1.7510	2.3390	4.2220	5.6010	7.2110	1.7600	2.4450	4.2600	5.6460	7.2510	1.7950	2.6680	4.3260	5.7220	7.3170
ω_{12}	2.7210	3.2290	4.4560	6.6660	8.6490	2.7240	3.2520	4.5090	6.6930	8.6830	2.7320	3.2940	4.5930	6.7380	8.7390
ω_{13}	3.4040	3.8690	4.9250	6.6850	8.8810	3.4060	3.8820	4.9610	6.7250	8.8990	3.4100	3.9060	5.0240	6.7910	8.9300
ω_{14}	3.9190	4.4070	6.0250	8.3400	10.830	3.9200	4.4210	6.0590	8.3720	10.850	3.9230	4.4460	6.1170	8.4260	10.900
ω_{15}	4.4320	5.2520	7.9130	11.120	14.480	4.4340	5.2710	7.9420	11.150	14.500	4.4360	5.3040	7.9900	11.190	14.540
(C) $R_I = 0.5$ nm and $L/R_O = 10$															
ω_1	...	0.1116	0.2832	0.4961	0.6799	...	0.3089	0.3086	0.5023	0.6841	...	0.5486	0.3828	0.5212	0.6937
ω_2	...	0.8370	1.2290	1.6280	1.9400	...	0.8661	1.2300	1.6290	1.9410	...	0.9723	1.2430	1.6330	1.9450
ω_3	...	1.2520	2.0270	2.4940	2.8730	...	1.3170	2.0140	2.4920	2.8740	...	1.4390	2.0150	2.4930	2.8750
ω_4	...	1.4610	2.4330	3.1350	3.4820	...	1.5710	2.5170	3.1280	3.4820	...	1.6720	2.5570	3.1230	3.4820
ω_5	...	1.5280	2.6650	3.5230	4.7230	...	1.7270	2.7000	3.5200	4.7270	...	2.0240	2.8540	3.5190	4.7350
ω_6	0.5801	1.8380	2.9610	3.6100	4.7900	1.1540	2.0010	3.0000	3.6960	4.8450	1.6960	2.0640	3.0090	3.8250	4.9330
ω_7	0.5813	1.9640	3.1200	4.4050	5.8540	1.1610	2.0940	3.2260	4.4780	5.9000	1.7390	2.4480	3.4150	4.5930	5.9740
ω_8	0.5836	2.1930	3.8020	5.6590	7.5330	1.1660	2.4020	3.9170	5.7130	7.5680	1.7450	2.7260	4.0850	5.8000	7.6250
ω_9	0.5862	2.7330	4.3390	6.0210	7.8110	1.1710	2.9560	4.4200	6.0930	7.8720	1.7540	3.1220	4.5600	6.2150	7.9720
ω_{10}	0.5886	2.9170	5.3080	7.3770	9.5580	1.1770	2.9860	5.3800	7.4360	9.6070	1.7630	3.2500	5.4950	7.5330	9.6900
ω_{11}	2.1850	3.9630	5.3750	7.9370	10.570	2.1970	3.9920	5.4340	7.9740	10.600	2.2440	4.0490	5.5370	8.0360	10.640
ω_{12}	3.3690	4.4790	6.7950	9.4600	12.290	3.3740	4.5980	6.8460	9.5060	12.330	3.3830	4.7840	6.9320	9.5830	12.390
ω_{13}	4.1580	4.7800	8.8890	13.250	17.250	4.1600	4.8060	8.9300	13.280	17.270	4.1660	4.8530	8.9970	13.340	17.320
ω_{14}	5.0370	6.2600	9.4350	13.320	17.760	5.0400	6.2870	9.4740	13.340	17.770	5.0450	6.3340	9.5380	13.380	17.790
ω_{15}	7.3600	10.050	15.7300	22.220	28.960	7.3630	10.070	15.760	22.240	28.980	7.3670	10.110	15.800	22.270	29.000

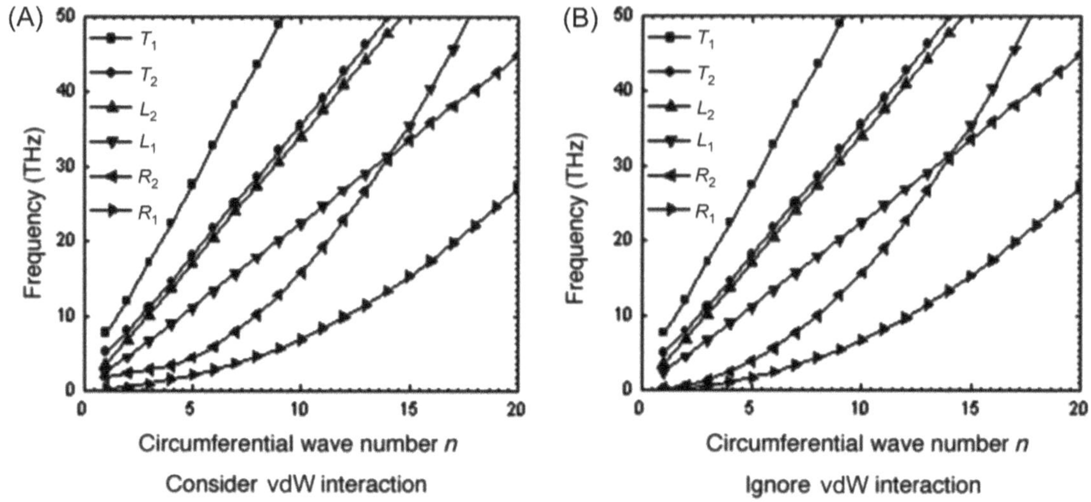

FIGURE 6.9

Effect of the vdW interaction on the natural frequencies of a DWCNT with $R_I = 0.65$ nm for $m = 1$ and various n. (A) Consider vdW interaction; (V) Ignore vdW interaction [24].

DWCNT, the TWCNT, and the five-walled MWCNT is almost the same, and thus the differences between the lowest and highest frequencies are almost the same for an MWCNT with different numbers of layers. This indicates that the more layers there are, the closer the natural frequencies will be.

Wang et al. [32] stated that the effect of vdW interaction is less pronounced for radial modes of small-radius MWCNTs due to the fact that the interlayer vdW interaction is characterized by a radius-independent vdW interaction coefficient. In contrast, our more refined model for the interlayer vdW interaction is characterized by a radius-dependent vdW interaction coefficient and the results for the natural frequencies did show that the effect of vdW interaction on the radial modes of small-radius MWCNTs is significant. Figs. 6.9 and 6.10 show the influence of vdW interaction on the natural frequencies of a DWCNT with small innermost radius $R_I = 0.65$ nm. Indeed, the effect of the vdW interaction on the frequencies which are associated with the torsional and longitudinal vibrations is very small and can be neglected. However, for the lower mode-order vibration such as $m = 1$ and $n < 5$ or $n = 1$ and $m < 10$, frequencies associated with the radial vibration when considering the vdW interaction are significantly different from those when ignoring the vdW interaction, indicating again that vdW interaction does play an important role in the radial vibration of the small-radius DWCNT.

To show the effect of the innermost radius on the lower mode-order frequencies of a DWCNT, frequencies associated with the radial vibration are calculated from Eqs. (6.55) and (6.56) and presented in Fig. 6.11 for the breathing vibration ($m = 1$, $n = 0$). The frequencies associated with the torsional, longitudinal, and radial vibrations ($m = 1$, $n = 1$) are calculated from Eq. (6.46) and presented in Fig. 6.12. It is observed from Figs. 6.11 and 6.12 that the frequencies are the function of the tube radius. With a small radius, say, less than 5 nm, the frequencies drop rapidly, indicating that the radius effect plays an important role in the frequencies of a small-radius MWCNT. For a DWCNT with a large radius, say, greater than 10 nm, the frequencies vary very slowly, showing

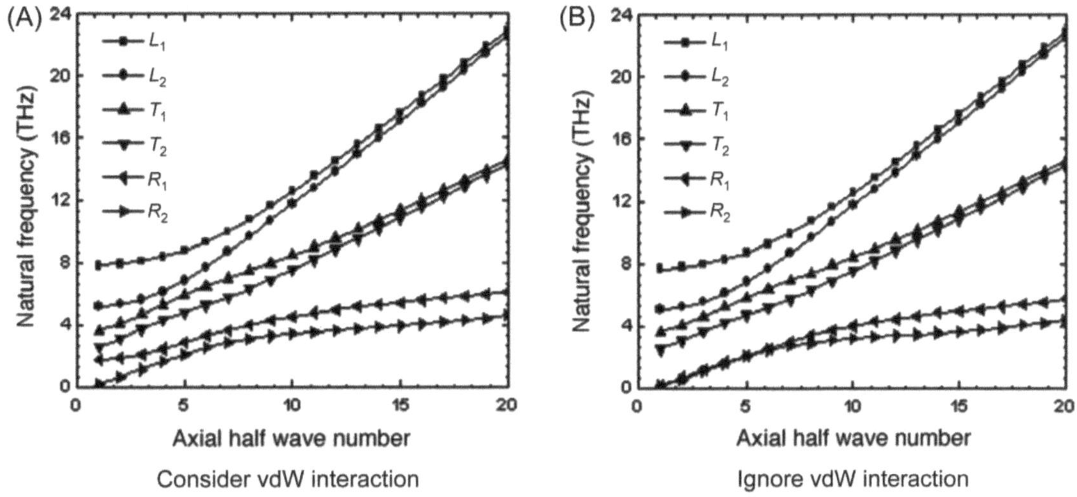

FIGURE 6.10

Effect of the vdW interaction on the natural frequencies of a DWCNT with $R_I = 0.65$ nm for $n = 1$ and various m. (A) Consider vdW interaction; (V) Ignore vdW interaction [24].

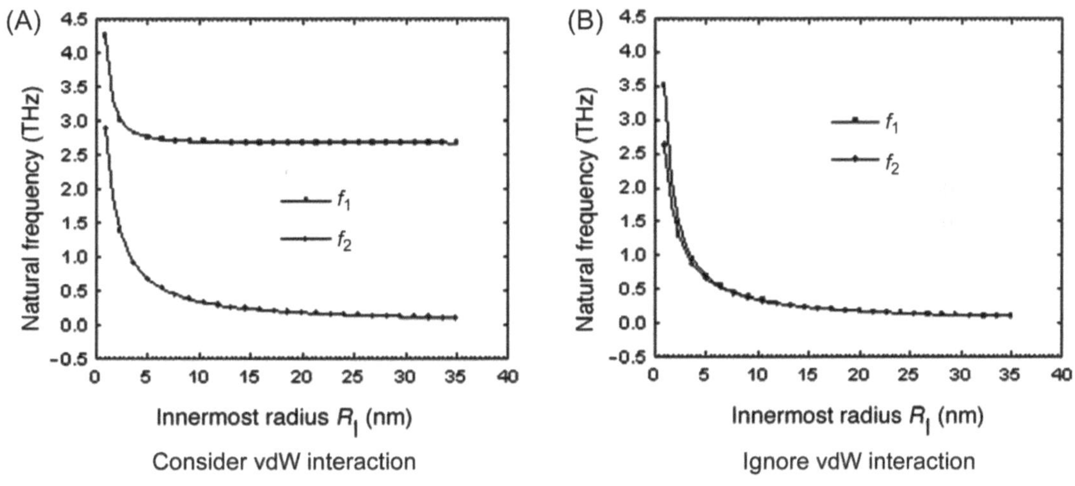

FIGURE 6.11

Effect of the size of CNTs on the breathing frequencies for a DWCNT with $m = 1$ and $n = 0$. (A) Consider vdW interaction; (V) Ignore vdW interaction [24].

that the dependence of the frequencies on the radius can be neglected when the tube radius is large enough. Also, the higher frequency associated with the radial vibration is almost a constant and independent of the tube radius when the innermost radius is greater than 10 nm, indicating again that the vdW interaction plays a critical role on the higher radial vibration of MWCNTs. Also, it is

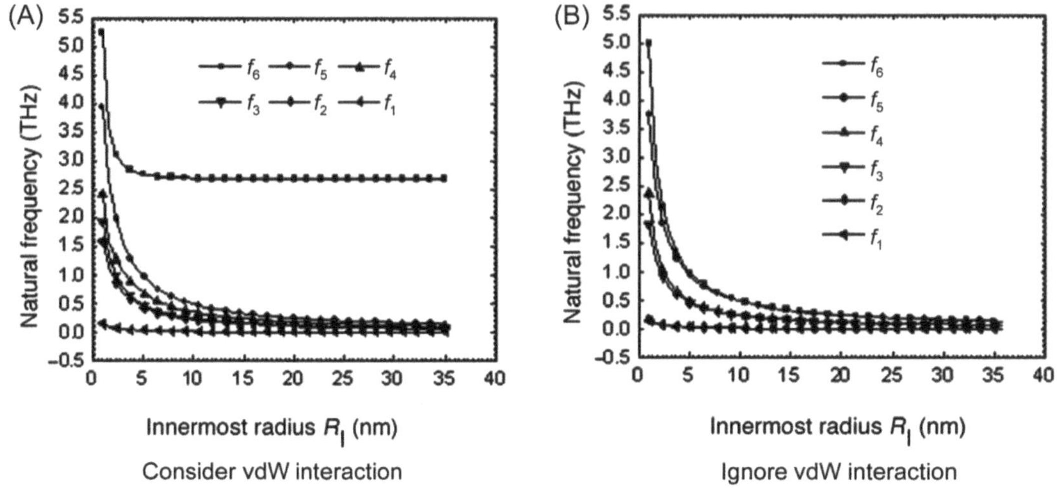

FIGURE 6.12

Effect of the size of CNTs on the natural frequencies for a DWCNT with $m=1$ and $n=1$. (A) Consider vdW interaction; (V) Ignore vdW interaction [24].

clear from Figs. 6.3 and 6.4 that the interlayer vdW interaction plays a critical role on the higher frequency associated with radial vibration of an MWCNT with arbitrary radius, but the effect of vdW interaction on the lower frequency associated with the radial vibration and on the frequencies associated with the torsional and the longitudinal vibrations is very small and is negligible for MWCNTs with radii from very small to very large.

The effect of the innermost radius on the higher mode order ($m=5$, $n=5$) frequencies of a DWCNT is shown in Fig. 6.13 for various innermost radii. When the interlayer vdW interaction is ignored, the mode shapes associated with f_1–f_6 do not change for different tube radii, as shown in Fig. 6.13B. However, when the vdW interaction is considered, the mode shapes which are associated with some frequencies will change from radial, torsional, and longitudinal mode shapes to other shapes as the innermost radius is varied from small to large radii. For example, in Fig. 6.13A, as the innermost radius increases, the mode shape associated with the frequency f_6 changes from torsional mode ($R_I < 7$ nm) to a mix of radial and torsional modes (7 nm $\leq R_I \leq$ 9 nm), and then to a radial mode ($R_I \geq 9$ nm), and the mode shape associated with frequency f_4 changes from longitudinal mode ($R_I \leq 4$ nm) to a mix of longitudinal and torsional modes (4 nm $\leq R_I \leq$ 8 nm), and then to a torsional mode ($R_I > 8$ nm). Similarly, the mode shape associated with frequency f_2 changes from radial mode ($R_I \leq 2$ nm) to a mix of radial and longitudinal modes (2 nm $\leq R_I \leq$ 4 nm), and then to a longitudinal mode ($R_I \leq 4$ nm). In contrast, the mode shapes associated with the frequencies f_3 and f_1 do not change for different tube radii.

As the radius of a tube in an MWCNT is increased the stiffening effect of that particular tube on the entire MWCNT assembly is reduced, and therefore a different physical combination among all the factors is obtained. This is similar to the vibrations of beams on elastic foundations, where the mode number is increased as the stiffness of the elastic foundation is increased. In the current problem, on top of this, also the coupling in the displacement field makes it more complex, and one can see in Fig. 6.13 the continuous change in the modal composition when the vdW interaction is considered.

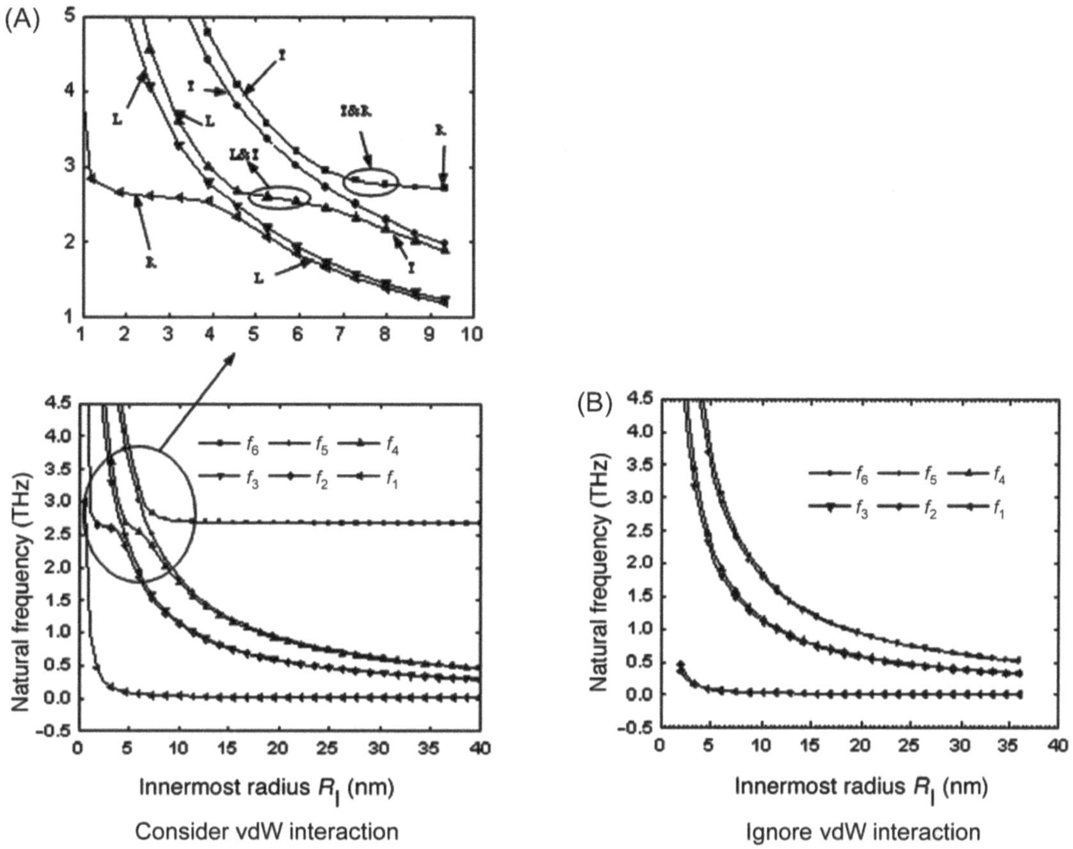

FIGURE 6.13

Effect of the size of CNTs on the natural frequencies for a DWCNT with $m = 5$ and $n = 5$. (A) Consider vdW interaction; (V) Ignore vdW interaction [24].

In summary, based on the more refined vdW interaction formulas, the free vibration of an MWCNT is investigated by using the Donnell shell model. In contrast to the work by Wang et al. [32] where they conclude that the effect of the vdW interaction is less significant for the radial vibration of small-radius MWCNTs, our numerical results reveal that the effect of the vdW interaction plays a significant role on the higher natural frequencies which are associated with radial vibration of small-radius MWCNTs. This is due to the fact that our continuum model is based on radius-dependent vdW interaction formulas, as compared to the radius-independent vdW interaction formulas that were used by Wang et al. [32] The effects of the tube radius and the interlayer vdW interaction are also examined on the mode shapes of MWCNTs. When ignoring the vdW interaction, the mode shapes of MWCNTs do not change as the radii of the MWCNTs are varied from very small to very large. However, when the vdW interaction is considered, it is observed that some of the mode shapes will change between radial, torsional, and longitudinal modes, due to the effect of the vdW interaction even for cases where the innermost radius is less than 2 nm, indicating again that the interlayer vdW interaction plays a significant role in the vibration of MWCNTs with small radius. In addition, the numerical simulations show that ultrahigh natural frequency can be achieved by using an MWCNT. For a given combination of m and n, the lowest natural frequencies are not influenced by the vdW interaction. However, other higher natural frequencies are significantly dependent on the vdW interaction. Especially, the vdW interaction plays a dominant role on the higher radial frequencies so that the effect of the mode order can be neglected.

REFERENCES

[1] Iijima S. Helical microtubules of graphitic carbon. Nature 1991;354(6348):56–8.

[2] Iijima S, Brabec C, Maiti A, et al. Structural flexibility of carbon nanotubes. J Chem Phys 1996;104:2089–92.

[3] Yakobson BI, Brabec CJ, Bernholc J. Nanomechanics of carbon tubes: instabilities beyond linear response. Phys Rev Lett 1996;76(14):2511–14.

[4] Yakobson BI, Campbell MP, Brabec CJ, Bernholc J. High strain rate fracture and C-chain unraveling in carbon nanotubes. Comp Mater Sci 1997;8(4):341–8.

[5] Hernandez E, Goze C, Bernier P, et al. Elastic properties of C and BxCyNz composite nanotubes. Phys Rev Lett 1998;80:4502–5.

[6] Sanchez-Portal D, Artacho E, Soler JM, Rubio A, Ordejon P. Ab initio structural, elastic, and vibrational properties of carbon nanotubes. Phys Rev B Condens Matter 1999;59:12678–88.

[7] Liew KM, Wong CH, He XQ, Tan MJ, Meguid SA. Nanomechanics of single and multi-walled carbon nanotubes. Physi Rev B 2004;69(11):115429.

[8] Liew KM, He XQ, Wong CH. On the study of elastic and plastic properties of multi-walled carbon nanotubes under axial tension using molecular dynamics simulation. Acta Mater 2004;52(9):2521–7.

[9] Govindjee S, Sackman JL. On the use of continuum mechanics to estimate the properties of nanotubes. Solid State Commun 1999;110:227–30.

[10] Li CY, Chou TW. A structural mechanics approach for the analysis of carbon nanotubes. Int J Solids Struct 2003;40(10):2487–99.

[11] Li C, Chou TW. Elastic moduli of multi-walled carbon nanotubes and the effect of van der Waals forces. Composites Sci Technol 2003;63:1517–24.

[12] Odegard GM, Gates TS, Nicholson LM, et al. Equivalent-continuum modeling of nano-structured materials. Composites Sci Technol 2002;62:1869–80.

[13] Pantano A, Parks DM, Boyce MC. Mechanics of deformation of single- and multi-wall carbon nanotubes. J Mechanics and Phys Solids 2004;52:789–821.

[14] Liu B, Huang Y, Jiang H, Qu S, Hwang KC. The atomic-scale finite element method. Comput Method Appl M. 2004;193(17-20):1849–64.

[15] Ru CQ. Effective bending stiffness of carbon nanotubes. Phys Rev B 2000;62:9973–6.

[16] Ru CQ. Effect of van der Waals forces on axial buckling of a double-walled carbon nanotube. J Appl Phys 2000;87(10):7227–31.

[17] Ru CQ. Elastic buckling of single-walled carbon nanotube ropes under high pressure. Phys Rev B 2000;62:10405–8.

[18] Ru CQ. Axially compressed buckling of a double-walled carbon nanotube embedded in an elastic medium. J Mech Phys Solids 2001;49:1265–79.

[19] Ru CQ. Degraded axial buckling strain of multi-walled carbon nanotubes due to interlayer slips. J Appl Phys 2001;89(6):3426–33.

[20] Wang CY, Ru CQ, Mioduchowski A. Axially compressed buckling of pressured multiwall carbon nanotubes. Int J Solids and Struct 2003;40(15):3893–911.
[21] Shen HS. Postbuckling prediction of double-walled carbon nanotubes under hydrostatic pressure. Int J Solids Struct 2004;41(9-10):2643–57.
[22] He XQ, Kitipornchai S, Liew KM. Buckling analysis of multi-walled carbon nanotubes: a continuum model accounting for van der Waals interaction. J Mech Phys Solids 2005;53(2):303–26.
[23] Liew KM, He XQ, Kitipornchai S. Buckling characteristics of embedded multi-walled carbon nanotubes. P Roy Soc a-Math Phy 2005;461(2064):3785–805.
[24] He XQ, Eisenberger M, Liew KM. The effect of van der Waals interaction modeling on the vibration characteristics of multiwalled carbon nanotubes. J Appl Phys 2006;100(12).
[25] He XQ, Kitipornchai S, Wang CM, Liew KM. Modeling of van der Waals force for infinitesimal deformation of multi-walled carbon nanotubes treated as cylindrical shells. Int J Solids Struct 2005;42(23):6032–47.
[26] Kitipornchai S, He XQ, Liew KM. Buckling analysis of triple-walled carbon nanotubes embedded in an elastic matrix. J Appl Phys 2005;97(11).
[27] Yoon J, Ru CQ, Mioduchowski A. Vibration of an embedded multiwall carbon nanotube. Composite Sci Technol 2003;63(11):1533–45.
[28] Mahan GD. Oscillations of a thin hollow cylinder: carbon nanotubes. Phys Rev B 2002;65(23).
[29] Rao AM, Chen J, Richter E, Schlecht U, Eklund PC, Haddon RC, et al. Effect of van der Waals interactions on the Raman modes in single walled carbon nanotubes. Phys Rev Lett 2001;86(17):3895–8.
[30] Zhang YQ, Liu GR, Xie XY. Free transverse vibrations of double-walled carbon nanotubes using a theory of nonlocal elasticity. Phys Rev B 2005;71(19).
[31] Li CY, Chou TW. Vibrational behaviors of multiwalled-carbon-nanotube-based nanomechanical resonators. Appl Phys Lett 2004;84(1):121–3.
[32] Wang CY, Ru CQ, Mioduchowski A. Free vibration of multiwall carbon nanotubes. J Appl Phys 2005;97(11).
[33] Wang CY, Ru CQ, Mioduchowski A. Applicability and limitations of simplified elastic shell equations for carbon nanotubes. J Appl Mech-T Asme 2004;71(5):622–31.
[34] Popov VN, Henrard L. Breathinglike phonon modes of multiwalled carbon nanotubes. Phys Rev B 2002;65(23).
[35] Natsuki T, Endo M, Tsuda H. Vibration analysis of embedded carbon nanotubes using wave propagation approach. J Appl Phys 2006;99(3).
[36] Lennard Jones JE. The determination of molecular fields: from the variation of the viscosity of a gas with temperature. Proc Roy Soc 1924;106A:441.
[37] Girifalco LA, Lad RA. Energy of cohesion, compressibility, and the potential energy functions of the graphite system. J Chem Phys 1956;25(4):693–7.
[38] Girifalco LA. Interaction potential for C60 molecules. J Phys Chem-Us 1991;95(14):5370–1.
[39] Saito R, Dresselhous G, Dresselhous MS. Physical properties of carbon nonotubes. London: Imperial College Press; 1998.
[40] Timoshenko SP, Gere JM. Theory of elastic stability. New York: McGraw-Hill; 1961.
[41] Saito R, Matsuo R, Kimura T, Dresselhaus G, Dresselhaus MS. Anomalous potential barrier of double-wall carbon nanotube. Chem Phys Lett 2001;348(3-4):187–93.
[42] Robertson DH, Brenner DW, Mintmire JW. Energetics of nanoscale graphitic tubules. Physical Review B 1992;45:12592–5.
[43] Gulseren O, Yildirim T, Ciraci S. Systematic ab initio study of curvature effects in carbon nanotubes. Phys Rev B 2002;65:153405.
[44] Allen HG, Bulson PS. Background to buckling. London: McGraw-Hill; 1980.
[45] Charlier J-C, Michenaud J-P. Energetics of multilayered carbon tubules. Phys Rev Lett 1993; 70(12):1858.

CHAPTER

NONLOCAL ELASTICITY THEORIES 7

7.1 INTRODUCTION

Carbon nanotubes (CNTs) have attracted worldwide attention [1–4] for their wide range of applications, including atomic-force microscopes, field emitters, frictionless nanoactuators, nanomotors, nanobearings, nanosprings, nanofillers for composite materials, and nanoscale electronic devices [4]. In addition to experimental endeavors, CNT modeling is classified into two main categories. The first is atomic modeling, which includes such techniques as molecular dynamics (MD) and TBMD and the density functional model [5–11]. Li and Chou [12] reported an atomistic simulation of single-walled CNT (SWCNT) subjected to harmonic waves. Atomic modeling is limited to systems with a small number of molecules and atoms and is therefore confined to small-scale modeling. The second category is continuum modeling [13–18], which includes classical (or local) beam and shell theories that are practical for analyzing CNTs for large-scale systems. Successful work has been conducted with continuum modeling, such as buckling analysis, dynamics studies, and mechanical property investigations of CNTs [12,19–39]. Dynamic mechanical behaviors of CNTs have been studied widely in recent years. Elastic wave propagation in CNTs was studied through continuum models and MD simulations [40–44]. Vibrations of CNTs are of importance in various applications, such as sensors, high frequency oscillators, and other nanoscale devices. Vibrations of CNTs have been investigated by continuum models and molecular mechanics methods [45–49]. It is important to apply accurate theoretical models for the natural frequencies and mode shapes of CNTs [50]. The modeling and simulation methods of vibrating CNTs were reviewed by Gibson et al. [50]. Static and dynamic mechanical deflections can be electrically induced in cantilevered, multi-walled CNTs (MWCNTs) in a transmission electron microscope [51]. Natsuki et al. [52] investigated the wave propagation in SWCNTs and double-walled CNTs (DWCNTs) conveying fluid through the use of the shell model. Insepov et al. [53] summarized a few ways to activate surface-traveling waves on the nanotube surface.

Although the classical continuum models are efficient in CNT mechanical analysis through their relatively simple formulae, their applicability to the identification of the small-scale effect on the mechanical behavior of CNTs is questionable. This limited applicability of the classical or local continuum models at small-length scales is partly due to the fact that classical modeling does not admit intrinsic size dependence in the elastic solutions of inclusions and inhomogeneities. Sun and Zhang [54] discussed the limited applicability of continuum models to nanotechnology and proposed a semicontinuum model to analyze nanomaterials. The values of the material properties were found to be completely dependent on the thickness of the plate structure. The modeling of such a size-dependent phenomenon has become an active and interesting subject of research in continuum modeling [8,55]. Wang [56] demonstrated the effective in-plane stiffness and bending rigidity of armchair and zigzag CNTs through analysis of the representative volume element of a graphene

Mechanical Behaviors of Carbon Nanotubes. DOI: http://dx.doi.org/10.1016/B978-0-323-43137-8.00007-X

layer with continuous elastic models. It was found that the bending rigidity of the CNTs had a size effect. Wang and Hu [57] studied flexural wave propagation in SWCNTs through the use of a nonlocal elastic Timoshenko beam model and an MD simulation based on the first-generation reactive empirical bond order (REBO) potential [58]. Liew et al. [59] studied the flexural wave propagation in SWCNTs through the use of an MD simulation based on the second-generation REBO [60]. The material microstructure at small size, such as the lattice spacing between individual atoms, becomes increasingly important and the discrete structure of the material can no longer be homogenized into a continuum. Therefore, the nonlocal elastic continuum model is expected to take into account the scale effect when tackling large-scale nanomaterial analysis. Size effects often become prominent at nanometer scales, and there is increasing interest in the general area of nanotechnology in explicitly addressing the cause of these effects [61].

The nonlocal elasticity theory [62,63] assumes that the stress state at a given reference point is a function of the strain field at every point in the body. Therefore, the scale effect can be considered in constitutive equations simply as a material parameter. However, the classical elasticity theory cannot account for the scale effect, as the stress state is uniquely dependent on the strain state at the same point. Peddieson et al. [64] expected the application of the nonlocal elasticity theory to have potential in the analysis of nanomaterials. They applied it to the formulation of a nonlocal version of the Euler–Bernoulli beam model and concluded that nonlocal continuum mechanics could potentially play a useful role in nanotechnology applications. Further applications of nonlocal continuum mechanics have been employed in the study of CNT buckling. Sudak [65] presented a multiple column model for the linearized column buckling of MWCNTs using the theory of nonlocal continuum mechanics. Zhang et al. [66] proposed a nonlocal multishell model for the axial buckling of MWCNTs under axial compression. The nonlocal elastic Euler–Bernoulli and Timoshenko beam models were applied to study wave propagation in CNTs [67]. Wang and Varadan [44] studied wave propagation in CNTs based on the nonlocal elastic shell theory. Both theoretical analyses and numerical simulations have explicitly revealed the small-scale effect on wave dispersion relations for different CNT wavenumbers in the longitudinal and circumferential directions and for different wavelengths in the circumferential direction. However, no MD simulations have been available for the validation of shell models and beam models considering the vdW interaction between adjacent tubes in studying transverse, torsional wave, and flexural propagation in CNTs.

This chapter aims to investigate the applicability of nonlocal beam models, nonlocal rod models, and nonlocal shell models for free transverse, longitudinal, and torsional vibrations and waves in SWCNTs and DWCNTs through verifications from MD simulations. Beam models and MD simulations are employed to study the transverse vibration of SWCNTs. The nonlocal rod model is used to predict the MD results for longitudinal and torsional vibrations of SWCNTs. The nonlocal shell model is applied to study transverse and torsional wave in SWCNTs and DWCNTs. The free transverse, longitudinal, and torsional vibrations of SWCNTs with two different boundary conditions are studied.

7.2 NONLOCAL ELASTIC BEAM MODEL

7.2.1 NONLOCAL BEAM AND ROD MODELS FOR VIBRATION OF SWCNTs

The Euler–Bernoulli beam model is usually employed in the analysis of longer one-dimensional beamlike structures, normally with ratio of the length to the thickness over 20 in engineering structures. If the cross-sectional dimensions are not small enough compared to the length of the beam, we need to consider the effects of rotary inertia and shear deformation using Timoshenko beam model [68]. To account for small-scale effects on CNT's mechanical behaviors, the nonlocal beam and shell models have been employed to overcome the limitation of the applicability of the classical or local continuum modes at small length scales [35,40,41,47,65,69].

The natural frequencies of the local Euler–Bernoulli beam are shown as [68]

$$\omega_n = (\beta_n l)^2 \sqrt{\frac{EI}{\rho A l^4}}, \quad (n = 1, 2...) \tag{7.1}$$

where $\beta_1 l = 1.875$ and 4.73 are for the fundamental frequency of a cantilevered beam and a clamped–clamped beam, respectively. l is the length of the beam.

The governing equation for free vibration of the nonlocal Euler–Bernoulli beam [67] is proposed to be:

$$EI\frac{\partial^4 u}{\partial x^4} + \rho A\left[\frac{\partial^2 u}{\partial t^2} - (e_0 a)^2 \frac{\partial^4 u}{\partial x^2 \partial t^2}\right] = 0 \tag{7.2}$$

The nonlocal Timoshenko model needs to be employed to account for the small scale effect and the effects of rotary inertia and shear deformation. The governing equations for the vibrations of nonlocal Timoshenko beam are as follows [47]:

$$\begin{aligned} &GA\kappa\left(\frac{\partial \varphi}{\partial x} + \frac{\partial^2 u}{\partial x^2}\right) + \rho A \omega^2 u = 0, \\ &EI\frac{\partial^2 \varphi}{\partial x^2} - GA\kappa\left(\frac{\partial u}{\partial x} + \varphi\right) + \rho I \omega^2 \varphi - (e_0 a)^2\left(\rho A \omega^2 \frac{\partial u}{\partial x} + \rho I \omega^2 \frac{\partial^2 \varphi}{\partial x^2}\right) = 0 \end{aligned} \tag{7.3}$$

where A is the beam cross-sectional area. $\rho =$ 2.27 g/cm^3 is the density of CNT. κ is Timoshenko's shear coefficient which depends on the shape of the cross-section. $\kappa = 0.5$ is used for SWCNT [57]. $G = E/(2(1 + \nu))$ is the shear modulus. $I = (\pi(D_O^4 - D_I^4))/64$ is the area moment of inertia, where D_O and D_I are the outside diameter and inside diameter of SWCNTs, respectively. The wall thickness of an SWCNT is 0.34 nm. The length of the C–C bond is $a = 0.142$ nm. The parameter $e_0 a$ is the scale coefficient revealing the small scale effect on the responses of structures of nanosize.

Eringen [63] proposed e_0 as 0.39 by matching the dispersion curves from the nonlocal theory for a plane waves with the Born–Karman model of lattice dynamics. However, there are no experiments conducted to determine the value of e_0 for CNTs. Wang and Hu proposed $e_0 = 1/\sqrt{12} \approx 0.288$ used in the nonlocal beam model [57]. Zhang et al. [35] estimated $e_0 \approx 0.82$ by matching the theoretical buckling strain obtained by the nonlocal thin shell model [66] to those from molecular mechanics simulations given by Sears and Batra [70]. A conservative estimate of the scale coefficient $e_0 a < 2.0$ nm for an SWCNT if the measured wave propagation frequency value for the SWCNT is greater than 10 THz [67]. Hu et al. [41] estimated the values of $e_0 = 0.6$ and 0.2 for the transverse and torsional wave propagation in CNTs, respectively, by matching the dispersions of CNTs observed from the MD results with the numerical results obtained from the nonlocal shell model. In this study, we will fix $a = 0.142$ nm which is the length of C–C bond and provide an estimation of parameter e_0 by matching the fundamental frequency of SWCNTs observed from MD simulations with the numerical results obtained from the nonlocal beam and rod models. The parametere_0 is postulated to be dependent on the diameter and boundary conditions, and independent on the length of SWCNT in this study.

The bending moment and shear force of a nonlocal Timoshenko beam are as follows [47]:

$$M = EI\frac{d\phi}{dx} - (e_0a)^2\left(\rho A\omega^2 u + \rho I\omega^2\frac{d\phi}{dx}\right) \tag{7.4}$$

$$Q = GA\kappa\left(\phi + \frac{du}{dx}\right) \tag{7.5}$$

The boundary conditions for a cantilevered nonlocal Timoshenko beam:

$$M|_{x=0} = Q|_{x=0} = 0, \quad u|_{x=l} = \varphi|_{x=l} = 0 \tag{7.6}$$

The boundary conditions for a clamped–clamped nonlocal Timoshenko beam:

$$u|_{x=0,l} = 0, \quad \varphi|_{x=0,l} = 0, \tag{7.7}$$

The free vibration solution for CNTs can be expressed as:

$$u(x,t) = U(x)e^{i\omega t}, \quad \varphi(x,t) = \phi(x)e^{i\omega t}, \tag{7.8}$$

where $i = \sqrt{-1}$. The natural frequencies and mode shapes of nonlocal Timoshenko beam can be calculated by the finite difference method.

Next, we will introduce nonlocal rod theory for the analysis of longitudinal and torsional vibrations of SWCNT. The nonlocal constitutive relation is as follows [63]:

$$\sigma(x) - (e_0a)^2\frac{d^2\sigma(x)}{dx^2} = E\varepsilon(x) \tag{7.9}$$

Based on Eq. (7.9), the governing equation for the longitudinal vibration of the nonlocal rod is given by

$$\frac{\partial^2 u}{\partial t^2} - (e_0a)^2\frac{\partial^4 u}{\partial x^2\partial t^2} = \frac{E}{\rho}\frac{\partial^2 u}{\partial x^2} \tag{7.10}$$

Substituting $u = U(x)e^{iwt}$ into (7.10) and using the boundary conditions, we can find the natural frequency of longitudinal and torsional vibrations of cantilevered and clamped–clamped nonlocal rods.

The natural frequency of longitudinal vibration of a cantilevered nonlocal rod is derived as

$$\omega_n = \frac{(2n-1)\pi}{2l}\sqrt{\frac{E}{\rho[1+(e_0a)^2((2n-1)\pi/(2l))^2]}}, \quad n = (1,\ 2\ldots) \tag{7.11}$$

The natural frequency of longitudinal vibration of a clamped–clamped nonlocal rod is derived as

$$\omega_n = \frac{n\pi}{l}\sqrt{\frac{E}{\rho[1+(e_0a)^2(n\pi/l)^2]}}, \quad n = (1,\ 2\ldots) \tag{7.12}$$

For torsional vibration of nonlocal rod, the Young's modulus E in the above formulas of natural frequencies should be replaced by shear modulus G. Poisson's ratio ν and Young's modulus E calculated from the MD simulations for the test of axial tension are listed in Table 7.1.

Table 7.1 Young's Modulus and Poisson's Ratio of SWCNT [40]

SWCNT	Young's Modulus (TPa)	Poisson's Ratio
(5, 5)	$E = 0.89$	$\nu = 0.25$
(10, 10),(15, 15)	$E = 0.83$	$\nu = 0.317$
(10, 0)	$E = 0.96$	$\nu = 0.21$
(19, 0)	$E = 0.81$	$\nu = 0.35$
(20, 20)	$E = 0.84$	$\nu = 0.312$

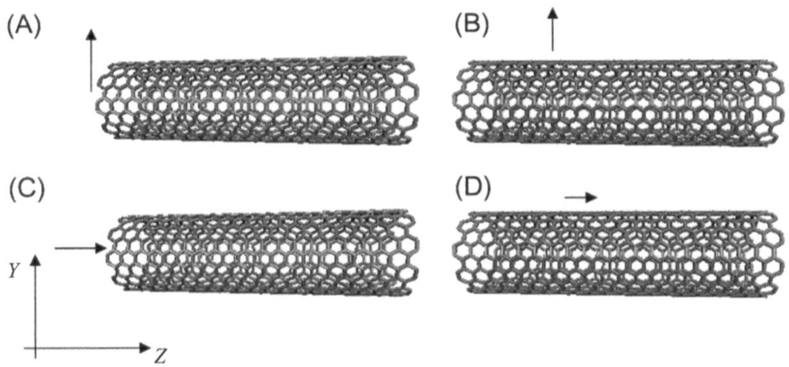

FIGURE 7.1

The atoms labeled red are fixed during the simulation of SWCNT vibration. (A) Transverse vibration of cantilevered SWCNT. (B) Transverse vibration of clamped–clamped SWCNT. (C) Longitudinal vibration of cantilevered SWCNT. (D) Longitudinal vibration of clamped–clamped SWCNT [71].

Reprinted with permission from Hu YG, Liew KM, Wang Q. Nonlocal continuum model and molecular dynamics for free vibration of single-walled carbon nanotubes. J Nanosci Nanotechno 2011;11(12):10401–7. Copyright © American Scientific Publishers.

MD simulations for SWCNTs are conducted to verify the applicability of the nonlocal beam and nonlocal rod models for free transverse, longitudinal, and torsional vibrations of SWCNTs [71]. The interaction of atoms is described by the second-generation REBO potential energy expression for hydrocarbons [60] and there is no temperature control in the MD simulations. The verlet algorithm in the velocity form with time step 1 fs is used to simulate the vibration of SWCNTs. The shortest and longest SWCNTs including (5, 5), (10, 10), (15, 15), and (20, 20) SWCNTs in the MD simulations are 1.2 and 9.0 nm, respectively.

The atoms at the end of SWCNT and the middle of SWCNT are moved away from their equilibrium positions in the Y direction as shown in Fig. 7.1A and B, respectively. Fig. 7.1C and D shows that the atoms will be moved in the Z direction to produce the longitudinal vibration of SWCNT. The torsional vibration of SWCNT can be realized through releasing the torsional SWCNT in MD simulation. For example, when SWCNT are stretched from its equilibrium position in the Y direction, the tension acts as the restoring force that brings the SWCNT to its original undeformed position. The deformed SWCNT with different boundary conditions can vibrate transversely at its natural frequency. The displacement d of a cross-section of the SWCNT is calculated as follows: $d = 1/n \sum_{i=1}^{n} y_i$, where n is the number of atoms at the cross-section of SWCNT. y_i is

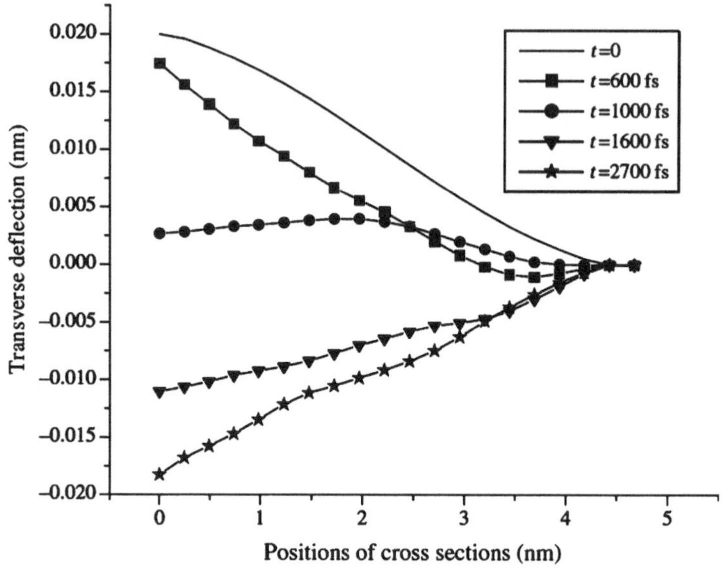

FIGURE 7.2

The transverse deflections of a cantilevered (10, 10) SWCNT at different times during MD simulation of its free vibration [71].

Reprinted with permission from Hu YG, Liew KM, Wang Q. Nonlocal continuum model and molecular dynamics for free vibration of single-walled carbon nanotubes. J Nanosci Nanotechno 2011;11(12):10401–7. Copyright © American Scientific Publishers.

the displacement of the i atom. Each atom at the cross-section passes through its respective equilibrium position simultaneously which is observed from the MD simulations. Fig. 7.2 shows that the fundamental vibration mode can be excited after a small deflection of cantilevered SWCNT is produced. The transverse deflections of SWCNT during the vibration at different time are presented. The vibrations of the end of the cantilevered (10, 10) SWCNT with different lengths are shown in Fig. 7.3. The periods and fundamental frequencies can be calculated from the MD simulations.

7.2.2 NONLOCAL ELASTIC BEAM MODELS FOR FLEXURAL WAVE PROPAGATION IN DWCNTs

In nonlocal elasticity [62], the stress at a reference point x is considered to be a function of the strain field at every point in the body. The basic equations for linear, homogeneous, isotropic, nonlocal elastic solid with zero body force are given as [63]

$$\begin{aligned} &\sigma_{ij,j} = 0, \\ &\sigma_{ij}(x) = \int \alpha(|x - x'|, \tau) C_{ijkl} \varepsilon_{kl}(x') dV(x'), \quad \forall x \in V, \\ &\varepsilon_{ij} = \frac{1}{2}(u_{i,j} + u_{j,i}), \end{aligned} \tag{7.13}$$

where σ_{ij} and ε_{ij} are stress and strain tensors, respectively, C_{ijkl} is the elastic modulus tensor in classical isotropic elasticity, u_i is the displacement vector, $\alpha(|x - x'|, \tau)$ is the nonlocal modulus or

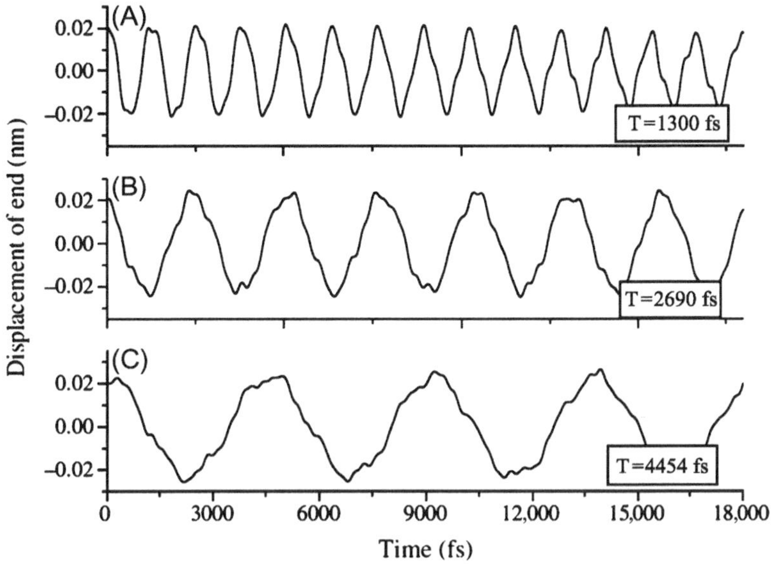

FIGURE 7.3

The vibration of the free end of cantilevered (10, 10) SWCNT with different lengths. (A) 1.7 nm long. (B) 2.9 nm long. (C) 4.1 nm long [71].

Reprinted with permission from Hu YG, Liew KM, Wang Q. Nonlocal continuum model and molecular dynamics for free vibration of single-walled carbon nanotubes. J Nanosci Nanotechno 2011;11(12):10401–7. Copyright © American Scientific Publishers.

attenuation function which incorporates the nonlocal effects at the reference point x produced by local strain at the source x' into the constitutive equations, and $|x - x'|$ is the Euclidean distance. In$\tau = e_0 a/l$, e_0 is a constant appropriate to each material, a is an internal characteristic length (e.g., length of C–C bond, lattice parameter, and granular distance), and l is an external characteristic length (e.g., crack length and wavelength).

Integral-partial differential equations of the above linear nonlocal elasticity are reduced to singular partial differential equations of a special class of physically admissible kernel [63]. Hence, Hooke's law for stress and strain relation in the polar coordinate system is thus expressed by

$$\sigma - (e_0 a)^2 \frac{\partial^2 \sigma}{\partial x^2} = E\varepsilon, \tag{7.14}$$

where σ is the axial stress and ε is the axial strain. The parameter $e_0 a$ is the scale coefficient revealing the small-scale effect on the responses of structures of nanosize. The parameter $e_0 = 0.39$ was given by Eringen [63], but there are no experiments conducted to determine the value of e_0 for CNT. Wang and Hu [57] proposed $e_0 = 1/\sqrt{12} \approx 0.028$ used in the nonlocal beam model. Zhang et al. [35] estimated $e_0 \approx 0.82$ by matching the theoretical buckling strain obtained by the nonlocal thin shell model [66] to those from molecular mechanics simulations given by Sears and Batra [70]. A conservative estimate of the scale coefficient $e_0 a < 2.0$ nm for an SWCNT if the measured wave propagation frequency value for the SWCNT is greater than 10 THz [67]. Hu et al. [41] gave the estimation of e_0 investigated through the comparison between the dispersion relations obtained from the numerical results of governing equations of shell and MD simulation. The value of e_0 which is a coefficient to identify the scale effect should be different for different issues. In this study, we will fix $a = 0.142$ nm and give the estimation of parameter e_0 by matching the dispersions of CNTs observed from the MD results with the numerical results obtained from the nonlocal double-beam model. The parameter a will be assumed to be independent of the length, diameter, and chirality of CNTs for the wave dispersion.

The governing equation for DWCNT's wave propagation by considering the vdW effect via the nonlocal Euler–Bernoulli beam model is given as [35]

$$\begin{aligned} EI_1\frac{\partial^4 u_1}{\partial x^4} + \rho A_1\frac{\partial^2 u_1}{\partial t^2} - (e_0a)^2\left(\rho A_1\frac{\partial^4 u_1}{\partial x^2\partial t^2} - \frac{\partial^2 p_1}{\partial x^2}\right) = p_1, \\ EI_2\frac{\partial^4 u_1}{\partial x^4} + \rho A_2\frac{\partial^2 u_1}{\partial t^2} - (e_0a)^2\left(\rho A_2\frac{\partial^4 u_1}{\partial x^2\partial t^2} - \frac{\partial^2 p_2}{\partial x^2}\right) = p_2. \end{aligned} \tag{7.15}$$

The wave propagation solution for CNTs can be expressed as

$$u_i(x,t) = U_i e^{i(kx-\omega t)} \tag{7.16}$$

where U_i is the amplitude of the wave motion, k is the wave number, and ω is the frequency of the wave motion.

The analytic solution can be obtained from the Euler–Bernoulli beam model,

$$\begin{aligned} v_1 = \sqrt{\frac{-B+\sqrt{B^2-4AC}}{2A}}, \\ v_2 = \sqrt{\frac{-B-\sqrt{B^2-4AC}}{2A}} \end{aligned} \tag{7.17}$$

where

$$\begin{aligned} &A = \rho^2 A_1 A_2 k^4\left[1+(e_0a)^2k^2\right]^2, \\ &B = -\rho k^2\left[1+(e_0a)^2k^2\right]\left[(A_1I_2+A_2I_1)Ek^4 + (e_0a)^2k^2(A_1c_{21}-A_2c_{12}) - A_1c_{21} - A_2c_{12}\right], \\ &C = EI_1I_2k^8 - EI_1k^4c_{21}\left[1-(e_0a)^2k^2\right] - EI_2k^4c_{12}\left[1+(e_0a)^2k^2\right] \end{aligned} \tag{7.18}$$

The governing equation for DWCNT's wave propagation by considering the vdW effect via the nonlocal Timoshenko beam model is given as [43]

$$\begin{aligned} &GA_1\kappa\left(\frac{\partial\varphi_1}{\partial x} - \frac{\partial^2 u_1}{\partial x^2}\right) + \rho A_1\frac{\partial^2 u_1}{\partial t^2} = p_1, \\ &GA_1\kappa\left[1-(e_0a)^2\frac{\partial^2}{\partial x^2}\right]\left(\frac{\partial u_1}{\partial x} - \varphi_1\right) + EI_1\frac{\partial^2\varphi_1}{\partial x^2} - \rho I_1\frac{\partial^2}{\partial t^2}\left[\varphi_1 - (e_0a)^2\frac{\partial^2\varphi_1}{\partial x^2}\right] = 0, \\ &GA_2\kappa\left(\frac{\partial\varphi_2}{\partial x} - \frac{\partial^2 u_2}{\partial x^2}\right) + \rho A_2\frac{\partial^2 u_2}{\partial t^2} = p_2, \\ &GA_2\kappa\left[1-(e_0a)^2\frac{\partial^2}{\partial x^2}\right]\left(\frac{\partial u_2}{\partial x} - \varphi_1\right) + EI_2\frac{\partial^2\varphi_2}{\partial x^2} - \rho I_2\frac{\partial^2}{\partial t^2}\left[\varphi_2 - (e_0a)^2\frac{\partial^2\varphi_2}{\partial x^2}\right] = 0. \end{aligned} \tag{7.19}$$

where p_k ($k = 1, 2$) is the vdW force between the adjacent tubes. The (inward positive) net radial pressure p_k on the tube k is given as follows:

$$p_1 = c_{12}(u_1 - u_2) \quad \text{and} \quad p_2 = c_{21}(u_2 - u_1), \tag{7.20}$$

where c_{ij} is the vdW interaction coefficient, which is estimated as [24]

$$c_{ij} = -\left[\frac{1001\ \pi\varepsilon\sigma^{12}}{3a^4}E_{ij}^{13} - \frac{1120\ \pi\varepsilon\sigma^6}{9a^4}E_{ij}^{17}\right]R_j \tag{7.21}$$

and

$$E_{ij}^m = (R_j + R_i)^{-m}\int_0^{\pi/2}\frac{d\theta}{\left[1 - K_{ij}\cos^2\theta\right]^{m/2}}, \tag{7.22}$$

in which

$$K_{ij} = \frac{4R_jR_i}{(R_j + R_i)^2}. \tag{7.23}$$

Nonlocal beam theory can conform to the local theory if the scale coefficient e_0 is set to zero.

In order to check the applicability of the dispersion relations by the nonlocal double-beam model, MD simulation for DWCNTs is also presented. The lengths of the DWCNTs are 19.7 nm. The interaction of atoms in each tube is described by the second-generation REBO potential energy expression for hydrocarbons [60], while the vdW force between inner and outer walls is described by the Lennard–Jones potential [72]. Poisson's ratio and Young's modulus calculated from the MD simulations for the test of axial tension are listed in Table 7.1.

7.3 NONLOCAL ELASTIC SHELL MODEL [73]

Based on Eq. (7.13), Hook's law for the stress and strain relation in the polar coordinate system in nonlocal elastic shell model is expressed by

$$\begin{aligned}
\sigma_x - (e_0a)^2\nabla^2\sigma_x &= \frac{E}{1-\upsilon^2}(\varepsilon_x + \upsilon\varepsilon_\theta),\\
\sigma_\theta - (e_0a)^2\nabla^2\sigma_\theta &= \frac{E}{1-\upsilon^2}(\varepsilon_\theta + \upsilon\varepsilon_x),\\
\sigma_{\theta x} - (e_0a)^2\nabla^2\sigma_{\theta x} &= \frac{E}{2(1+\upsilon)}\gamma_{x\theta},\\
\sigma_{x\theta} - (e_0a)^2\nabla^2\sigma_{x\theta} &= \frac{E}{2(1+\upsilon)}\gamma_{x\theta},
\end{aligned} \tag{7.24}$$

where E is the Young's modulus of the material; υ is the Poisson's ratio; x and θ are the longitudinal and angular circumferential coordinates; σ_x, σ_θ, $\sigma_{\theta x}$, and $\sigma_{x\theta}$ are the normal and shear stresses; and ε_x, ε_θ, and $\gamma_{x\theta}$ are the normal and shear strains. The Laplace operator in the polar coordinate system is given by $\nabla^2 = \frac{\partial^2}{\partial x^2} + \frac{\partial^2}{r^2\partial\theta^2}$, and r is the radius measured from the mid-plane of the cross-section in the following CNT analysis.

The nonlocal elastic cylindrical shell theory, which is based on the Flugge shell theory [74], was developed by Wang and Varadan [44]. The application of the nonlocal elastic cylindrical shell theory to each tube of an armchair–armchair (10, 10)@(15, 15) DWCNT yields six equations that govern its free vibrations, as follows:

$$r_k{}^2\frac{\partial^2 u_k}{\partial x^2}+\frac{1}{2}(1-\upsilon)\frac{\partial^2 u_k}{\partial \theta^2}+\frac{r_k}{2}(1+\upsilon)\frac{\partial^2 v_k}{\partial x\partial\theta}-\upsilon r_k\frac{\partial w_k}{\partial x}+(1-\upsilon^2)\frac{D}{Ehr_k{}^2}\left(\frac{1}{2}(1-\upsilon)\frac{\partial^2 u_k}{\partial\theta^2}+r_k{}^3\frac{\partial^3 w_k}{\partial x^3}-\frac{r_k}{2}(1-\upsilon)\frac{\partial^3 w_k}{\partial x\partial\theta^2}\right)$$
$$=\frac{\rho h}{Eh}(1-\upsilon^2)r_k{}^2(1-(e_0a)^2\nabla^2)\frac{\partial^2 u_k}{\partial t^2},$$

$$\frac{r_k}{2}(1+\upsilon)\frac{\partial^2 u_k}{\partial x\partial\theta}+\frac{r_k{}^2}{2}(1-\upsilon)\frac{\partial^2 v_k}{\partial x^2}+\frac{\partial^2 v_k}{\partial\theta^2}-\frac{\partial w_k}{\partial\theta}+(1-\upsilon^2)\frac{D}{Eh}\left(\frac{3}{2}(1-\upsilon)\frac{\partial^2 v_k}{\partial x^2}+\frac{1}{2}(3-\upsilon)\frac{\partial^3 w_k}{\partial x^2\partial\theta}\right)$$
$$=\frac{\rho h}{Eh}(1-\upsilon^2)r_k{}^2(1-(e_0a)^2\nabla^2)\frac{\partial^2 v_k}{\partial t^2},$$

$$\upsilon r_k\frac{\partial u_k}{\partial x}+\frac{\partial v_k}{\partial\theta}-w_k-(1-\upsilon^2)\frac{D}{Ehr_k{}^2}\left(r_k{}^4\nabla^2\nabla^2 w_k+r_k{}^3\frac{\partial^3 u_k}{\partial x^3}-\frac{r_k}{2}(1-\upsilon)\frac{\partial^3 u_k}{\partial x\partial\theta^2}+\frac{r_k{}^2}{2}(3-\upsilon)\frac{\partial^3 v_k}{\partial x^2\partial\theta}+w_k+2\frac{\partial^2 w_k}{\partial\theta^2}\right)$$
$$=\frac{1}{Eh}(1-\upsilon^2)r_k{}^2(1-(e_0a)^2\nabla^2)(\frac{\partial^2 w_k}{\partial t^2}\rho h-p_k),$$

(7.25)

where $p_k(k=1,\ 2)$ is the vdW force between the adjacent tubes. The (inward positive) net radial pressure p_k on tube k is given by (7.20).

The wave propagation solution for CNTs can be expressed as

$$u(x,t)=Ue^{i(lx+m\theta-\omega t)},\quad v(x,t)=Ve^{i(lx+m\theta-\omega t)},\quad w(x,t)=We^{i(lx+m\theta-\omega t)},\tag{7.26}$$

where U, V, and W denote the longitudinal, circumferential, and radial displacement amplitudes of tube k, respectively. l is the wavenumber in the longitudinal direction, ω is the frequency of the wave motion. The proposed nonlocal Flugge shell theory in equations (7.25) can conform to the local theory if the scale coefficient, $e_0a=0$, is set to zero.

An important case results from considering motion independent of θ. Thus $\frac{\partial}{\partial\theta}=0$.

$$r^2\frac{\partial^2 u}{\partial x^2}-\upsilon r\frac{\partial w}{\partial x}+(1-\upsilon^2)\frac{D}{Ehr^2}r^3\frac{\partial^3 w}{\partial x^3}=\frac{\rho h}{Eh}(1-\upsilon^2)r^2(1-(e_0a)^2\nabla^2)\frac{\partial^2 u}{\partial t^2},$$
$$\frac{r^2}{2}(1-\upsilon)\frac{\partial^2 v}{\partial x^2}+(1-\upsilon^2)\frac{D}{Eh}\frac{3}{2}(1-\upsilon)\frac{\partial^2 v}{\partial x^2}=\frac{\rho h}{Eh}(1-\upsilon^2)r^2(1-(e_0a)^2\nabla^2)\frac{\partial^2 v}{\partial t^2},\tag{7.27}$$
$$\upsilon r\frac{\partial u}{\partial x}-w-(1-\upsilon^2)\frac{D}{Ehr^2}\left(r^4\nabla^2\nabla^2 w+r^3\frac{\partial^3 u}{\partial x^3}\right)=\frac{1}{Eh}(1-\upsilon^2)r^2(1-(e_0a)^2\nabla^2)\left(\frac{\partial^2 w}{\partial t^2}\rho h-p\right).$$

The second equation of (7.27) is uncoupled from the other two equations and represents the purely torsional motion of the shell. The pure torsional wave propagation solution for CNTs can be expressed as $v(x,t) = Ve^{i(kx-\omega t)}$. Substituting the expression to the equation (7.25) leads to the velocity of torsion wave:

$$v = \frac{1}{1-(e_0a)^2k^2}\left(\frac{E}{2\rho(1+\upsilon)} + \frac{3D(1-\upsilon)}{2h\rho r^2}\right) \tag{7.28}$$

The material parameters of CNTs are proposed as: in-plane stiffness $Eh = 360\ \mathrm{J/m^2}$[13], mass density $\rho h = (2.27\ \mathrm{g/cm^3}) \times 0.34\ \mathrm{nm}$.

Adams et al. [75] suggested the bending rigidity, $D = 1.62$ eV, by applying the quantum MD methods using both empirical and local density functional methods to evaluate the energies of a number of ball-shaped and tubular fullerenes of various sizes. The bending rigidity $D = 1.46$ eV was evaluated based on ab initio energy calculations by Kudin et al. [76]. The value of bending rigidity, $D = 2$ eV, was proposed in a study in which the results of the phonon dispersion relations of a CNT were compared with those based on the local elasticity theory for a cylindrical medium [77]. Wang and Liew [11] proposed the bending rigidity 1.78 eV using Molecular mechanics calculations for relatively large tubes and graphite sheets. $D = 1.78$ eV will be used in this study. The substitution of Eqs. (7.27) into Eqs. (7.25) ultimately yields a generalized eigenvalue problem, as follows.

$$M\{(l,m),\omega\}_{6\times 6}\begin{Bmatrix} U_1 \\ V_1 \\ W_1 \\ U_2 \\ V_2 \\ W_2 \end{Bmatrix} = 0, \tag{7.29}$$

where the entries of matrices M are functions of the wavenumbers l and ω and the scale coefficient e_0a. The condition for the existence of nonzero solutions U_k and W_k $(k = 1, 2)$ is

$$\det M = 0. \tag{7.30}$$

The six dispersion relations can be obtained by solving equation (7.30) for a given wavenumber $m = 2$. The numerical results are shown in Fig. 7.4. One of them will be used to predict the dispersion of the transverse wave in an armchair–armchair (10, 10)@(15, 15) DWCNT.

To check the applicability of the dispersion relations produced by the nonlocal cylindrical shell model to SWCNTs and DWCNTs, MD simulations are presented. The transverse and torsional wave can be produced when the atoms at the end of CNT are excited in a way showed in Fig. 7.5A and B, respectively. The lengths of SWCNTs and DWCNTs are about 19.7 nm. Using the second-generation REBO potential, we computed the Poisson's ratio $\nu = 0.317$ and Young's modulus $E = 0.83$ TPa for the armchair (10, 10) and (15, 15) CNTs that were derived from the MD simulation for the test of axial tension. The Poisson's ratio $\nu = 0.347$ and Young's modulus $E = 0.8$ TPa are calculated for the zigzag (20, 0) CNT.

7.4 VIBRATION CHARACTERISTICS OF CNTs [71]

Free transverse, longitudinal, and torsional vibrations of SWCNTs with different lengths and radii are investigated. Applicability of nonlocal models is verified by MD simulations.

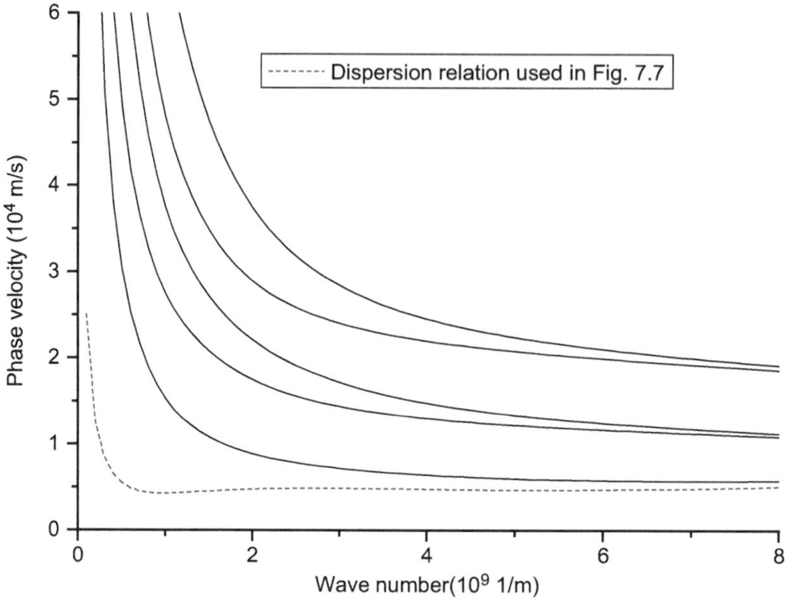

FIGURE 7.4

Wave velocity versus wavenumber of the armchair–armchair (10, 10)@(15, 15) DWCNT at wavenumber $m = 2$ [73].

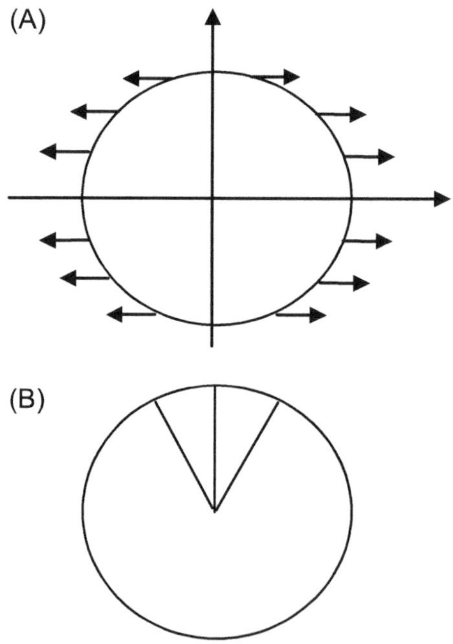

FIGURE 7.5

(A) The atoms at the end of the CNT are subjected to harmonic excitation to move in different direction at the same time. (B) Every atom like atom A at the end of the CNT swings through a small angle after exit [73].

Figs. 7.6 and 7.7 show the fundamental frequencies of four cantilevered SWCNT with different lengths and radii. The fundamental frequency becomes lower when an SWCNT is longer. The Timoshenko beam model gives a better prediction of the fundamental frequency of shorter SWCNTs than the Euler–Bernoulli beam model does. Both beam models are appropriate for relative longer SWCNTs. For example, the two beam models can predict the fundamental frequencies of SWCNTs well when the length of a (5, 5) SWCNT is larger than 3.5 nm.

The MD results and beam models for free transverse vibration of clamped–clamped SWCNT are showed in Fig. 7.8. The fundamental frequency of the clamped–clamped SWCNT is higher than a cantilevered SWCNT with the same length. The fundamental frequency of transverse vibration of the clamped–clamped SWCNT can reach 1.06 terahertz for a (5, 5) SWCNT with the length of 2.91 nm.

Figs. 7.6–7.8 indicate that the nonlocal Timoshenko beam model will not be necessary when the length of a CNT is longer than 3.5 nm. But the Timoshenko beam model is not effective to predict the frequencies from verifications by the MD results when the SWCNT is shorter, e.g., 1.7 nm long in Figs. 7.6–7.8 The values of parameter e_0 in the nonlocal Timoshenko beam and nonlocal Euler–Bernoulli beam models for different SWCNTs are list in Table 7.2. The boundary conditions have a significant effect on the value of parameter e_0. The parameter e_0 should be 52 to match the fundamental frequency obtained from MD simulation for cantilevered (5, 5) SWCNTs with length shorter than 3.5 nm.

The parameter e_0 should be 5.2 for (5, 5) SWCNTs with both ends clamped. The nonlocal Euler–Bernoulli beam model fails to predict the fundamental frequency of short cantilevered SWCNTs because the fundamental frequency will not decrease with the increased parameter e_0.

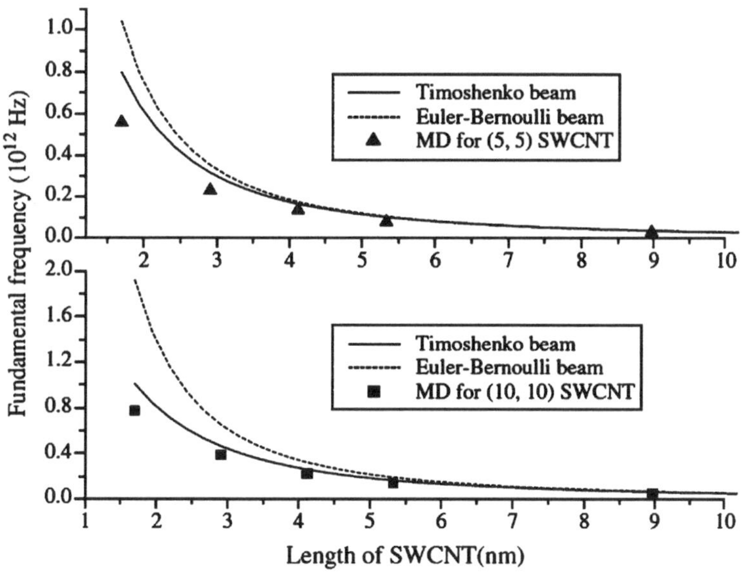

FIGURE 7.6

Fundamental frequencies of cantilevered (5, 5) and (10, 10) SWCNTs with different lengths [71].

Reprinted with permission from Hu YG, Liew KM, Wang Q. Nonlocal continuum model and molecular dynamics for free vibration of single-walled carbon nanotubes. J Nanosci Nanotechno 2011;11(12):10401–7. Copyright © American Scientific Publishers.

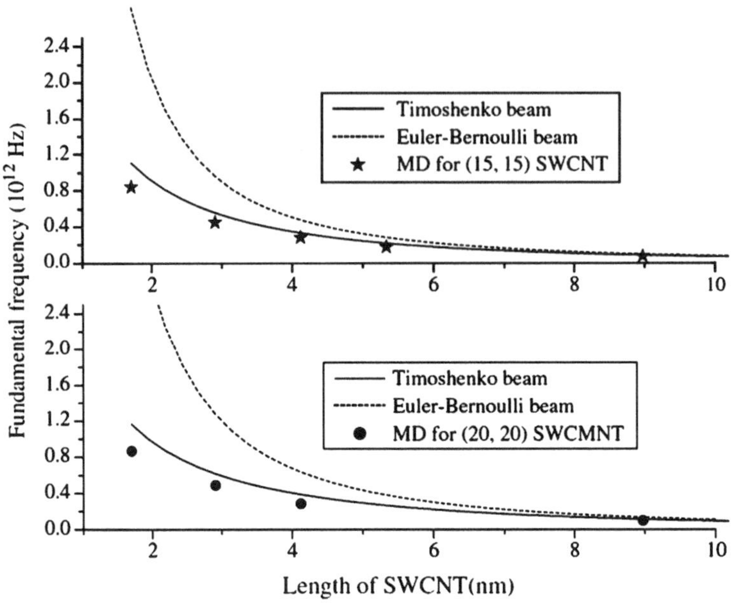

FIGURE 7.7

Fundamental frequencies of cantilevered (15, 15) and (20, 20) SWCNTs with different lengths [71].

Reprinted with permission from Hu YG, Liew KM, Wang Q. Nonlocal continuum model and molecular dynamics for free vibration of single-walled carbon nanotubes. J Nanosci Nanotechno 2011;11(12):10401–7. Copyright © American Scientific Publishers.

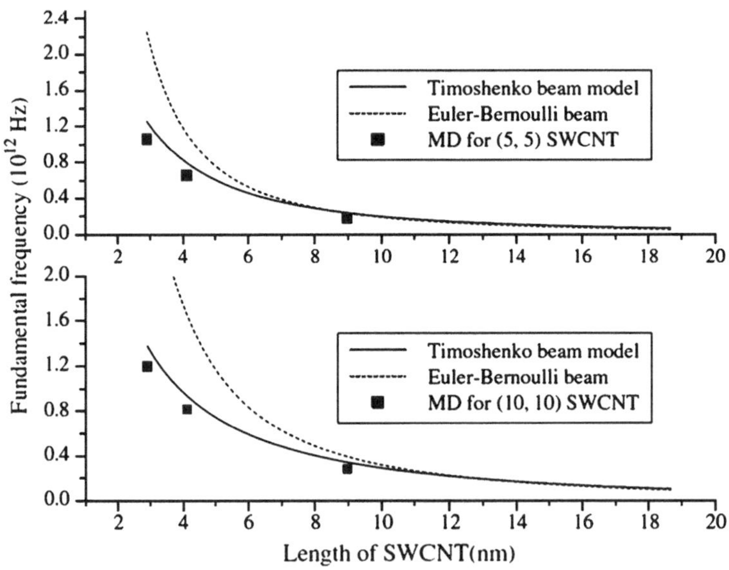

FIGURE 7.8

Fundamental frequencies of clamped–clamped (5, 5) and (10, 10) SWCNTs with different lengths [71].

Reprinted with permission from Hu YG, Liew KM, Wang Q. Nonlocal continuum model and molecular dynamics for free vibration of single-walled carbon nanotubes. J Nanosci Nanotechno 2011;11(12):10401–7. Copyright © American Scientific Publishers.

Table 7.2 The Parameter e_0 in Nonlocal Timoshenko (NT) Beam and Nonlocal Euler (NE) Beam Models for Transverse Vibration of SWCNTs With Two Different Boundary Conditions [71]

(5, 5)		(10, 10)		(15, 15)		(20, 20)	
Cantilevered CNT							
52		34.1		30.8		29.7	
Clamped–clamped NT and NE							
NT	NE	NT	NE	NT	NE	NT	NE
5.5	10.5	7.8	19.5	10.1	30.5	12.3	44.1

Reprinted with permission from Hu YG, Liew KM, Wang Q. Nonlocal continuum model and molecular dynamics for free vibration of single-walled carbon nanotubes. J Nanosci Nanotechno 2011;11(12):10401–7. Copyright © American Scientific Publishers.

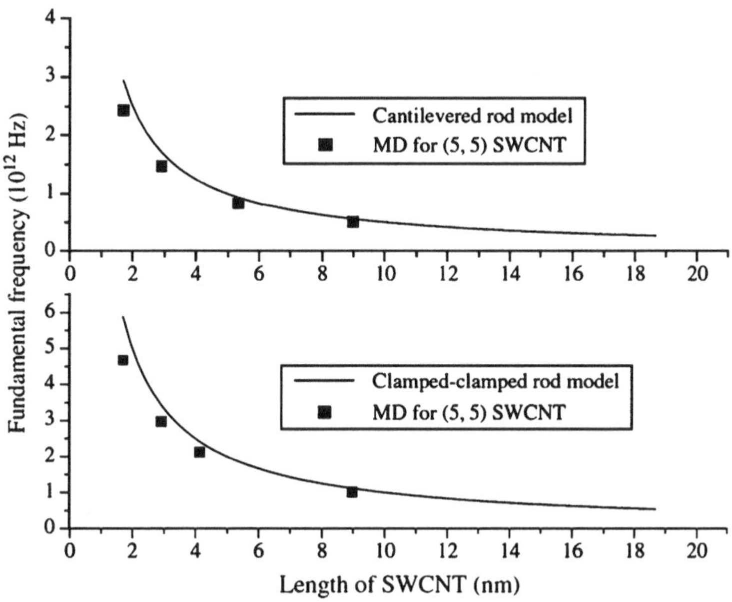

FIGURE 7.9

Fundamental frequencies of longitudinal vibrations of cantilevered and clamped–clamped (5, 5) SWCNTs with different lengths [71].

Reprinted with permission from Hu YG, Liew KM, Wang Q. Nonlocal continuum model and molecular dynamics for free vibration of single-walled carbon nanotubes. J Nanosci Nanotechno 2011;11(12):10401–7. Copyright © American Scientific Publishers.

But the fundamental frequency of a clamped–clamped nonlocal Euler–Bernoulli beam decreases with the parameter e_0. In this case, the parameter should be 10.5 for a clamped–clamped nonlocal Euler–Bernoulli beam to match the MD results of (5, 5) SWCNT with length shorter than 3.5 nm.

Fig. 7.9 shows the fundamental frequencies of longitudinal vibration of SWCNT with different lengths and boundary conditions. The rod model can be used to predict the MD results for long SWCNT. The fundamental frequency of longitudinal vibration of clamped–clamped SWCNT can reach 2.96 terahertz for (5, 5) SWCNT with 2.91 nm long.

Fig. 7.10 shows the torsional vibration of a cantilevered and clamped–clamped (5, 5) SWCNT. The fundamental frequencies can be predicted well by the rod model for long SWCNT. It is necessary to use the nonlocal rod model to predict the natural frequencies for short SWCNT, such as 1.7 nm long. The values of parametere_0 in nonlocal rod model for SWCNTs with lengths smaller than 3.5 nm are listed in Table 7.3. When the length is larger than 3.5 nm, the classical rod model can provide good predictions of fundamental frequency of free longitudinal and torsional vibrations of SWCNT.

Table 7.4 shows the fundamental frequencies of SWCNTs with different boundary conditions and the length of 4.12 nm. The boundary conditions have a significant effect on the fundamental

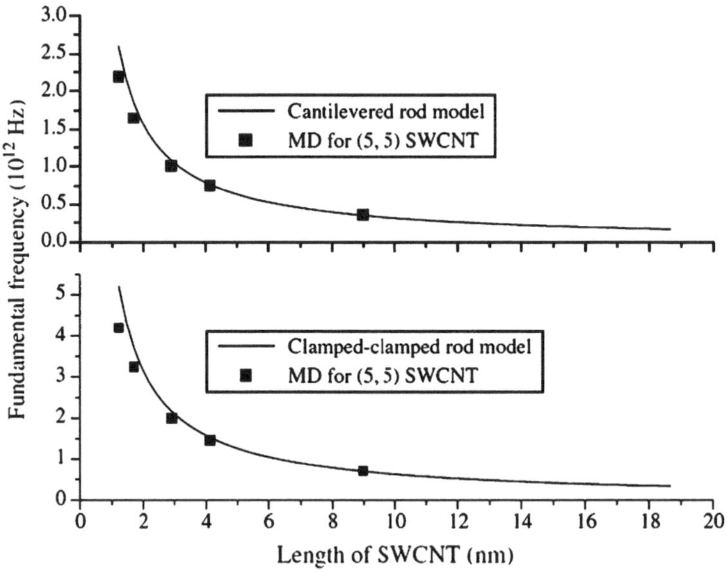

FIGURE 7.10

Fundamental frequencies of torsional vibrations of cantilevered and clamped–clamped (5, 5) SWCNTs with different lengths [71].

Reprinted with permission from Hu YG, Liew KM, Wang Q. Nonlocal continuum model and molecular dynamics for free vibration of single-walled carbon nanotubes. J Nanosci Nanotechno 2011;11(12):10401–7. Copyright © American Scientific Publishers.

Table 7.3 The Parameter e_0 in Nonlocal Rod Models for Longitudinal and Torsional Vibrations of SWCNTs With Two Different Boundary Conditions [71]

	Longitudinal Vibration		Torsional Vibration	
SWCNT	Cantilevered	Clamped–Clamped	Cantilevered	Clamped–Clamped
(5, 5)	5.8	3	4	2.2
(10, 10)	5.8	3.1	3.8	2.3
(15, 15)	7.3	3	3.4	2.7
(20, 20)	7.7	2.3	3.3	2.8

Reprinted with permission from Hu YG, Liew KM, Wang Q. Nonlocal continuum model and molecular dynamics for free vibration of single-walled carbon nanotubes. J Nanosci Nanotechno 2011;11(12):10401–7. Copyright © American Scientific Publishers.

Table 7.4 MD Results of the Fundamental Frequencies (10^{12} Hz) of SWCNTs With Different Boundary Conditions and the Same Length 4.12 nm [71]

SWCNT	Transverse (C–F)	Transverse (C–C)	Longitudinal (C–F)	Longitudinal (C–C)	Torsion (C–F)	Torsion (C–C)
(5, 5)	0.136	0.66	1.07	2.12	0.75	1.46
(10, 10)	0.22	0.82	1.07	2.05	0.68	1.42
(15, 15)	0.29	0.83	1.02	1.99	0.67	1.41
(20, 20)	0.32	0.83	1.01	1.83	0.7	1.45

Reprinted with permission from Hu YG, Liew KM, Wang Q. Nonlocal continuum model and molecular dynamics for free vibration of single-walled carbon nanotubes. J Nanosci Nanotechno 2011;11(12):10401–7. Copyright © American Scientific Publishers.

frequency. The symbols C–F and C–C in the table represent cantilevered and clamped–clamped boundary conditions, respectively. The clamped–clamped SWCNT's fundamental frequency is higher than the cantilevered one. The fundamental frequency of longitudinal vibration of a clamped–clamped SWCNT is higher than others for the same SWCNT in Table 7.4.

In summary, free transverse, longitudinal, and torsional vibrations of cantilevered and clamped–clamped SWCNTs are studied through nonlocal beam models, nonlocal rod models, and MD simulations. Nonlocal beam and nonlocal rod models for the free vibrations of SWCNTs are verified by MD simulations. The nonlocal Timoshenko beam model can predict MD results better than the classical beam model does for short SWCNTs. The nonlocal Euler–Bernoulli model can also predict the fundamental frequency of short clamped–clamped SWCNTs. The classical (local) beam and rod models can give good predictions of fundamental frequencies of long SWCNTs when the length is larger than 3.5 nm.

7.5 WAVE PROPAGATION OF CNTs [40]

7.5.1 NONLOCAL ELASTIC BEAM MODELS FOR FLEXURAL WAVE PROPAGATION

The phase velocity and wavenumber can be determined from the flexural vibration of two arbitrary sections of a CNT that is simulated using MD. The transverse displacement of some section of an armchair–armchair (10, 10)@(15, 15) DWCNT can be obtained by computing the average displacements of the atoms in the section. The harmonic deflection is achieved by shifting the edge atoms at one end of the CNT shown in Fig. 7.11. For example, the atoms that are located at the end of the armchair–armchair (10, 10)@(15, 15) DWCNT denoted by 0 are assumed to be subjected to a harmonic deflection of period $T = 500$ fs, as shown in Fig. 7.12. The corresponding angular frequency is $\omega = 2/T \approx 1.26 \times 10^{13}$ rad/s while the other end is kept free. Fig. 7.12B and C shows the flexural vibrations of 1 at $x_1 = 2.46$ nm and 2 at $x_2 = 4.92$ nm, respectively, of the DWCNT that is simulated using MD. If the transient deflection of the first two periods is neglected, then the propagation duration Δt of the wave from 1 to 2 can be estimated as

$$\Delta t \approx \frac{(t_{23} - t_{13}) + (t_{24} - t_{14}) + \cdots + (t_{2n} - t_{1n})}{n - 2} \tag{7.31}$$

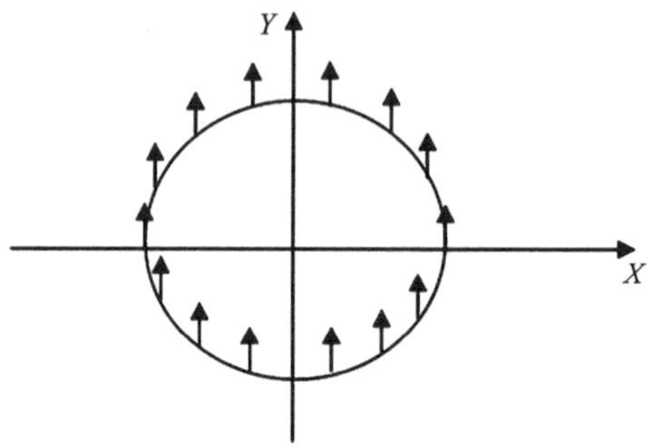

FIGURE 7.11

Atoms exited at the end of the inner and outer tube in DWCNT during MD simulation [40].

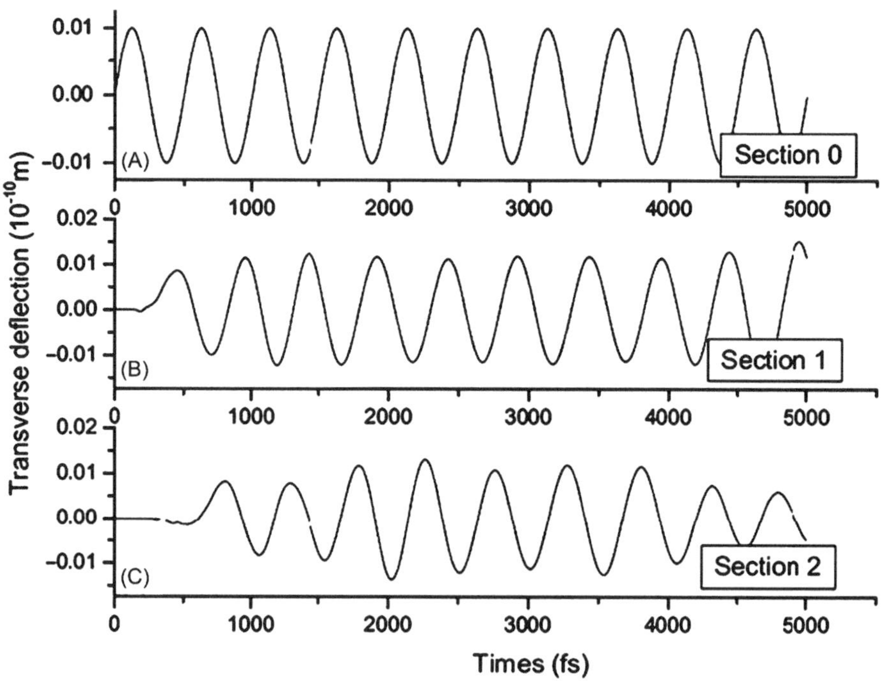

FIGURE 7.12

Time histories of the deflection of atoms at different sections of the armchair–armchair (10, 10)@(15, 15) DWCNT: (A) sinusoidal wave of period $T = 500$ fs input at section 0; (B) deflection of section 1 which is 2.46 nm ahead of section 0; and (C) deflection of section 2 which is 4.92 nm ahead of section 0 [40].

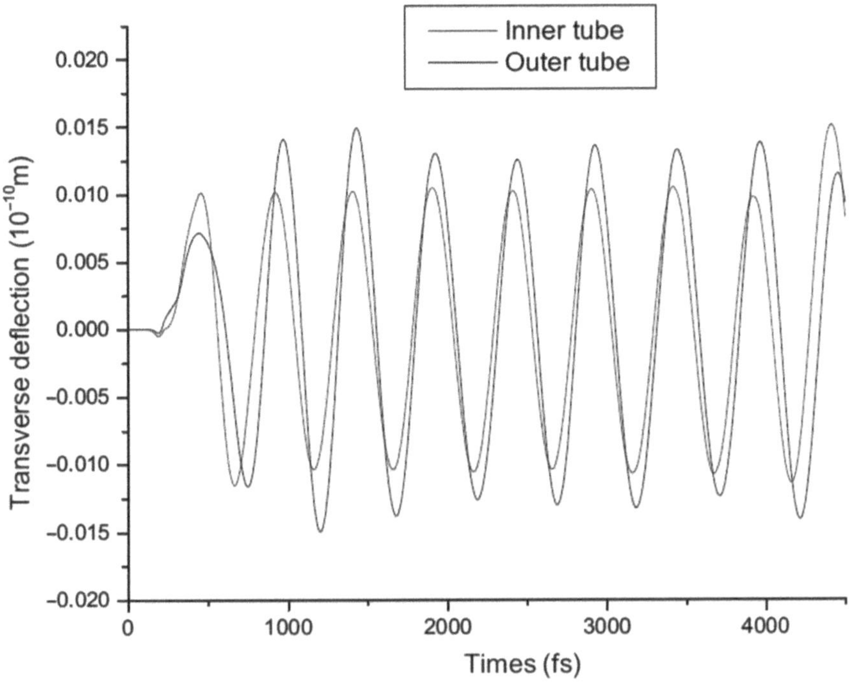

FIGURE 7.13

Time histories of the deflection of the section which is 2.46 nm ahead of the end of the inner and outer tubes for the armchair–armchair (10, 10)@(15, 15) DWCNT [40].

The phase velocity and wave number are expressed as follows, respectively,

$$c = \frac{x_2 - x_1}{\Delta t}, \quad k = \frac{2\pi}{\lambda} = \frac{\omega T}{\lambda} = \frac{\omega}{c}. \tag{7.32}$$

One section of the inner tube and outer tube at period $T = 500$ fs is presented in Fig. 7.13. The inner tube and outer tube are observed to vibrate coaxially in the figure.

Fig. 7.14 illustrates the dispersion relations between the phase velocity c and the wave number k of the flexural wave in the (10, 10)@(15, 15) DWCNT. The flexural wave dispersion relation derived from the above governing equation for nonlocal double Euler–Bernoulli and Timoshenko beam models are showed in Fig. 7.14. The classical (or local) beam theories cannot reveal the small effect on flexural wave dispersion. The scale coefficients, $e_0 = 6.5$ and $a = 0.142$ nm, are used in the nonlocal elastic double Timoshenko beam model. When the wave number is approximately smaller than 1×10^9 1/m, the phase velocities given by local or nonlocal beam models are close to each other and they can all predict the MD simulations well. When the wave number is larger than 1.5×10^9 1/m, the phase velocity decreases, and it is found that only the nonlocal elastic double Timoshenko beam model can predict the flexural wave dispersion of the DWCNT. Figs. 7.15 and 7.16 show the dispersion relation of flexural wave in (5, 5)@(10, 10) DWCNT and (10, 0)@(19, 0) DWCNT. It is clearly seen that only the nonlocal elastic double Timoshenko beam model can

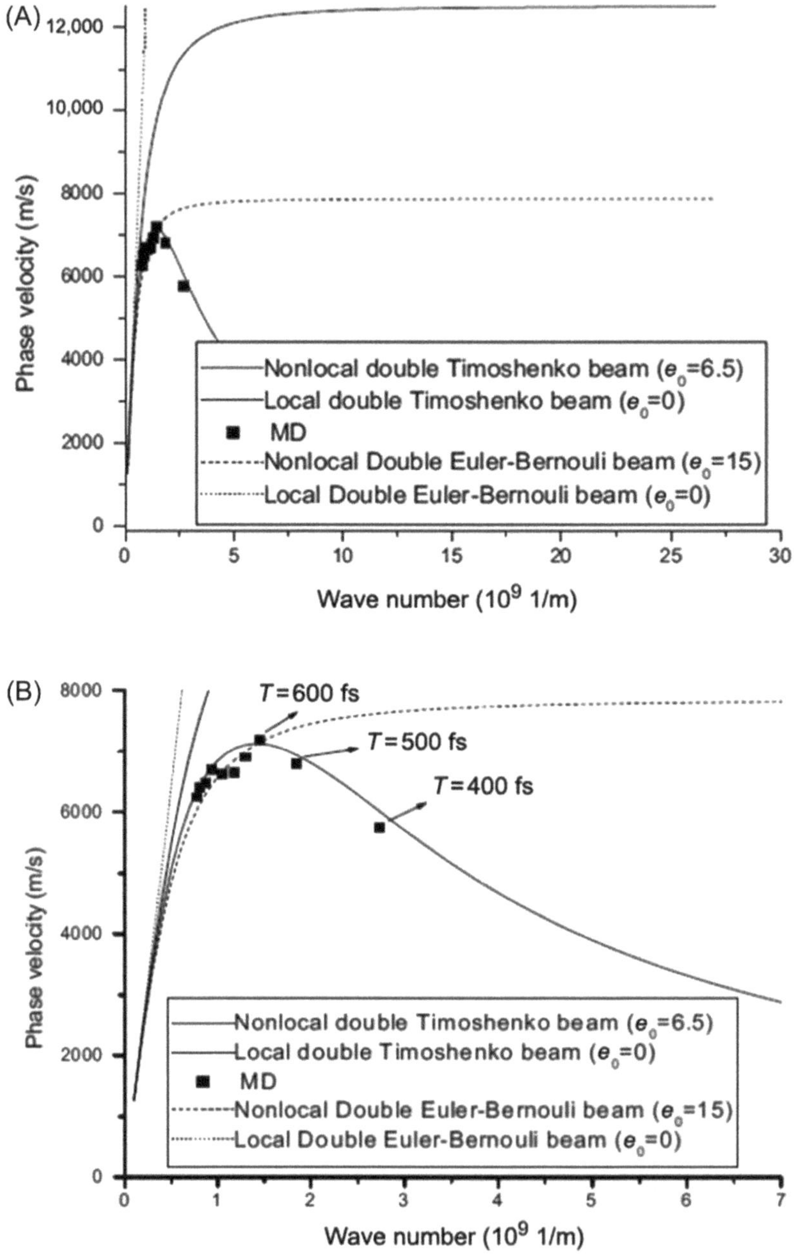

FIGURE 7.14

Dispersion relation of the (10, 10)@(15, 15)DWCNT [panel (B) is the zoom of (A)] [40].

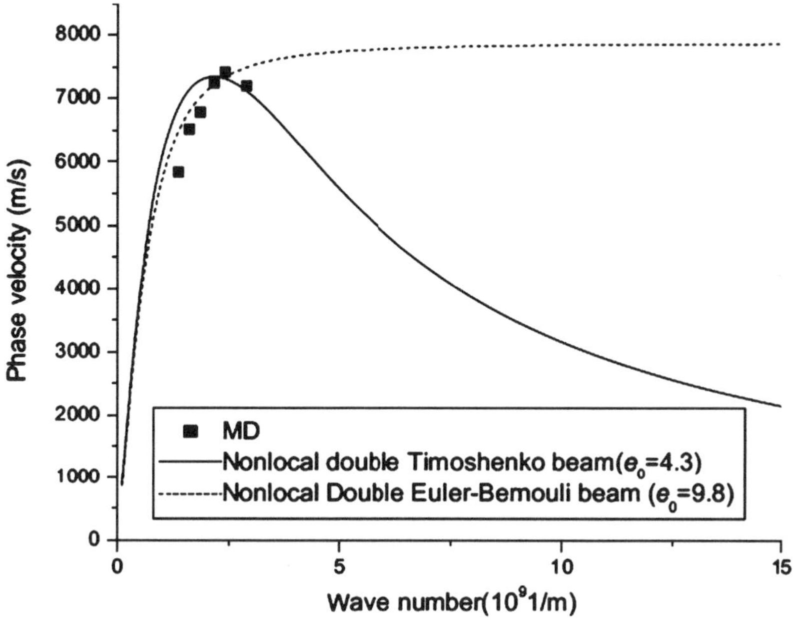

FIGURE 7.15

Dispersion relation of the (5, 5)@(10, 10) DWCNT [40].

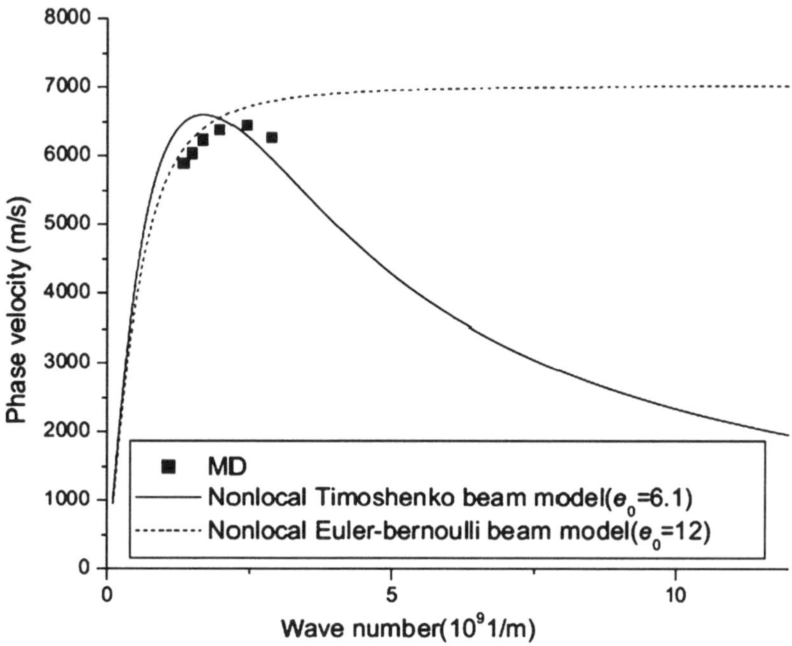

FIGURE 7.16

Dispersion relation of the (10, 0)@(19, 0) DWCNT [40].

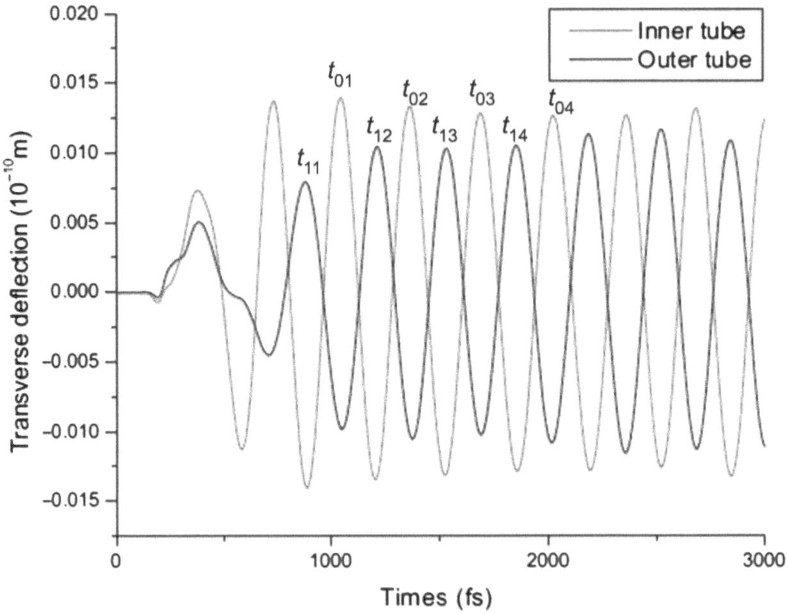

FIGURE 7.17

Time histories of the deflection of atoms at the section which is 2.46 nm ahead of the end of the armchair–armchair (10, 10)@(15, 15) DWCNT (the sinusoidal wave of period $T = 330$ fs input at the end of the DWCNT) [40].

predict the flexural wave dispersion of the DWCNTs when the wave numbers are larger than 2.4×10^9 1/m and 2.44×10^9 1/m for (5, 5)@(10, 10) DWCNT and (10, 0)@(19, 0) DWCNT, respectively. The nonlocal Euler beam model fails to describe wave propagation in the DWCNTs. The values of the scale coefficient $e_0 = 4.3$ and 6.1 are used in the nonlocal double Timoshenko beam model for the two DWCNTs. In addition, it is found that there are little differences among nonlocal models with different scale parameters e_0 when the wave number is small. The good agreement between the dispersion relation induced from the nonlocal elastic double Timoshenko beam model and the MD simulations verifies the effectiveness of this model.

Fig. 7.17 provides the vibration of the (10, 10)@(15, 15) DWCNT at the period $T = 330$ fs and the corresponding angular frequency is $\omega = 2/T \approx 1.9 \times 10^{13}$ rad/s. From the figure, it shows that when the frequency is larger than 1.9×10^{13} rad/s, a noncoaxial vibration of the DWCNT will occur. The average phase difference between the inner tube and the outer tube can be estimated from Fig. 7.17 as $\bar{t} = 1/n \sum_{i=1}^{n} (t_{0i} - t_{1i}) \approx 165.3$ fs. In order to check the effect of the vdW force on the noncoaxial vibration of the DWCNT, MD simulations for flexural wave in SWCNTs are conducted. Fig. 7.18 indicates that the (10, 10) and (15, 15) SWCNTs have different phases at the period $T = 330$ fs without considering the vdW forces between them. The average phase difference between the (10, 10) and (15, 15) SWCNTs can also be estimated from Fig. 7.18 $\bar{s} = 1/n \sum_{i=1}^{n} (s_{0i} - s_{1i}) \approx 153.3$ fs. The other average phase differences at different periods are shown in Table 7.5. The phase difference of the vibration becomes increasingly obvious with the

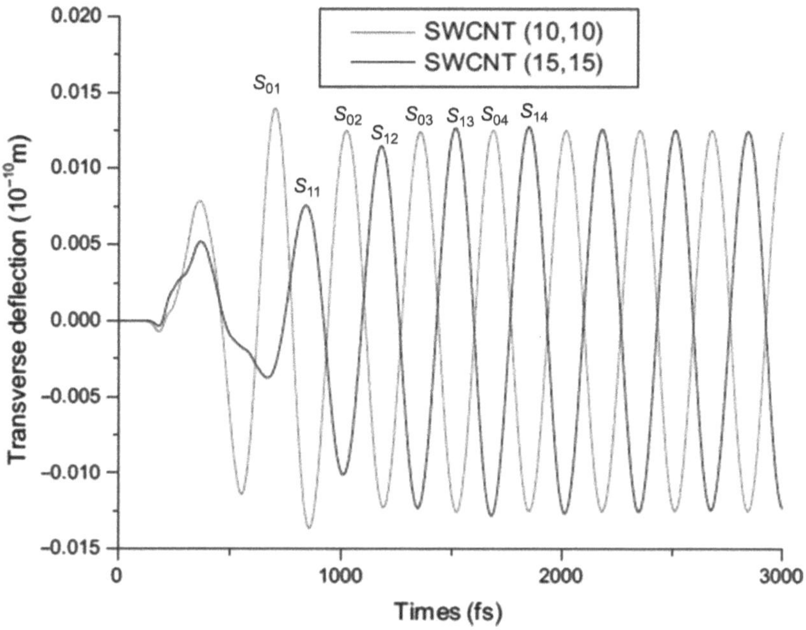

FIGURE 7.18

Time histories of the deflection of the section which is 2.46 nm ahead of the end of the SWCNTs [40].

Table 7.5 Average Phase Differences of the Vibration of the Section Which Is 2.46 nm Ahead of the End of the DWCNT (10, 10)@(15, 15) and the SWCNTs (10, 10) and (15, 15) [40]

Period (fs)	DWCNT(10, 10)@(15, 15) (fs)	SWCNTs (10, 10) and (15, 15) (fs)	$\lvert\bar{s}-\bar{t}\rvert$ (fs)
$T=330$	$\bar{t}=165.3$	$\bar{s}=153.3$	12
$T=400$	$\bar{t}=91$	$\bar{s}=82$	9
$T=500$	$\bar{t}=27$	$\bar{s}=28$	1
$T=600$	$\bar{t}=11.8$	$\bar{s}=12.5$	0.7

decrease in the period. The phase difference increases from 11.8 to 165.3 fs as the period decreases from 600 to 330 fs for DWCNTs. Similarly, the phase difference increases from 12.5 to 153.3 fs as the period decreases from 600 to 330 fs for SWCNTs. The values of $\bar{s}$ and $\bar{t}$ are very close at the periods $T=500$ fs and $T=600$ fs, at which the phase velocity also starts to decrease, as indicated in Fig. 7.14. It is concluded that the different small-scale effects of SWCNTs with different diameters have a significant effect on the noncoaxial flexural vibration of DWCNTs at high frequency. The vdW forces make no significant contribution to the phase differences, as the noncoaxial vibration of DWCNTs would occur without consideration of the vdW forces between the adjacent tubes. The continuum mechanics model is inapplicable in the analysis of flexural wave in DWCNT when the noncoaxial vibration of DWCNTs happens at high frequency, especially when the angular frequency is larger than 1.9×10^{13} rad/s.

In summary, the study of flexural wave dispersion in DWCNTs in a wide range of wavenumbers by the nonlocal elastic double-beam theory and MD simulations is presented. The results indicate that only the nonlocal elastic double Timoshenko beam model is able to predict the small effect on flexural wave dispersion when the wavenumber becomes so large that the CNT microstructure plays an important role in the flexural wave. These small-scale effects have a significant effect on the noncoaxial flexural vibration of DWCNTs at high frequency. The vdW interaction has little effect on the noncoaxial flexural vibration of DWCNT, and the nonlocal elastic Timoshenko beam theory is inapplicable in modeling the noncoaxial wave propagation in CNTs.

7.5.2 NONLOCAL SHELL MODEL FOR ELASTIC WAVE PROPAGATION

It is quite straightforward to determine the phase velocity and wavenumber from the transverse vibration of atoms in two arbitrary sections of a CNT that was simulated using MD. The harmonic deflection of some atoms was achieved by shifting the atoms at one end of the nanotube while the other end was kept free, as shown in Fig. 7.5A. For example, some atoms that are located at the end of the armchair (15, 15) SWCNT denoted by Section 0 were assumed to be subject to the harmonic deflection of period $T = 500$ fs, as shown in Fig. 7.19A. The corresponding angular

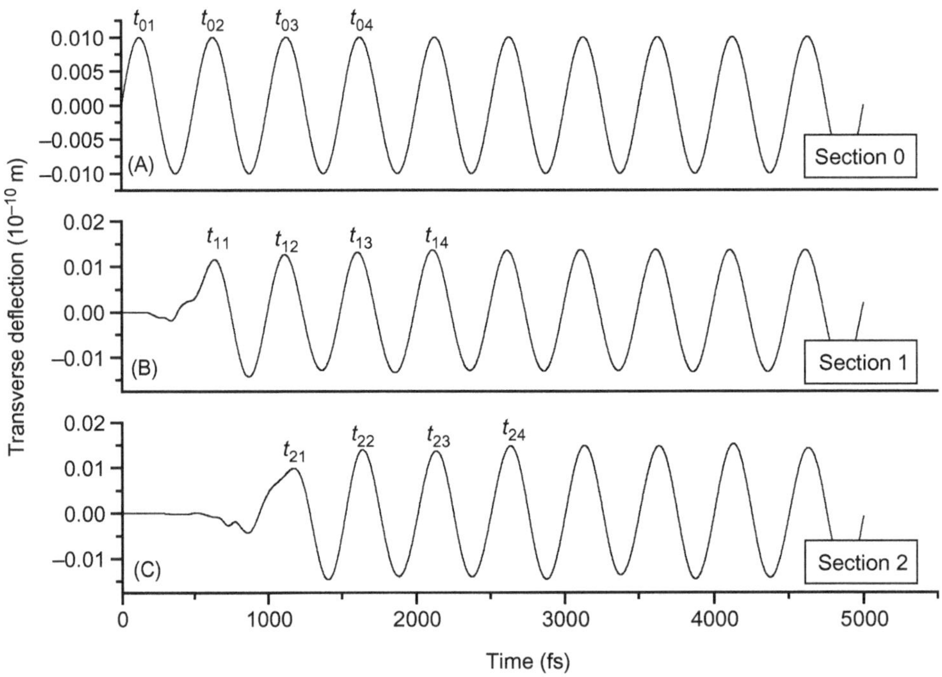

FIGURE 7.19

Time histories of the deflection of atoms at different sections of the armchair (15, 15) CNT. (A) The sinusoidal wave of period $T = 500$ fsinput at section 0. (B) The deflection of section 1, 2.46 nm ahead of section 0. (C) The deflection of section 2, 4.92 nm ahead of section 0 [73].

frequency is $\omega = 2\pi/T \approx 1.26 \times 10^{13}$ rad/s. Fig. 7.19B and C show the transverse vibrations of the atom in Section 1 at $x_1 = 2.46$ nm and Section 2 at $x_2 = 4.92$ nm, respectively. If the transient deflection of the first two periods is neglected, then the propagation duration Δt of the wave from Section 1 to Section 2 can be estimated as

$$\Delta t \approx \frac{(t_{23} - t_{13}) + (t_{24} - t_{14}) + \cdots + (t_{2n} - t_{1n})}{n - 2} \tag{7.33}$$

The phase velocity and wavenumber are, respectively,

$$c = \frac{x_2 - x_1}{\Delta t}, \quad k = \frac{2\pi}{\lambda} = \frac{\omega T}{\lambda} = \frac{\omega}{c} \tag{7.34}$$

When the atoms at the end of CNT are subjected to the harmonic excitation indicated in Fig. 7.5B, the torsional wave will propagate in SWCNTs as shown in Fig. 7.20. The torsional wave velocity can be calculated through the method for calculating transverse wave velocity.

Fig. 7.21 illustrates the dispersion relations between the phase velocity c and the wavenumber k of the transverse wave in the armchair (15, 15) SWCNT and zigzag (20, 0) SWCNT. Fig. 7.21 shows the nonlocal shell model can predict the MD results better than the local shell model can when the wave lengths are less than 2.36×10^{-9} m and 1.88×10^{-9} m for (15, 15) SWCNT and (20, 0) SWCNT, respectively. The value of the parameter $e_0 = 0.6$ is used for transverse wave in armchair (15, 15) and zigzag (20, 0) CNT.

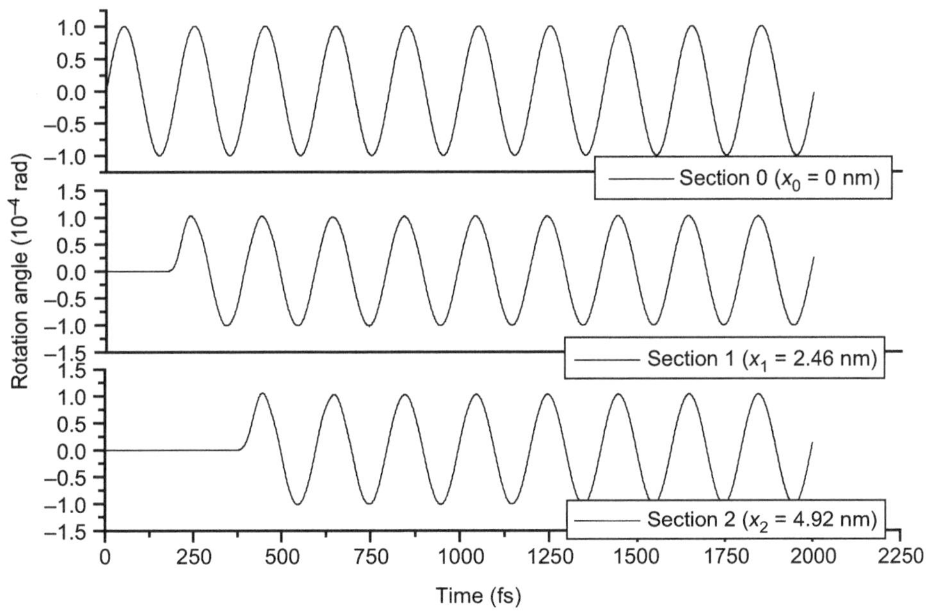

FIGURE 7.20

The time histories of the rotation angle of different sections of the armchair (10, 10) SWCNT. (A) The sinusoidal wave of period $T = 200$ fs input at section 0. (B) The deflection of section 1, 2.46 nm ahead of section 0. (C) The deflection of section 2, 4.92 nm ahead of section 0 [73].

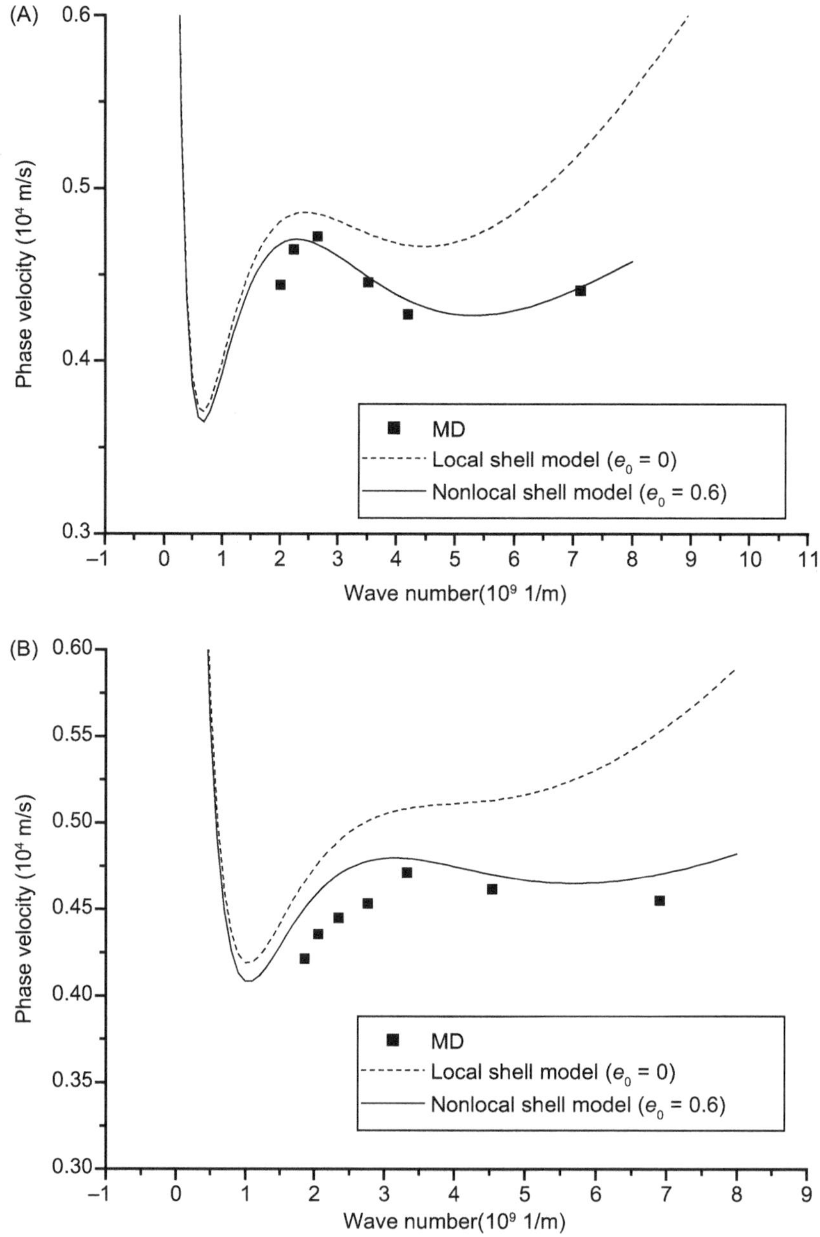

FIGURE 7.21

(A) Dispersion relation of transverse wave in the armchair (15, 15) carbon nanotube. (B) Dispersion relation of transverse wave in the zigzag (20, 0) carbon nanotube [73].

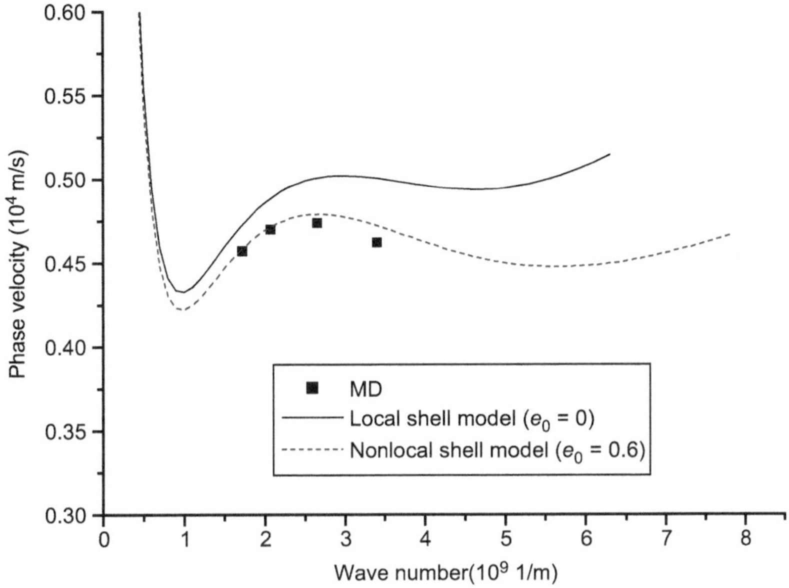

FIGURE 7.22

Dispersion relation of the armchair–armchair (10, 10)@(15, 15) DWCNT [73].

Fig. 7.22 shows that the nonlocal elastic cylindrical double-shell model can predict the dispersion relation of the transverse wave better than local double-shell model when the wavenumber is large enough. This is because the interlayer van der Waals interaction of the DWCNT and the small-scale effect on the dispersion of the transverse wave of the SWCNT are considered. The phase velocity decreases when the wavenumber is approximately larger than 2.65×10^9 1/m, or the wavelength is approximately less than 2.35×10^{-9}m. The nonlocal elastic double-shell model can predict the wave dispersion better than local shell model when the wavelength is approximately smaller than 2.35×10^{-9}m. Fig. 7.23 indicates that the nonlocal shell model can predict dispersion relation obtained from the MD simulation of torsional wave in (10, 10) SWCNT and (15, 15) SWCNT better than local shell model can when the wavelength is smaller than 0.95×10^{-9}m and 0.9×10^{-9} m, respectively. The good agreement between the nonlocal elastic cylindrical shell models results and the MD simulations verifies the effectiveness of this model. The value of the parameter $e_0 = 0.6$ is adopted for (10, 10)@(15, 15) DWCNTs to predict the dispersion of a transverse wave in CNTs, and $e_0 = 0.2$ and 0.23 are used for a torsional wave in SWCNT (10, 10) and SWCNT (15, 15), respectively. The estimates of e_0 are investigated through the comparison between the dispersion relations obtained from numerical results of equations and MD simulation. The value of e_0 which is a coefficient to identify the scale effect should be different for different issues, such as dispersions of different elastic waves in SWCNT and DWCNT. The scale coefficient is dependent on SWCNTs and DWCNTs for transverse wave, and CNTs for torsional wave. When the value of e_0 becomes smaller, the scale effect will be weaker. The torsional wave motion is only related to the displacement in tangential direction, while the transverse wave motion is involved in the

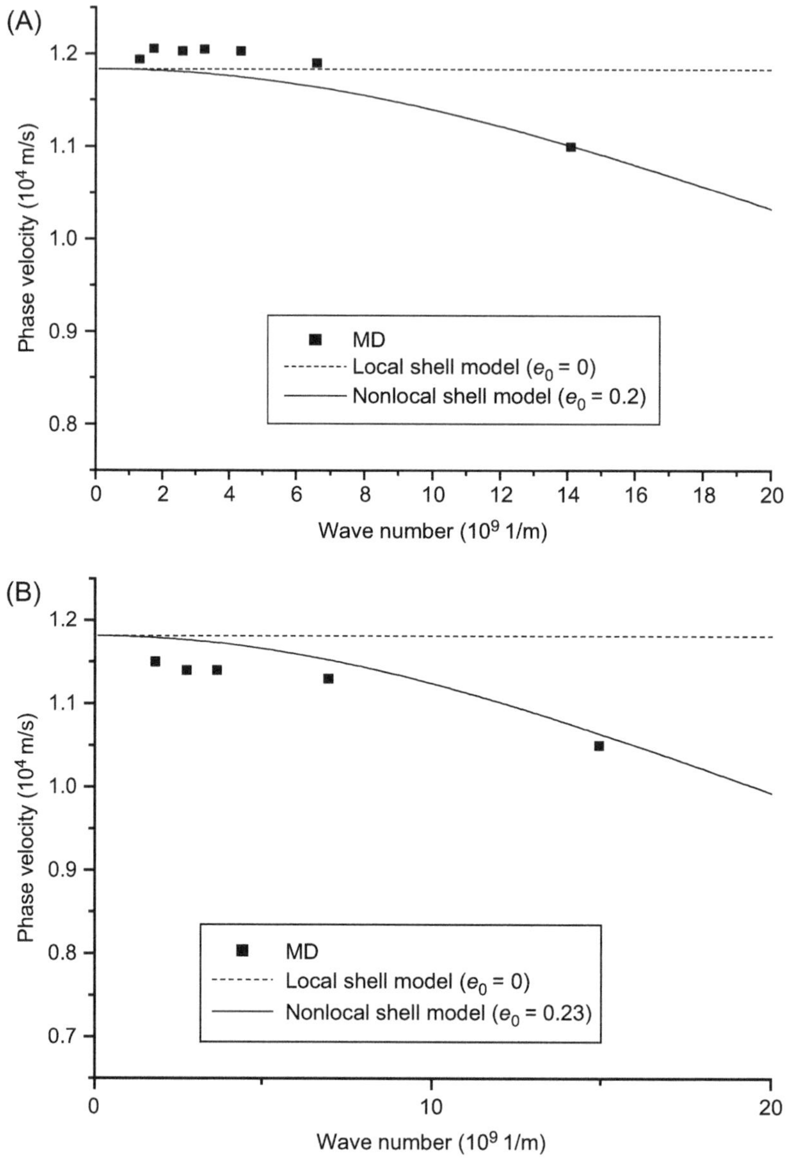

FIGURE 7.23

(A) Dispersion relation of torsional wave in the armchair (10, 10) SWCNT. (B) Dispersion relation of torsional wave in the armchair (15, 15) SWCNT [73].

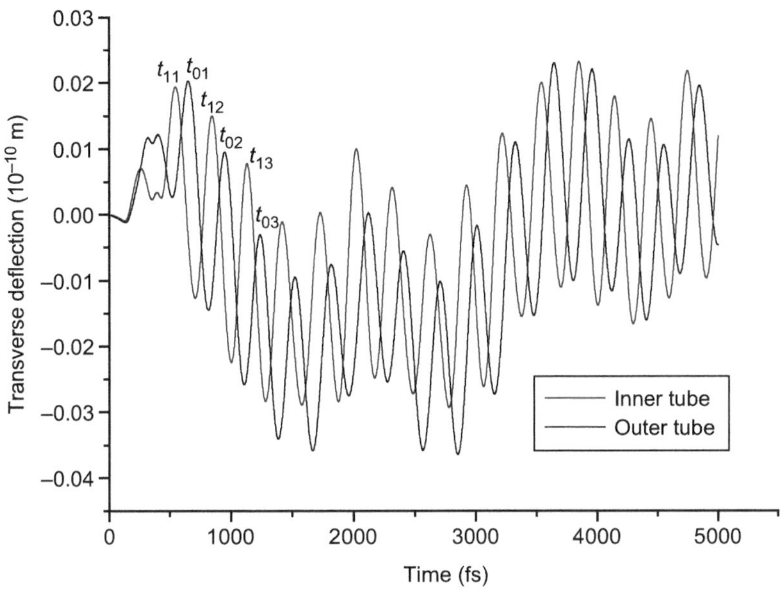

FIGURE 7.24

The time histories of the deflection of atoms at the section, 2.46 nm ahead of the end of the armchair–armchair (10, 10)@(15, 15) DWCNT. The sinusoidal wave of period $T = 300$ fs input at the end of the DWCNT [73].

displacements in longitudinal, radial, and tangential direction. The carbon atoms more complicated vibration in the transverse wave propagation in CNT leads to the stronger scale effect on the dispersion of the transverse wave. This is the main reason why the vale of $e_0 = 0.2$ for the torsional wave in SWCNT is much smaller than the values of $e_0 = 0.6$ used for transverse wave in CNTs. The structure of CNTs, such as chirality, diameter, and layers, may also be the factors to affect the value of e_0. But for the same problems, such as transverse wave or torsional wave, in CNT, these structure differences lead to a relatively small change of the value of e_0 compared to the changes of values for different problems.

Fig. 7.24 indicates that the obvious phase difference of inner and outer tube of the (10, 10)@(15, 15) DWCNT will occur at the period $T = 300$ fs. The corresponding angular frequency is $\omega = 2\pi/T \approx 3.14 \times 10^{13}$ rad/s. The average phase difference between the inner tube and the outer tube can be estimated from Fig. 7.8 as $\bar{t} = \frac{1}{n}\sum_{i=1}^{n}(t_{0i} - t_{1i}) \approx 103.2$ fs. Fig. 7.25 indicates that the SWCNTs (10, 10) and (15, 15) have different phases at the period $T = 300$ fs without considering the van der Waals forces between them. The average phase difference between the SWCNTs (10, 10) and (15, 15) can also be estimated from Fig. 7.25: $\bar{s} = \frac{1}{n}\sum_{i=1}^{n}(s_{0i} - s_{1i}) \approx 101.2$ fs. The other average phase differences at different periods are shown in Table 7.6. The phase difference of the vibration becomes increasingly obvious with the decrease of the period. The phase difference increases from 43.3 to 103.2 fs as the period decreases from 700 to 300 fs for DWCNTs. Similarly,

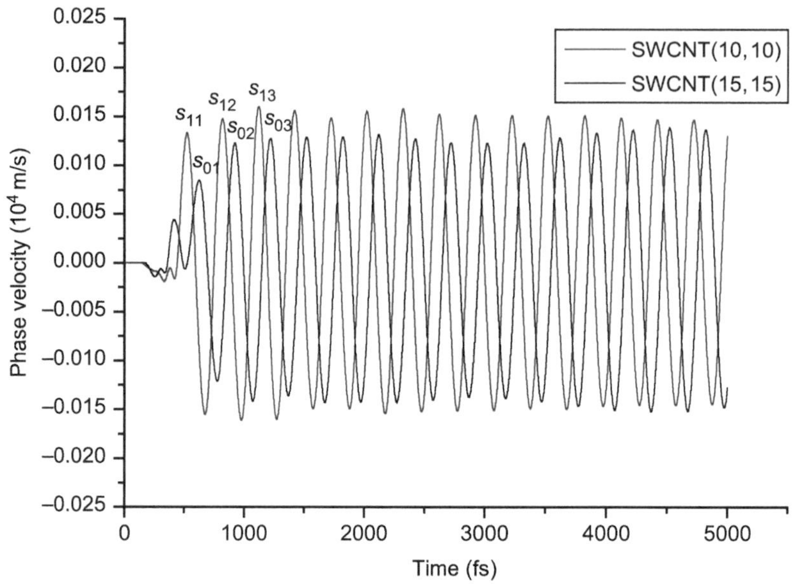

FIGURE 7.25

The time histories of the deflection of atoms at the section, 2.46 nm ahead of the end of the SWCNT (10, 10) and SWCNT (15, 15). The sinusoidal wave of period $T = 300$ fs input at the end of the SWCNTs [73].

Table 7.6 The Average Phase Differences of the Vibrations of Atom at the Section, 2.46 nm Ahead of the End of the DWCNT (10, 10)@(15, 15) and the SWCNTs (10, 10) and (15, 15) [73]

Period (fs)	DWCNT (10, 10)@(15, 15) (fs)	SWCNTs (10, 10) and (15, 15) (fs)	$\lvert\bar{s}-\bar{t}\rvert$ (fs)
$T = 300$	$\bar{t} = 103.2$	$\bar{s} = 101.2$	2
$T = 400$	$\bar{t} = 80.3$	$\bar{s} = 74.1$	6.2
$T = 500$	$\bar{t} = 59.8$	$\bar{s} = 38.8$	21
$T = 600$	$\bar{t} = 54$	$\bar{s} = 29$	25
$T = 700$	$\bar{t} = 43.3$	$\bar{s} = 40$	3.3

the phase difference increases from 40 to 101.2 fs as the period decreases from 700 to 300 fs for SWCNTs. The van der Waals forces make no significant contribution to the phase differences of the transverse wave in the inner and outer tube, as the phase differences would occur without consideration of the van der Waals forces between the adjacent tubes. This indicates that the small-scale effects can change the phase when the frequency is extremely high.

In summary, the study of transverse and torsional wave dispersion in SWCNTs and DWCNTs on the basis of the nonlocal elastic cylindrical shell theory and the use of MD simulations for SWCNT and an armchair–armchair (10, 10)@(15, 15) DWCNT in a wide range of wavenumbers are presented. The value of parameter e_0 is estimated based on the MD result to predict the dispersion of a transverse wave in CNTs through using the nonlocal shell models. The results of this study indicate that the nonlocal elastic cylindrical shell model is able to offer a better prediction for transverse and torsional wave dispersion in CNTs than the local elastic shell model when the wavenumber is large enough. The nonlocal shell models are required when the wavelengths are approximately less than 2.36×10^{-9} m and 0.95×10^{-9} m for a transverse wave in armchair (15, 15) SWCNTs and a torsional wave in armchair (10, 10) SWCNTs, respectively. These small-scale effects have a significant effect on the phase difference of a transverse wave in the inner and outer tubes of DWCNTs at high frequency.

REFERENCES

[1] Iijima S. Helical microtubules of graphitic carbon. Nature 1991;354(6348):56–8.
[2] Ball P. Roll up for the revolution. Nature 2001;414(6860):142–4.
[3] Baughman RH, Zakhidov AA, de Heer WA. Carbon nanotubes—the route toward applications. Science 2002;297(5582):787–92.
[4] Lau KT. Interfacial bonding characteristics of nanotube/polymer composites. Chem Phys Lett 2003; 370(3–4):399–405.
[5] Iijima S, Brabec C, Maiti A, Bernholc J. Structural flexibility of carbon nanotubes. J Chem Phys 1996;104(5):2089–92.
[6] Hernandez E, Goze C, Bernier P, Rubio A. Elastic properties of C and B x C y N z composite nanotubes. Phys Rev Lett 1998;80(20):4502.
[7] Sánchez-Portal D, Artacho E, Soler JM, Rubio A, Ordejón P. Ab initio structural, elastic, and vibrational properties of carbon nanotubes. Phys Rev B 1999;59(19):12678–88.
[8] Yakobson BI, Campbell MP, Brabec CJ, Bernholc J. High strain rate fracture and C-chain unraveling in carbon nanotubes. Comp Mater Sci 1997;8(4):341–8.
[9] Liew KM, Wong CH, He XQ, Tan MJ, Meguid SA. Nanomechanics of single and multiwalled carbon nanotubes. Phys Rev B 2004;69(11):115429.
[10] Liew KM, Wei JX, He XQ. Carbon nanocones under compression: buckling and post-buckling behaviors. Phys Rev B 2007;75(19).
[11] Wang Q, Liew KM. Molecular mechanics modeling for properties of carbon nanotubes. J Appl Phys 2008;103(4).
[12] Li CY, Chou TW. Elastic wave velocities in single-walled carbon nanotubes. Phys Rev B 2006;73(24).
[13] Yakobson BI, Brabec CJ, Bernholc J. Nanomechanics of carbon tubes: instabilities beyond linear response. Phys Rev Lett 1996;76(14):2511–14.
[14] Krishnan A, Dujardin E, Ebbesen T, Yianilos P, Treacy M. Young's modulus of single-walled nanotubes. Phys Rev B 1998;58(20):14013.
[15] Ru CQ. Effective bending stiffness of carbon nanotubes. Phys Rev B 2000;62(15):9973–6.
[16] Parnes R, Chiskis A. Buckling of nano-fibre reinforced composites: a re-examination of elastic buckling. J Mech Phys Solids 2002;50(4):855–79.
[17] Ru CQ. Elastic buckling of single-walled carbon nanotube ropes under high pressure. Phys Rev B 2000;62(15):10405–8.
[18] Liew KM, Sun YZ. Elastic properties and pressure-induced structural transitions of single-walled carbon nanotubes. Phys Rev B 2008;77(20):205437.
[19] Wong EW, Sheehan PE, Lieber CM. Nanobeam mechanics: elasticity, strength, and toughness of nanorods and nanotubes. Science 1997;277(5334):1971–5.

[20] Lourie O, Cox D, Wagner H. Buckling and collapse of embedded carbon nanotubes. Phys Rev Lett 1998;81(8):1638.

[21] Govindjee S, Sackman JL. On the use of continuum mechanics to estimate the properties of nanotubes. Solid State Commun 1999;110(4):227–30.

[22] Liew KM, He XQ, Wong CH. On the study of elastic and plastic properties of multi-walled carbon nanotubes under axial tension using molecular dynamics simulation. Acta Mater 2004;52(9):2521–7.

[23] Liew KM, Wong CH, Tan MJ. Tensile and compressive properties of carbon nanotube bundles. Acta Mater 2006;54(1):225–31.

[24] He X, Kitipornchai S, Liew K. Buckling analysis of multi-walled carbon nanotubes: a continuum model accounting for van der Waals interaction. J Mech Phys Solids 2005;53(2):303–26.

[25] Ru C. Column buckling of multiwalled carbon nanotubes with interlayer radial displacements. Physical Review B 2000;62(24):16962.

[26] Wang Q. Instability analysis of double-walled carbon nanotubes subjected to axial compression. J Appl Phys 2008;104(3):036102.

[27] Wang Q. Torsional buckling of double-walled carbon nanotubes. Carbon N Y 2008;46(8):1172–4.

[28] Wang Q, Duan W, Liew K, He X. Inelastic buckling of carbon nanotubes. Appl Phys Lett 2007; 90(3):033110.

[29] Wang Q, Liew KM, He X, Xiang Y. Local buckling of carbon nanotubes under bending. Appl Phys Lett 2007;91(9):093128.

[30] Zhang Y, Wang C, Duan W, Xiang Y, Zong Z. Assessment of continuum mechanics models in predicting buckling strains of single-walled carbon nanotubes. Nanotechnology 2009;20(39):395707.

[31] Wang Q, Liew K. Application of nonlocal continuum mechanics to static analysis of micro-and nano-structures. Phys Lett A 2007;363(3):236–42.

[32] Yoon J, Ru C, Mioduchowski A. Sound wave propagation in multiwall carbon nanotubes. J Appl Phys 2003;93(8):4801–6.

[33] Mitra M, Gopalakrishnan S. Vibrational characteristics of single-walled carbon-nanotube: time and frequency domain analysis. J Appl Phys 2007;101(11):114320.

[34] Natsuki T, Ni Q-Q, Endo M. Wave propagation in single-and double-walled carbon nanotubes filled with fluids. J Appl Phys 2007;101(3):034319.

[35] Zhang Y, Liu G, Xie X. Free transverse vibrations of double-walled carbon nanotubes using a theory of nonlocal elasticity. Physical Review B 2005;71(19):195404.

[36] Yoon J, Ru C, Mioduchowski A. Timoshenko-beam effects on transverse wave propagation in carbon nanotubes. Composites Part B: Eng 2004;35(2):87–93.

[37] Yoon J, Ru C, Mioduchowski A. Noncoaxial resonance of an isolated multiwall carbon nanotube. Phys Rev B 2002;66(23):233402.

[38] Natsuki T, Endo M, Tsuda H. Vibration analysis of embedded carbon nanotubes using wave propagation approach. J Appl Phys 2006;99(3):034311.

[39] Liew KM, Wang Q. Analysis of wave propagation in carbon nanotubes via elastic shell theories. Int J Eng Sci 2007;45(2-8):227–41.

[40] Hu YG, Liew KM, Wang Q. Nonlocal elastic beam models for flexural wave propagation in double-walled carbon nanotubes. J Appl Phys 2009;106(4):044301.

[41] Hu Y-G, Liew KM, Wang Q, He X, Yakobson B. Nonlocal shell model for elastic wave propagation in single-and double-walled carbon nanotubes. J Mech Phys Solids 2008;56(12):3475–85.

[42] Wang L, Hu H, Guo W. Validation of the non-local elastic shell model for studying longitudinal waves in single-walled carbon nanotubes. Nanotechnology 2006;17(5):1408.

[43] Wang Q, Varadan V. Wave characteristics of carbon nanotubes. Inter J Solids Struct 2006; 43(2):254–65.

[44] Wang Q, Varadan VK. Application of nonlocal elastic shell theory in wave propagation analysis of carbon nanotubes. Smart Mater Struct 2007;16(1):178–90.
[45] Li C, Chou T-W. Single-walled carbon nanotubes as ultrahigh frequency nanomechanical resonators. Phys Rev B 2003;68(7):073405.
[46] Wang C, Tan V, Zhang Y. Timoshenko beam model for vibration analysis of multi-walled carbon nanotubes. J Sound Vibration 2006;294(4):1060–72.
[47] Wang C, Zhang Y, He X. Vibration of nonlocal Timoshenko beams. Nanotechnology 2007; 18(10):105401.
[48] Yoon J, Ru C, Mioduchowski A. Vibration of an embedded multiwall carbon nanotube. Composites Sci Technol 2003;63(11):1533–42.
[49] Yoon J, Ru C, Mioduchowski A. Terahertz vibration of short carbon nanotubes modeled as Timoshenko beams. J Appl Mechan 2005;72(1):10–17.
[50] Gibson RF, Ayorinde EO, Wen Y-F. Vibrations of carbon nanotubes and their composites: a review. Composites Sci Technol 2007;67(1):1–28.
[51] Poncharal P, Wang Z, Ugarte D, De Heer WA. Electrostatic deflections and electromechanical resonances of carbon nanotubes. Science 1999;283(5407):1513–16.
[52] Natsuki T, Ni QQ, Endo M. Wave propagation in single- and double-walled carbon nanotubes filled with fluids. J Appl Phys 2007;101(3).
[53] Insepov Z, Wolf D, Hassanein A. Nanopumping using carbon nanotubes. Nano Lett 2006;6(9):1893–5.
[54] Sun CT, Zhang HT. Size-dependent elastic moduli of platelike nanomaterials. J Appl Phys 2003;93 (2):1212–18.
[55] Sheehan PE, Lieber CM. Nanotribology and nanofabrication of MoO3 structures by atomic force microscopy. Science 1996;272(5265):1158–61.
[56] Wang Q. Effective in-plane stiffness and bending rigidity of armchair and zigzag carbon nanotubes. Int J Solids Struct 2004;41(20):5451–61.
[57] Wang LF, Hu HY. Flexural wave propagation in single-walled carbon nanotubes. Phys Rev B 2005; 71(19):195412.
[58] Brenner DW. Empirical potential for hydrocarbons for use in simulating the chemical vapor deposition of diamond films. Phys Rev B 1990;42:9458.
[59] Liew KM, Hu YG, He XQ. Flexural wave propagation in single-walled carbon nanotubes. J Comput Theor Nanos 2008;5(4):581–6.
[60] Brenner DW, Shenderova OA, Harrison JA, Stuart SJ, Ni B, Sinnott SB. A second-generation reactive empirical bond order (REBO) potential energy expression for hydrocarbons. J Phys Condensed Matter 2002;14(4):783.
[61] Sharma P, Ganti S, Bhate N. Effect of surfaces on the size-dependent elastic state of nano-inhomogeneities. Appl Phys Lett 2003;82(4):535–7.
[62] Eringen A. Nonlocal polar field models. New York: Academic; 1976.
[63] Eringen AC. On differential equations of nonlocal elasticity and solutions of screw dislocation and surface waves. J Appl Phys 1983;54(9):4703–10.
[64] Peddieson J, Buchanan GR, McNitt RP. Application of nonlocal continuum models to nanotechnology. Int J Eng Sci 2003;41(3–5):305–12.
[65] Sudak LJ. Column buckling of multiwalled carbon nanotubes using nonlocal continuum mechanics. J Appl Phys 2003;94(11):7281–7.
[66] Zhang YQ, Liu GR, Wang JS. Small-scale effects on buckling of multiwalled carbon nanotubes under axial compression. Phys Rev B 2004;70(20).
[67] Wang Q. Wave propagation in carbon nanotubes via nonlocal continuum mechanics. J Appl Phys 2005;98(12).

[68] Wood W. Practical time-stepping schemes. Oxford: Oxford University Press; 1990.
[69] Hsieh JY, Lu JM, Huang MY, Hwang CC. Theoretical variations in the Young's modulus of single-walled carbon nanotubes with tube radius and temperature: a molecular dynamics study. Nanotechnology 2006;17(15):3920–4.
[70] Sears A, Batra R. Macroscopic properties of carbon nanotubes from molecular-mechanics simulations. Phys Rev B 2004;69(23):235406.
[71] Hu YG, Liew KM, Wang Q. Nonlocal continuum model and molecular dynamics for free vibration of single-walled carbon nanotubes. J Nanosci Nanotechno 2011;11(12):10401–7.
[72] Jones JE. On the determination of molecular fields. I. From the variation of the viscosity of a gas with temperature. Proceedings of the Royal Society of London A: Mathematical, Physical and Engineering Sciences: The Royal Society; pp. 441–462.
[73] Hu YG, Liew KM, Wang Q, He XQ, Yakobson BI. Nonlocal shell model for elastic wave propagation in single- and double-walled carbon nanotubes. J Mech Phys Solids 2008;56(12):3475–85.
[74] Flugge W. Stresses in shells. Berlin, Heidelberg: Springer-Verlag; 1960.
[75] Adams GB, Sankey OF, Page JB, O'Keeffe M, Drabold DA. Energetics of large fullerenes: balls, tubes, and capsules. Science (New York, NY) 1992;256(5065):1792–5.
[76] Kudin KN, Scuseria GE, Yakobson BI. C2F, BN, and C nanoshell elasticity from ab initio computations. Phys Rev B 2001;64(23):235406.
[77] Wang CY, Ru CQ, Mioduchowski A. Axisymmetric and beamlike vibrations of multiwall carbon nanotubes. Phys Rev B 2005;72(7):075414.

CHAPTER

TECHNOLOGICALLY RELEVANT APPLICATIONS

8

8.1 INTRODUCTION

Carbon nanotubes (CNTs) hold significant promise for potential applications such as fluid transport, fluid storage, biosensors, and drug-delivery devices, in view of their perfect hollow cylindrical structure and superior elastic properties. CNTs offer high diffusion coefficients for liquids as well as gases due to the natural smoothness of their inner walls [1–5]. The simulations and experimental results in recent years have proven that the molecular transport through CNTs is far in excess of what is predicted by continuum hydrodynamics theories [6–9]. Studies of the elastic properties of CNTs are numerous in the literature, whereas studies on fluid conveying CNTs [3,6,8,10–12] and fluid flow around CNTs [13,14] are relatively scarce. The behavior of the fluid inside CNTs is expected to be fundamentally different from the behavior of the fluid in macro/microsystems at low velocities because of the very small diameter of CNTs, and therefore it offers itself as a significant and challenging research topic in nanoelectromechanical systems (NEMS) [10,15]. Dynamic analysis has been conducted by researchers to study the behavior of CNTs under a wide range of fluid flow velocities from 0 to 100,000 m/s using elastic beam models and molecular dynamics (MD) simulations [8,10,11,16]. The simulations give insight into the direction of fluid flow measurements in single-walled CNTs (SWCNTs).

Water molecules confined in nanoscale pores usually have different transport behaviors from those of bulk system [17–21]. The molecular transport in nanoporous media raises many interesting theoretical problems [22]. Understanding the transport of water molecules through CNT is an important practical issue in biological science [23–28], because the transport behavior of water through CNT is very similar to the flow of water in the cellular membrane of biological systems. Hummer and coworkers [19] observed the spontaneous and continuous filling of CNT with a one-dimensional ordered chain of water. Joseph and Aluru [29] used MD simulation to investigate water flow through different types of tubes and showed that the enhanced flow rate was due to the smooth surface of the nanotube. Rossi et al. [30] studied the flow and condensation of water inside CNT by means of environmental scanning electron microscopy (SEM). He suggested that the slow flow of liquid is a result of the confinement of water and its strong interaction with tube walls. Although much progress has been achieved in both the theoretical and experimental studies on the nanofluidic properties of water through CNTs, some fundamental questions like the nano-hydromechanics of water flow through CNTs remain unanswered. Hence, it is valuable to design a functional device [31,32] to control the transport of water molecules through CNTs. Pressure gradients were imposed on a nanoscale system to drive the water transport through CNTs [33], where pressure difference of two ends of SWCNT is up to 60 atm. Zambrano et al. [34] used thermal gradients to induce water transport through SWCNTs, in which the temperature difference imposed on two ends of SWCNT reached up to 140K. High pressure and temperature gradient imposed on the

Mechanical Behaviors of Carbon Nanotubes. DOI: http://dx.doi.org/10.1016/B978-0-323-43137-8.00008-1

given systems would change the physical and chemical properties of water (even phase transition), which affect the accuracy of the research results. Electric current applied to the SWCNT is a good method to drive water flow through nanotubes [35]. However, this experiment is difficult to perform because the required apparatus is complicated. Moreover, the water molecules are possible to be electrolyzed by electric current. Manipulation of the flow of water molecules through CNTs is a challenging issue. Until now, this issue has not been well tackled. Few research results have been reported on the transport of water molecules driven by methane molecules.

In this chapter, the possibility of fluid flow measurement in fluid-conveying SWCNTs is investigated using the free vibration analysis. MD simulation is performed to investigate how the methane molecules manipulate the transport of water molecules through SWCNTs and diameter-gradient SWCNTs with junctions.

8.2 CONVEYING FLUID

The free vibration analysis of fluid-conveying SWCNTs is performed by assuming a frictionless contact between the fluid and SWCNT wall. The objective is to examine the effects of an internal moving fluid on the free vibration natural frequency, mode shape, and various forces, i.e., stiffness, inertia, Coriolis, and centrifugal, as well as the structural instability of SWCNTs.

The SWCNT in the present analysis is treated as a perfect hollow cylindrical tube having an equivalent bending rigidity EI with a mass m_c per unit length, as shown in Fig. 8.1. Let v be the uniform mean flow velocity of the fluid and m_f be the mass of fluid per unit length of the SWCNT. The governing differential equation of motion for the fluid conveying tube can be expressed as [37]

$$(m_f + m_c)\frac{\partial^2 u}{\partial t^2} + 2m_f v\frac{\partial^2 u}{\partial x \partial t} + m_f v^2\frac{\partial^2 u}{\partial x^2} + EI\frac{\partial^4 u}{\partial x^4} = 0 \tag{8.1}$$

where the four terms on the left side correspond to the inertial, Coriolis/gyroscopic, centrifugal, and stiffness forces, respectively, the last term is entirely related to the structure SWCNT, the second and third terms are entirely related to the fluid flow, while the first term is related to both structure and fluid. The bending rigidity EI in Eq. (8.1) is an input parameter for a given fluid and size of a SWCNT. One needs to know the exact values of Young's modulus and wall thickness of a SWCNT to compute the bending rigidity. It is unclear as to how one can choose appropriate values in view of wide scattered data reported in the literature. Hence, a different approach is

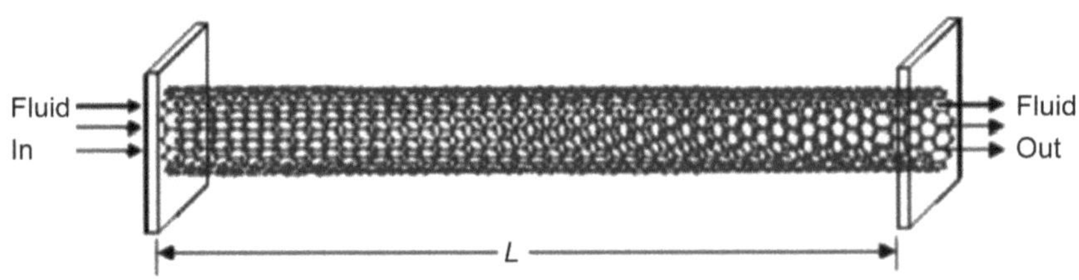

FIGURE 8.1

SWCNT conveying fluid of mass m_f per unit length with velocity v [36].

followed here: an equivalent bending rigidity is computed instead of using the assumed values of Young's modulus and wall thickness from literature. The stiffness matrix of the SWCNT is first computed using the finite difference method. Next, a free vibration analysis is performed using the obtained stiffness matrix and the lumped mass matrix of carbon atoms to compute the fundamental natural frequencies, where the mass of the fluid is set to be zero in Eq. (8.1). The relationship between the fundamental natural frequency f and the bending rigidity *EI* can be expressed as $f = (\beta^2/2\pi L^2)\sqrt{EI/m_c}$ for a fixed–fixed SWCNT, where $\beta = 1.506\pi$ for the fundamental mode. The bending rigidity is therefore obtained as $m_c(2\pi L^2 f/\beta^2)^2$. The computed *EI* values for various diameters of SWCNTs are listed in Table 8.1.

Eq. (8.1) represents a linear homogeneous system that can be solved for eigenvalues and eigenvectors by recasting it into an equivalent state–space form. The computed natural frequencies depend on the size of the SWCNT and the mass flow rate of the fluid. Nondimensional parameters, i.e., the mass ratio m_r and the dimensionless velocity v_n are introduced for convenience, i.e., $m_r = m_f/(m_f + m_c)$ and $v_n = \sqrt{m_f/(EI)}vL$. The natural frequencies of the fluid conveying SWCNT are computed by varying these nondimensional parameters. The fluid being conveyed is assumed to be water. It is also assumed that no fluid is outside the SWCNT. The approach presented here is not limited to water-conveying SWCNTs and can be extended to any fluid. Fig. 8.2 shows the variation of the natural frequency with respect to the dimensionless velocity of the fluid for the lowest three natural modes of a fluid conveying SWCNT. The computed values of the first, second, and third frequencies at zero velocity of the fluid are $3.2564 \times 10^{10}, 8.9765 \times 10^{10}$, and 1.7598×10^{11}Hz respectively. The centrifugal force increases and the resultant stiffness decreases as the velocity of the fluid increases, this is because the centrifugal force acts in the opposite direction to the stiffness force, and as a result the natural frequencies decrease with an increase in the fluid velocity. The dimensionless velocity at which the fundamental natural frequency becomes zero is defined as the critical velocity v_{cr} and is equal to 2π, as shown in Fig. 8.2. The structure is unstable at this velocity because the resultant stiffness is zero.

The magnitudes of inertial, Coriolis, centrifugal, and stiffness forces are computed along the length of the SWCNT at various fluid velocities. The distribution of these forces normalized with respect to the stiffness force is plotted against the nondimensional length of the SWCNT, as shown in Fig. 8.3 at 20% of the critical flow velocity. It is observed from Fig. 8.3 that the inertial force curve follows the fundamental mode shape, and the stiffness force curve is slightly perturbed from the fundamental mode shape due to the centrifugal force. The Coriolis force curve showed that the force acts in one direction in the first half of the length of the SWCNT and in the opposite direction in another half. The stiffness forces are supposed to be zero at the ends of the SWCNT for a free

Table 8.1 μ1 Computed Bending Rigidity Values for SWCNTs [36]

SWCNT Type	Diameter (nm)	Bending Rigidity *EI* ($N \cdot m^2$)
(5, 0)	0.415	4.7102×10^{-27}
(8, 0)	0.65	1.9770×10^{-26}
(12, 0)	0.961	7.7374×10^{-26}
(14, 0)	1.12	1.1122×10^{-25}
(16, 0)	1.28	1.6867×10^{-25}

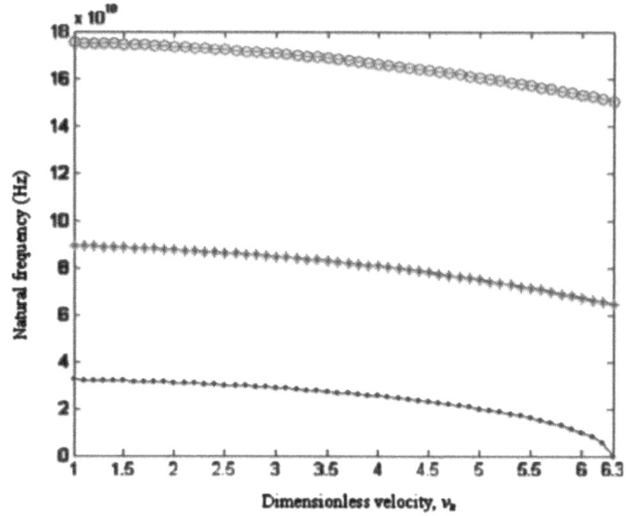

FIGURE 8.2

Variation of dimensionless frequencies with dimensionless velocity for the lowest three modes. Dots, stars, and circles represent first, second, and third mode natural frequencies, respectively [36].

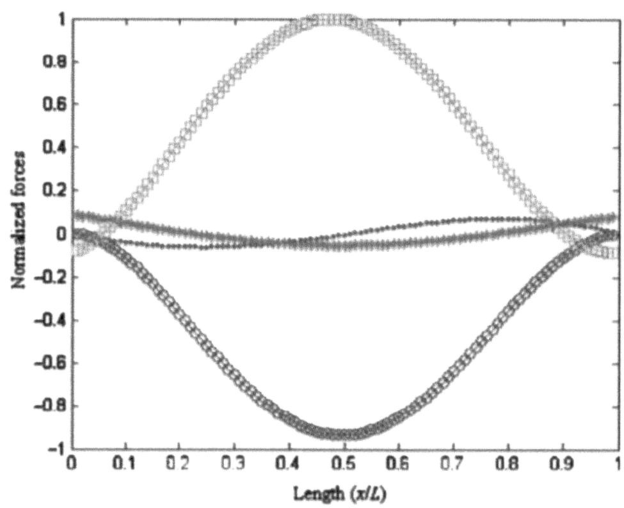

FIGURE 8.3

Variation of normalized forces with the length of the CNT. Circles, squares, dots, and stars represents inertial, stiffness, Coriolis, and centrifugal forces, respectively [36].

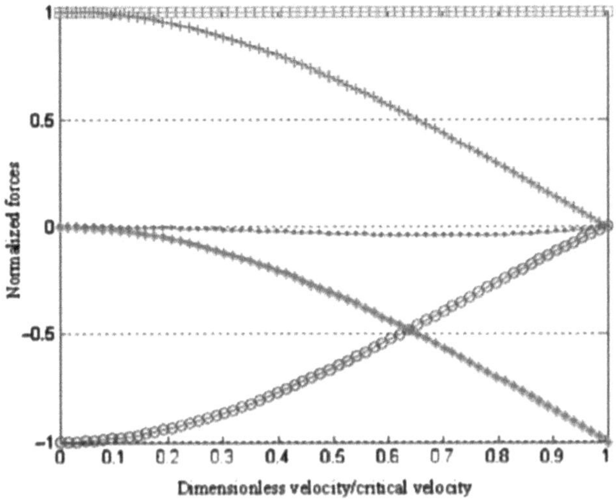

FIGURE 8.4

Variation of forces with fluid velocity at $0.5L$ from one end of the SWCNT. Circles, squares, dots, stars, and plus symbols represent inertial stiffness, Coriolis, centrifugal, and effective stiffness forces, respectively [36].

vibration mode without fluid flow, but are seen to be nonzero here because there is a centrifugal force acting in the opposite direction and the resultant of these two forces is zero.

The magnitude of various forces is studied at a midsection, $0.5L$ from the fixed end, of the SWCNT by increasing the velocity ratio, defined as the ratio of the dimensionless velocity to the critical velocity, in the range of 0–1 and is shown in Fig. 8.4. All forces are normalized with respect to the stiffness force. The inertial and stiffness forces are equal in magnitude and opposite in direction when the fluid velocity is zero, and the Coriolis and centrifugal forces are zero. Perturbation forces increase as the velocity of the fluid increases, which cause a flow induced structural instability in the SWCNT. The effective stiffness force, i.e., the resultant of the stiffness force and centrifugal force, converges to zero as the velocity of the fluid tends to be the critical velocity, which leads to the instability of the structure. The Coriolis force increases linearly with the velocity well below the critical velocity. This force is used for the mass flow measurement in the Coriolis mass flowmeters.

The change in the mass flow rate in Coriolis mass flowmeters can be determined by measuring the phase difference using two electromagnetic detectors mounted at appropriate locations on the pipe [38]. Flow measurement in NEMS devices is a difficult and challenging task in practice, and flow measurement techniques in nanodevices are yet to be developed. The solution of Eq. (8.1) contains complex eigenvalues and eigenvectors. These complex eigenvectors can be used to compute the mass flow rate. The phase $\varnothing$ is the inverse tangent of the ratio of the imaginary value to the real value of the complex eigenvector corresponding to the fundamental mode. Mass flow rate can be determined by measuring the phase difference between two appropriate points on the SWCNT [38]. The computed phase varies linearly along the length of the SWCNT [39]. Two appropriate points for phase measurement in the present study are chosen to be $0.25L$ and $0.75L$

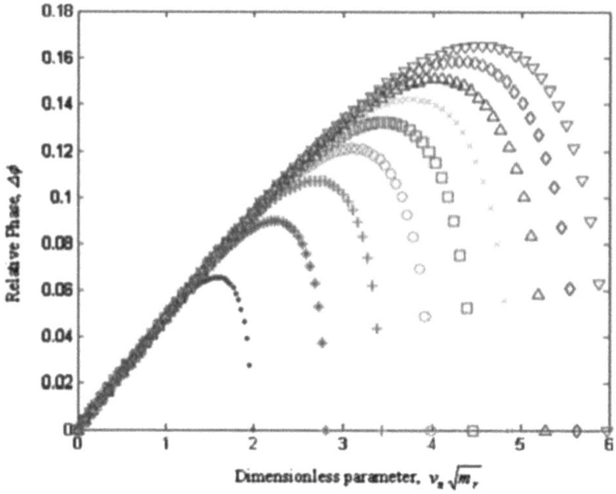

FIGURE 8.5

Relative phase variation with the dimensionless parameter $vn\sqrt{m_r}$. The symbols from dot to downward triangle represent the mass ratio m_r varying 0.1–0.9, respectively [36].

from one end of the SWCNT. The magnitude of relative phase, $\Delta\varnothing$ between these two points depends on the mass and velocity of the flowing fluid. Numerical experiments are conducted to compute the relative phase by changing the density of the fluid, i.e., m_r is varied from 0.1 to 0.9, and changing the velocity of the fluid, i.e., v_n is varied from 0 to 2π. It is observed that the computed relative phase has a linear relationship with the dimensionless parameter $v_n\sqrt{m_r}$ for low fluid velocities, as shown in Fig. 8.5, and this linear range can be exploited for the fluid flow measurement. The relationship between the relative phase and the dimensionless parameter for low fluid velocities can be expressed using a linear curve fit as $\Delta\varnothing = 0.0483 v_n\sqrt{m_r}$ and $\Delta\varnothing$ can be used to measure the mass flow rate of the fluid in the range of 0–40% of its critical flow velocity.

8.2.1 DRIVING WATER MOLECULES ALONG AN SWCNT

The realization of an effective transport technique may guide us to explore its application for a nanoscale pump. A nanoscale pump incorporated into the CNT systems would be superior to other porous materials systems in many aspects due to the flexibility, homogeneity, and controllability of the structural parameters, as well as the unusual transport properties in SWCNTs [24,6].

MD simulations are performed using the MATERIALS STUDIO molecular modeling software packages. The COMPASS force field is adopted to model the interaction between molecules in system [40]. The coordinates of the CNTs are fixed. The type of nanotube is zigzag (n, 0) SWCNTs with indices $n = 12$, 14, 16, and 18, whose diameters are 9.39, 10.96, 12.53, and 14.09 Å, respectively. The length of all nanotubes is 17.04 Å. The numbers of methane molecules are designated as 40, 50, 60, 70, 80, and 120, respectively. The cut-off distance is set to be 10.0 Å. The time-step is 1 fs. Data are recorded at each 0.5 ps for further analysis.

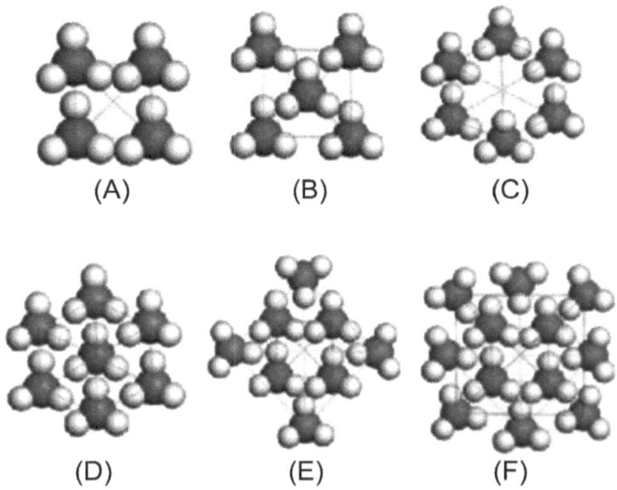

FIGURE 8.6

Top views of arrays of different numbers of methane molecules. (A) 40 methane molecules; (B) 50 methane molecules; (C) 60 methane molecules; (D) 70 methane molecules; (E) 80 methane molecules; (F) 120 methane molecules [41].

Firstly, 20 water molecules are inserted randomly into the interior of the SWCNT, and then 5000 time-steps are run to relax the initial configuration in order to obtain the lowest energy structure. Subsequently, methane molecules are arranged at one gate of the SWCNT in an array regularly and symmetrically to draw these water molecules out from the inner of the SWCNT. Fig. 8.6 shows the schematic drawing of the arrays. The center lines of arrays of methane molecules are identical with the axis of the SWCNT. As for 40 methane molecules, molecules are arranged in a $2 \times 2 \times 10$ array illustrated in Fig. 8.6A. Based on 40 methane molecules, 10 methane molecules are inserted into the center of the array of 40 methane molecules to construct the array of 50 methane molecules. Similarly, other arrays of 60, 70, 80, and 120 methane molecules are also fabricated according to the Fig. 8.6C–F.

Fig. 8.7 shows overall pictures of the moving process of water molecules extracted from the SWCNTs caused by 40 methane molecules. Water molecules in the nanotube start to move to an end of the SWCNT. Simultaneously, the initial ordered configuration of methane molecules becomes disordered because of diffusion. The water molecules are drawn out to a certain distance L as shown in Fig. 8.7B. It takes 180 ps to draw all water molecules out completely from the SWCNT (as shown in Fig. 8.7C). Finally, the nanotube is empty. The methane molecules are responsible for the driving force, therefore, the effect of methane molecules on the translocation of water molecules becomes important. It is observed that the transport behavior of water molecules driven by 60 methane molecules as shown in Fig. 8.8 is much different from that driven by 40 methane molecules as shown in Fig. 8.7.

At the same simulation time 55 ps, 60 methane molecules make the water molecules move 11.798 Å shown in Fig. 8.8A while 40 methane molecules make them move 6.794 Å shown in

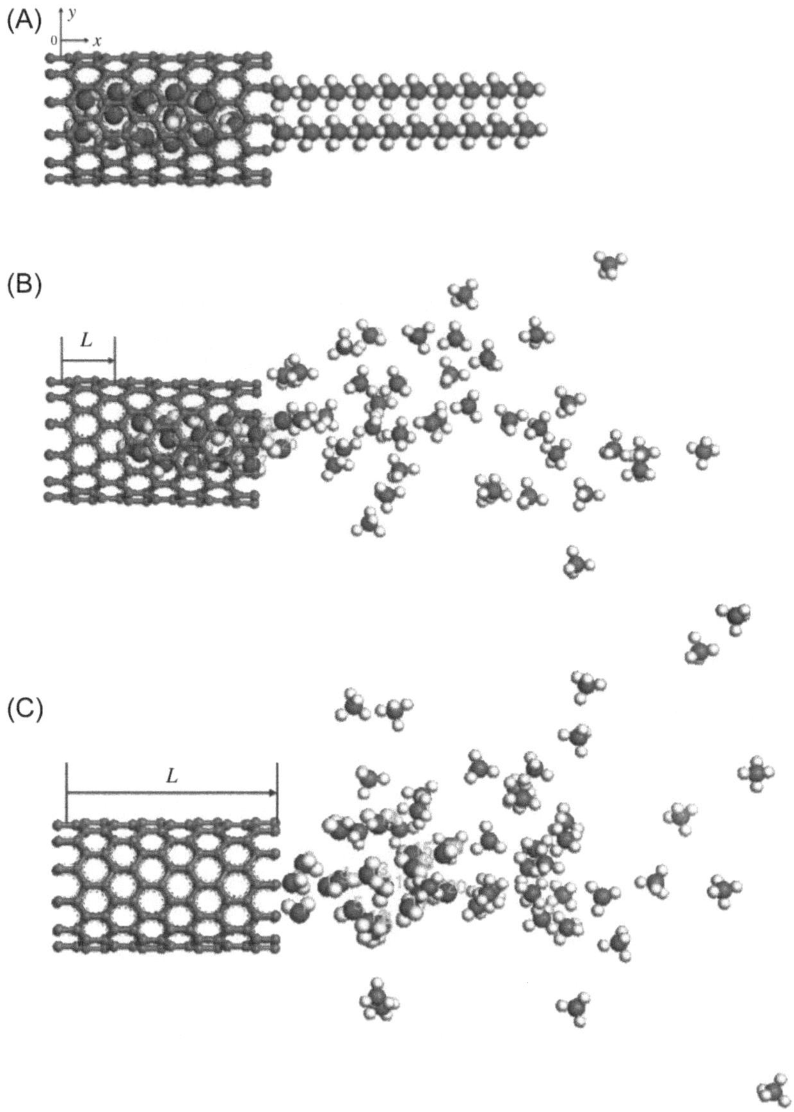

FIGURE 8.7

Snapshots of water molecules drawn out from (12, 0) SWCNT by 40 methane molecules. (A) $t = 0$ ps; (B) $t = 55$ ps; (C) $t = 180$ ps. More clearly, SWCNT is colored in gray, hydrogen atoms in white spheres, oxygen atoms in black (red in web version), and carbon atoms of methane in dark grey (blue in web version). Water molecules are labeled by numbers in grey (green in web version) for the trajectory observation. L is the displacement distance for water molecules from the initial site 0 [41].

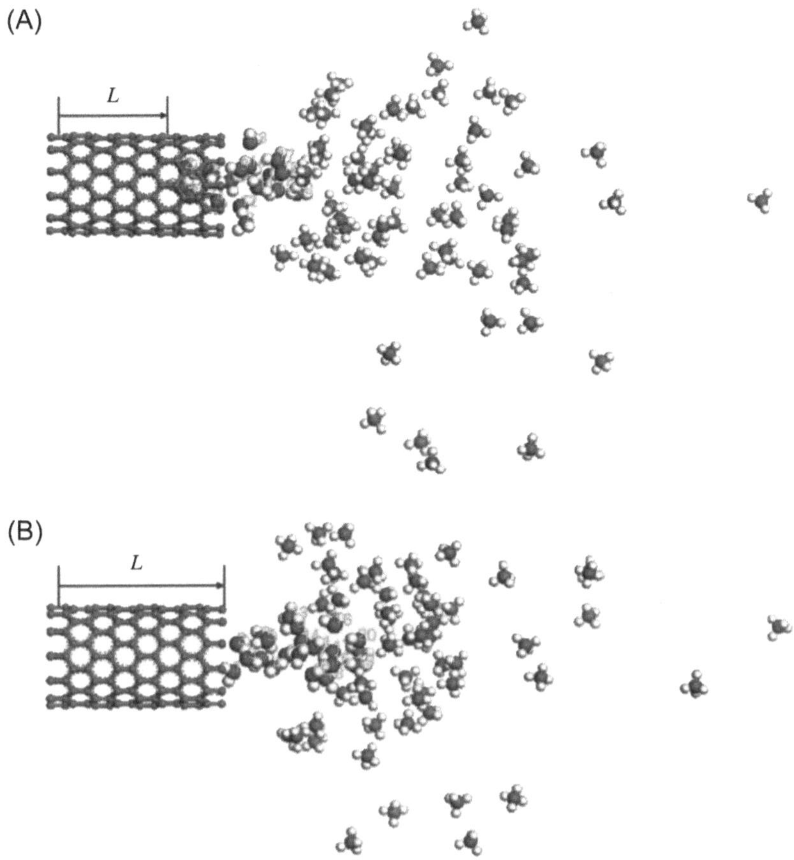

FIGURE 8.8

Simulation snapshots obtained for water molecules drawn out from (12, 0) SWCNT by 60 methane molecules at different time (A) $t = 55$ ps; (B) $t = 125$ ps [41].

Fig. 8.7B. The average translocation velocity is 0.2145 and 0.1235 Å/ps. Therefore, we suggest that the driving force generated by 60 methane molecules is larger than that by 40 methane molecules. An increase of methane molecules favors the reduction of the extracting time of water molecules from the CNT. As shown in Fig. 8.9, the displacement $L(t)$ of water molecules displays roughly linear dependence on time t for different numbers of methane molecules. The translocation rate of water molecules is found to increase with the number of methane molecules. When there are 80 methane molecules, velocity can reach up to 4.817 Å/ps, almost more than 40 orders of magnitude greater than in the case of 40 methane molecules.

As for a simple SWCNT–water system, water molecules are relatively stable in the CNT. It is impossible for water molecules to escape from the interior of CNT automatically without any additional driving force. Whether the water molecules could be drawn out or not is decided by the result of competition between the methane–water attractive force and the SWCNT–water

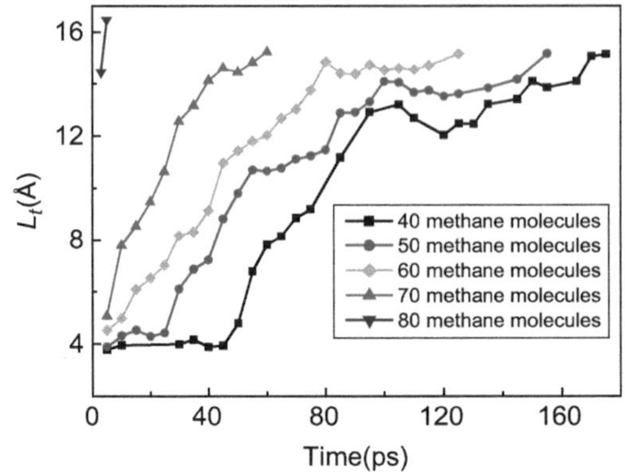

FIGURE 8.9

Time evolution of the translocation displacement $L(t)$ of water molecules driven by different numbers of methane molecules [41].

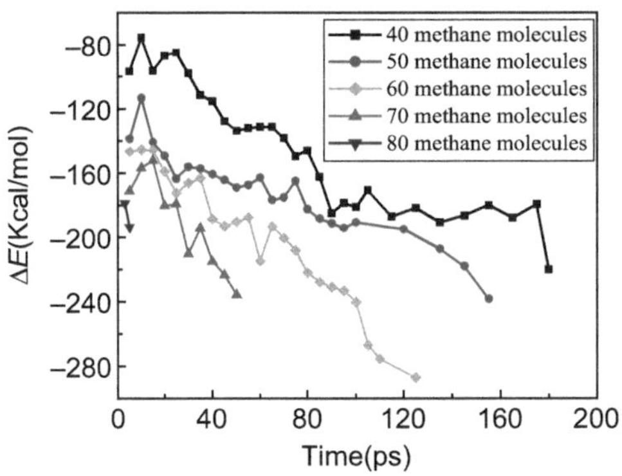

FIGURE 8.10

ΔE ($\Delta E = E_2 - E_1$) as a function of time t with the number of methane molecules. E_1 is the SWCNT—water attractive potential energy and E_2 the methane—water attractive potential energy [41].

attractive force. When the methane—water attractive force is dominant and the SWCNT—water attractive force is less dominant, the water molecules could be drawn out successfully. On the contrary, when the SWCNT—water attractive force is stronger and the methane—water attractive force is weaker, the water molecules would fail to be drawn out. To ascertain this hypothesis, we perform a series of energy computations. As shown in Fig. 8.10, the difference of E_2 and E_1, i.e., ΔE ($\Delta E = E_2 - E_1$) is a function of the time t with different numbers of methane molecules, in which ΔE increases with time. We estimate the attractive potential energy of SWCNT—water and methane—water, defined as E_1 and E_2. Fig. 8.10 shows that, as time goes, methane—water attractive interaction is getting stronger than SWCNT—water attractive interaction, and methane—water attractive

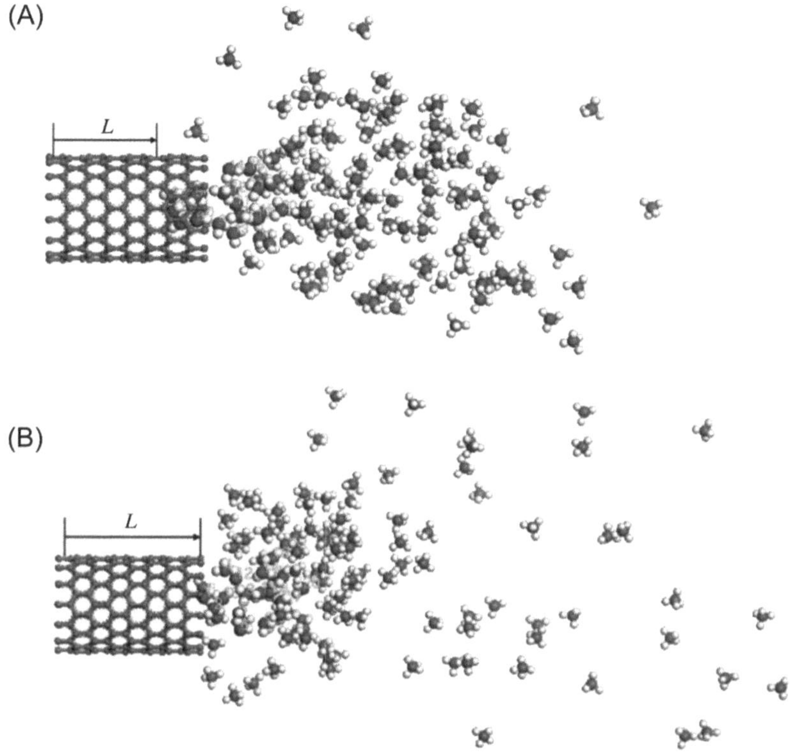

FIGURE 8.11

Snapshots of water molecules drawn out from (14, 0) SWCNT (diameter 10.96 Å) by 120 methane molecules at different time (A) $t = 185$ ps; (B) $t = 325$ ps. L is the displacement distance for water molecules from the initial site [41].

interaction rises as the number of the methane molecules increases. That is to say, more methane molecules would generate a stronger attractive force on the water molecules. It is the reason why more methane molecules can draw water molecules out from the SWCNTs with shorter time.

As reported [22,42], the water fluidity in CNTs increases as the diameter of the SWCNT increases. The question is how water molecules transport under an extra driving force when confined in a CNT. In Fig. 8.11, the water molecules can be easily drawn out from the (14, 0) SWCNT (diameter 10.96 Å) completely; however, it is hard to draw the water molecules out from the (16, 0) and (18, 0) SWCNTs with diameters of 12.53 and 14.09 Å in Figs. 8.12 and 8.13. This suggests that an increased diameter of the SWCNT decreases the transport velocity of water molecules. It is worth pointing out that the water molecules eventually reach up to the maximum distance in the nanotube and bounce back for some distance due to the unstable interaction force, which is the result of the competition of attractive interaction between the SWCNT–water and methane–water systems.

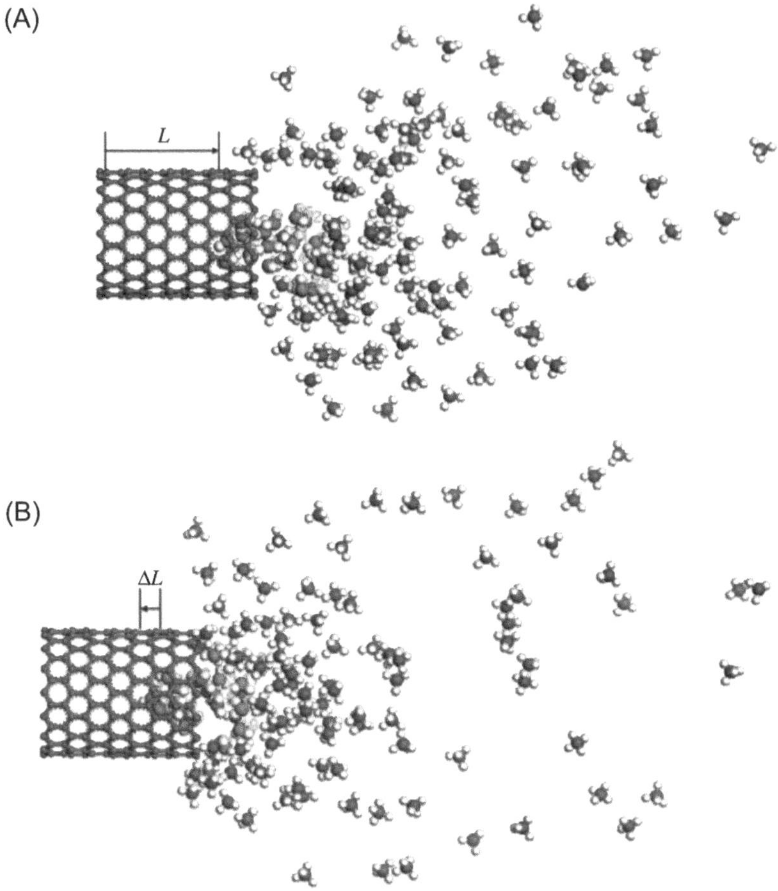

FIGURE 8.12

Snapshots of water molecules drawn out from (16, 0) SWCNT (diameter 12.53 Å) by 120 methane molecules at different time (A) $t = 195$ ps; (B) $t = 605$ ps. L is the maximum translocation distance of water molecules inside of the SWCNT at $t = 195$ ps. ΔL is the "bounce back" distance for water molecules from $t = 195$ ps to $t = 605$ ps [41].

The "bounce back" distances ΔL of water molecules in Figs. 8.12 and 8.13 are 1.625 and 1.749 Å, which illustrate that the nanotube with a bigger diameter would cause a more serious "bounce back" phenomenon than that with a smaller diameter. The translocation displacement $L(t)$ of water molecules as a function of time t with diameters of SWCNTs is illustrated in Fig. 8.14, where the curves display the opposite tendency to the Fig. 8.9.

The axial translocation velocity of water molecules in SWCNT increases sharply as the diameter of the SWCNT decreases. Obviously, two curves for SWCNTs with diameters of 12.53 and 14.09 Å have their peak values, denoted by A and B. The peak value A is higher than B. They correspond to the maximum translocation distance of water molecules inside of the SWCNTs. The water molecules in the SWCNT with diameter of 12.53 Å will take less time to reach the peak

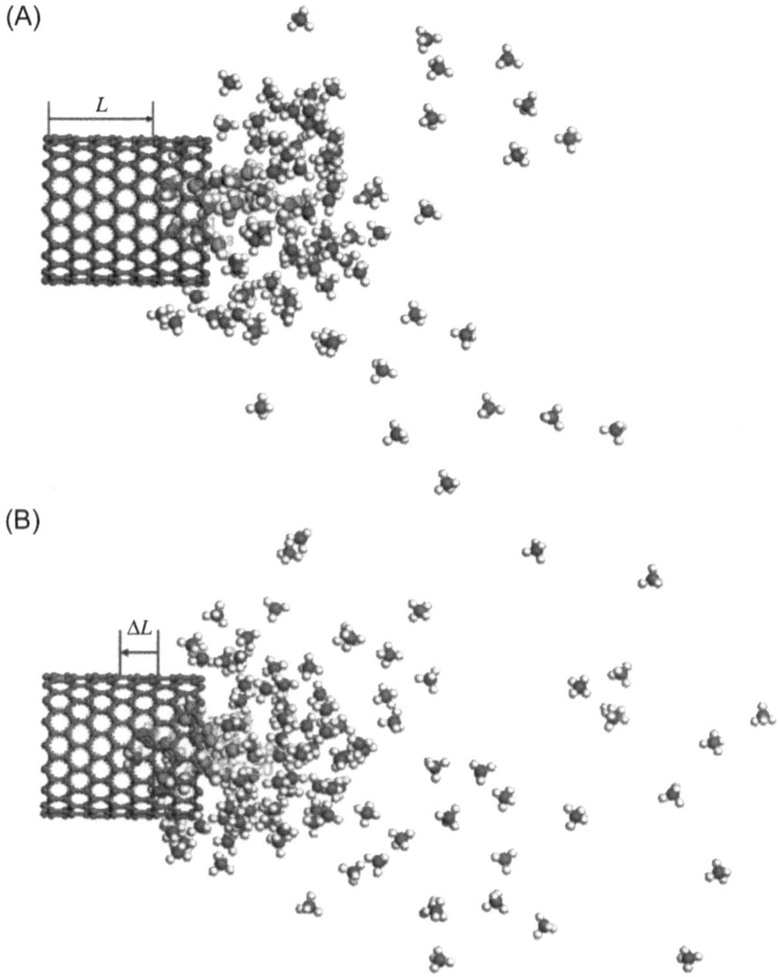

FIGURE 8.13

Simulation snapshots of water molecules drawn out from (18, 0) SWCNT (diameter 14.09 Å) by 120 methane molecules during different periods of time (A) $t = 385$ ps and (B) $t = 885$ ps. L the maximum translocation distance of water molecules inside of the SWCNT at $t = 385$ ps. ΔL is the "bounce back" distance for water molecules from $t = 385$ ps to $t = 885$ ps [41].

value than that with diameter of 14.09 Å. Subsequently, as the maximum peak is reached, the curves begin to drop down due to the "bounce back" phenomenon illustrated in Figs. 8.12 and 8.13. The "bounce back" is due to the SWCNT–water attractive interaction being greater than the methane–water attractive interaction.

In a word, the water molecules confined in the SWCNT can be drawn out by methane molecules. The competition of attractive force between methane–water and SWCNT–water systems

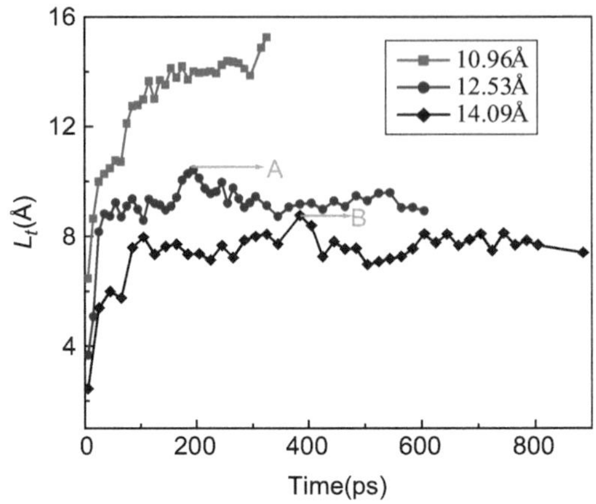

FIGURE 8.14

Dependence of translocation displacement $L(t)$ of water molecules on time t in different diameters of SWCNTs. Points A and B are the peak values of two curves for SWCNTs with diameters of 12.53 and 14.09 Å [41].

dominates the moving direction of water molecules. The number of methane molecules has a positive effect on the transport of water molecules through the SWCNT. The translocation velocity of water molecules increases sharply with the increasing number of methane molecules. The diameter of the SWCNT is however not in favor of the transport of water molecules. The transport velocity of water molecules decreases with the increasing diameter of the SWCNT. The "bounce back" of water molecules is clearly observed in SWCNTs as diameters get bigger.

The result may arouse interest in fabrication of molecular-selective CNT devices which may act through mechanisms similar to those of biological transmembrane channels. The filling/emptying transition in functionalized nanotubes can be potentially used as wetting–dewetting nanoscale devices and would have wide applications in nanotechnology and biotechnology.

8.2.2 DRIVING WATER MOLECULES ALONG A DIAMETER-GRADIENT SWCNT

Water usually travels through many passages connected by junctions of different diameters, rather than single-diameter tubes, which can cause barriers to water transport. Although much work focuses on the transport of water confined in a single-diameter SWCNT [43,44], little research has been conducted on the transport behavior of water molecules in a diameter-gradient SWCNT with many junctions. The manipulation of the flow of water molecules through diameter-gradient SWCNTs connected by junctions with different diameters is a challenging issue and has not been well tackled.

MD simulation is also performed to investigate how methane molecules pull water molecules through a diameter-gradient SWCNT with junctions. The model is highly appropriate for studying the influence of junctions on the transport of water through SWCNTs with different diameters. Understanding the mechanism of water transport in these CNT-based membranes is crucial in the

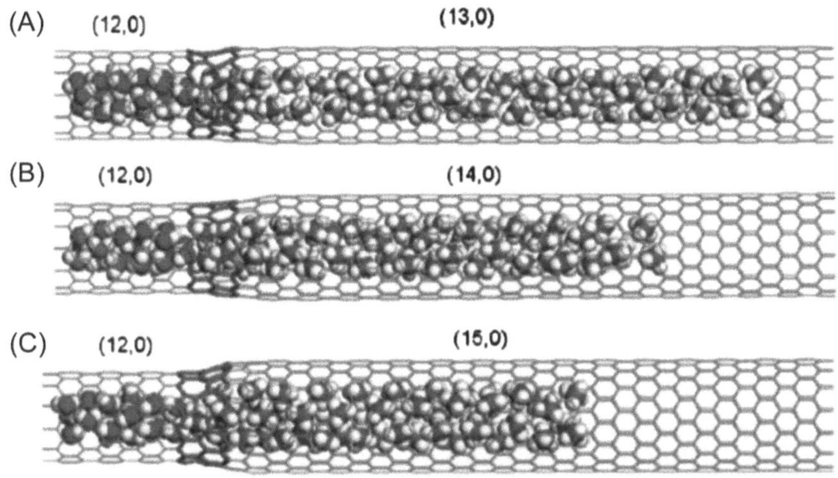

FIGURE 8.15

Snapshots of the initial configurations of three bottle-like SWCNT systems (A) a (12, 0) SWCNT connected with (13, 0); (B) a (12, 0) SWCNT connected with (14, 0); and (C) a (12, 0) SWCNT connected with (15, 0) SWCNT. The SWCNTs are gray, the hydrogen atoms are represented by white spheres, the oxygen atoms are in black (red in web version), and the carbon atoms of methane are dark grey (blue in web version) [45].

design of nanomachinery and synthetic membranes, and for the development of novel drug-delivery devices.

Each system consists of coaxial CNTs with different diameters, which are linked by a junction region with a pentagon–heptagon (5–7) topological structure, as illustrated in Fig. 8.15. Such connections have been directly observed by Iijima and others with transmission electron microscopy [46,47]. The nanotubes are zigzag (n, 0) SWCNTs with indices $n = 12$, 13, 14, 15, 16, and 17, the diameters of which are 9.39, 10.18, 10.96, 11.74, 12.53, and 13.31 Å, respectively. Both bottle-like and terrace-like SWCNTs are used in the simulation. Fig. 8.15 shows the bottle-like SWCNTs, which comprise a (12, 0) SWCNT connected to (13, 0), (14, 0), and (15, 0) SWCNTs, respectively. Fig. 8.16 shows the terrace-like SWCNTs, in which (12, 0), (13, 0), (14, 0), and (15, 0) SWCNTs are connected with smooth junctions formed by five- and seven-membered rings to form a (12, 0)–(13, 0)–(14, 0)–(15, 0) series-connected SWCNT system. In the same way, other series-connected SWCNT systems are also fabricated successfully. The length of all the series-connected SWCNT systems is 83.6 Å. The coordinates of the SWCNTs are fixed. The quantity of methane molecules is chosen as 50. The cut-off distance is set to 10.0 Å, and the time-step is 1 fs. Data are recorded every 0.5 ps for further analysis. Initially, 20 water molecules are inserted randomly into the interior of the (12, 0) SWCNT, and 5000 time-steps are run to relax the initial configuration to obtain the lowest energy structure. Finally, the methane molecules are arranged in another nanotube connected with a (12, 0) SWCNT to be used to drag the water molecules from the inside of the (12, 0) SWCNT.

Fig. 8.17 shows the translocation displacement L of water molecules dragged by methane molecules through three bottle-like SWCNTs at 6000 ps. Fig. 8.17C demonstrates that the displacement

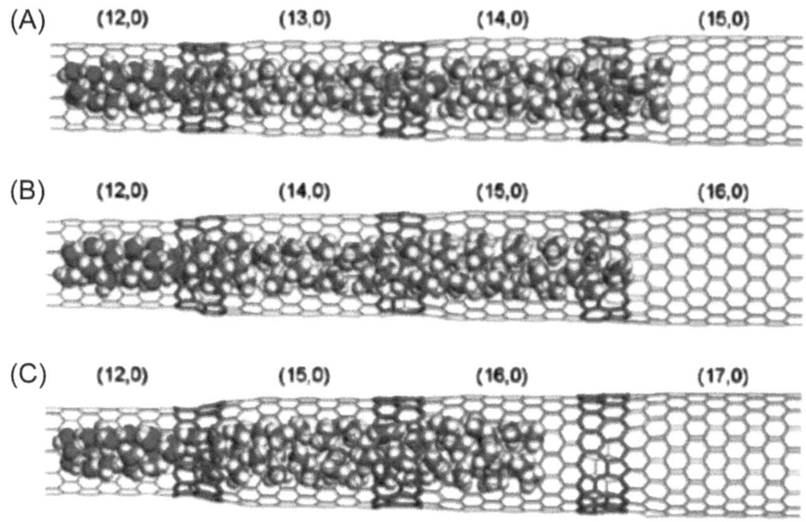

FIGURE 8.16

Snapshots of the initial configurations of three terrace-like SWCNT systems (A) (12, 0)–(13, 0)–(14, 0)–(15, 0); (B) (12, 0)–(14, 0)–(15, 0)–(16, 0); and (C) (12, 0)–(15, 0)–(16, 0)–17, 0) [45].

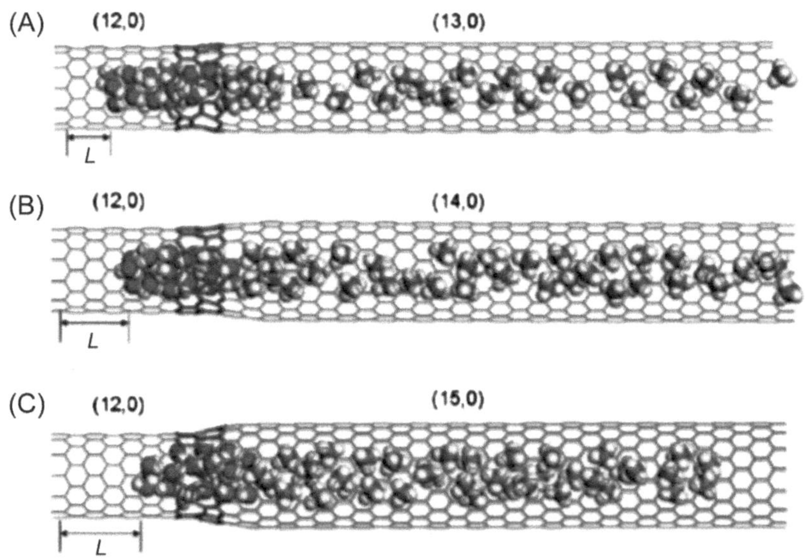

FIGURE 8.17

Snapshots of water molecules dragged by methane molecules in three bottle-like systems at $t = 6000$ ps obtained by MD simulation [45].

distance L of water molecules in the (12, 0)−(15, 0) system is the longest and differs from the behavior of the system in the (12, 0)−(13, 0) SWCNT shown in Fig. 8.17A. Obviously, the motion of the water molecules is attributed to the methane molecules that generate the driving force to make the water molecules flow through the SWCNT. Because methane molecules confined in the SWCNT are not at the lowest energy status, they continuously diffuse with an initial acceleration. Subsequently, the water molecules will follow the methane molecules through the SWCNT due to the van der Waals attraction between the methane molecules and the water molecules. The methane molecules are responsible for the driving force, and therefore the translocation of the water molecules is dominated by the methane molecules.

When the water molecules reach a junction region, they oscillate due to the potential barrier of the junction, which counteracts the attractive force of the methane molecules. If the attractive force of the methane molecules overcomes the potential barrier caused by the junction, then the methane molecules are able to drag the water molecules to another compartment; otherwise, they become jammed in the junction region. Fig. 8.18 shows the evolution of the potential energy in the three bottle-like systems with time. It is worthy of note that the magnitudes of the potential energy of the three systems decrease as the diameter of the SWCNTs increases. This is because the water molecules are more stable inside the bigger SWCNT than they are inside the smaller SWCNT, as demonstrated in previous work [48]. The potential energy barriers in the (12, 0)−(13, 0) SWCNT, the (12, 0)−(14, 0) SWCNT, and the (12, 0)−(15, 0) SWCNT are 75, 100, and 168 kJ, respectively, which indicates that a slight change in the diameter of a junction causes significant variation in the potential energy. In the (12, 0)−(15, 0) system, it takes the water molecules 19,330 ps to cross the junction region, whereas in the (12, 0)−(13, 0) SWCNT, it takes them about 20,550 ps to cross the junction region. The systems tend to equilibrate once the water molecules have crossed the

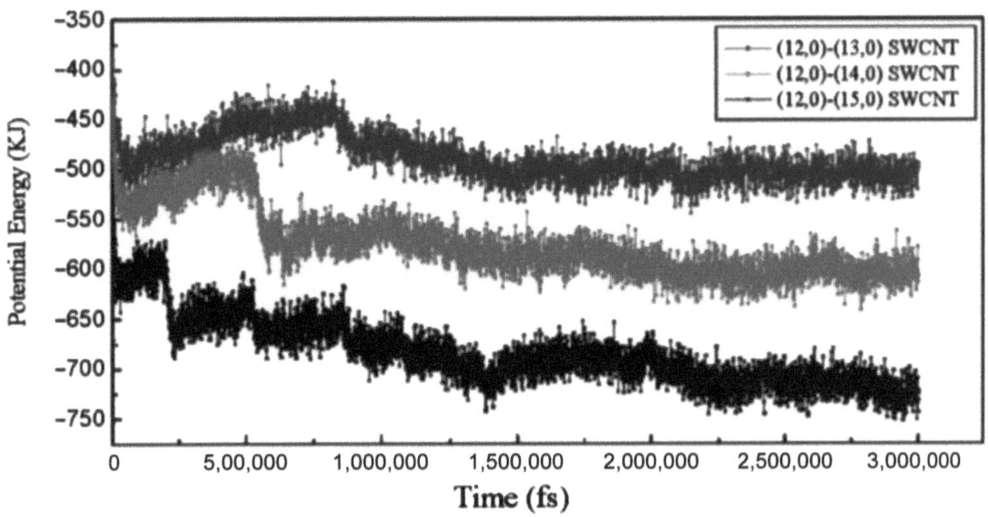

FIGURE 8.18

Evolution curves of the potential energy of the three systems with time t [45].

junction region. When the methane molecules leave the interior of the SWCNTs, the attractive interaction between the methane molecules and water molecules weakens. Finally, the methane–-water attractive interaction is equal to that of SWCNT–water and SWCNT–methane, which means that the whole system reaches equilibrium. Hence the number of the methane molecules in the SWCNT also affects the transport behavior of the water molecules through the nanotube. More methane molecules mean a larger driving force, and therefore it is easier for the water molecules to cross the junction region. As mentioned in Fig. 8.18, the potential energy curves display clear fluctuations around the average value, which is caused by the vibration of the water molecules in the interior of the SWCNTs at their equilibrium position in the axial direction. The final configurations of the three systems can be seen in Fig. 8.19. The maximum value of the translocation displacement of the water molecules in the (12, 0)–(13, 0), (12, 0)–(14, 0), and (12, 0)–(15, 0) SWCNTs is 18.037, 23.708, and 34.703 Å, respectively. The corresponding average velocities of the three systems are $9 \times 10{-}4$, $1.15 \times 10{-}3$, and $1.7 \times 10{-}3$ Å/ps. As shown in Fig. 8.20, the displacement $L(t)$ of the water molecules displays a roughly linear dependence on time t for the different diameters of SWCNTs to which the (12, 0) SWCNT is connected. Hence, the translocation velocity of the water molecules increases with the diameter of the SWCNTs. However, for a simple SWCNT–water system without the methane molecules, as shown in Fig. 8.21, there is still a quite small displacement L of water moving toward the next compartment, which is much less than those driven by the methane molecules. Due to the interaction of the long-lasting hydrogen bonds between water molecules, it is a collective motion behavior for the confined water molecules, which is consistent with the ballistic-type mechanism of Striolo's results [49]. When the water molecules reach a junction region, they oscillate backward and forward along the axial direction, but finally they would not spontaneously enter the next compartment due to the barrier of the junction. This is different from Hummer's calculation in which it was found that a one-dimensionally

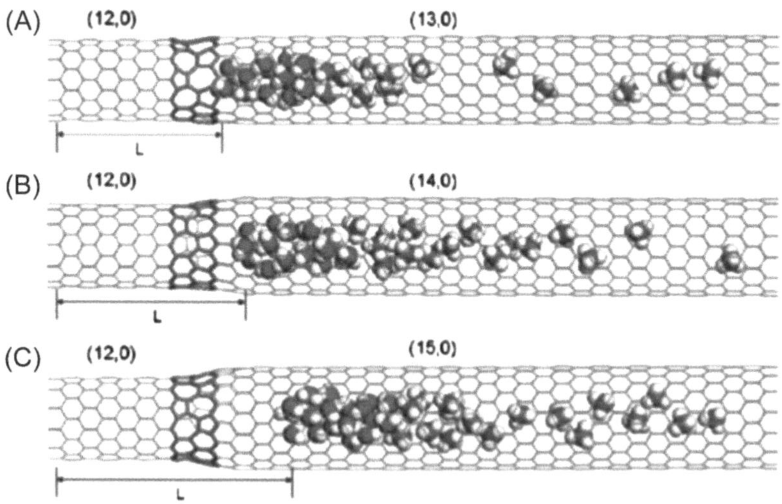

FIGURE 8.19

Sequential snapshots of water molecules drawn by methane molecules in the three systems at time $t = 20640$ ps [45].

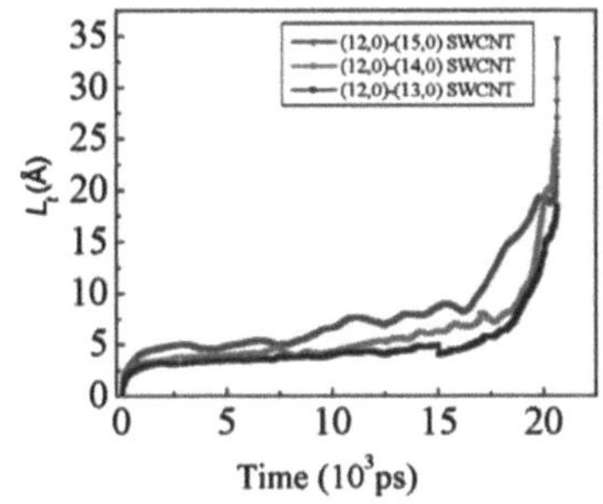

FIGURE 8.20

Translocation displacement $L(t)$ in the axial direction as a function of time t of water molecules dragged by methane molecules in three bottle-like SWCNTs [45].

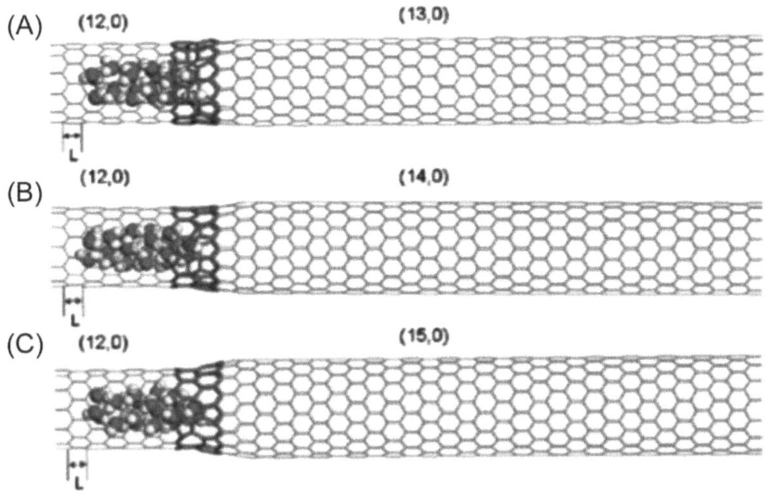

FIGURE 8.21

Simulation snapshots of water molecules in three bottle-like SWCNTs without methane molecules at $t = 20$ 640 ps [45].

ordered chain of water can fill a nonpolar CNT spontaneously and continuously with little resistance [19]. Likewise, our finding differs greatly from the previous investigations of fast water transport through a single-diameter CNT reported by Holt [6], Majumder [50], and Thomas [51].

To further examine the effect of a series-connected SWCNT system on the transport of water molecules, a system with more junctions, such as the terrace-like SWCNT, is designed to perform another simulation. As shown in the inset in Fig. 8.22, during the first 100 ps in the simulation, the translocation velocity of the water molecules increases with the pore size, as in Fig. 8.20. The reason is that, during the first 100 ps, the translocation displacement of water molecules in the (12, 0)

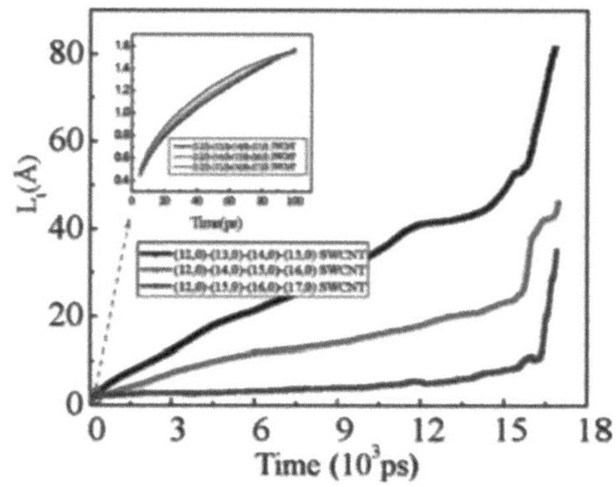

FIGURE 8.22

Evolution of the translocation displacement *L*(*t*) of water molecules driven by methane molecules in three different terrace-like SWCNTs. The inset shows the dependence of the translocation displacement *L*(*t*) of the water molecules on time ($t = 0-100$ ps) [45].

(13, 0)–(14, 0)–(15, 0), (12, 0)–(14, 0)–(15, 0)–(16, 0), and (12, 0)–(15, 0)–(16, 0)–(17, 0) SWCNTs is 1.545, 1.564, and 1.573 Å, respectively, less than those in the (12, 0)–(13, 0), (12, 0)–(14, 0), and (12, 0)–(15, 0) SWCNTs (2.106, 2.360, and 2.863 Å). This indicates that the methane molecules in the terrace-like SWCNTs just start to diffuse. More junctions of the terrace-like system do not affect the diffusion of the methane molecules. The transport behavior of the water molecules in the terrace-like SWCNTs is similar to that in the bottle-like SWCNTs. Over time, an unexpected opposite trend is observed in Fig. 8.22, in that the axial translocation velocity of the water molecules in the terrace-like SWCNTs increases sharply with the decrease of the diameters of the SWCNTs connected to the (12, 0) SWCNT. This indicates that the narrower terrace-like SWCNTs give the methane molecules a greater driving force. Consequently, the translocation velocity of the water molecules increases slowly as the diameters of the SWCNTs connected to the (12, 0) SWCNT increase during the initial 100 ps but decreases rapidly with the increase in diameter after 100 ps. Fig. 8.23 shows the final snapshots of the three terrace-like systems at *t*) 18,000 ps. It is worth noting that the water molecules escape completely from the (12, 0)–(13, 0)–(14, 0)–(15, 0) SWCNT. The translocation displacement *L* (length 81.296 Å) of the water molecules is the longest in this SWCNT (as shown in Fig. 8.23A), whereas the translocation displacement *L* (length 35.201 Å) of the water molecules in the (12, 0)–(15, 0)–(16, 0)–(17, 0) SWCNT is the shortest (as shown in Fig. 8.23C). These results conflict with those in Fig. 8.19. Moreover, the water molecules in the terrace-like SWCNTs move further than those in the bottle-like SWCNTs over the same interval. A comparison of the results for the transport of water molecules in the bottle-like SWCNTs with those for the terrace-like SWCNTs confirms that the greater number of junctions in the terrace-like SWCNTs gives the methane molecules a more sustained driving force, which favors the rapid transport of the water molecules along the SWCNTs.

In summary, methane molecules can overcome the potential barrier caused by the junctions and successfully pull the water molecules across the junction regions from one compartment to the next. In the bottle-like SWCNTs, the junctions are detrimental obstacles to the transport of water

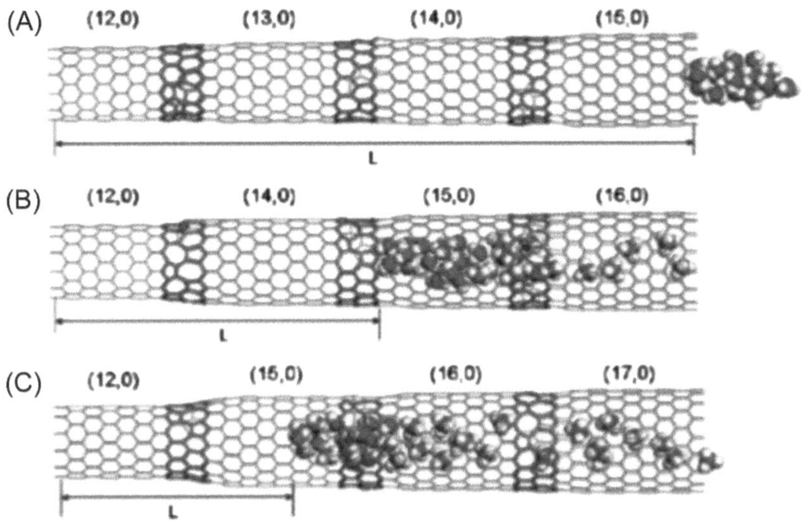

FIGURE 8.23

Simulation snapshots of three terrace-like SWCNTs at time $t = 1.8 \times 104$ ps [45].

through the nanotube, especially when the CNTs have small diameters. In contrast, for the terrace-like SWCNTs with more junctions, the junctions become a driving force that accelerates the diffusion of the methane molecules. Thus, the translocation displacement of water molecules in the terrace-like SWCNTs is much longer than that in the bottle-like SWCNTs. These results have implications for the design of nanoscale molecular diffusive drivers as motors for the transport of liquid or gas molecules. This study also indicates that the filling/emptying transition in functionalized nanotubes can potentially be used as a wetting–dewetting nanoscale device, which would have wide application in nanotechnology and biotechnology.

8.3 HYDROGEN STORAGE [52,53]

Gas adsorption properties of CNTs [54–63] and related carbon materials, such as graphitic nanofibers [64,65], activated carbons [66–73], fullerene, and nanotanks [74–79] have been extensively studied over the last decade. However, the reported storage capacities for these materials are considerably scattered, ranging from around 2–4 wt.% [80,81] to significant 5–10 wt.% [82] values. This makes it necessary to apply theoretical quantum simulations to understand the storage mechanisms. To date, theoretical quantum calculations on the adsorption process have been carried out on many aspects, including the adsorption sites of the H_2/H locating on the carbon materials [83,84], the coverage capacity of H atoms chemisorbed onto the CNTs [85,86], and the relations between hydrogen uptake capacity with the arrangements of the nanotubes [87,88], with nanotube diameter [52,85,89,90], with types of CNTs (such as zigzag, armchair, chiral, and coiled CNTs) [90–92], and with different metal doped CNTs [93–97]. Only recently has the chemisorption of hydrogen on SWCNTs been observed [98–102]. However, there is a void of information on the transient kinetics of the storage process that involves the interaction between hydrogen atoms and the CNT, and the corresponding thermodynamic properties. To obtain this important information, it is necessary to apply theoretical molecular simulation. The problem is frequently formulated in terms of potential functions, or in other words, MD or ab inito parameters. Classical potentials in MD are not always very accurate, and they usually have a simple analytical form so that numerical computations can be made faster. On the other hand, quantum mechanical ab initio methods are supposed to be highly accurate and, therefore, suitable to monitor the progress of a molecule translocating from the exterior of the CNTs into the inner space or onto the surface of the CNT tube. Although the choice of *ab initio* calculation methods can provide accurate results for simple molecules, the extreme computation effort makes it prohibitively difficult to produce accurate outcomes for complex molecules and macromolecules.

Additional assumptions are often required to investigate the chemical reaction process of a large cluster of molecules. Frequently applied assumptions for the H_2/H chemical reaction with the graphene/CNT include treating a group of selective molecules as "fixed atoms" or simply clustering a group of molecules as environment atoms. Both methods have been utilized by many research groups. First, as regards the "fixed atoms," it was utilized by Arellano et al. [100], who kept the CNT geometry unchanged and only varied the distances and angles of the hydrogen atoms with reference to the CNT. Verónica et al. [101] applied a similar idea by permitting the hydrogen atom to approach only from the center of the nanotube (inner wall) and imposing immobility on all the carbon atoms. Lee et al. [103] artificially set the radial component of the hydrogen atom at each step while moving the hydrogen atoms along the CNT. Although these assumptions lead to a significant reduction in computation time, it is achieved at the cost of neglecting the interactive role of the environment atoms/molecules at the vicinity of the chemical reaction sites, whose configurations often change due to the nature of the newly bonded atoms when the chemical reaction occurs.

Another method frequently observed in many reported theoretical works is the concept of a finite cluster of atoms, where atoms are cut (or taken) from the actual surface with the distance between the atoms forming the cluster kept at their bulk values [104]. This idea was applied by Han et al. [102,105], who studied the reaction pathway of hydrogen molecule reacting with CNT by applying density functional theory (DFT) calculation, with the cluster size of 100 carbon atoms. Due to the assumption of invariable carbon atoms in their work, it was not able to reflect the phenomena of C–C bond severance in the CNT sidewall caused by hydrogen interaction, which is important for the explication of CNTs coalescence observed by Nikolaev et al. [105]. One fundamental problem in the "cluster modeling" is that the convergence of the adsorption energy and the corresponding properties are associated with the cluster size. Generally, a very large cluster is required. However, if more than a few molecules are explicitly described by quantum calculation, computational costs rapidly become prohibitive.

In order to balance the efficiency and accuracy in the study of large systems, one of the popular and versatile hybrid approaches, the ONIOM (our own *N*-layered integrated molecular orbital and molecular mechanics (MM)), was developed by Morokuma and coworkers [106–112]. The basic concept of the ONIOM method is to divide a large system (the real system) into several layers, with each layer treated with distinct computational methods. The partition can be based on identifying the reaction center molecule. The target calculation for the relative energy is a high-level quantum mechanics (QM) treatment for the entire real system, which is usually computationally demanding and impractical. Therefore, instead of evaluating the energy of the real system at the high level, the ONIOM scheme can extrapolate it from independent less demanding calculations. The chemically important region is dealt with using the accurate high level QM method while the rest of the layers are treated with the computationally less intensive lower level QM (e.g., semiempirical) or MM methods. To date, this hybrid framework has been mostly applied to large organic molecules and organometallic complexes.

In the present work, a systematic study of the adsorption process of hydrogen atoms onto the (5, 5) CNT sidewall will be presented by taking on the two-layer ONIOM2 scheme. The corresponding thermodynamics properties are also investigated. A moderately sized active site region is described using high level theory, and the extended framework/environment atoms considered using a low-level method. With the theoretical level combination as suggested by Morokuma et al. [106–112], the integrated system energy is obtained from three independent calculations, E^{QM}_{model}, E^{MM}_{real}, and E^{MM}_{model}

$$E^{ONIOM2} = E^{QM}_{model} + E^{MM}_{real} - E^{MM}_{model}$$

where, the "real" system denotes all the complex geometry structures, which contains all the framework atoms and is calculated at the MM level. The "model" system refers to the chemical reaction sites, and its energy is treated at both the QM and MM levels, respectively. Fictitious hydrogen link atoms are used to cap those dangling bonds caused by the cutting of the covalent bonds between the QM and the MM regions. More specifically, in this study, the small model system includes the hydrogen and the necessary carbon atoms to represent the chemical activation sites, while other carbon atoms are used to simulate the local environment of the active sites during the adsorption process.

The conventional DFT calculations are directed toward obtaining electronic structure properties at absolute temperature. Therefore, in order to obtain refined and efficient simulation results at finite temperature, it is necessary to derive the corresponding thermodynamics properties from the mathematic framework that the DFT provides. The Boltzmann factor and partition functions are incorporated with DFT calculation to establish the temperature dependence of the Gibbs free energy of the electrochemical system. The transition states of the reaction are also studied in order to determine the energy barrier of the reaction and the reaction rate. It is important to note that the attractive feature of this method is that it connects seamlessly with conventional phenomenological equations for modeling macroscopic phenomena, on the basis of the intermolecular forces that the DFT uses for microscopic structure.

8.3.1 COMPUTATIONAL METHODOLOGY AND PHYSICAL MODELS

The model consists of a relatively narrow (5, 5) armchair CNT constituting of 180 carbon atoms with tube length of 22.6 Å, as shown in Fig. 8.24. The reason for the selection of relative smaller radii nanotubes lies in that it is much more energetically preferable than larger ones, as the C–C bonds of smaller diameter nanotubes are under greater strain [113,114]. The tube length used in this work is consistent with many earlier studies on the addition of hydrogen atoms to CNTs [84–86,115,116]. Clearly the tube caps possess higher potential energy fields and stronger attracting forces than the tube walls [86], as the electronic clouds of the carbon atoms at the caps overlap more significantly than those at the walls. In order to focus on the reaction sites at the sidewall of

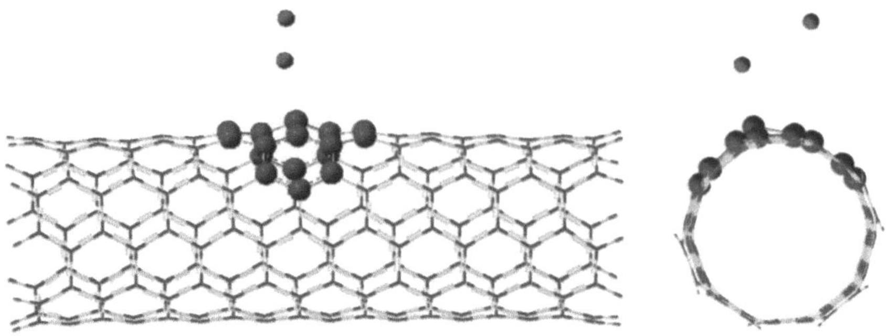

FIGURE 8.24

Model for the study of two hydrogen atoms chemisorbed onto the CNT sidewall with 180 carbon atoms, using ONIOM2 (B3LYP(6−31G + (d,p)):UFF) hybrid simulation. The initial state begins with two hydrogen atoms physisorbed at the exterior of the CNT due to van der Waals attraction. The ball-and-stick atoms represent the model system that are treated with B3LYP, and the line type atoms are the real regions treated with UFF [53].

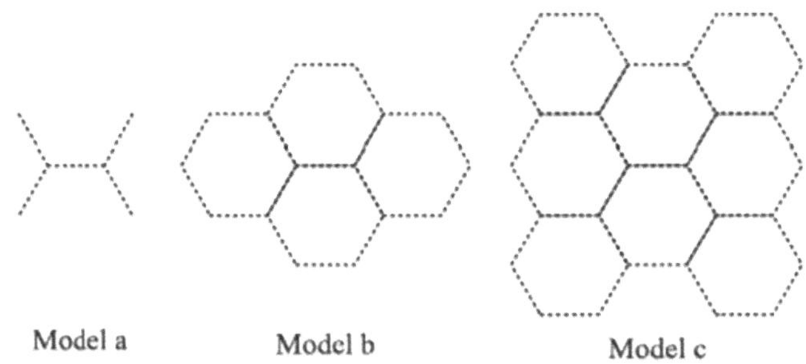

FIGURE 8.25

The number of carbon atoms used in the ONIOM scheme for testing necessary size of the small model system. Left: 6; middle: 16; right: 28 carbon atoms. The hydrogen atoms, which are part of this small model system, are not shown in the figure [53].

CNT and discount the effects caused by the tube caps, it is necessary to remove the caps during the simulation. The corresponding dangling bonds at the tube ends due to the removal are tied off with hydrogen atoms.

Our study begins with two hydrogen atoms which are physisorbed at the exterior of the CNT due to vdW attraction, see Fig. 8.25. The subsequent chemical reaction of hydrogen atoms with the sidewall of CNT, and the corresponding molecular thermodynamics properties, are then examined. In this work, the frequencies (the second derivative of energy with respect to the atomic coordinates) for the optimized intermediate and transition states are calculated to verify the correct transition structures (TSs) obtained.

The accuracy of the ONIOM scheme depends considerably on the choice of the calculation levels for the "model" and "real" regions. In this study, the small model system is treated via high-level DFT, specifically, the hybrid functional B3LYP method (Becke's three-parameter hybrid method [117–122] with the exchange functional of Lee et al. [123]). The remaining of the framework connecting the active site is handled via the low-level universal force field (UFF) MM to reduce the computational time and to represent the proper "mechanical" confinement effect for the embedded DFT part. The UFF force field replaces the interactions of the electrons with each other and with the nuclei by explicitly approximating all these interactions with a sum of refined energy expressions. Due to the fact that the UFF force field is based on vdW parameters to approximate interatomic parameters from the atomic parameter set, it can reasonably account for vdW interactions which have been reported to have significant contributions to most absorption processes. Overall this B3LYP:UFF combination provides small geometry deviation from the target calculation, and has been suggested/recommended by several researchers [84,86,115,116].

Another factor that determines the accuracy of the hybrid ONIOM method is the physical selection of the high- and low-level regions for calculation. Although the amount for each model system and the theoretical level employed for every problem varies to some extent, it is important to properly choose the region sizes for the model system. In the case of small molecules or atoms, such as the H_2 or H in this work, the size for the reactant species should include those in the proximity of the surrounding atoms, which can at least interact with the reactant through vdW forces. The appropriate dimension of the "model system" is brought out more clearly by carrying out a proper convergence test for progressively different model systems. The amount of carbon atoms of the small model system was varied from 6 to 16, and then to 28, as shown in Fig. 8.25. Comparison of the adsorption energy results obtained from the 6-carbon and 16-carbon small model systems showed significant discrepancies. The subsequent comparison of the 16-carbon and 28-carbon small model systems revealed very small discrepancy of less than 0.1%, and we can, therefore, verify that the solutions have converged when using the 16-carbon small model. Thus all subsequent calculations were carried out using the 16-carbon small model.

8.3.2 REACTION PATHWAY OF ATOMIC HYDROGEN INTERACTION WITH CNT

The basis set of 6–31 + G(d,p) is used for the small model system simulation, which emphasizes the polarization and diffusion function effects in the chemical reactions of the hydrogen atoms interacting with the CNT. The adsorption energies of the hydrogenated (5, 5) CNT systems are obtained according to Eq. (8.2). For the purpose of determining the activation barriers and understanding the chemisorptions mechanism, most of the effort here is focused on characterizing not only the stable intermediate (SI) but also the TSs states, where the latter follows the potential energy surface (PSE) path down to the SI. In principle, the transition state can be obtained from the maximum in the free energy surface projected onto the reaction coordinate, while the stable states is obtained from the minimum. In practice, there are many structural transformations (reaction pathways) that one can envision for the transformation of the same reactants into the same products. When plotting energies versus bond distances that involve bond breakage and formation to illustrate the chemical reaction process including the reactants, TS structure, intermediates, and products, the potential surfaces are often combined in three or more dimensions. In other words, the chemisorptions can proceed through a multiplicity of routes with an ensemble of TS structures. Therefore, it is more appropriate to say that we are exploring the potential energy transformation, when exploring the reaction outcome or possible mechanisms for some chemical transformations [124].

Considering the importance of SI and transition states structures in the understanding of the chemical transformation, the search, optimization and verification of the geometry for these structures were first performed with the ONIOM2 integrated schemes as part of the GAUSSIAN03 computational package. To verify the structures for the SI and TS, and to obtain the corresponding thermal properties for these structures, one more step is necessary, namely, performing the frequency analysis of the optimized structures. This vibrational frequency calculation is required to be carried out at the same B3LYP/6−31 + G (d,p) level as that in the geometry optimization. From the output of these calculations, we can establish that they are true viable SI and TS structures when only one imaginary frequency (negative force constant) exists for each TS structure, while an all-positive frequency condition is required for the SI structures. The underlying theoretical basis for this is due to the TSs corresponding to the first-order saddle points on the PSE. Such a saddle point is a point where there is a minimum in all dimensions but one. The SI states are located at the minimum points on the PSE. To further verify the simulation results for the TS and SI states, the auxiliary Gauss view, which is a Graphical User Interface program, was used to visually examine structures at all steps throughout the calculations. All stable structures were verified to be geometry minima by performing quadratic potential frequency calculations and examining the frequency list for imaginary values. By computing the frequencies of the transition state structures, it was also shown that the transition states are characterized by one imaginary frequency.

Once the information on geometry optimization is obtained, it is then possible to calculate the adsorption energy for the hydrogen, which is defined according to

$$E_{ad} = E_{CNT-2H} - E_{CNT} - E_{2H} \tag{8.2}$$

Where E_{CNT-2H}, E_{CNT}, and E_{2H} are the total energies of the fully optimized CNT−2*H* system, the stand-alone nanotube, and the hydrogen atoms, respectively. According to this definition, the negative binding energies obtained imply that the system is stable.

To investigate the activation energy of the hydrogen chemisorptions process, we first predict the SI and TS states of the reaction. The connections between reactants, intermediates, TSs, and products are shown in Fig. 8.26, where the contour line represents the entire chemical adsorption pathway of the two hydrogen atoms being chemisorbed onto the (5, 5) CNT. The adsorption energies are calculated according to Eq. (8.2), and relative reaction energy barriers are listed in the figure as well. This figure indicates that the two hydrogen atoms first form C−H bond and eventually sever the C−C bond of the CNT to form two H−C σ-bonds, and the overall chemical transformation is an exothermal reaction. The detailed configurations at each stable and transition states can be found in Fig. 8.24, Figs. 8.27−8.29. In addition to these structures, it is interesting to examine how the geometry of the system changes along the reaction pathway. Looking at the geometrical changes, one can clearly observe the principal movements during the hydrogen chemisorption. The present calculations begin with two hydrogen atoms which are initially physisorbed at the exterior of the CNT. Following geometry optimization, they attain the SI state I (SI I) see Fig. 8.24. Experimentally, the SI I physical state can be achieved under high-pressure conditions, or by injecting hydrogen atoms with high kinetic energies into the reaction area [100]. It can be observed that, in the stable SI state structure, both of the hydrogen atoms are attracted to the exterior surface of CNT due to the vdW attraction from the exterior of the CNT, and both are directly above two carbon atoms, along the radial direction of the CNT, see Fig. 8.24. Clearly, the hydrogen atom (H_1) which is nearer to the surface of the CNT weakens the C−C π-bond of the CNT, and subsequently

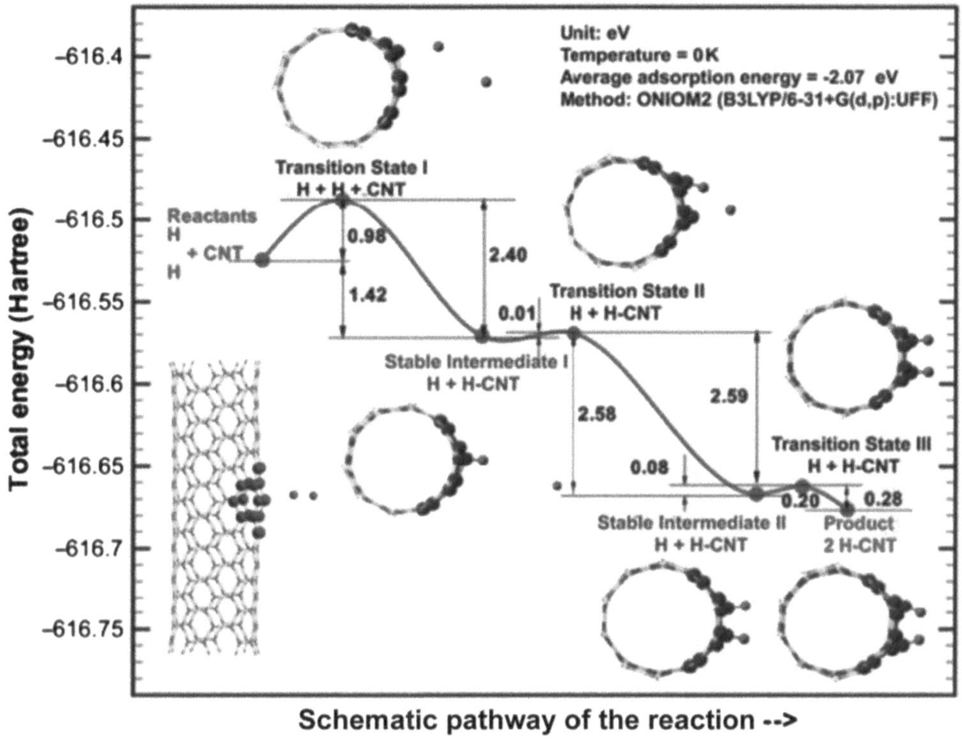

FIGURE 8.26

Schematic pathway for the chemisorption of two hydrogen atoms on a (5, 5) CNT. Three transition states are found during the reaction [53].

forms the C—H σ-bond at the exterior of the CNT in SI state II (SI II), see Fig. 8.27. For convenience, we shall name the carbon atom in the C—H bond as C_1. Following the reaction pathway, we found the other hydrogen atom (H_2) experiences the same reaction procedure as the H_1 atom. It is attracted and eventually chemisorbed with another carbon atom (C_2) that is located on the same circumferential layer (of the CNT) as the first carbon atom (C_1). Our results further show that the adsorption of the second hydrogen atom (H_2) on the CNT wall leads to an obvious distortion of the CNT structure, the corresponding configurations are shown in Fig. 8.28. Thus, the reaction from the physisorption of the hydrogen atoms to their chemisorption has greatly changed the bonds adjacent to the reaction sites. To confirm the bond breakage, the C—C bond length change from SI I to SI III was calculated. The results show that the distance between C_1 and C_2 extends from 1.55 Å in SI I to 1.70 Å in SI II and subsequently to 2.49 Å in the final products of SI III, see Fig. 8.29. This distance variation clearly indicates that the C—C bond is severed when both hydrogen atoms are chemisorbed at the exterior of the CNT, in the process of attaining the most stable SI III configuration.

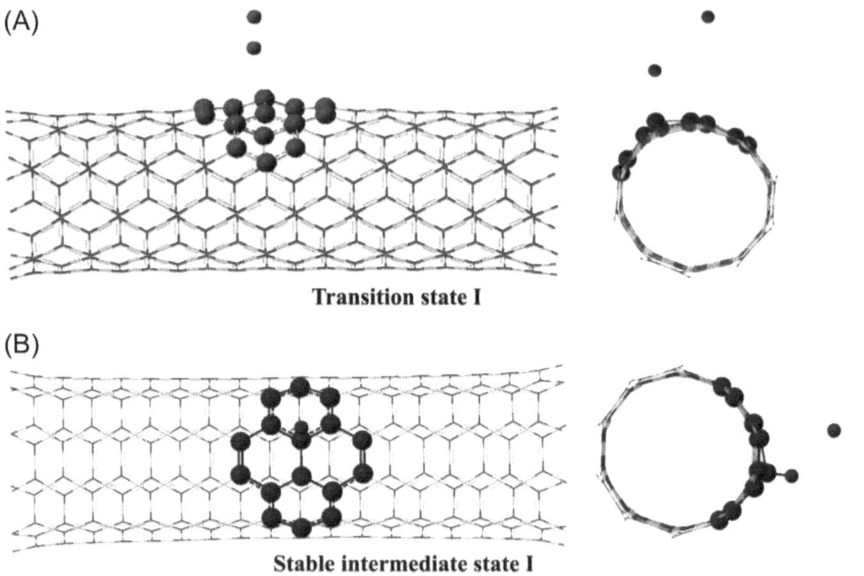

FIGURE 8.27

Optimized structure for transition state I (TS I) and stable intermediate state I (SI I) [53].

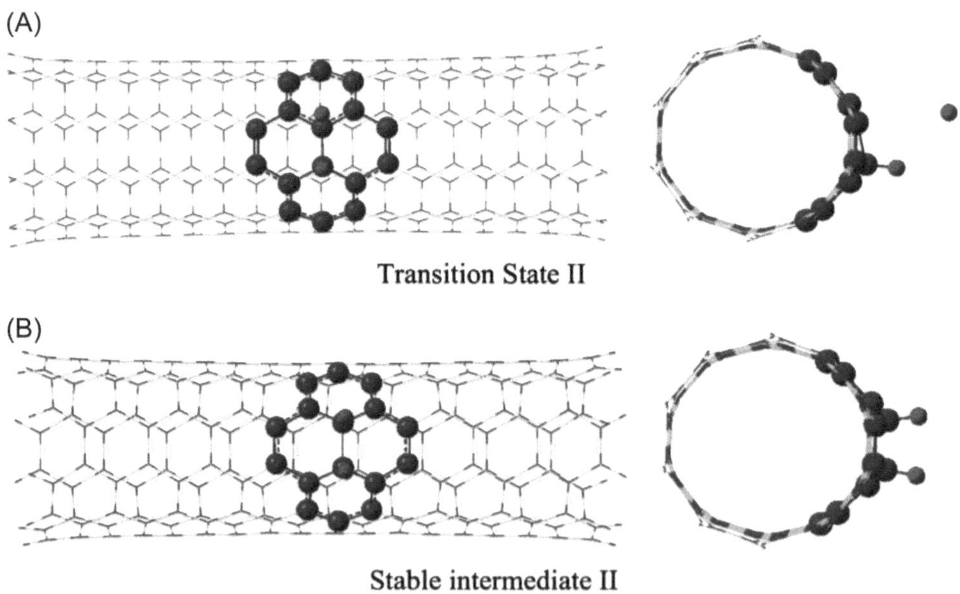

FIGURE 8.28

Optimized structure for transition state II (TS II) and stable intermediate II (SI II) [53].

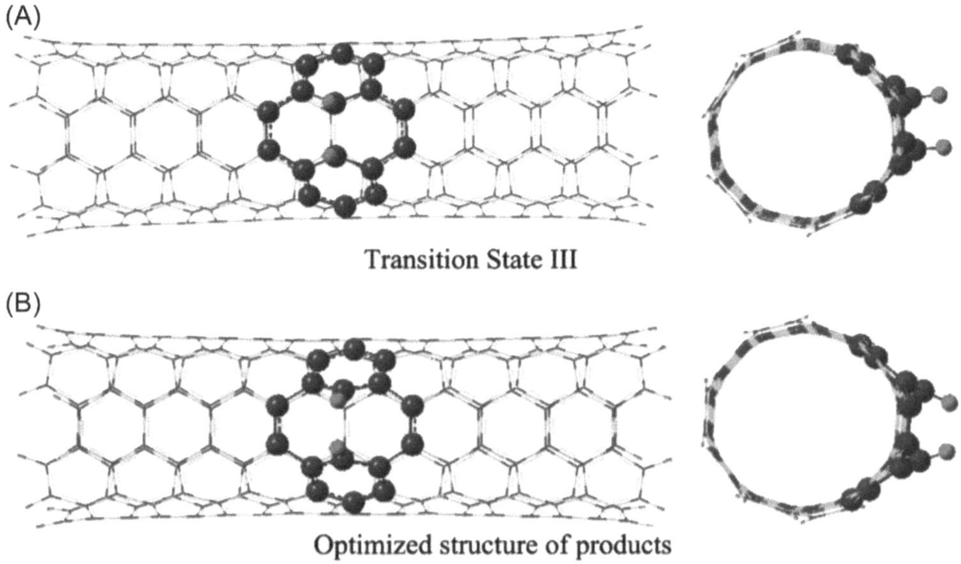

FIGURE 8.29

Optimized structure for transition state III (TS III) and the final stable product of 2H-CNT [53].

We have discussed the relative geometry variation for each state along the chemical pathway. Now we move on to examine the energy changes, which we have calculated, as the reaction progresses from each state to the next. It can be observed from Fig. 8.26 that the adsorption processes for both of the hydrogen atoms onto the sidewall of the tube are exothermic reactions. For the first hydrogen atom (H_1), with the H–C bonding energy of -1.42 eV (-32.746 kcal/mol), it is evident that the reactants achieve an energetically favorable state following the chemical reaction between one hydrogen atom and the CNT. For the second hydrogen atom (H_2), with -2.58 eV released from the reaction, it reaches the stable state where both H_1 and H_2 are chemisorbed at the exterior of the CNT, but the C–C bond remains intact, see Fig. 8.28. However, it is not the most favorable state until the C–C bond is finally severed and this severance process is still energetically favorable with -0.20 eV released. The present results of this exothermic reaction between hydrogen atoms and the CNT are different from that of the report for H_2 molecules with CNT, by Han et al. [102], which indicated that the reaction is an endothermic process with 0.796 eV (18.352 kcal/mol) being absorbed by the system. The difference in the results may be due to the reason that more energy is required to break the H–H bond. To substantiate this, we note that it has been reported that about 4.565 eV (105.269 kcal/mol) will be absorbed to dissociate one hydrogen molecule into two hydrogen atoms [125].

8.3.3 ENTHALPIES AND FREE ENERGIES OF THE REACTION

Thus far, for the various states of the chemisorptions process, we have obtained the energy and geometry structure of atoms and molecules from first principle calculations. However, all the above arguments are carried out at absolute temperature. A practical question thus arises as to how temperature effects will influence these stable and transition states, and how the associated molecular vibrations vary over these energy states at a given temperature.

The thermodynamic state quantities are accessible by means of mathematical relationships given by molecular thermodynamics theory [126]. The central themes to solve this problem are the Boltzmann factor P_j and the partition function $Q(N, V, \beta)$. Once the ground state geometry is obtained, we are then able to extend the investigation further by studying the chemical reaction as a function of temperature T. Following the finite temperature study, the effects of Gibbs free energy, entropy, and enthalpy on the bond formation are explored in this section. Still the ONIOM scheme is applied to obtain the frequencies, while the Boltzmann factor and partition function are employed to establish the temperature dependence of the Gibbs free energy of the reaction.

In molecular thermodynamics theory, the Boltzmann factor P_j expresses the "probability" of a state at energy state E_j relative to the probability of a state of zero energy. For a system with energy states E_1, E_2, E_n, the probability of P_j in the state E_j depends exponentially on the energy of that state

$$P_j = \frac{\exp(-\beta E_j)}{\sum_i \exp(-\beta E_i)} \tag{8.3}$$

The temperature T (in Kelvin) is introduced into the quantum level formulation by $\beta = (k_B T)^{-1}$, where k_B is the Boltzmann constant. We are then able to calculate the average energy of a system in the ensemble (NVT) by

$$\langle E \rangle = \sum_j P_j E_j(N,V) = \sum_j \frac{E_j(N,V)\exp\left[-\beta E_j(N,V)\right]}{Q(N,V,\beta)} \tag{8.4}$$

where $Q(N, V, \beta)$ is the partition function of the system.

$$Q(N,V,\beta) = \sum_{j=0}^{\infty} \exp\left[-\beta E_j(N,V)\right] = \frac{1}{1-\exp\left[-\beta E_j(N,V)\right]} \tag{8.5}$$

Applying the partition function, the molecular partition energy can be written as

$$q(V,T) = q_{trans}\, q_{rot}\, q_{vib}\, q_{elec} \tag{8.6}$$

The energy of a molecule E can be written as the sum of the translational, rotational, vibrational, and electronic motion energies

$$E = E_{trans} + E_{rot} + E_{vib} + E_{elec} \tag{8.7}$$

From Eqs. (8.6) and (8.7), certain thermodynamic parameters, such as the Gibbs free energy G, enthalpy H, entropy S, and the heat capacities, can be investigated. The deduction for these parameters can be found in McQuarrie [126].

To enhance our understanding of the feasibility and rates of the chemisorption process in this study, it is necessary to consider two thermodynamic properties of the reaction. These are the free energy difference, ΔG that is between the products (the final state) and reactants (the initial state), and $\Delta G^{\ddagger}$ that describes the energy difference for initiating the conversion of reactants to the transition states. The former determines whether the reaction will be a spontaneous process, while the latter determines the rate of the reaction. The latter is commonly known as the Gibbs free energy of activation.

Based on the earlier results obtained in the present work, we note that it is easier for the hydrogen atom to react with the tube than the H_2 molecules. The adsorption of the former is energetically favorable, suggesting that the chemisorption process occurs spontaneously. We begin with a discussion on the thermochemical properties of entropy, enthalpy, and Gibbs free energy for the reaction process from 20K to 1000K, and the results from the reactants to the final products are presented in Fig. 8.30. It is clearly observed from this figure that as temperature T increases, the enthalpy (ΔH) and entropy (ΔS) decreases, while the Gibbs free energy ΔG increases. The maximum value of the free energy is still negative, at about −33 kcal/mol at 1000K. This indicates that the chemisorption process of the hydrogen atoms onto the outside wall of the CNT is a spontaneous one, when the temperature ranges from 20K to 1000K. The negative Gibbs free energy (ΔG) also indicates an exergonic process, where the system releases energy to its surroundings during the chemisorption process. Throughout the chemical transformation, the reactants change into the products, release the stored energy, and gradually the system reaches a stable condition of equilibrium. The magnitude of the free energy change is a measure of the "driving force" behind the reaction. The larger this magnitude is, which in this case occurs at lower temperatures, indicates that the reaction process is more likely to take place. Thus, the reaction with $\Delta G = -95$ kcal/mol at 20K has a higher probability of occurring than that of $\Delta G = -33$ kcal/mol at 1000K, see Fig. 8.30. The magnitudes

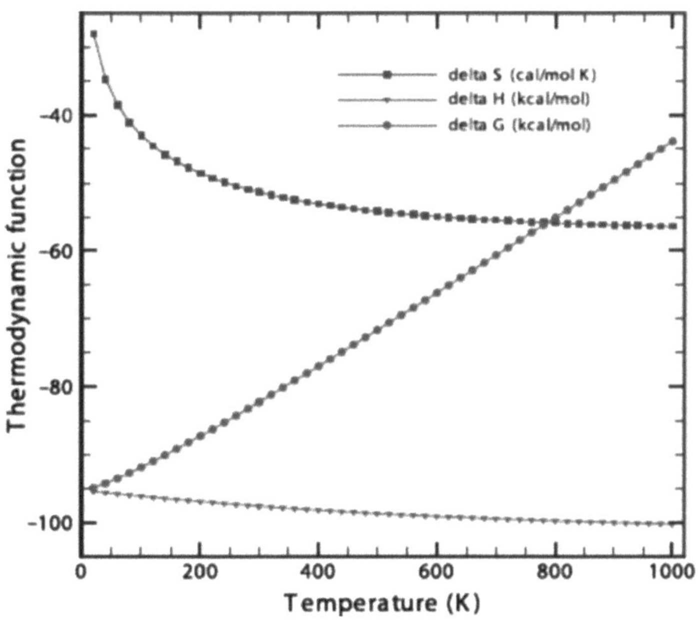

FIGURE 8.30

Variation of thermodynamic properties with temperatures for the reaction of two hydrogen atoms chemisorbed on the sidewall of (5, 5) CNT, from reactants to products [53].

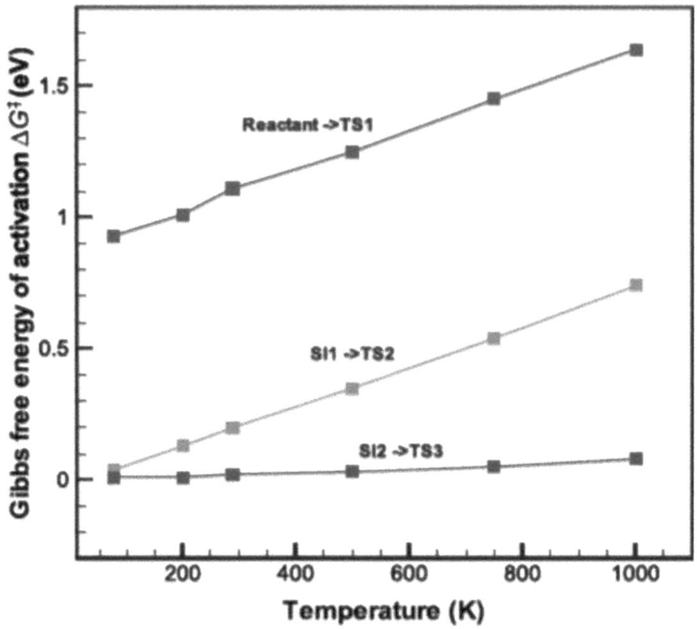

FIGURE 8.31

Gibbs free energies calculated for the reactants, products, the stable intermediates and the transition states [53].

of the enthalpy and entropy are decreasing as well, and the spontaneous change in these properties is often referred to as structural relaxation.

The next issue that naturally arises is whether we can explain the reaction rate in terms of the thermos dynamic properties. The free energy difference between reactants and products accounts for the equilibrium of the reaction, and a negative ΔG only provides information that a reaction can occur spontaneously, but it does not reveal whether it will proceed at a perceptible rate. In other words, the reaction speed is largely unrelated to the ΔG of the reaction. To determine how quickly the equilibrium state is attained, the Gibbs free energy of activation ($\Delta G^{\ddagger}$) which regulates the rate of the reaction has to be considered. The free energy of activation $\Delta G^{\ddagger}$ is sometimes simply called the activation energy, which can be obtained from the difference of the free energy between the transition-state intermediate and the reactant. As $\Delta G^{\ddagger}$ generally has a very large positive value, only a small fraction of the reactant molecules will at any one time acquire this free energy, and the overall rate of the reaction will be limited by the rate of formation of the TS. Thus, we have to consider not only the end points of the reaction, but also the chemical pathway between the end points.

Fig. 8.31 shows the variation of $\Delta G^{\ddagger}$ at typical temperatures, and the corresponding activation energies are illustrated in the plots of Fig. 8.32 which chart the changes in ΔG. These curves reveal that the reaction is more accelerated at lower temperatures due to the resulting lower activation energy between the reactants and its transition state intermediate. To summarise, a pristine CNT can interact with hydrogen atoms at ambient temperatures since the reaction for the chemisorption

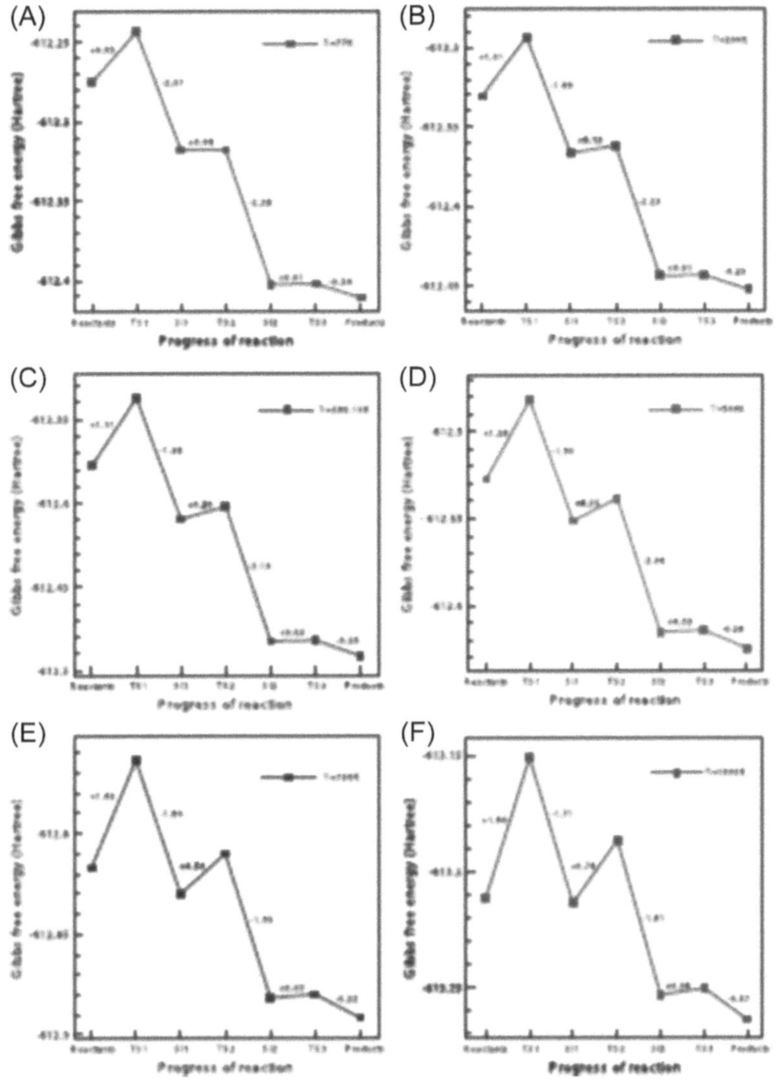

FIGURE 8.32

Reaction process and Gibbs free energy changes at different temperatures [53].

occurs spontaneously, but to improve (increase) the reaction rate, the reaction temperature should preferably be lower since the activation energy is found to be smaller at lower temperatures. This can be clearly seen in Fig. 8.32, where at 77K the activation energies from the reactants and SI states to the respective subsequent TS states are lower than at other higher temperatures.

We reach a similar conclusion for the reaction rates. The rate coefficient for the overall reaction rate can be calculated according to the Arrhenius equation

$$k(T) = \frac{k_B T}{hc^o} e^{-\Delta G_+^{\ddagger}/RT} \tag{8.8}$$

where c° represents the concentration constants. Eq. (8.8) provides an illustration of the physical time-scales involved. Comparing the reaction rates at 77K and 1000K for overcoming the energy of activation from the reactants to the first SI state, we have $k(77\text{K}){:}k(1000\text{K}) = 1{:}133$. It shows that an increase of 0.71 eV in $\Delta G^{\ddagger}$ leads to a 133-fold decrease in the reaction rate. Thus relatively small changes in $\Delta G^{\ddagger}$ can lead to large changes in the overall rate of the reaction. Again, we can see that the reaction at higher temperatures is indeed slower, as one would now come to expect.

8.4 MASS DETECTION

Nanomechanical resonators provide methods for highly sensitive mass detection of virus, gas, and other molecules. Generally, the sensitivity depends on how the resonance frequency or surface stress of a resonator changes when additional mass is adsorbed onto its surface [127–129]. Carbon nanostructures supposedly have the potential for ultra sensitive mass detection due to their extremely small size with combined extraordinary stiffness and low mass density. The vibration characteristics of carbon nanostructures have been studied in much experimental [130–133] and theoretical research [127,134–141]. With fundamental frequencies of the order of terahertz [134,140], cantilevered or bridged SWCNTs may have a mass sensitivity of up to 10^{-21} g when used for mass detection [127]. Is there any method which leads to more sensitive mass detection? The answer is Yes. Here we suggest a new strategy for ultra sensitive mass detection via SWCNC resonators. Since the experimental discovery in 1994 [142], CNCs with all five apex angles have been synthesized [143]. Recently, CNCs have also attracted much attention due to their unusual form. To the best of the knowledge of the authors, although a few publications have reported vibration characteristics of CNCs [144,145], studies focusing on mass sensitivity of CNC resonators are still inadequate. Hu et al. [144] simulated free transverse vibrations of open-ended single-wall CNCs (SWCNCs) using MD and Timoshenko beam model, and compared fundamental frequencies of SWCNCs with those of equivalent SWCNTs. Their MD results reveal that fundamental frequencies of SWCNCs with an apex angle of 19.2 degrees and a top radius of 0.34 nm are several times larger than those of equivalent SWCNTs. A higher resonant frequency implies a higher mass sensitivity. Besides, another natural characteristic of CNCs, its damping or quality factor, determines the energy storage time, which also plays an important role in mass detection. It should not be neglected that the quality factor decreases significantly due to surface effects as the mechanical resonator scales down. Jiang et al. [146] theoretically predicted that the favorable properties for CNTs include a high Young's modulus and low density, giving rise to a ultrahigh quality factor with the order of 2×10^5 via classical MD. Later, Hüttel et al. [147] experimentally measured the quality factors for CNTs and they observed gate-tunable resonances with resonance frequencies of approximately 350 MHz and quality factors of over 10^5. Such a large quality factor makes CNTs extremely suitable for mass detection. CNCs and CNTs are both carbon based nanomaterials with the same crystal lattice microstructures and thus have similar natural characteristics. For example, it is found that SWCNCs have high Young's moduli of the same order as the equivalent SWCNTs [148,149]. Therefore, it is reasonable to suggest that CNCs possess similar high quality factors as CNTs. The above studies clearly confirm the suitability of using carbon nanocones for the ultra sensitive mass detection.

So far, there are no studies of mass sensitivity of SWCNC-based mass detectors. The target of this study is to explore frequency shifts of cantilevered SWCNC resonators when extra mass attached at the free end using higher-order gradient theory. Various cantilevered SWCNC and SWCNT resonators with different top radii and lengths are simulated to exhibit the outstanding performance of SWCNC resonators in mass detection.

A SWCNC is treated as being formed by rolling up a fan-shape graphite sheet and connecting its two sides together into a seamless hollow conical structure. In the hexagonal crystal lattice microstructure, each atom connects it to three neighboring atoms by covalent bonds and we adopt this as a representative cell. The unique conical structure leads to a monotonically increasing radius along the longitudinal direction which gives rise to a change in the deformation at different positions. In addition, each atom in the SWCNC possesses a different chirality even at the same level [148,150].

The deformation process of an initial equilibrium SWCNC mapped from an initial equilibrium graphite sheet is divided into two steps. Firstly, the SWCNC is just rolled rigorously from the initial equilibrium fan-shaped graphite sheet which can be precisely described by geometrical conditions:

$$\begin{aligned} x_1 &= \sqrt{X_1^2 + X_2^2}\cos\left(\frac{\alpha}{2}\right) \\ x_2 &= R\sin\left(2\pi\frac{\beta}{\phi}\right) \\ x_3 &= -R\cos\left(2\pi\frac{\beta}{\phi}\right) \end{aligned} \tag{8.9}$$

where $\mathbf{x}_c = \mathbf{x}_c(x_1, x_2, x_3)$ is nodal coordinate of the formed SWCNC, ϕ is the angle of the graphite sector, α is the apex angle of the cone, and R is the radius of the corresponding section. These parameters are determined by:

$$\begin{aligned} \phi &= \frac{i\pi}{3}, \quad i = 1, 2, 3, 4, 5 \\ \alpha &= 2\arcsin\left(\frac{\phi}{2\pi}\right) \\ R &= \frac{\phi}{2\pi}\sqrt{X_1^2 + X_2^2} \\ \beta &= \arctan\left(\frac{X_2}{X_1}\right) \end{aligned} \tag{8.10}$$

$\mathbf{F}(\mathbf{X}) = \frac{\partial \mathbf{x}}{\partial \mathbf{X}}$ and $\mathbf{G}(\mathbf{X}) = \frac{\partial^2 \mathbf{x}}{\partial \mathbf{X}^2}$ are the first- and second-order deformation gradients tensors in the rolled process. The achievement of the minimum-potential configuration of a SWCNC in the second step will be introduced as follows.

8.4.1 THE INITIAL EQUILIBRIUM SWCNC

In the investigation of resonant frequency shifts of SWCNCs when attached to extra mass, the initial equilibrium configuration with the consideration of edge effect is obtained by the following two steps. The deformation of a SWCNC is first mapped rigidly from the initial equilibrium graphite sheet and then follows an optimized process of minimizing the potential of the system to obtain the initial equilibrium SWCNC. Therefore, the first- and second-order deformation gradients are calculated by the following two terms:

$$\mathbf{F} = \mathbf{F}_1 + \mathbf{F}_2, \quad \mathbf{G} = \mathbf{G}_1 + \mathbf{G}_2 \tag{8.11}$$

where $\mathbf{F}_1$ and $\mathbf{G}_1$ denote the gradients arrived from the process of rolling rigidly the planar graphite sheet to a SWCNC in the first step. $\mathbf{F}_2$ and $\mathbf{G}_2$ in the second step are approximated by MK interpolation:

$$\mathbf{F}_2 = \sum_{i=1}^{n} \phi_{i,\mathbf{X}} u_i, \quad \mathbf{G}_2 = \sum_{i=1}^{n} \phi_{i,\mathbf{XX}} u_i \tag{8.12}$$

where n is the total number of nodes covered in the compact support domain. The MK interpolation is built in the two-dimensional sheet and the quadratic basis is adopted.

Consequently, we obtain the incremental equation of the system:

$$\int_{\Omega} [(\mathbf{L}\cdot\boldsymbol{\Phi})^T \cdot \mathbf{M} \cdot (\mathbf{L}\cdot\boldsymbol{\Phi})] \Delta \mathbf{U}\, dV = -\int_{\Omega} (\mathbf{L}\cdot\boldsymbol{\Phi})^T \cdot \sigma_n\, dV + \int_{\partial\Omega} \boldsymbol{\Phi}^T \mathbf{t}_0^P\, dS + \int_{\partial\Omega} (N_1\boldsymbol{\Phi}_1 + N_2\boldsymbol{\Phi}_2)^T \cdot \mathbf{t}_0^Q\, dS \tag{8.13}$$

The initial equilibrium configuration of the SWCNC with the consideration of edge effect is achieved by iteratively solving the incremental equation until the nonequilibrium force (the right of Eq. (8.13)) reaches zero. Similarly, the initial equilibrium configuration of the SWCNT with the consideration of edge effect can be obtained using the above approach.

8.4.2 ANALYSIS OF RESONANT FREQUENCY AND FREQUENCY SHIFT

Hamilton's principle is employed to study the resonant frequency and frequency shift of cantilevered carbon nanostructure-based mass detectors. The real solution of history of motion during period $t_1 \sim t_2$ is determined by minimizing the Lagrangian functional:

$$\delta \int_{t_1}^{t_2} (E(t) - W(t) + L(t))dt = 0 \tag{8.14}$$

where $E(t)$ and $W(t)$ denote kinetic energy and elastic energy stored in carbon nanostructures, and $L(t)$ is total work done by the external force on the body and is not considered due to being free of external force. Variations of kinetic energy of the system and elastic energy stored in nanostructures are:

$$\delta \int_{t_1}^{t_2} E(t)\, dt = \delta \int_{t_1}^{t_2} \int_V \frac{1}{2} \rho \hat{\mathbf{U}}'(t)^T \hat{\mathbf{U}}'(t)\, dVdt = -\int_{t_1}^{t_2} \delta(\mathbf{U}(t)^T) \int_V \rho \boldsymbol{\Phi}^T \boldsymbol{\Phi} \mathbf{U}''(t)\, dVdt \tag{8.15}$$

$$\delta \int_{t_1}^{t_2} W(t)\, dt = \delta \int_{t_1}^{t_2} \int_V \frac{1}{2} \mathbf{grad}^T \cdot \mathbf{M} \cdot \mathbf{grad}\, dVdt = \int_{t_1}^{t_2} \delta(\mathbf{U}(t)^T) \int_V (\mathbf{L}\boldsymbol{\Phi})^T \mathbf{M}\mathbf{L}\boldsymbol{\Phi}\mathbf{U}(t)\, dVdt \tag{8.16}$$

where ρ is the area mass density calculated by $\frac{m_c}{\Omega_I}$ with m_c being 1.9943×10^{-23} g.

Substituting Eqs. (8.16) and (8.15) into Eq. (8.14), we can derived the dynamic equation of the system as follows:

$$\int_V [\rho \boldsymbol{\Phi}^T \boldsymbol{\Phi} \mathbf{U}''(t) + (\mathbf{L}\boldsymbol{\Phi})^T \mathbf{M} \mathbf{L} \boldsymbol{\Phi} \mathbf{U}(t)]\, dV = 0 \tag{8.17}$$

Considering the field function $\mathbf{U}(t) = \mathbf{A}\exp(i\omega t)$, Eq. (8.17) can be rewritten as:

$$\int_V [(\mathbf{L}\boldsymbol{\Phi})^T \mathbf{M} \mathbf{L} \boldsymbol{\Phi} - \omega^2 \rho \boldsymbol{\Phi}^T \boldsymbol{\Phi}]\mathbf{U}\, dV = 0 \tag{8.18}$$

where $\omega = 2\pi f$ is angular frequency. For simplification, we denote $\mathbf{K} = \int_V (\mathbf{L}\boldsymbol{\Phi})^T \mathbf{M}\mathbf{L}\boldsymbol{\Phi} dV$, and $\mathbf{M}_A = \int_V \rho \boldsymbol{\Phi}^T \boldsymbol{\Phi} dV$. Solutions of resonant frequencies become a matter of solving eigenvalue problems. The resonant frequency is sensitive to the resonator mass matrix $\mathbf{M}_A$ and thus a change in extra mass attached at the free end of cantilevered carbon nanostructure resonator results in a variation in resonant frequency. The critical issue of mass detection is the magnitude of frequency shift corresponding to the attached mass. A larger frequency shift of resonators with the same added mass implies greater sensitivity for mass detection. That is to say, the mass sensitivity is determined by the magnitude of resonant frequency shift.

The higher-order gradient theory is adopted to calculate the resonant frequency and frequency shift of cantilevered SWCNC resonators, according to the attached mass at the free end. This theory can give a good prediction of fundamental frequencies of carbon nanostructures. To illustrate this point clearly, we select cantilevered 19.2 degrees SWCNC and SWCNT as examples. The initial equilibrium SWCNC has a height of 4.33 nm and a top radius of 0.34 nm. The fundamental frequency of this cantilevered SWCNC obtained using the present method is 0.393 THz, which is quite close to that of Hu et al.'s result obtained using molecular dynamic simulation (approximately 0.4 THz) [144]. In addition, the obtained fundamental frequency of a cantilevered 6 nm long (7, 7) SWCNT using the present method and atomistic simulation are, respectively, 9.006×10^{-2} THz and 8.989×10^{-2} THz, with only an 0.19% difference. Therefore, the above comparative studies confirm the accuracy of the present approach.

To investigate the mass sensitivity, numerical simulation of kinds of SWCNC-based mass detectors with apex angles of 19.2 and 38.9 degrees is carried out, along with the equivalent SWCNT-based mass detectors. Fig. 8.33 depicts the schematic of cantilevered SWCNC- (A) and SWCNT-based mass detectors (B) with attached mass at the free end.

Fig. 8.34 demonstrates fundamental frequencies of cantilevered SWCNCs and (7, 7) SWCNT resonators with the same top radius of 0.48 nm. It clearly reveals that the two kinds of SWCNCs with apex angles of 19.2 and 38.9 degrees exhibit far higher fundamental frequencies than equivalent SWCNTs. To further illustrate this feature, we select the cantilevered 19.2 degrees SWCNC and SWCNT resonators with the same length of 6 nm as an example. The fundamental frequency of 19.2 degrees SWCNC resonator is 0.282 THz, while that of the SWCNT resonator is only 0.090 THz, indicating a greater potential of SWCNC resonators to be used for high-performance mass detection than SWCNT resonators. The curves of these two kinds of cantilevered SWCNC resonators demonstrate that the apex angle has a slight decreasing effect on fundamental frequencies. It is worthwhile to note that the heights of these two kinds of nanostructures distinctively affect fundamental frequencies. Fundamental frequencies decrease sharply as heights of SWCNCs increase. A decline of 51.75% is observed in the fundamental frequency of 19.2 degrees SWCNC

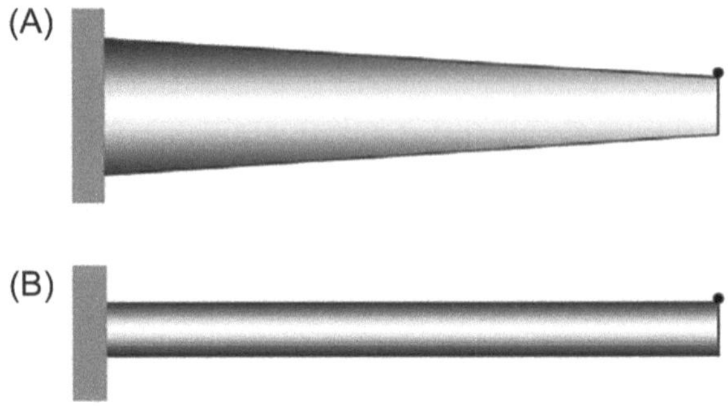

FIGURE 8.33

The schematic of cantilevered SWCNC (A) and SWCNT resonators (B) with a mass attached at the free end [151].

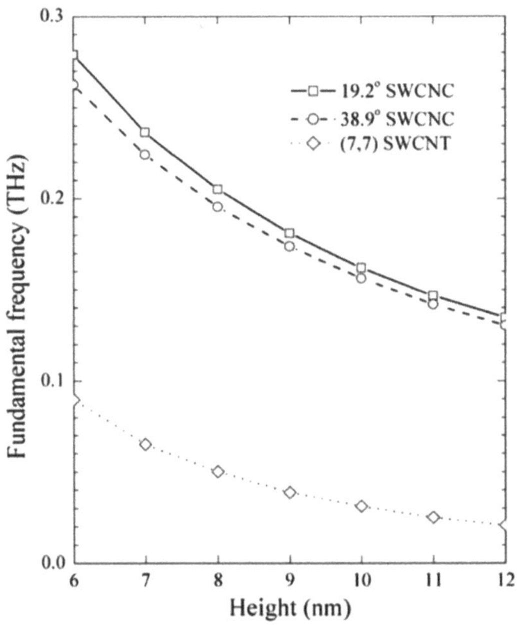

FIGURE 8.34

Fundamental frequencies of two kinds of cantilevered SWCNCs with the same top radius of 0.48 nm and equivalent SWCNTs for different heights [151].

resonator as the height increases by twofold. This suggests that reduction of the size of resonators can markedly enhance the mass detection ability.

Fig. 8.35A demonstrates resonant frequencies of cantilevered (7, 7) SWCNT resonators with attached mass at the free end. The frequency-attached mass curves of SWCNT resonators show a clear variation of resonant frequency when the attached mass is larger than 10^{-21} g, implying the

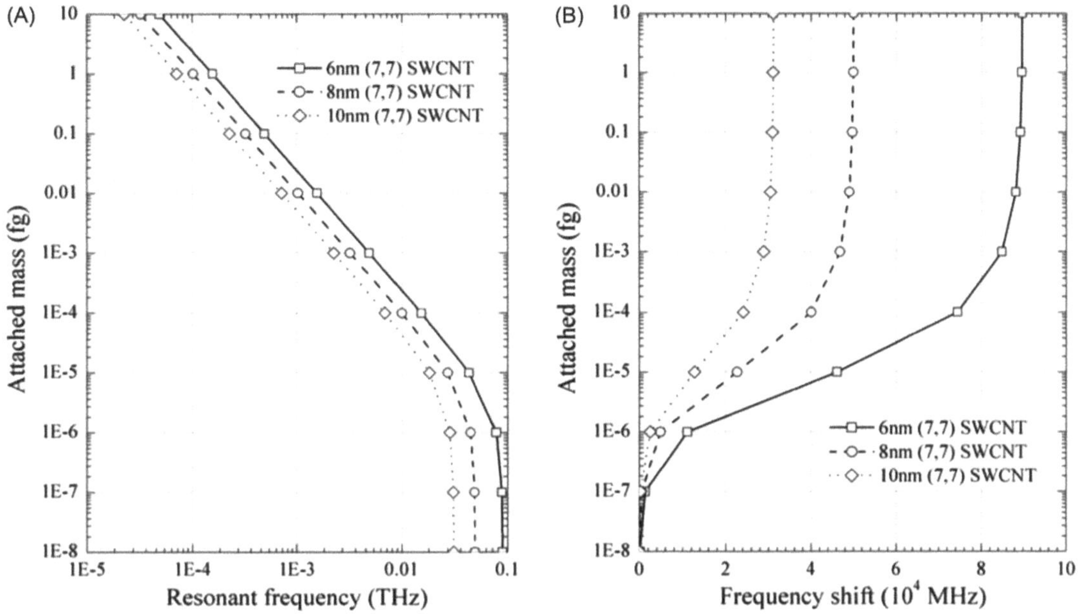

FIGURE 8.35

Resonant frequencies (A) and frequency shifts (B) of cantilevered (7, 7) SWCNT resonators with and different heights for different attached mass [151].

mass sensitivity of SWCNT resonators can reach a level of 10^{-21} g. This conclusion is in agreement with the conclusion of Li and Chou using the molecular structural mechanics method [127]. Frequency shifts of cantilevered SWCNT resonators by attached mass are shown in Fig. 8.35B. Apparently, the shortest SWCNT resonator has an evident change of frequency shift especially when the attached mass changes from 10^{-20} to 10^{-21} g. This confirms the aforementioned suggestion that reducing the size of resonators provides a way of improving the mass detection ability.

Logarithmically linear functional relationships exist between resonant frequencies of cantilevered SWCNT resonators and the attached mass if the attached mass is larger than 10^{-20} g. This conclusion is also applicable to the cantilevered SWCNC resonators. Fig. 8.36A shows the comparison of resonant frequencies of cantilevered SWCNC and (7, 7) SWCNT resonators with the same height of 6 nm and top radius of 0.48 nm by attached different mass. It reveals that the resonant frequency of cantilevered 38.9 degrees SWCNC resonator has a parallel trend as that of the (7, 7) SWCNT resonator and much higher than the 19.2 degrees SWCNC resonator in most cases although it possesses a lower fundamental frequency. In other word, the cantilevered 38.9 degrees SWCNC resonator is not as sensitive as the 19.2 degrees SWCNC resonator when attached to the same extra mass at the free end. It is noteworthy that the same conclusion can be even extended to 19.2 degrees SWCNC resonators if the attached mass is larger than 10^{-21} g which is a magnitude more sensitive than other two resonators. On the other hand, curves shown in Fig. 8.36B display clearly that both two kinds of cantilevered SWCNC-based mass detector exhibit much higher

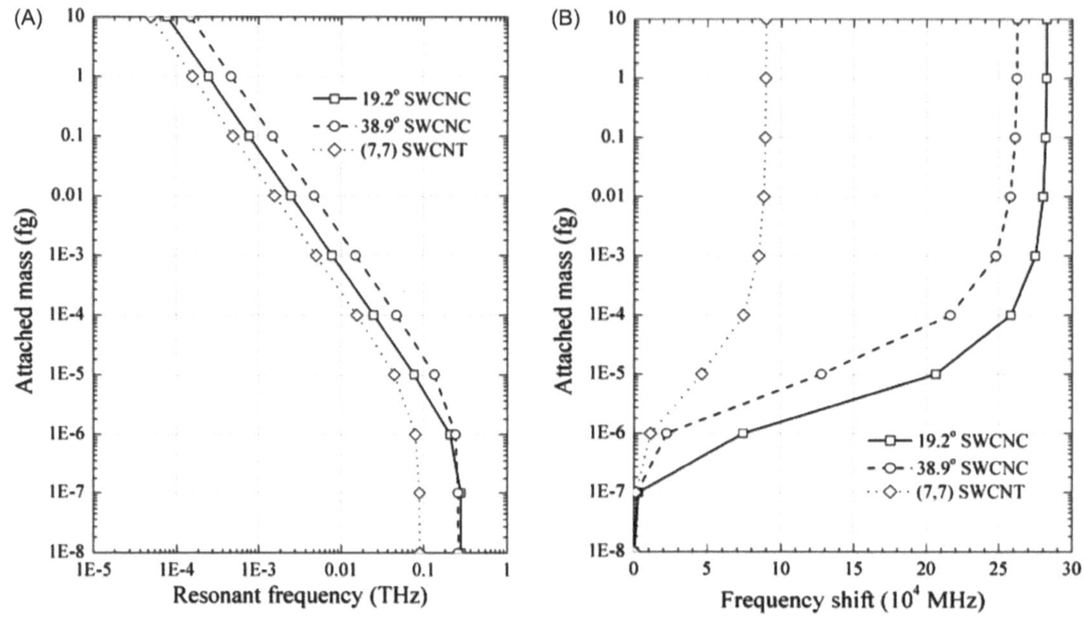

FIGURE 8.36

Resonant frequencies (A) and frequency shifts (B) of cantilevered SWCNC and SWCNT resonators with the same height of 6 nm for different attached mass [151].

frequency shifts than cantilevered SWCNT-based ones when attached a extra mass smaller than 10^{-20} g. Nevertheless, the changes in frequency shifts of cantilevered 19.2 degrees SWCNC-based mass detector are far larger than SWCNT-based ones and also the 38.9 degrees SWCNC-based mass detectors when the attached mass is smaller than 10^{-20} g. The frequency shift of a cantilevered 19.2 degrees SWCNC-based mass detector with an attached mass of 10^{-21} g is 74.1464 GHz, while that of the 38.9 degrees SWCNC- and the equivalent SWCNT-based mass detector are only 22.022 and 11.096 GHz, respectively. A larger resonant frequency shift is much easier to be detected, which means 19.2 degrees SWCNC resonators are most sensitive than the others. When the attached mass changes from 10^{-20} to 10^{-21} g and from 10^{-21} to 10^{-22} g, the changes of frequency shifts of the 19.2 degrees SWCNC-based mass detector are 132.256 and 71.220 GHz, respectively; while the changes in 38.9 degrees SWCNC- and SWCNT-based mass detectors are, 105.673 and 19.838 GHz, 35.078 and 9.807 GHz, respectively, for the same changes in attached mass. Therefore, we can concluded that the mass sensitivity of the cantilevered 19.2 degrees SWCNC-based mass detector can reach a level of 10^{-22} g from the comparison of the changes in the frequency shifts when the attached mass changes from 10^{-21} to 10^{-22} g.

Figs. 8.37–8.40 present the resonant frequency and frequency shift of cantilevered SWCNC-based mass detectors with different apex angles, top radii, and heights obtained using the present method. Fig. 8.37 shows the resonant frequency and frequency shift of the cantilevered 19.2 degrees SWCNC-based mass detector with top radius of 0.48 nm and different height by attached extra mass at the free end. An interesting phenomenon shown in Fig. 8.37A is that the height of SWCNC

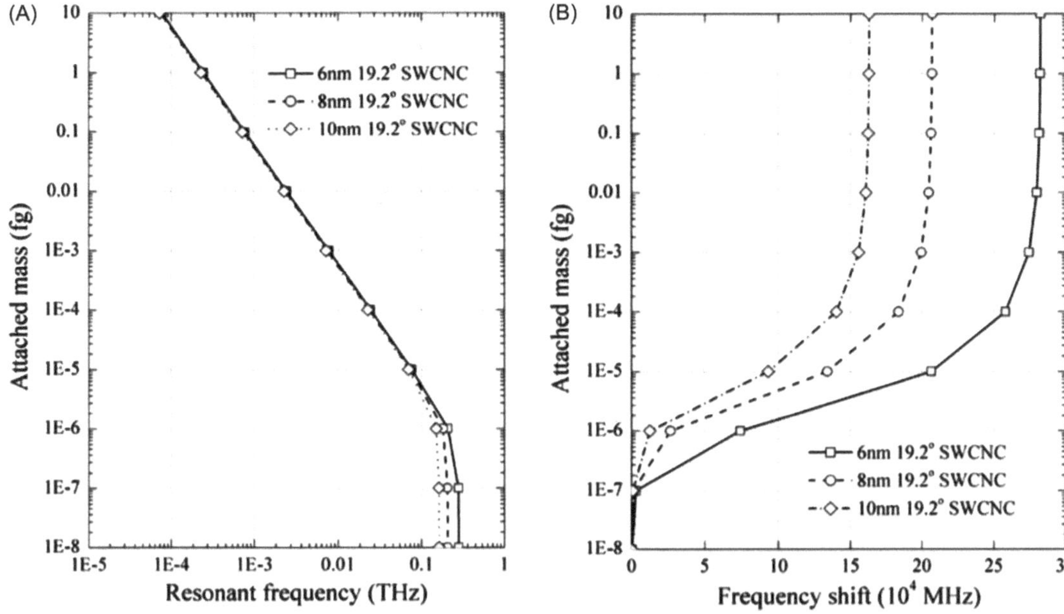

FIGURE 8.37

Resonant frequencies (A) and frequency shifts (B) of cantilevered 19.2 degrees SWCNC resonators with the same top radius of 0.48 nm and different heights for different attached mass [151].

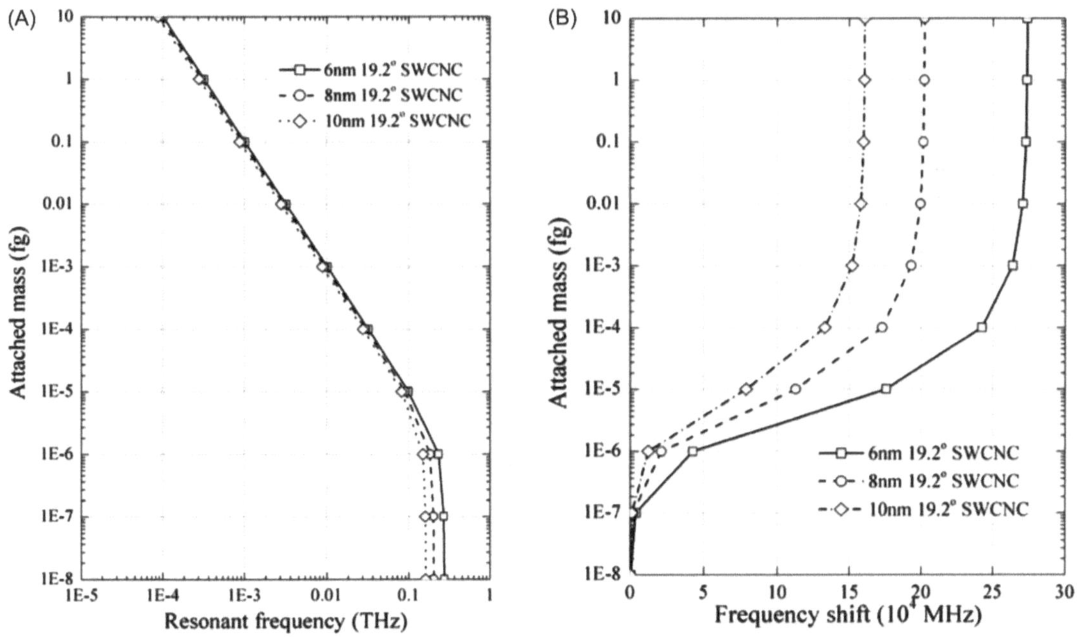

FIGURE 8.38

Resonant frequencies (A) and frequency shifts (B) of cantilevered 19.2 degrees SWCNC resonators with the same top radius of 0.34 nm and different heights for different attached mass [151].

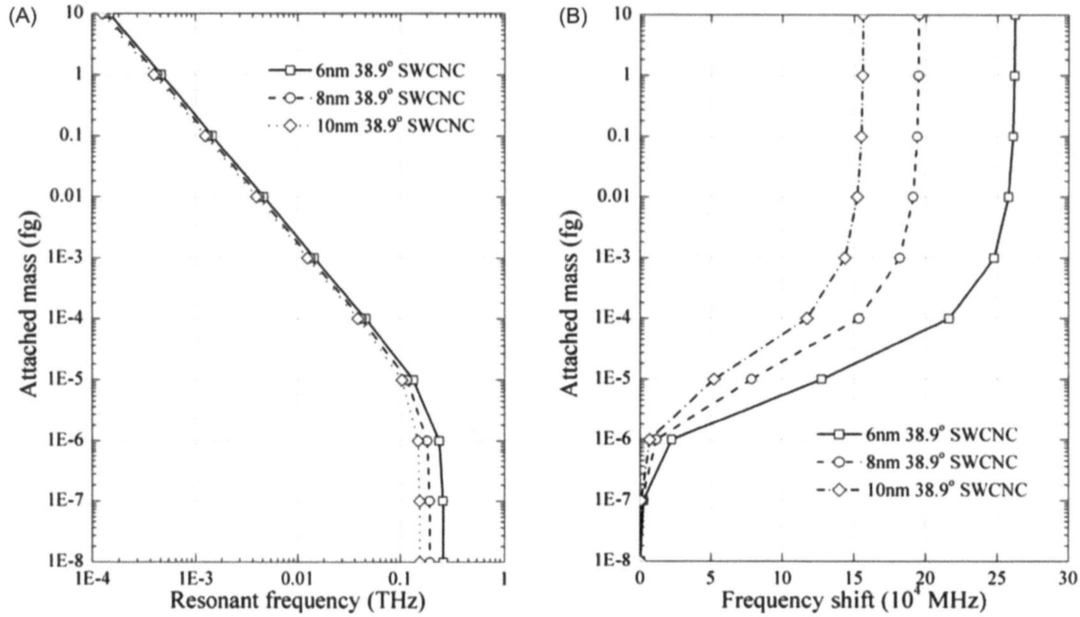

FIGURE 8.39

Resonant frequencies (A) and frequency shifts (B) of cantilevered 38.9 degrees SWCNC resonators with the same top radius of 0.48 nm and different heights for different attached mass [151].

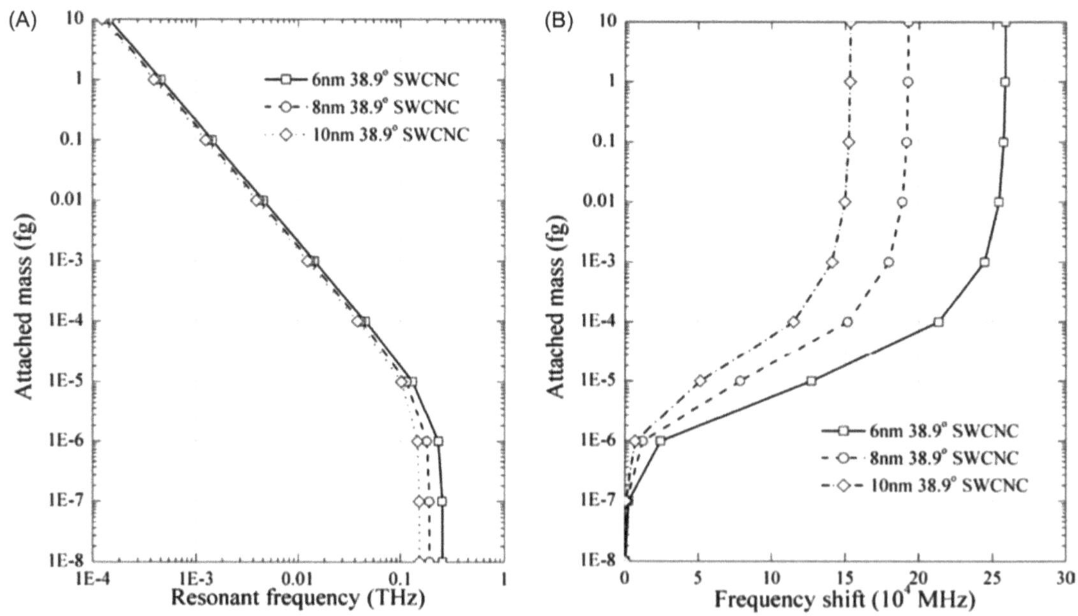

FIGURE 8.40

Resonant frequencies (A) and frequency shifts (B) of cantilevered 38.9 degrees SWCNC resonators with the same top radius of 0.34 nm and different heights for different attached mass [151].

resonators plays an insignificant role on the resonant frequency when the attached mass is larger than 10^{-20} g. Correspondingly, Fig. 8.37B shown the three curves are quite parallel to each other when the attached mass is larger than 10^{-20} g. It reveals that the changes of frequency shift of the three kinds of 19.2 degrees SWCNC-based mass detectors, owing to the variation of attached mass, are the same. This special characteristic offers better practicability in practical application because the fabrication process of a 19.2 degrees SWCNC-based mass detector is not as demanding as a SWCNT-based mass detector. A deviation in the size of SWCNC resonators causes a slight loss of accuracy of precision.

The resonant frequency and frequency shift of cantilevered 19.2 degrees SWCNC-based mass detectors with a top radius of 0.34 nm and different heights shown in Fig. 8.38A demonstrates that as the top radius of SWCNC resonator decreasing, the difference of resonant frequency between SWCNC resonators with different height becomes larger.

Figs. 8.39 and 8.40 indicate that as the apex angle increases, the cantilevered SWCNC-based mass detectors become similar to the equivalent SWCNT-based mass detector. The apex angle has a decreasing effect on the mass sensitivity of a SWCNC-based mass detector.

The effect of top radii of SWCNCs on frequency shifts is also examined. The two kinds of SWCNC resonators with different top radii and apex angles are simulated and computational results

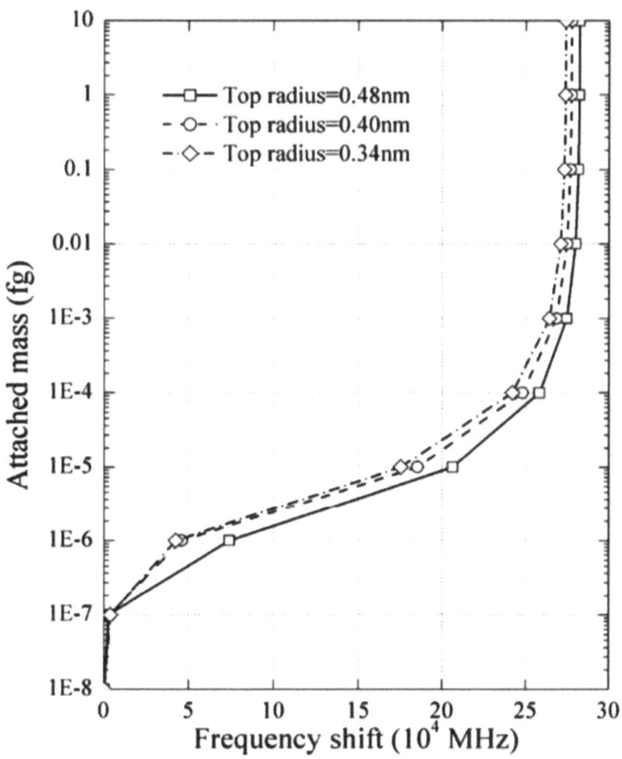

FIGURE 8.41

Resonant frequency shifts of cantilevered 19.2 degrees SWCNCs with the same height of 6 nm and different top radii for different attached mass [151].

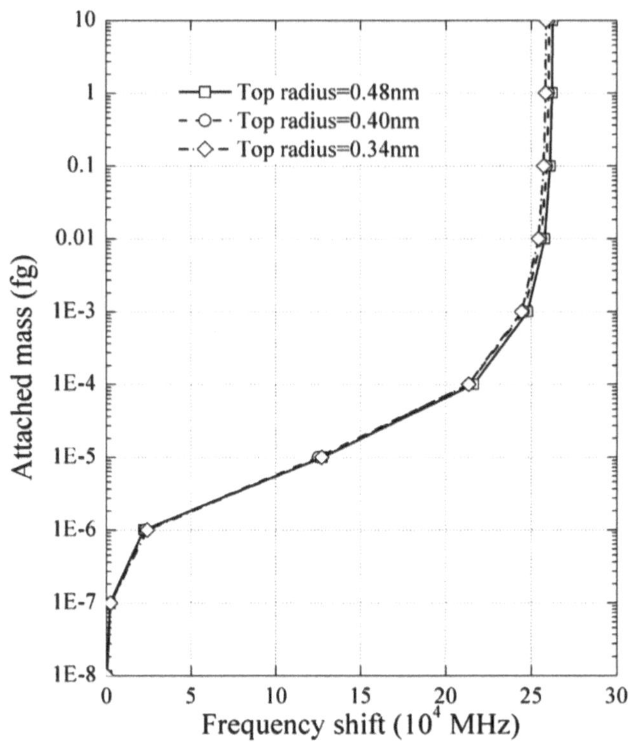

FIGURE 8.42

Resonant frequency shifts of cantilevered 38.9 degrees SWCNCs with the same height of 6 nm and different top radii for different attached mass [151].

are showed in Figs. 8.41 and 8.42. The change of the top radius gives rise to a slight deviation of the frequency shift-attached mass curve and as the apex angle increases, the change becomes even smaller. It indicates that the frequency shift is slightly dependent on the top radius of SWCNC-based mass detectors.

We further investigated the high-performance of SWCNC resonators and SWCNT resonators in mass detection, numerical simulations of extra masses attached at arbitrary places. The axial distance between the attached place and free end is denoted as d. Here we introduce a geometrical parameter λ which is defined as $\lambda = \frac{d}{L}$, where L is the total length of mechanical resonators and the lengths of samples are selected as 6 nm. Figs. 8.43 and 8.44 show the variations of resonant frequency shift for cantilevered SWCNTs and SWCNCs with extra masses of 10^{-22} and 10^{-21} g versus different attached positions on the surface. Keeping the mass constant, it is observed that the frequency shifts for both SWCNC and SWCNT resonators decrease sharply as λ increases. With an extra mass of 10^{-22} g, the frequency shift for the 19.2 degrees SWCNC resonator decreases more than two times and sixteen times when the extra mass moved from the free end to the positions $\lambda = 0.32$ and $\lambda = 0.64$. This indicates that the sensitivity is largely dependent on the

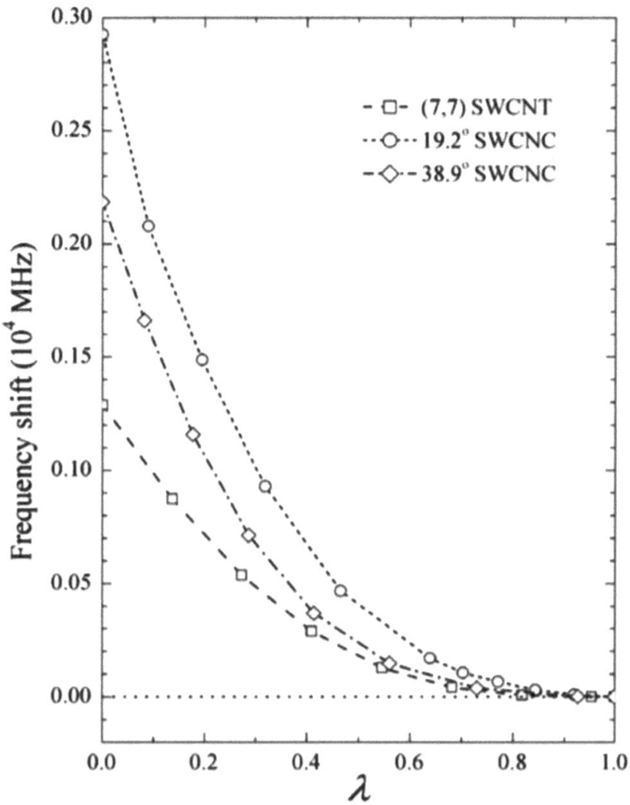

FIGURE 8.43

The variations of resonant frequency shift of cantilevered SWCNT and SWCNCs with an extra mass of 10^{-22} g versus different attached position on the surface [151].

position of the mass. The free end is the most sensitive position. The 19.2 degrees SWCNC resonator has the highest sensitivity. Keeping the same top radius, when the geometrical parameter λ is larger than 0.6, only a small discrepancy in frequency shift for 38.9 degrees SWCNC and (7, 7) SWCNT is observed.

In a word, the potential of cantilevered SWCNC resonators for use in mass detection was examined using the higher-order gradient theory under a mesh-free computational framework based on MK interpolation. Since resonant frequencies of SWCNC resonators are several times higher than those of equivalent SWCNT resonators, SWCNC resonators are more promising for application in mass detection. The shifts of resonant frequencies of SWCNC-based mass detectors are far larger than those of SWCNT-based mass detectors when the same mass is attached at the free end, implying the former has a sensitivity higher than the latter. Comparison of shifts of frequencies of carbon nanostructure-based mass detectors shows that mass sensitivity of 19.2 degrees SWCNC-based mass detectors can at least reach a level of 10^{-22} g. The examination of the effect of top radii on

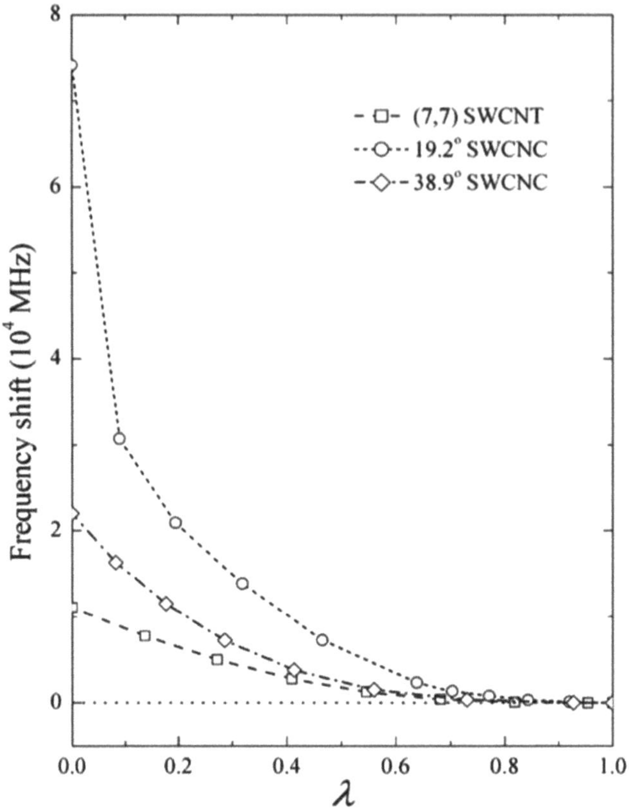

FIGURE 8.44

The variations of resonant frequency shift of cantilevered SWCNT and SWCNCs with an extra mass of 10^{-21} g versus different attached positions on the surface [151].

frequency shifts demonstrates that mass sensitivity of SWCNC-based mass detectors is slightly dependent on top radii and decreases as the apex angle increases.

REFERENCES

[1] Mao ZG, Sinnott SB. A computational study of molecular diffusion and dynamic flow through carbon nanotubes. J Phys Chem B 2000;104(19):4618–24.
[2] Mao Z, Sinnott SB J. Phys Chem B 2001;106:6916.
[3] Marti J, Gordillo MC. Chem Phys Lett 2002;354:227.
[4] Alberto S. Nano Lett 2006;6:633.
[5] Skoulidas AI, Ackerman DM, Johnson JK, Sholl DS. Phys Rev Lett 2002;89:185901.
[6] Holt JK, Park HG, Wang YM, Stadermann M, Artyukhin AB, Grigoropoulos CP, et al. Fast mass transport through sub-2-nanometer carbon nanotubes. Science 2006;312(5776):1034–7.

[7] Hummer G, Rasaiah JC, Noworyta JP. Naturr(London) 2001;414:188.

[8] Yoon J, Ru CQ. Compos Sci Technol. 2005;65:1326.

[9] Majumder M, Chopra N, Andrews R, Hinds BJ. Nature(London) 2005;438:44.

[10] Tuzun RE, Noidy DW, Sumpter BG, Merkle RC. Nanotechnology 1996;7:241.

[11] Mao Z, Sinnott SB. Phys Rev Lett 2002;89:278301.

[12] Solares SD, Blanco M, Goddard WA. Nanotechnology 2004;15:1405.

[13] Walther JH, Werder T, Jaffe RL, Koumoutsakos P. Phys Rev E 2004;69:062201.

[14] Ghosh S, Sood AK, Kumar N. Science 2003;299:1042.

[15] Yarin AL, Yazicioglu AG, Megaridis CM, Rossi MP, Gogotsi Y. J. Appl. Phys. 2005;97:124309.

[16] Supple S, Quirke N. Phys Rev Lett 2003;90:214501.

[17] Cicero G, Grossman JC, Schwegler E, Gygi F, Galli G. Water confined in nanotubes and between graphene sheets: a first principle study. J Am Chem Soc 2008;130(6):1871–8.

[18] Striolo A. The mechanism of water diffusion in narrow carbon nanotubes. Nano Lett 2006;6(4):633–9.

[19] Hummer G, Rasaiah JC, Noworyta JP. Water conduction through the hydrophobic channel of a carbon nanotube. Nature 2001;414(6860):188–90.

[20] de Groot BL, Grubmuller H. The dynamics and energetics of water permeation and proton exclusion in aquaporins. Curr Opin Struc Biol 2005;15(2):176–83.

[21] Vidossich P, Cascella M, Carloni P. Dynamics and energetics of water permeation through the aquaporin channel. Proteins 2004;55(4):924–31.

[22] Liu YC, Wang Q, Zhang L, Wu T. Dynamics and density profile of water in nanotubes as one-dimensional fluid. Langmuir 2005;21(25):12025–30.

[23] Lozano-Castello D, Cazorla-Amoros D, Linares-Solano A, Quinn DF. Micropore size distributions of activated carbons and carbon molecular sieves assessed by high-pressure methane and carbon dioxide adsorption isotherms. J Phys Chem B 2002;106(36):9372–9.

[24] Skoulidas AI, Ackerman DM, Johnson JK, Sholl DS. Rapid transport of gases in carbon nanotubes. Phys Rev Lett 2002;89(18):185901.

[25] Sun L, Crooks RM. Single carbon nanotube membranes: a well-defined model for studying mass transport through nanoporous materials. J Am Chem Soc 2000;122(49):12340–5.

[26] Koga K, Gao GT, Tanaka H, Zeng XC. Formation of ordered ice nanotubes inside carbon nanotubes. Nature 2001;412(6849):802–5.

[27] Walther JH, Jaffe RL, Kotsalis EM, Werder T, Halicioglu T, Koumoutsakos P. Hydrophobic hydration of C-60 and carbon nanotubes in water. Carbon 2004;42(5-6):1185–94.

[28] Tang DS, Ci L, Zhou WY, Me SS. Effect of H2O adsorption on the electrical transport properties of double-walled carbon nanotubes. Carbon 2006;44(11):2155–9.

[29] Joseph S, Aluru NR. Why are carbon nanotubes fast transporters of water? Nano Lett 2008;8(2):452–8.

[30] Rossi MP, Ye HH, Gogotsi Y, Babu S, Ndungu P, Bradley JC. Environmental scanning electron microscopy study of water in carbon nanopipes. Nano Lett 2004;4(5):989–93.

[31] Werder T, Walther JH, Jaffe RL, Halicioglu T, Noca F, Koumoutsakos P. Molecular dynamics simulation of contact angles of water droplets in carbon nanotubes. Nano Lett 2001;1(12):697–702.

[32] Supple S, Quirke N. Rapid imbibition of fluids in carbon nanotubes. Phys Rev Lett 2003;90(21):214501.

[33] Thomas JA, McGaughey AJH. Reassessing fast water transport through carbon nanotubes. Nano Lett 2008;8(9):2788–93.

[34] Zambrano HA, Walther JH, Koumoutsakos P, Sbalzarini IF. Thermophoretic motion of water nanodroplets confined inside carbon nanotubes. Nano Lett 2009;9(1):66–71.

[35] Zhao YC, Song L, Deng K, Liu Z, Zhang ZX, Yang YL, et al. Individual water-filled single-walled carbon nanotubes as hydroelectric power converters. Adv Mater 2008;20(9):1772.

[36] Reddy CD, Lu C, Rajendran S, Liew KM. Free vibration analysis of fluid-conveying single-walled carbon nanotubes. Appl Phys Lett 2007;90(13).

[37] Chen SS. Flow-induced Vibration of Cylindrical Structures. Hemisphere, Washington, 1987.

[38] Sultan G, Hemp J. J. Sound Vib. 1989;132:473.

[39] Reddy CD. Ph.D. thesis. Nanyang Technical University, Singapore, 2006.

[40] Battezzati L, Pisani C, Ricca F. Equilibrium conformation and surface motion of hydrocarbon molecules physisorbed on graphit. J Chem Soc, Faraday Trans 2 1975;71:1629–39.

[41] Yu HQ, Li H, Zhang JX, Liu XF, Liew KM. Possibility of driving water molecules along a single-walled carbon nanotube using methane molecules. Carbon 2010;48(2):417–23.

[42] Jepps OG, Bhatia SK, Searles DJ. Wall mediated transport in confined spaces: exact theory for low density. Phys Rev Lett 2003;91(12):126102.

[43] Maniwa Y, Matsuda K, Kyakuno H, Ogasawara S, Hibi T, Kadowaki H, et al. Nature 2007;6:135–41.

[44] Kalra A, Hummer G, Garde S. J Phys Chem B 2004;108:544–9.

[45] Yu HQ, Li YF, Li H, Zhang K, An CG, Liu XF, et al. Methane molecules drive water molecules along diameter-gradient SWCNTs with junctions. J Phys Chem B 2010;114(26):8676–9.

[46] Iijima S, Ichihashi T, Ando Y. Nature 1992;356:776–8.

[47] Zhen Y, Henk WCP, Leon B, Cees D. Nature 1999;40:273–6.

[48] Maniwa Y, Kataura H, Abe M, Udaka A, Suzuki S, Achiba Y, et al. Chem Phys Lett 2005;401:534–8.

[49] Striolo A. Nano Lett. 2006;6:633–9.

[50] Majumder M, Chopra N, Andrews R, Hinds BJ. Nature. 2005;438:44.

[51] Thomas JA, McGaughey AJH. Nano Lett 2008;8:2788–93.

[52] Ren YX, Ng TY, Liew KM. State of hydrogen molecules confined in C-60 fullerene and carbon nanocapsule structures. Carbon 2006;44(3):397–406.

[53] Ng TY, Ren YX, Liew KM. Adsorption of hydrogen atoms onto the exterior wall of carbon nanotubes and their thermodynamics properties. Int J Hydrogen Energ 2010;35(10):4543–53.

[54] Chen X, Zhang Y, Gao XP, Pan GL, Jiang XY, Qu JQ, et al. Electrochemical hydrogen storage of carbon nanotubes and carbon nanofibers. Int J Hydrogen Energy 2004;29(7):743–8.

[55] Chen P, Wu X, Lin J, Tan KL. High H_2 uptake by alkali-doped carbon nanotubes under ambient pressure and moderate temperatures. Science 1999;285:91–3.

[56] Fernández PS, Castro EB, Real SG, Martins ME. Electrochemical behaviour of single walled carbon nanotubes-hydrogen storage and hydrogen evolution reaction. Int J Hydrogen Energy 2009;34(19):8115–26.

[57] Zheng QR, Gu AZ, Lu XS, Lin WS. Adsorption equilibrium of supercritical hydrogen on multi-walled carbon nanotubes. J Supercrit Fluids 2005;34(1):71–9.

[58] Dillon AC, Gennett T, Alleman JL, Jones KM, Parilla PA, Heben MJ. Carbon nanotube materials for hydrogen storage. In: Proceedings of the 2000 US DOE Hydrogen Program Review, vol II. 2000.

[59] Dillon AC, Gennett T, Alleman JL, Jones KM, Parilla PA, Heben MJ. Carbon nanotube materials for hydrogen storage. In: Proceedings of the 1999 US DOE Hydrogen Program Review, vol II. 1999.

[60] Hirscher M, Becher M, Haluska M, Dettlaff-Weglikowska U, Quintel A, Duesberg GS, et al. Hydrogen storage in sonicated carbon materials. Appl Phys B Lasers Opt 2001;72(2):129–32.

[61] Hirscher M, Becher M, Haluska M, Quintel A, Skakalova V, Choi YM, et al. Hydrogen storage in carbon nanostructures. J Alloys Compd 2002;330–332:654–8.

[62] Li X, Zhu H, Ci L, Xu C, Mao Z, Wei B, et al. Hydrogen uptake by graphitized multi-walled carbon nanotubes under moderate pressure and at room temperature. Carbon 2001;39(13):2077–9.

[63] Ritschel M, Uhlemann M, Gutfleisch O, Leonhardt A, Graff A, Taschner C, et al. Hydrogen storage in different carbon nanostructures. Appl Phys Lett 2002;80(16):2985–7.

[64] Ahn CC, Ye Y, Ratnakumar BV, Whitlam C, Bowman RC, Fultz B. Hydrogen desorption and adsorption measurements on graphite nanofibers. Appl Phys Lett 1998;73(23):3378–80.

[65] Chambers A, Park C, Baker RTK, Rodriguez NM. Hydrogenstorage in graphite nanofibers. J Phys Chem B 1998;102:4253–6.
[66] Chahine R, Bose TK. Low-pressure adsorption storage ofhydrogen. Int J Hydrogen Energy 1994;19 (2):161–4.
[67] Zhou L, Zhou Y, Sun Y. A comparative study of hydrogen adsorption on superactivated carbon versus carbon nanotubes. Int J Hydrogen Energy 2004;29(5):475–9.
[68] Amankwah KAG, Noh JS, Schwarz JA. Hydrogen storage on superactivated carbon at refrigeration temperatures. Int J Hydrogen Energy 1989;14(7):437–47.
[69] Kiyobayashia T, Takeshitaa HT, Tanakaa H, Takeichia N, Züttelb A, Schlapbachb L, et al. Hydrogen adsorption in carbonaceous materials—how to determine the storage capacity accurately. J Alloys Compd 2002;330-332:666–9.
[70] Strobel R, Jorissen L, Schliermann T, Trapp V, Schutz W, Bohmhammel K, et al. Hydrogen adsorption on carbon materials. J Power Sources 1999;84(2):221–4.
[71] Kojima Y, Suzuki N. Hydrogen adsorption and desorption by potassiumÒ»doped superactivated carbon. Appl Phys Lett 2004;84(20):4113–15.
[72] Panella B, Hirscher M, Roth S. Hydrogen adsorption in different carbon nanostructures. Carbon 2005;43 (10):2209–14.
[73] Blackman JM, Patrick JW, Arenillas A, Shi W, Snape CE. Activation of carbon nanofibres for hydrogen storage. Carbon 2006;44(8):1376–85.
[74] Kong J, Franklin NR, Zhou C, Chapline MG, Peng S, Cho K, et al. Nanotube molecular wires as chemical sensors. Science 2000;287(5453):622–5.
[75] Türker L. AM1 treatment of (Li + nH2)n = 0-5@C60 systems. Int J Hydrogen Energy 2003;28 (2):223–8.
[76] Naghshineh N, Hashemianzadeh M. First-principles study of hydrogen storage on Si atoms decorated C60. Int J Hydrogen Energy 2009;34(5):2319–24.
[77] Türker L. Hydrogen storage capability of Se@C120 system. Int J Hydrogen Energy 2007;32 (12):1933–8.
[78] Talyzin AV, Jacob A. Hydrogen adsorption by ball milled C60. J Alloys Compd 2005;395(1-2):154–8.
[79] Türker L. Hydrogen storage capacity of Be@C115 system. J Mol Struct THEOCHEM 2005;723(1-3):105–10.
[80] Orimoa S, Züttelb A, Schlapbachb L, Majer G, Fukunaga T, Fujii H. Hydrogen interaction with carbon nanostructures: current situation and future prospects. J Alloys Compd 2003;356-357:716–19.
[81] Zuttel A. Materials for hydrogen storage. Materials Today 2003;6(9):24–33.
[82] Dillon AC, Jones KM, Bekkedahl TA, Kiang CH, Bethune DS, Heben MJ. Storage of hydrogen in single-walled carbon nanotubes. Nature 1997;386:377–9.
[83] Lee SM, Lee YH. Hydrogen storage in single-walled carbon nanotubes. Appl Phys Lett 2000;76 (20):2877–9.
[84] Froudakis GE. Hydrogen interaction with single-walledcarbon nanotubes: a combined quantum-mechanics/molecular-mechanics study. Nano Lett 2001;1(4):179–82.
[85] Bauschlicher Jr CW. High coverages of hydrogen on a (10,0) carbon nanotube. Nano Lett 2001; 1(5):223–6.
[86] Bauschlicher Jr CW, So CR. High coverages of hydrogen on (10,0), (9,0) and (5,5) carbon nanotubes. Nano Lett 2002;2:337–41.
[87] Wang Q, Johnson JK. Optimization of carbon nanotube arrays for hydrogen adsorption. J Phys Chem B 1999;103(23):4809–13.
[88] Dodziuk H, Dolgonos G. Molecular modeling study of hydrogen storage in carbon nanotubes. Chem Phys Lett 2002;356(1-2):79–83.

[89] Ma Y, Xia Y, Zhao M, Ying M. Hydrogen storage capacity in single-walled carbon nanotubes. Phys Rev B Condens Matter 2002;65.

[90] Cheng H, Cooper AC, Pez GP, Kostov MK, Piotrowski P, Stuart SJ. Molecular dynamics simulations on the effects of diameter and chirality on hydrogen adsorption in single-walled carbon nanotubes. J Phys Chem B 2005;109(9):3780–6.

[91] Gayathri V, Devi NR, Geetha R. Hydrogen storage in coiled carbon nanotubes. Int J Hydrogen Energy 2010;35(3):1313–20.

[92] Furuya Y, Hashishin T, Iwanaga H, Motojima S, Hishikawa Y. Interaction of hydrogen with carbon coils at low temperature. Carbon 2004;42(2):331–5.

[93] López-Corral I, German E, Volpe MA, Brizuela GP, Juan A. Tight-binding study of hydrogen adsorption on palladium decorated graphene and carbon nanotubes. Int J Hydrogen Energy 2010;35 (6):2377–84.

[94] Iyakutti K, Kawazoe Y, Rajarajeswari M, Surya VJ. Aluminum hydride coated singleÒ»walled carbon nanotube as a hydrogen storage medium. Int J Hydrogen Energy 2009;34(1):370–5.

[95] Oku T, Kuno M, Narita I. Hydrogen storage in boron nitride nanomaterials studied by TG/DTA and cluster calculation. J Phys Chem Solids 2004;65(2-3):549–52.

[96] Oku T, Narita I. Calculation of H_2 gas storage for boron nitride and carbon nanotubes studied from the cluster calculation. Phys B: Condens Matter 2002;323(1-4):216–18.

[97] Wu X, Yang J, Hou JG, Zhu Q. Deformation-induced site selectivity for hydrogen adsorption on boron nitride nanotubes. Phys Rev B Condens Matter 2004;69(15):153411.

[98] Lee SM, Park KS, Choi YC, Park YS, Bok JM, Bae DJ, et al. Hydrogen adsorption and storage in carbon nanotubes. Synth Met 2000;113(3):209–16.

[99] Gang L, Henry S, Nicholas K. Hydrogen-induced unzipping of single-walled carbon nanotubes. Phys Rev B Condens Matter 2003;68(20):201402(R).

[100] Alonso JA, Arellano JS, Molina LM, Rubio A, Lopez MJ. Interaction of molecular and atomic hydrogen with single-wall carbon nanotubes. IEEE Trans Nanotechnol 2004;3(2):304–10.

[101] Verónica B, Jochen H, Gustavo ES. Interaction of atomic hydrogen with single-walled carbon nanotubes: a density functional theory study. J Chem Phys 2004;120(15):7169–73.

[102] Han SS, Lee HM. Adsorption properties of hydrogen on (10,0) single-walled carbon nanotube through density functional theory. Carbon 2004;42(11):2169–77.

[103] Lee SM, An KH, Kim WS, Lee Y, Park YS, Seifert G, et al. Hydrogen storage in carbon nanotubes. Synth Met 2001;121(1–3):1189–90.

[104] Jeloaica L, Sidis V. DFT investigation of the adsorption of atomic hydrogen on a cluster-model graphite surface. Chem Phys Lett 1999;300(1-2):157–62.

[105] Nikolaev P, Thess A, Rinzler AG, Colbert DT, Smalley RE. Diameter doubling of single-wall nanotubes. Chem Phys Lett 1997;266(5-6).

[106] Dapprich S, Komáromi I, Byun KS, Morokuma K, Frisch MJ. A new ONIOM implementation in gaussian98. Part I. The calculation of energies, gradients, vibrational frequencies and electric field derivatives. J Mol Struct THEOCHEM 1999;461–462:1–21.

[107] Kerdcharoen T, Morokuma K. ONIOM-Xs: an extension of the ONIOM method for molecular simulation in condensed phase. Chem Phys Lett 2002;355(3-4):257–62.

[108] Svensson M, Humbel S, Froese R, Matsubara T, Sieber S, Morokuma K. ONIOM: a multilayered integrated MO + MM method for geometry optimizations and single point energy predictions. A test for Diels–Alder reactions and Pt(P(t-Bu)3)2 + HZ oxidative addition. J Phys Chem 1996;100 (50):19357–63.

[109] Tschumper GS, Morokuma K. Gauging the applicability of ONIOM (mo/mo) methods to weak chemical interactions in large systems: hydrogen bonding in alcohol dimers. J Mol Struct Theochem 2002;592(1-3):137–47.
[110] Vreven T, Morokuma K. The ONIOM (our own n-layered integrated molecular orbital + molecular mechanics) method for the first singlet excited (S sub 1) state photoisomerization path of a retinal protonated schiff base. J Chem Phys 2000;113(8):2969–75.
[111] Vreven T, Morokuma K. On the application of the IMOMO (integrated molecular orbital + molecular orbital) method. J Comput Chem 2000;21(16):1419–32.
[112] Vreven T, Morokuma K, Farkas O, Schlegel HB, Frisch MJ. Geometry optimization with QM/MM, ONIOM, and other combined methods. I. Microiterations and constraints. J Comput Chem 2003;24 (6):760–9.
[113] Sanchez-Portal D, Artacho E, Soler JM, Rubio A, Ordejon P. Ab initio structural, elastic, and vibrational properties of carbon nanotubes. Phys Rev B Condens Matter 1999;59:12678–88.
[114] Blase X, Benedict LX, Shirley EL, Louie SG. 72:. Hybridization effects and metallicity in small radius carbon nanotubes. Phys Rev Lett 1994;1994(72):1878–81.
[115] Bauschlicher CW. Hydrogen and fluorine binding to the sidewalls of a (10,0) carbon nanotube. Chem Phys Lett 2000;322(3-4):237.
[116] Froudakis GE. Hydrogen interaction with carbon nanotubes: a review of ab initio studies. J Phys Condens Matter 2002;14:8453–65.
[117] Becke AD. Density-functional thermochemistry I. The effect of the exchange-only gradient correction. J Chem Phys 1992;96(3):2155–60.
[118] Becke AD. Density-functional thermochemistry II. The effect of the PerdewĊWang generalized-gradient correlation correction. J Chem Phys 1992;97(12):9173–7.
[119] Becke AD. Density-functional thermochemistry III. The role of exact exchange. J Chem Phys 1993;98 (7):5648–52.
[120] Becke AD. Density-functional thermochemistry IV. A new dynamical correlation functional and implications for exact-exchange mixing. J Chem Phys 1996;104(3):1040–6.
[121] Becke AD. Density-functional thermochemistry V. Systematic optimization of exchang-correlation functionals. J Chem Phys 1997;107(20):8554–60.
[122] Becke AD. A new mixing of HartreeĊFock and local density functional theories. J Chem Phys 1993;98 (2):1372–7.
[123] Lee C, Yang W, Parr RG. Development of the Colle-Salvetti correlation-energy formula into a functional of the electron density. Phys Rev B: Condens Matter 1988;37:785–9.
[124] Lowry TH, Richardson KS. Mechanism and theory in organic chemistry. 3rd ed. New York: Harper&Row; 1987.
[125] Lide DR. CRC handbook of chemistry and physics. Boca Raton, FL: CRC Press; 1995.
[126] McQuarrie DA. Molecular thermodynamics. Sausalito, CA: University Science Books; 1999.
[127] Li CY, Chou TW. Mass detection using carbon nanotube-based nanomechanical resonators. Appl Phys Lett 2004;84(25):5246–8.
[128] Ren Q, Zhao YP. Influence of surface stress on frequency of microcantilever-based biosensors. Microsyst Technol 2004;10(4):307–14.
[129] Zhang Y, Ren Q, Zhao YP. Modelling analysis of surface stress on a rectangular cantilever beam. J Phys D Appl Phys 2004;37(15):2140–5.
[130] Garcia-Sanchez D, Paulo AS, Esplandiu MJ, Perez-Murano F, Forro L, Aguasca A, et al. Mechanical detection of carbon nanotube resonator vibrations. Phys Rev Lett 2007;99(8).
[131] Chiu HY, Hung P, Postma HWC, Bockrath M. Atomic-scale mass sensing using carbon nanotube resonators. Nano Lett 2008;8(12):4342–6.

[132] Sazonova V, Yaish Y, Ustunel H, Roundy D, Arias TA, McEuen PL. A tunable carbon nanotube electromechanical oscillator. Nature 2004;431(7006):284–7.

[133] Wang ZL, Poncharal P, de Heer WA. Measuring physical and mechanical properties of individual carbon nanotubes by in situ TEM. J Phys Chem Solids 2000;61(7):1025–30.

[134] Li CY, Chou TW. Single-walled carbon nanotubes as ultrahigh frequency nanomechanical resonators. Phys Rev B 2003;68(7):073405.

[135] Li CY, Chou TW. Vibrational behaviors of multiwalled-carbon-nanotube-based nanomechanical resonators. Appl Phys Lett 2004;84(1):121–3.

[136] Wang LF, Hu HY. Flexural wave propagation in single-walled carbon nanotubes. Phys Rev B 2005;71(19):195412.

[137] Kahn D, Lu JP. Vibrational modes of carbon nanotubes and nanoropes. Phys Rev B 1999;60(9):6535–40.

[138] Yoon J, Ru CQ, Mioduchowski A. Noncoaxial resonance of an isolated multiwall carbon nanotube. Phys Rev B 2002;66(23):233402.

[139] Wu DH, Chien WT, Chen CS, Chen HH. Resonant frequency analysis of fixed-free single-walled carbon nanotube-based mass sensor. Sensor Actuat a-Phys 2006;126(1):117–21.

[140] Yoon J, Ru CQ, Mioduchowski A. Terahertz vibration of short carbon nanotubes modeled as Timoshenko beams. J Appl Mech-T Asme 2005;72(1):10–17.

[141] He XQ, Kitipornchai S, Liew KM. Resonance analysis of multi-layered graphene sheets used as nanoscale resonators. Nanotechnology 2005;16(10):2086–91.

[142] Ge MH, Sattler K. Observation of fullerene cones. Chem Phys Lett 1994;220(3-5):192–6.

[143] Krishnan A, Dujardin E, Treacy MMJ, Hugdahl J, Lynum S, Ebbesen TW. Graphitic cones and the nucleation of curved carbon surfaces. Nature 1997;388(6641):451–4.

[144] Hu YG, Liew KM, He XQ, Li ZL, Han J. Free transverse vibration of single-walled carbon nanocones. Carbon 2012;50(12):4418–23.

[145] Firouz-Abadi RD, Fotouhi MM, Haddadpour H. Free vibration analysis of nanocones using a nonlocal continuum model. Phys Lett A 2011;375(41):3593–8.

[146] Jiang H, Yu MF, Liu B, Huang Y. Intrinsic energy loss mechanisms in a cantilevered carbon nanotube beam oscillator. Phys Rev Lett 2004;93(18).

[147] Huttel AK, Steele GA, Witkamp B, Poot M, Kouwenhoven LP, van der Zant HSJ. Carbon nanotubes as ultrahigh quality factor mechanical resonators. Nano Lett 2009;9(7):2547–52.

[148] Yan JW, Liew KM, He LH. Predicting mechanical properties of single-walled carbon nanocones using a higher-order gradient continuum computational framework. Compos Struct 2012;94(11):3271–7.

[149] Wei CY, Srivastava D. Nanomechanics of carbon nanofibers: Structural and elastic properties. Appl Phys Lett 2004;85(12):2208–10.

[150] Compernolle S, Kiran B, Chibotaru LF, Nguyen MT, Ceulemans A. Ab initio study of small graphitic cones with triangle, square, and pentagon apex. J Chem Phys 2004;121(5):2326–36.

[151] Yan JW, Liew KM, He LH. Ultra-sensitive analysis of a cantilevered single-walled carbon nanocone-based mass detector. Nanotechnology 2013;24(12) 125703-1-12

CHAPTER

2-D GRAPHENE AND WHITE GRAPHENE

9

9.1 INTRODUCTION

Intense interest has been aroused in graphene in fields such as physics, mechanics, chemistry, and materials science. Graphene is a two-dimensional sp^2-hybridized carbon sheet only one-atom-thick and with a repetitive hexagonal honeycomb crystal lattice structure. It is an exciting material [1] with an extra-large theoretical specific surface area (2630 m^2/g), high intrinsic mobility (200,000 $cm^2/v/s$) [2,3], ultrahigh Young's modulus ($\sim$1.0 TPa) [4], excellent thermal conductivity ($\sim$5000 W/m/K) [5], high optical transmittance ($\sim$97.7%), and good electrical conductivity. Naturally, graphene is not in a free state of equilibrium [6] and will transform into other important allotropes [1] such as stacked three-dimensional graphite, rolled one-dimensional carbon nanotubes (CNTs), and warped zero-dimensional fullerenes. This particular atomic structure yields remarkable material properties with a low mass density, making graphene a good material candidate for silicon materials [7]. In recent years, graphene has engendered the intensive interest of many theoretical studies and more recently has become an exciting area for experimentalists in terms of its potential application.

Previous studies have focused on fewer and fewer layers of graphite [8]. In 2004, the first single-layer samples isolated from graphite by Geim and coworkers [9] at Manchester University were recognized as a milestone in this field (see Fig. 9.1). This great discovery sparked an explosion of material innovation and aroused an explosion of investigation due to the thought of thermodynamical unstability of two-dimensional crystals at finite temperatures [10,11]. Molecular beam epitaxy was typically used for quasi-two-dimensional films growth on a supporting substrate, with an appreciable influence on electrical properties [12]. On the contrary, the mechanical exfoliation technique exfoliated the two-dimensional single-layer crystals from three-dimensional graphite. The yielding single- and few-layer flakes were stuck on the substrate only according to vdW forces and could be freestanding by means of etching away the substrate [2,13–15]. Such a method can maximize the reduction of induced effects and allow scientists to explore the intrinsic properties of graphene.

The experimentally isolated single-layer graphene opens a new era of nanomaterials involving a large amount of interesting physics [1,16]. In the original studies, the intrinsic properties of graphene involving graphene's ambipolar field effect [9], the quantum Hall effect at room temperature [17–20], extremely high carrier mobility [2,21], and even the single molecule adsorption [22,23] were explored. Countless devices were expected to be prepared to take full advantage of these properties including the next generation of high-speed and radio frequency logic devices, nano-electromechanical systems [24], thermally and electrically conductive reinforced composites, sensors, resonators [25], gas detection [26], solar cells, and transparent electrodes for displays.

Mechanical Behaviors of Carbon Nanotubes. DOI: http://dx.doi.org/10.1016/B978-0-323-43137-8.00009-3

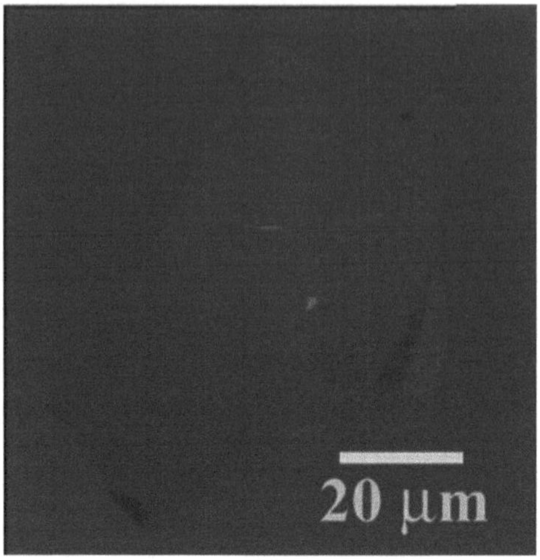

FIGURE 9.1

Single-layer graphene was first observed by Geim and others at Manchester University. Here a few-layer flake is shown, with optical contrast enhanced by an interference effect at a carefully chosen thickness of oxide [9].

Although great successes have been achieved by device physicists, the extensive utilization of graphene is still a serious challenge. A principal reason is that it is extremely difficult to prepare reliably high-quality samples, especially in large-scale production [27]. Since the performance of graphene is not only closely related to the number of layers but also the overall quality of crystal lattice, the challenge is double [3,13,28,29]. The original top-down approach of mechanical exfoliation has so far synthesized the highest quality production. Nevertheless, it is confined to only low throughput. vdW force between the neighboring layers must be exactly overcome to exfoliate a single sheet but with no disturbance on other subsequent sheets. Consequently, several alternative approaches have been developed to prepare single layers and a few of them have been used to implement promising proof-of-concept devices.

In general, there are three alternative approaches to mechanical exfoliation, including chemical process to exfoliate and stabilize individual sheets in solution [30–33], bottom-up methods for direct graphene growth on the basis of organic precursors [34–36], and catalyze growth in situ on a substrate [37–41]. Each kind of these approaches has its own function and its limitation as well. For example, extensive modification of the 2D crystal lattice has been demanded for chemically exfoliated graphene and thus degrades device performance [33,42]. Nevertheless, bottom-up techniques encounter many challenges when preparing large and uniform single-layer graphenes. Since macromolecules are insoluble and have accompanying side reactions, total organic syntheses have been limited to small scale production [43]. Substrate-based growth techniques by means of chemical vapor deposition (CVD) or the reduction of silicon carbide are dependent on the ability to walk a narrow thermodynamic tightrope [44]. It is noteworthy that conditions are required to be carefully controlled in order to prevent crystal growth from seeding additional second layers or forming grain boundaries when a sheet is nucleated. Compared with the abovementioned alternatives, mechanical exfoliation with cellophane tape is still the best way of high-quality graphene flakes preparation.

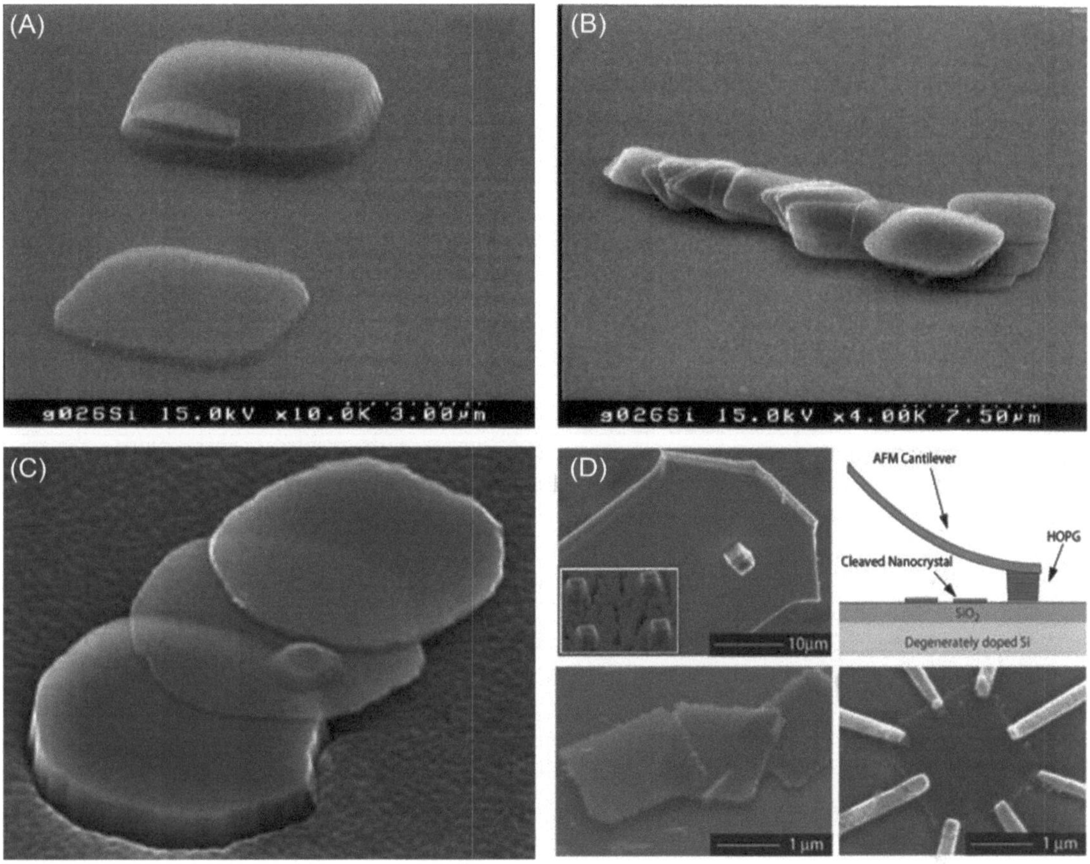

FIGURE 9.2

Scanning electron microscope images of early attempts at mechanical exfoliation using graphite pillars. (A and B) Ruoff's group peeled away layers with an AFM tip [8]. (C and D) Kim's group transferred the pillars to a tipless cantilever and deposited thin slabs onto other substrates in tapping mode. A series of scanning electron microscope images show thin samples cleaved onto the Si/SiO2 substrate and a typical mesoscopic device [45].

9.2 PREPARATION METHODS AND TESTING TECHNOLOGIES

9.2.1 PREPARATION METHODS

Mechanical exfoliation techniques have been adopted for some time in order to produce thin samples from layered compounds. In 1999, one such approach was proposed to tailor uniformly small pillars from patterned highly oriented pyrolytic graphite by reactive ion etching (see Fig. 9.2) by using an AFM tip [8]. The observed thinnest slabs were approximately 200 nm thick. Later in 2005, an improved method by transferring the pillars to a tipless cantilever on SiO2 was presented to manipulate a thin slab as thin as 10 nm [45]. Other work focused on

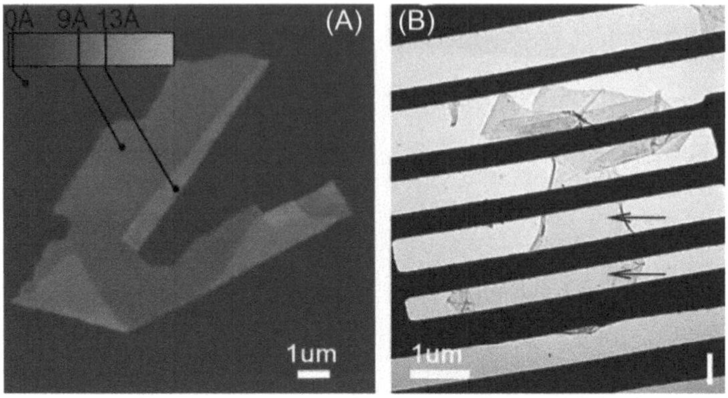

FIGURE 9.3

Mechanical exfoliation produced the very first single-layer graphene flakes. (A) An atomic force microscopy image shows the substrate–graphene step height of <1 nm and a folded step height of 0.4 nm [15]. (B) TEM image of a freestanding graphene film after etching of the underlying substrate [13].

graphene included using a high temperature around 1600°C for conversion of nanometer-sized graphene from nanodiamonds [46].

Among all these mechanical exfoliation methods for producing thin samples, a much simpler approach was ultimately proposed in 2004 by which the first isolation of single-layer graphene has been manipulated by Geim (see Fig. 9.3) at Manchester University [9]. In this peeling method, a cellophane tape is commonly utilized to successively remove layers from a graphite flake. The tape is pressed down against a substrate and then peeled away to remove a sample (see Fig. 9.1). However, the flakes deposited on the tape are often much thicker than one layer. A single sheet can be delaminated and isolated due to vdW attraction to the substrate when the tape is then lifted away. Ultimately, high-quality crystallites with a size of more than 100 μm^2 can be prepared by this repetitive peeling work, which can be more than 100 μm^2 in size.

However, perhaps the most important issue for isolating single-layer graphene was how to identify the atomically thin specimen. The measured optical absorbance of graphene is only 2.3%, so that it is extremely difficult to visually observe (see Fig. 9.4) [47,48]. For the sake of visual observation of single flakes, the Manchester group enhanced the optical contrast under white-light illumination by virtue of the interference effect at a specially chosen thickness (300 nm) of SiO_2 on Si [49]. On the basis of this idea, a great deal of work on imaging graphene on a variety of substrates has been done by adapting the same effect [49–52].

9.2.2 CHARACTERIZING GRAPHENE FLAKES

Since the two-dimensional graphene was discovered, characterization techniques have been urgently required to be developed for deposited flakes. Typical optical microscopy which utilizes the interference effect offers a good way for identifying thin candidates. Nevertheless, it is not capable of furnishing adequate evidence to distinguish the number of layers for a given flake. In actual fact, many interesting properties of graphene are closely related to the crystalline thickness.

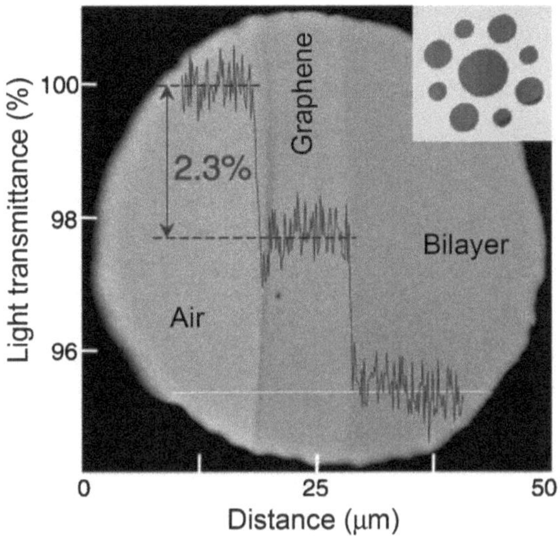

FIGURE 9.4

A single and bilayer sample suspended on a porous membrane. Optical absorbance is measured at 2.3% per layer. The inset shows the sample design with several apertures [47].

Therefore, conclusive distinction of the number of layers is an important issue and reliable identification techniques become imperative. So far, the most obvious way is measuring electronic band structure. A single-layer graphene is a semiconductor or semimetal with a zero band gap in which the highest occupied molecular orbital touches the lowest unoccupied one at a single Dirac point, while thicker flakes exhibit metallic properties due to the stacking behaviors of multiple layers, which leads to some overlap of their carrier wave functions.

9.2.3 SCANNING PROBE MICROSCOPY

Scanning probe microscopy (SPM) was commonly selected to verify the crystallite thickness. The characterization by this method costs much time, but the 0.34 nm step height for each successive layer is well within the detection limits for modern AFMs. It is indeed difficult to resolve the substrate–graphene height profile because of the differences in tip attraction/repulsion between the insulating substrate and semimetallic graphene. Moreover, the preferential adsorption of a thin water film on graphene exacerbates this issue under ambient conditions. Due to the complications, Novoselov et al. [9] reported that the thickness of a single-layer was typically in the range of 0.6–1.0 nm for substrate–graphene height profiles by AFM.

Since there is no alteration in material related with the location of the step, the folded edges of graphene provides a much more reliable and accurate way of measuring thickness by AFM. Using this method, the Manchester group originally reported a single-layer step height of ~0.4 nm (see Fig. 9.3). Whether admitting it or not, folds commonly appear during mechanical exfoliation to achieve an energetic minimum on account of the sizeable vdWs attraction between a sheet and itself.

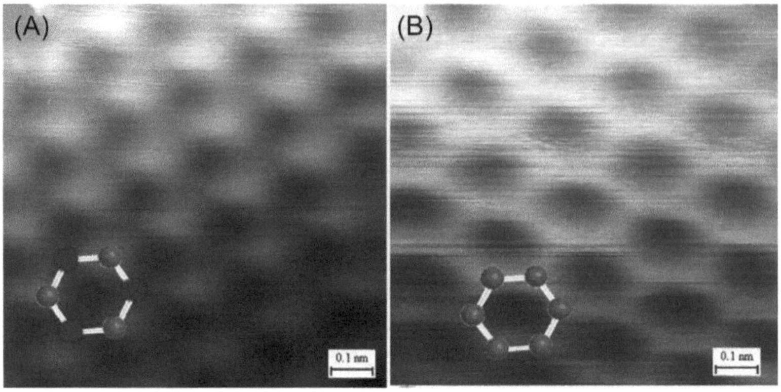

FIGURE 9.5

(A) STM image of graphite showing only the three carbons that eclipse a neighbor in the sheet directly below. (B) In contrast, all six carbons are equivalent and thus visible in mechanically exfoliated single-layer graphene [56].

The electronic topography of graphite is generally observed by scanning tunneling microscopy (STM) [53–55], and only three carbon atoms of the hexagonal rings are visible since graphite is in the AB stacking form (see Fig. 9.5) [56]. Electron density in this atomic arrangement is extremely high for the three α-carbons, allowing them visualization by STM. This abnormality is contrary to the expectation that the six carbons in the single-layer graphene are in completely equivalent status with equal electron intensity, which has been indeed confirmed by Stolyarova et al. [56] by using ultrahigh vacuum STM.

9.2.4 RAMAN SPECTROSCOPY

SPM can be used preferably for the investigation of the graphene's layered structure. Additional methods are required to determine the specimen thickness in a high-throughput style because of the sample preparation time and substrate requirements. Though it can be used instead of Raman spectroscopy, it was not regarded as a directly topographical technique. Meanwhile, it is found to be the most practical method to probe the thickness of mechanically exfoliated flakes. The changes in electronic structure could be reflected directly by the features of graphite and graphene from the stacking of successive layers. Although less than obvious, this approach can definitely make good sense [57–61]. In the Raman spectrum, the number of layers can be inferred by the observation of gradual changes (up to the screening length) in a "fingerprint" style (as shown in Fig. 9.6).

The G band and the G′ band at ~ 1584 and $\sim 2700\ cm^{-1}$, respectively, are the major features of graphite and graphene in the Raman spectra. The G band is on the basis of the E_{2g} vibrational mode; meanwhile, the G′ band is characterized as a second-order two-phonon mode. If the symmetry is damaged in samples or by edges with obvious defects, a third feature can be observed, the D band at $\sim 1350\ cm^{-1}$, though it is not Raman active for pristine graphene. The number of layers present for a given flake can be reflected by the changes in the positions and relative peak heights of the G and G′ bands. For single-layer graphene, the location of the G peak is 3–5 cm^{-1} higher than that for bulk graphite. However, the intensities of single-layer graphene and bulk graphite are

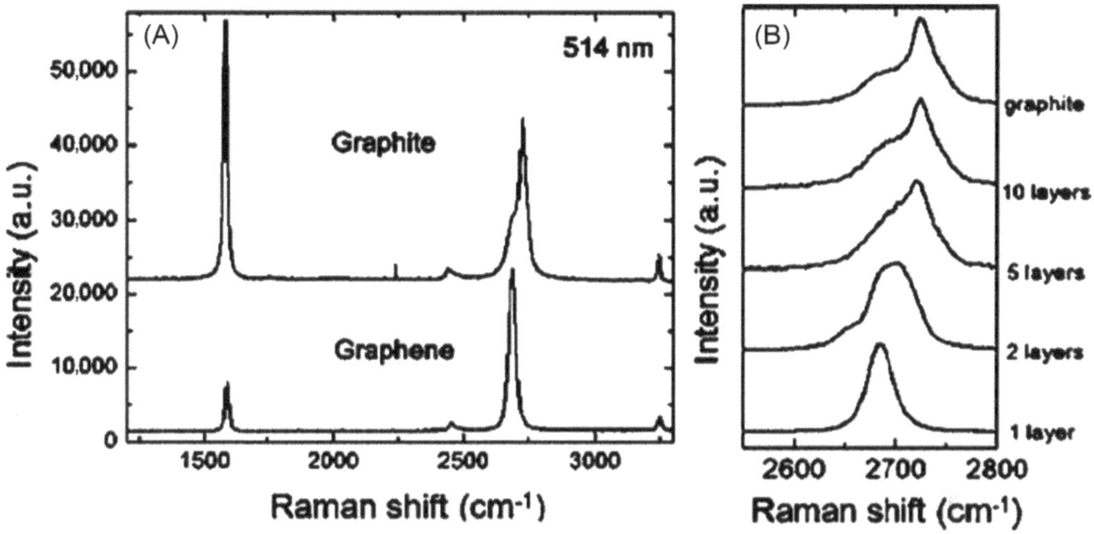

FIGURE 9.6

Raman spectroscopy is a powerful diagnostic tool for the study of graphene. Both the G (near 1584 cm^{-1}) and G2 (near 2700 cm^{-1}) bands undergo significant changes due to the thickness of AB stacked flakes, as produced by mechanical exfoliation [57].

roughly the same. An obvious change in both intensity and shape can be observed by the G′ peak when the number of layers decreases. The G′ band consists of two major components in bulk graphite. The intensities of the components are approximately 1/2 and 1/4 that of the G peak, respectively, for the high and low shifts. The G′ band is a single sharp peak at the lower shift for single-layer graphene; meanwhile, the intensity is generally four times that of the G peak. Eventually, the identity of mechanically exfoliated flakes is trustworthily determined by scientists using the fitting to these trends.

9.3 FUNDAMENTAL PROPERTIES AND GENERAL BEHAVIORS

9.3.1 ELECTRONIC PROPERTIES

The existence of massless Dirac quasiparticles in graphene was confirmed by the experimental observation of the cyclotron mass dependence on the square root of electronic density [19,21]. As a zero band gap semiconductor material, an ambipolar electric field effect was exhibited in graphene and continuous tuning of charge carriers can be realized between electrons and holes in a concentration of up to 10^{13} cm^{-2}, with a mobility of up to 15,000 cm^2/V/s (Fig. 9.7A) at room temperature [9,19,21]. Moreover, it is observed that the temperature has a weak effect on the mobility in graphene which suggests a new way of realizing an ultrahigh mobility [3]. By removing the impurity, an exceptionally high mobility of over 200,000 cm^2/V/s was achieved in suspended graphene

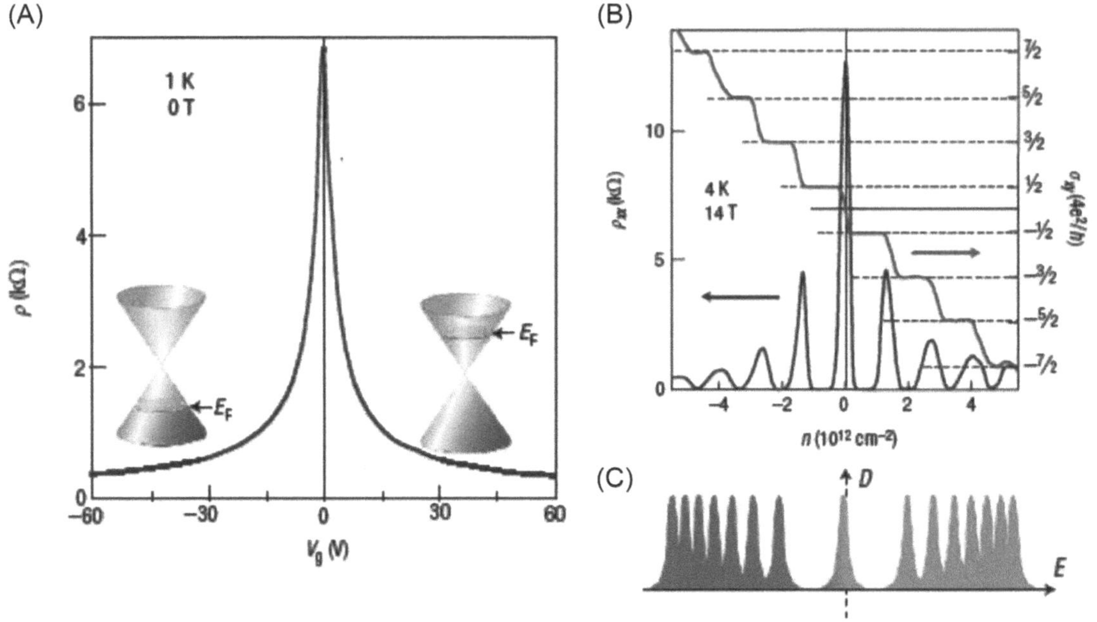

FIGURE 9.7

(A) Ambipolar electric field effect in monolayer graphene. The insets show the changes in the position of the Fermi energy E_F with changing gate voltage V_g; (B) The hallmark of massless Dirac fermions is QHE plateau in σ_{xy} at half integers of $4e^2/h$; (C) Landau levels in the density of states D in graphene are described by $EN \propto \sqrt{N}$ for massless Dirac fermions[1] (A and C) and [21] (B).

[2,62]. Such a high mobility in graphene maintaining even at high carrier density in both electrically and chemically doped devices [22] affirms ballistic transport at submicrometer level [63]. Another overarching concern of graphene is measuring the electronic quality in order to find out if there is the quantum Hall effect at room temperature [20].

9.3.2 MECHANICAL PROPERTIES

Mechanical properties of single-layer graphene mainly focus on Young's modulus and fracture strength and numerical simulations such as classical MD simulations have been applied for the investigation [64–66]. For few-layer graphene, the force–displacement curve has been experimentally measured by AFM on a strip of graphene suspended over trenches and hence the Young's modulus has been extracted [67]. Mechanical properties of circular few-layer graphene have also been obtained by force–volume measurements by AFM [68]. Lee et al. [4] measured the elastic properties of freestanding monolayer graphene by way of an nanoindentation test using an AFM (Fig. 9.8A and B), and extracted a Young's modulus of approximately 1 TPa and an intrinsic fracture strength of 130 GPa, while Frank et al. [67] obtained a Young's modulus of 0.5 TPa by measuring the effective spring constant of stacks of graphene.

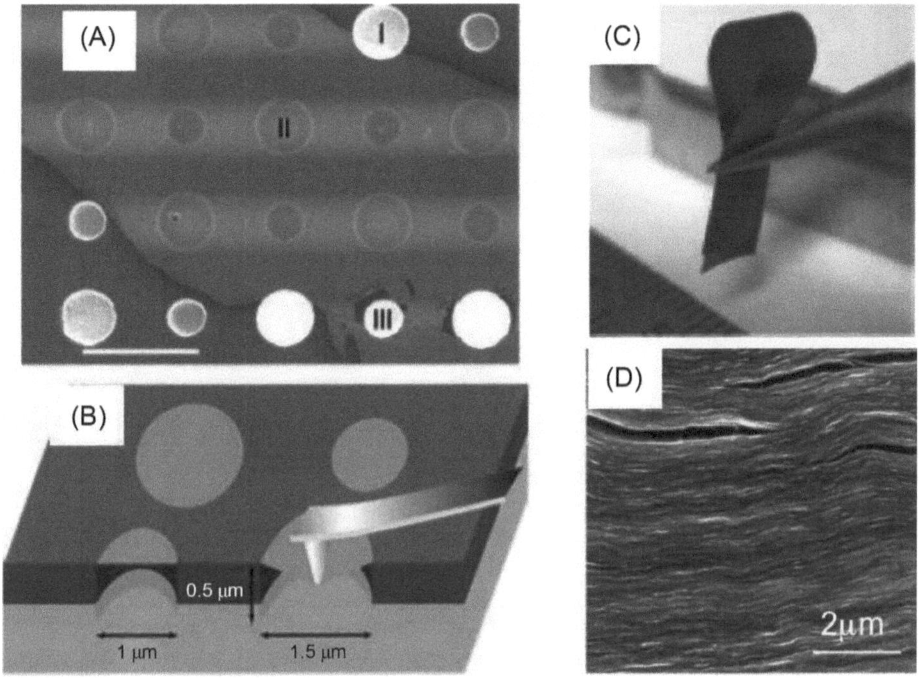

FIGURE 9.8

(A) SEM image of a graphene flake spanning an array of circular holes (scale bar, 3 μm) and (B) Schematic illustration of nanoindentation on membranes; (C) and (D) show graphene oxide paper and its cross-section in SEM (A and B) [4] and (C and D) [69].

9.3.3 OPTICAL PROPERTIES

On the basis of the infrared through the visible range of the spectrum, an expression of $\pi e^2/2h$ has been proposed to express the high-frequency conductivity for Dirac fermions in graphene [70,71]. The optical transmittance T and optical reflectance R are then expressed as $T = (1 + 1/2\pi\alpha)^{-2}$ and $R = 1/4\pi^2\alpha^2 T$, respectively, for normal incidence light, in which $\alpha = 2\pi e^2/hc \approx 1/137$, e is the electron charge, c is the light speed, and h is Planck's constant, and thus the opacity is $(1 - T) \approx \pi\alpha \approx$ 2.3%. Analyzed from the aforementioned expressions of T and R, no material parameters are directly involved and thus constants T and R can be stated to be the results of electronic properties and structures of graphene [72]. Experimental results shown in Fig. 9.9 demonstrate that the constant transparence of graphene in the visible range is approximately 97.7% and linearly decreases as the number of layers increases [47].

9.3.4 THERMAL PROPERTIES

The electronic contribution to thermal conductivity (Wiedemann–Franz law) can be neglected due to the relatively low carrier density of nondoped graphene. Therefore, phonon transport, including

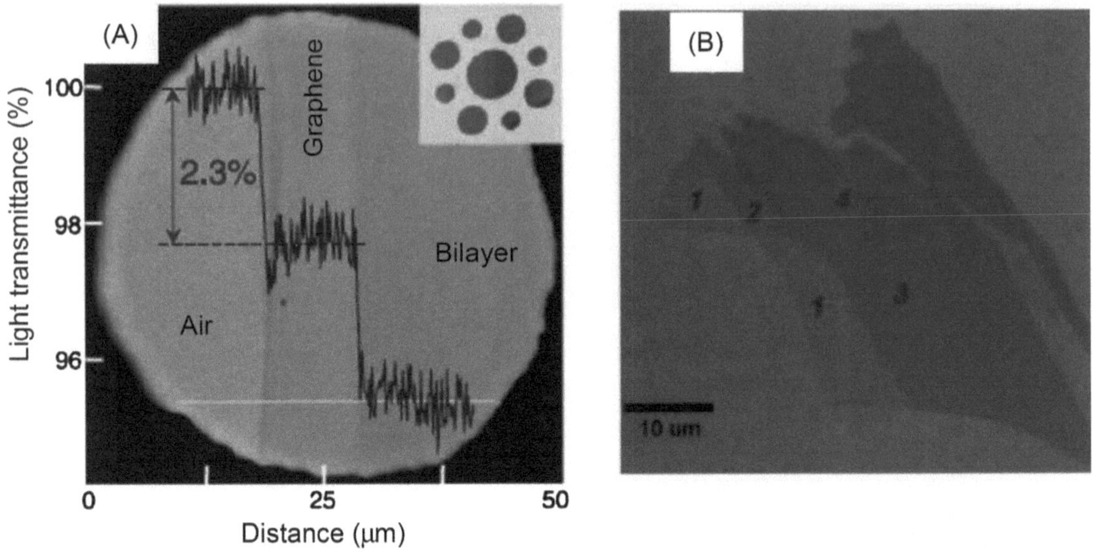

FIGURE 9.9

(A) Photograph of a 50 μm aperture partially covered by graphene and its bilayer. The line scan profile shows the intensity of transmitted white light along the gray line (yellow line in web version). Inset shows the sample design: a 20 μm thick metal support structure has apertures 20, 30, and 50 μm in diameter with graphene flakes deposited over them; (B) Optical image of graphene flakes with one, two, three, and four layers on a 285 nm thick SiO_2-on-Si substrate [47] (A) and [73] (B).

diffusive conduction at high temperature and ballistic conduction at sufficiently low temperature, plays a predominant role in the thermal conductivity (κ) of graphene [74].

Thermal properties of graphene have been investigated by using molecular dynamics (MD) simulations based on the Green–Kubo approach [75]. Numerical results demonstrated that thermal conductivity $\kappa \propto 1/T$ for nondefective graphene as temperature T increases beyond ~100K. A graphene exhibits a closely similar thermal transport behavior as an isolated CNT and the results showed a room temperature thermal conductivity of ~6000 W/m/K for a graphene which is much higher than that of pure diamond [75]. Based on the Boltzmann equation, it was found that when the diffusive conduction plays a predominate role, the thermal conductivity was particularly dependent on the width d of graphene nanoribbons (GNRs) and the "roughness" of the edges [76]. The dependence of κ for GNRs on different edge shapes including length, width, and strain was studied by using nonequilibrium MD simulation [77]. The results revealed that κ is a power law function of the length L ($\kappa \propto L^{\beta}$), in which β varies from 0.3 to 0.5 at room temperature. The strong dependence of thermal conductivity on length L may imply very long phonon mean free paths in GNRs.

For a suspended mechanically exfoliated single-layer graphene, a thermal conductivity of ~5000 W/m/K was predicted by optical measurement using the shift in the Raman G band [5,78]. The graphene is suspended on a trench as shown in Fig. 9.10A and B laser beam is focused on its center. The heat then radially flows from the center of the graphene to the support. Compared to heat conduction in the graphene, the heat loss by air conduction was stated to be negligible [80].

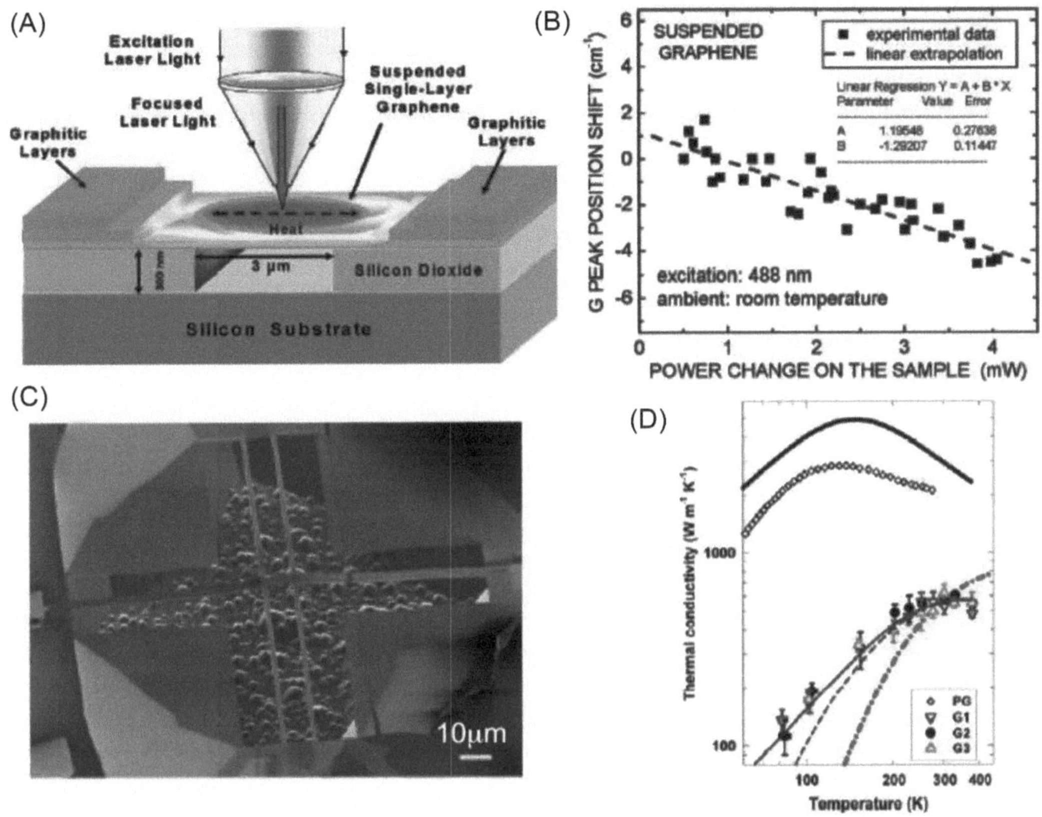

FIGURE 9.10

(A) Schematic of an experiment showing the excitation laser light focused on a graphene flake suspended across a trench. The focused laser light creates a local hot spot and generates a heat wave inside graphene propagating toward heat sinks; (B) Shift in Raman G peak position versus change in total dissipated power. The sample was excited at 488 nm and spectra were recorded at room temperature in the backscattering configuration. (C) SEM image of the suspended device for the thermal conductivity measurement of graphene. (D) Measured thermal conductivity values of graphene devices G1, G2, and G3, the highest reported values of pyrolytic graphite (PG), along with the Boltzmann transport equation (BTE) calculation results of suspended graphene (thick black solid line) and supported graphene (thin gray solid line (blue solid line in web version) and gray dashed line (blue dashed line in web version)), as well as the relaxation time approximation (RTA) calculation result (thick gray dash dot line (red dash dot line in web version)) for supported graphene (A and B; [5]) and (C and D; [79]).

The temperature rise in the laser-heating graphene gives rise to a red-shift of the Raman G peak owing to bond softening. The red-shift of the G peak for graphene is linearly dependent on the increase of sample temperature when a relatively low laser power is used [61]. The frequency of the Raman G peak was found to be a function of excitation power (Fig. 9.10B), in which the thermal conductivity, up to ~5000 W/m/K, was extracted from the slope of trend line.

However, a relatively lower thermal conductivity of ~2500 W/m/K (at 350K) was measured for CVD grown graphene which was deposited onto a thin silicon nitride membrane coating with a thin layer of gold for better thermal contact [81]. The recent report [79] showed a thermal conductivity of 600 W/m/K for micromechanically exfoliated graphene deposited on a SiO_2 substrate (Fig. 9.10C and D). This result is much lower than that obtained for suspended graphene because of the phonons leaking across the graphene–support interface and strong interface scattering of flexural modes [79].

9.4 RECENT RESEARCH ADVANCE IN 2-D GRAPHENE AND WHITE GRAPHENE

As mentioned before, the discovery of monolayer graphene with two-dimensional sp^2-hybridized carbon atoms has aroused intensive attention not only in both scientific research but also engineering designs [82,83]. The high carrier mobility of graphene, behaving as massless Dirac fermions [21], has attracted considerable interest. Ballistic charge transport has been also observed in graphene [84], which has already led to many unique physical phenomena, including the ambipolar electric field effect and half-integer quantum Hall effect [9]. Nevertheless, pristine graphene exhibits a semiconduction of which the zero bandgap feature is an obstacle to applications in the fabrication of electronic nanodevices. Therefore, how to effectively open the bandgap is an extreme technological foundation to indeed realize its engineering applications. In order to solve this issue, there have been many proposed strategies for tuning the gap of graphene, including fabricating GNRs [85,86], applying built-in strain or bias voltage into bilayer graphene [28,87–89], and doping periodic heteroatoms on graphene [90–92].

Apart from the abovementioned strategies, an unprecedented degree of modulation of electronic properties is available through the intercalation stack of graphene with other wide-gap semiconductors as well as relative orientation of component layers. Heterostructures [93,94] made from two-dimensional atomic crystals [9,15] have already led to the observation of new physical phenomena, such as the metal–insulator transition [95] and Coulomb drag [96], and to the realization of functional devices. A new graphene cousin, hexagonal boron nitride (h-BN) sheets, which alternately substitute boron and nitrogen atoms for carbon atoms entirely, has analogous crystallographic parameters of graphene: in-plane lattice constants are 0.2504 and 0.246 nm [97]. Since these BN sheets are composed of boron and nitrogen, they are expected to be far more oxidation resistant than graphene cousin. For example, pronounced higher resistance of BN nanotubes (BNNTs) to oxidation at high temperatures than CNTs have been reported [98]. BNNTs remain stable at 700°C in air and perfect multiwalled BNNTs with small diameters can survive even up to 900°C. It has also been found that the onset temperature for oxidation of BNNTs is 800°C which is much higher than the 400°C for CNTs under the same conditions. Porous BN sheets exhibit excellent sorption performance, reaching 33 times own weight, for oils, organic solvents, and dyes but which repel water have already been reported [99]. Due to its strong resistance to oxidation, the BN sheet can be cleaned and reused by means of heating and burning in air. Furthermore, BN is also inactive and does not react with iron and steel. The thermal and chemical stability of BN nanostructures implies an expectation of them being better candidates for specific applications such as protective shields to prevent encapsulated nanomaterials from getting destroyed when performing at high temperature and in hazardous environments [100].

Specifically, h-BN sheet is a semiconductor with a wide bandgap of 5.92 eV [101] and thus is suggested to be an attractive material to develop sandwiched structures. For example, a novel kind of field effect transistor with tunneling between two graphene electrodes controlled by the gate voltage was possible to be realized by using a third (gate) electrode [102,103]. Direct graphene/h-BN sandwiched structure growth has been experimentally synthesized. In order to avoid the difficult control and operation of exfoliation and heteroepitaxial techniques, a facile and scalable direct CVD growth of h-BN sheet on graphene and a controllable large-scale growth of graphene/h-BN stacked layers has been proposed [104]. These technical progresses have begun a new era of engineering novel electronic devices and interconnects on the basis of laminated C/BN heterostructures with tunable electrical performances. Mishchenko et al. [105] reported that a careful alignment of the crystallographic orientation graphene/h-BN sandwiched structure in a transistor device, two graphene electrodes separated by a layer of h-BN sheet, can realize resonant tunneling with conservation of electron energy, momentum, and chirality. A tunable radiofrequency oscillatory current can be induced by the resonance peak and negative differential conductance in the device characteristics. From the exploration of geometric and electronic characteristics of C/BN bilayer and multilayer heterostructures subjected to hydrogenation, it was found that the rearrangement of electrons at the interface would give rise to a built-in potential difference, playing a decisive role in the bandgap of the resulting system [106]. By controlling the potential difference, a continuously and linearly tuning method for the bandgap of hydrogenated C/BN bilayer from a semiconducting state to a metallic one was successfully developed. This linear tuning technique of bandgap implies a great potential of C/BN heterostructures films in electronic and optical electronics [107,108], such as tunnel transistors [102,109] and photovoltaic sensors [110].

9.5 APPLICATION PROSPECTS

9.5.1 FIELD EFFECT TRANSISTORS

The unique band structure of graphene allows a continuous tunability by using a gate electrical field for the bipolar carriers in graphene [9]. Many publications [9,19,21,111] have been focused on the investigation of graphene field effect transistor (FET) devices with a single back gate. The experimentally measured field effect mobility of graphene is about one order of magnitude higher than that of Si. A mass production method for graphene transistors based on graphene growth on Cu films has been presented to furnish large-scale uniformly electrical transistor arrays [112]. Lin et al. [113] used top gate geometry to fabricate a high frequency graphene FET working in a 26 GHz frequency condition. The device mobility could be further improved by introducing an organic polymer buffer layer between graphene and conventional gate dielectrics [114]. Based on epitaxial graphene which was created on a 2-inch SiC wafer, a cutoff frequency up to 100 GHz has been achieved [115]. Using graphene quantum dots created by electron beam lithography, Ponomarenko et al. [116] reported a single electron transistor (Fig. 9.11). Such a graphene single electron transistor has the similar behavior as conventional larger size ones including periodic Coulomb blockade

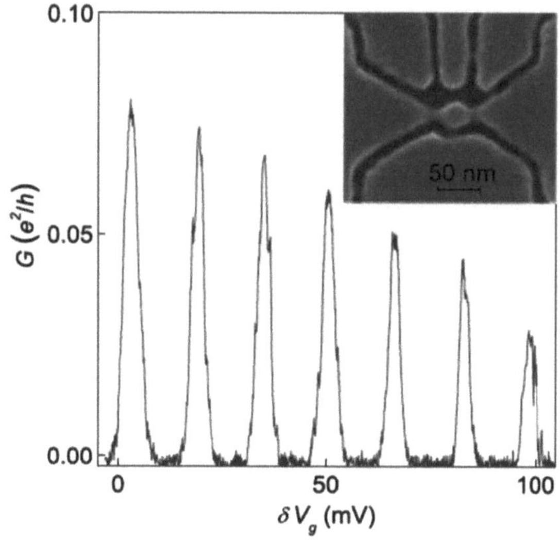

FIGURE 9.11

A graphene-based SET. Conductance G of a device with the central island of 250 nm in diameter and distant side gates as a function of V_g in the vicinity of +15 V; $T = 0.3$K. The inset shows one of the smaller devices illustrating the high resolution lithography "that allows features down to 10 nm" [116].

peaks. When the quantum dot is smaller than 100 nm, it was found that the peaks became strongly nonperiodic because of the major contribution of quantum confinement [116].

The finite minimum conductance for graphene with a zero gate voltage gives rise to relatively low I_{on}/I_{off} ratios [9] for FET devices. There have been several developed routes for inducing and controlling the energy gap in graphene to shut off the current. In order to clearly illustrate this issue, take the epitaxially grown graphene on the SiC substrate as an example. Because of the interaction with the SiC substrate, the band splitting can be up to 0.26 eV [117]. It was found that the sample thickness had a decreasing effect on the gap and if the number of layers was larger than four, the gap would approach zero through angle-resolved photoemission spectroscopy studies. In addition, an insulating state was allowed in bilayer graphene FETs through an additional top gate method [118]. Under low temperature condition, a small gap of ~10 meV was created for a perpendicular electric field. While at room temperature, a continuous tunable band gap for bilayer graphene FETs was realized by using a similar double-gated structure and infrared microspectroscopy [89].

9.5.2 SENSORS

The change in conductance of graphene is a function of the extent of surface adsorption, large specific surface area, and low Johnson noise [1,19,21]. In recent years, experimental [22,119–122] and theoretical [123–125] researches have been conducted to explore the potential application of monolayer graphene in detecting molecules, including gases and biomolecules. The existence of charge transfer between adsorbed molecules and graphene is stated to be a chemical reaction. The

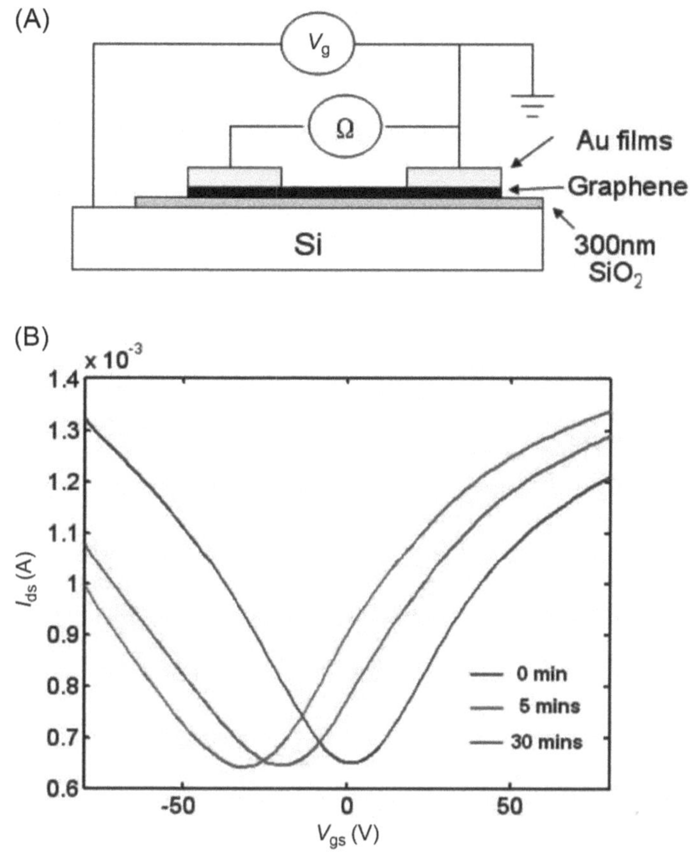

FIGURE 9.12

(A) Schematic of a graphene FET gas sensor device; (B) Evolution of I_{ds}-V_{gs} curves with the exposure to NH_3 of the graphene FET for different durations [126].

adsorbed molecules on the surface of graphene sustain a charge transfer as a donor or acceptor, which results in a change in the Fermi level, carrier density, and electrical resistance of the monolayer graphene. Fig. 9.12A illustrates a typical schematic of a graphene FET gas sensor device. Source-drain current (I_{ds}) versus gate voltage (V_{gs}) in the dynamic evolution over time could be recorded when the sensor device was exposed to gas such as NH_3 (Fig. 9.12B) [126]. The initial Dirac point (V_D) approaches the back gate bias of 0 V and then become -20 V after 5 min of exposure with a slow shift to its final status of about -30 V. These changes in Dirac point imply an adsorption of NH_3 on the graphene surface. Given the charge transfer rate and Dirac point shift, the concentration of the adsorbed molecules on the surface of graphene can be finally determined. As observed from the recent researches, NH_3 and CO molecules usually serve as donors while H_2O and NO_2 are acceptors [22,125].

9.5.3 CLEAN ENERGY DEVICES

The high theoretical surface area of 2630 m^2/g and capacity of facilitating electron or hole transfer along its two-dimensional surface allow graphene to be a promising electrode material. Several reports have been focused on the exploration of graphene-based electrodes for both rechargeable lithium ion batteries and electrochemical double layer capacitors. In rechargeable lithium ion batteries, graphite with a maximum "lithium insertion" capacity of 372 $mAhg^{-1}$, corresponding to a LiC_6 stoichiometry, is the most commonly used anode material [127]. For anode material consisting of reduced graphene oxide platelets, the capacity could increase to 540 $mAhg^{-1}$, while those of CNT- and C_{60}-based anode materials would increase to 730 and 784 $mAhg^{-1}$, respectively (Fig. 9.13) [128]. The increase in storage capacity results from the increase in interlayer distance between the graphene platelets when introducing CNTs or C_{60} molecules. Edges and defects have an increasing effect on the capacity and as a result, a higher reversible capacity of 784 $mAhg^{-1}$ after 15 cycles for graphene nanosheets, prepared by electron beam irradiation of graphene oxide suspensions, was achieved [129]. In contrast to powder made up of reduced graphene oxide platelets, relatively flexible graphene paper electrodes generally exhibit poor electrochemical performance although they possess good conductivity and mechanical stability [130]. Nevertheless, a large storage capacity of ~2200 $mAhg^{-1}$ after 50 cycles and ~1500 $mAhg^{-1}$ after 200 cycles was achieved when mixing silicon nanoparticles with reduced graphene oxide platelets to form a paper-like material in the recent report [131]. The uniform dispersion of silicon particles as well as the

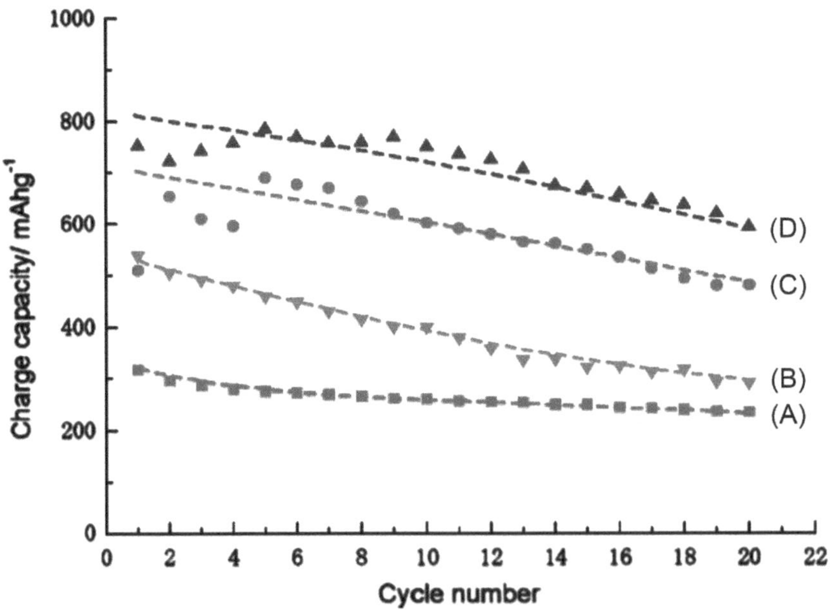

FIGURE 9.13

Lithium insertion/extraction properties of graphene nanosheets (GNS): Charge/discharge cycle performance of (A) graphite; (B) GNS; (C) GNS + CNT; (D) GNS + C_{60} [128].

reconstitution of graphene platelets to form a continuous, highly conducting 3D network is deemed to play a significant role in enhancing the storage capacities.

9.5.4 GRAPHENE–POLYMER NANOCOMPOSITES

Graphene serving as fillers are promising in polymer nanocomposites [132]. Other graphite derivatives, including graphene oxide and graphite intercalation compounds, have been used as precursors to bulk scale production of graphite nanoplatelets [133]. Graphene-based nanoplatelets have been filled in a variety of polymer matrix composites, including polystyrene [30,134], polymethyl methacrylate [135], polyvinyl alcohol [136,137], polypropylene [138], epoxy [139], polyester [140], silicone foam [141], polyurethane [142], polyvinyldiene fluoride [143], and polycarbonate [144]. It is noteworthy that graphene-based fillers are required to be exfoliated and well dispersed in the polymer matrix. Such a method offers a high degree of probability in enhancing the material property of nanocomposites [145]. Rapid heating and ultrasonication have been comprehensively adopted for highly exfoliated platelet production [146,147]. Several approaches, including solution mixing, melt blending, or in situ polymerization, can be utilized to disperse these fillers into polymers and in situ polymerization of fillers have the best dispersion [138].

Polymer nanocomposites with graphene-based filler have exhibited prominent enhancements in their material properties, including Young's modulus, tensile strength, electrical conductivity, and thermal stability. Moreover, interestingly, it is observed that a low loading of graphene-based filler in the nanocomposites gives rise to these evident improvements which are a result of their large interfacial area and high aspect ratio [135,148]. Liang et al.'s [137] experimental results revealed an increase of 76% and 62% in tensile strength and Young's modulus, respectively, for 0.7 wt% loading of graphene oxide in a solution-mixed polyvinyl alcohol nanocomposite. The large enhancement in the mechanical property results from interfacial hydrogen bonding to effectively load transfer to the graphene oxide filler. Reduced graphene oxide filler in polyvinyl alcohol can be used to make electrically conductive composites with a percolation threshold below 1 wt% and leads to a large change in glass transition temperature [136]. Owing to the onset of rheological percolation and corrugated morphology of the high-quality exfoliated nanoplatelets, a large increase in Young's modulus as well as a shift of 30°C in glass transition temperature is observed for only 0.05 wt% loading of a thermally expanded graphite oxide-polymethyl methacrylate composite [135]. Nanocomposites of platelets on the basis of thermally expanded graphite oxide exhibited a higher stiffness and less than or equal electrical percolation thresholds than carbon black- and CNT-based filler nanocomposites [149]. The coefficient of thermal expansion decreases in a low loading of exfoliated graphite oxide platelet filler in epoxy [150] while the critical buckling strength increases [151].

REFERENCES

[1] Geim AK, Novoselov KS. The rise of graphene. Nat Mater 2007;6(3):183–91.

[2] Bolotin KI, Sikes KJ, Jiang Z, Klima M, Fudenberg G, Hone J, et al. Ultrahigh electron mobility in suspended graphene. Solid State Commun 2008;146(9-10):351–5.

[3] Morozov SV, Novoselov KS, Katsnelson MI, Schedin F, Elias DC, Jaszczak JA, et al. Giant intrinsic carrier mobilities in graphene and its bilayer. Phys Rev Lett 2008;100(1):016602.
[4] Lee C, Wei XD, Kysar JW, Hone J. Measurement of the elastic properties and intrinsic strength of monolayer graphene. Science 2008;321(5887):385–8.
[5] Balandin AA, Ghosh S, Bao WZ, Calizo I, Teweldebrhan D, Miao F, et al. Superior thermal conductivity of single-layer graphene. Nano Lett 2008;8(3):902–7.
[6] Fasolino A, Los JH, Katsnelson MI. Intrinsic ripples in graphene. Nat Mater 2007;6(11):858–61.
[7] Schwierz F. Graphene transistors. Nat Nanotechnol 2010;5(7):487–96.
[8] Lu XK, Yu MF, Huang H, Ruoff RS. Tailoring graphite with the goal of achieving single sheets. Nanotechnology 1999;10(3):269–72.
[9] Novoselov KS, Geim AK, Morozov SV, Jiang D, Zhang Y, Dubonos SV, et al. Electric field effect in atomically thin carbon films. Science 2004;306(5696):666–9.
[10] Landau L. Zur Theorie der phasenumwandlungen II. Phys Z Sowjetunion 1937;11:26–35.
[11] Peierls R. Quelques propriétés typiques des corps solides. Annales de l'institut Henri Poincaré; p. 177-222.
[12] Stangl J, Holy V, Bauer G. Structural properties of self-organized semiconductor nanostructures. Rev Mod Phys 2004;76(3):725–83.
[13] Meyer JC, Geim AK, Katsnelson MI, Novoselov KS, Booth TJ, Roth S. The structure of suspended graphene sheets. Nature 2007;446(7131):60–3.
[14] Bolotin KI, Sikes KJ, Hone J, Stormer HL, Kim P. Temperature-dependent transport in suspended graphene. Phys Rev Lett 2008;101(9):096802.
[15] Novoselov KS, Jiang D, Schedin F, Booth TJ, Khotkevich VV, Morozov SV, et al. Two-dimensional atomic crystals. P Natl Acad Sci USA 2005;102(30):10451–3.
[16] Katsnelson MI. Graphene: carbon in two dimensions. Mater Today 2007;10(1-2):20–7.
[17] Novoselov KS, McCann E, Morozov SV, Fal'ko VI, Katsnelson MI, Zeitler U, et al. Unconventional quantum Hall effect and Berry's phase of 2 pi in bilayer graphene. Nat Phys 2006;2(3):177–80.
[18] Jiang Z, Zhang Y, Tan YW, Stormer HL, Kim P. Quantum Hall effect in graphene. Solid State Commun 2007;143(1-2):14–19.
[19] Zhang YB, Tan YW, Stormer HL, Kim P. Experimental observation of the quantum Hall effect and Berry's phase in graphene. Nature 2005;438(7065):201–4.
[20] Novoselov KS, Jiang Z, Zhang Y, Morozov SV, Stormer HL, Zeitler U, et al. Room-temperature quantum hall effect in graphene. Science 2007;315(5817) 1379-
[21] Novoselov KS, Geim AK, Morozov SV, Jiang D, Katsnelson MI, Grigorieva IV, et al. Two-dimensional gas of massless Dirac fermions in graphene. Nature 2005;438(7065):197–200.
[22] Schedin F, Geim AK, Morozov SV, Hill EW, Blake P, Katsnelson MI, et al. Detection of individual gas molecules adsorbed on graphene. Nat Mater 2007;6(9):652–5.
[23] Novoselov K, Geim A. Graphene detects single molecule of toxic gas. Mater Technol 2007;22 (3):178–9.
[24] Bunch JS, van der Zande AM, Verbridge SS, Frank IW, Tanenbaum DM, Parpia JM, et al. Electromechanical resonators from graphene sheets. Science 2007;315(5811):490–3.
[25] He XQ, Kitipornchai S, Liew KM. Resonance analysis of multi-layered graphene sheets used as nanoscale resonators. Nanotechnology 2005;16(10):2086–91.
[26] Arash B, Wang Q, Duan WH. Detection of gas atoms via vibration of graphenes. Phys Lett A 2011;375 (24):2411–15.
[27] Ruoff R. Graphene: Calling all chemists. Nat Nanotechnol 2008;3(1):10–11.
[28] Castro EV, Novoselov KS, Morozov SV, Peres NMR, Dos Santos JMBL, Nilsson J, et al. Biased bilayer graphene: semiconductor with a gap tunable by the electric field effect. Phys Rev Lett 2007;99 (21):216802.

[29] Yan J, Henriksen EA, Kim P, Pinczuk A. Observation of anomalous phonon softening in bilayer graphene. Phys Rev Lett 2008;101(13):136804.
[30] Stankovich S, Dikin DA, Dommett GHB, Kohlhaas KM, Zimney EJ, Stach EA, et al. Graphene-based composite materials. Nature 2006;442(7100):282−6.
[31] Stankovich S, Dikin DA, Piner RD, Kohlhaas KA, Kleinhammes A, Jia Y, et al. Synthesis of graphene-based nanosheets via chemical reduction of exfoliated graphite oxide. Carbon 2007;45(7):1558−65.
[32] Li D, Muller MB, Gilje S, Kaner RB, Wallace GG. Processable aqueous dispersions of graphene nanosheets. Nat Nanotechnol 2008;3(2):101−5.
[33] Gilje S, Han S, Wang M, Wang KL, Kaner RB. A chemical route to graphene for device applications. Nano Lett. 2007;7(11):3394−8.
[34] Muller M, Kubel C, Mullen K. Giant polycyclic aromatic hydrocarbons. Chem-Eur J 1998;4(11):2099−109.
[35] Tyutyulkov N, Madjarova G, Dietz F, Mullen K. Is 2-D graphite an ultimate large hydrocarbon? 1. Energy spectra of giant polycyclic aromatic hydrocarbons. J Phys Chem B 1998;102(50):10183−9.
[36] Yang XY, Dou X, Rouhanipour A, Zhi LJ, Rader HJ, Mullen K. Two-dimensional graphene nanoribbons. J Am Chem Soc 2008;130(13) 4216
[37] Berger C, Song ZM, Li TB, Li XB, Ogbazghi AY, Feng R, et al. Ultrathin epitaxial graphite: 2D electron gas properties and a route toward graphene-based nanoelectronics. J Phys Chem B 2004;108 (52):19912−16.
[38] Berger C, Song ZM, Li XB, Wu XS, Brown N, Naud C, et al. Electronic confinement and coherence in patterned epitaxial graphene. Science 2006;312(5777):1191−6.
[39] Reina A, Jia XT, Ho J, Nezich D, Son HB, Bulovic V, et al. Large area, few-layer graphene films on arbitrary substrates by chemical vapor deposition. Nano Lett. 2009;9(1):30−5.
[40] Kim KS, Zhao Y, Jang H, Lee SY, Kim JM, Kim KS, et al. Large-scale pattern growth of graphene films for stretchable transparent electrodes. Nature 2009;457(7230):706−10.
[41] Sutter PW, Flege JI, Sutter EA. Epitaxial graphene on ruthenium. Nat Mater 2008;7(5):406−11.
[42] Gomez-Navarro C, Weitz RT, Bittner AM, Scolari M, Mews A, Burghard M, et al. Electronic transport properties of individual chemically reduced graphene oxide sheets. Nano Lett 2007;7(11):3499−503.
[43] Wu JS, Pisula W, Mullen K. Graphenes as potential material for electronics. Chem Rev 2007;107 (3):718−47.
[44] Hass J, de Heer WA, Conrad EH. The growth and morphology of epitaxial multilayer graphene. J Phys-Condens Mat 2008;20(32):323202.
[45] Zhang YY, Tadigadapa S. Thermal characterization of liquids and polymer thin films using a microcalorimeter. Appl Phys Lett 2005;86(3):073104.
[46] Affoune AM, Prasad BLV, Sato H, Enoki T, Kaburagi Y, Hishiyama Y. Experimental evidence of a single nano-graphene. Chem Phys Lett 2001;348(1-2):17−20.
[47] Nair RR, Blake P, Grigorenko AN, Novoselov KS, Booth TJ, Stauber T, et al. Fine structure constant defines visual transparency of graphene. Science 2008;320(5881) 1308
[48] Stauber T, Peres NMR, Geim AK. Optical conductivity of graphene in the visible region of the spectrum. Phys Rev B 2008;78(8):085432.
[49] Blake P, Hill EW, Castro Neto AH, Novoselov KS, Jiang D, Yang R, et al. Making graphene visible. Appl Phys Lett 2007;91(6):063124.
[50] Blake P, Brimicombe PD, Nair RR, Booth TJ, Jiang D, Schedin F, et al. Graphene-based liquid crystal device. Nano Lett. 2008;8(6):1704−8.
[51] Jung I, Pelton M, Piner R, Dikin DA, Stankovich S, Watcharotone S, et al. Simple approach for high-contrast optical imaging and characterization of graphene-based sheets. Nano Lett 2007;7(12):3569−75.
[52] Ni ZH, Chen W, Fan XF, Kuo JL, Yu T, Wee ATS, et al. Raman spectroscopy of epitaxial graphene on a SiC substrate. Phys Rev B 2008;77(11):115416.

[53] Batra IP, Garcia N, Rohrer H, Salemink H, Stoll E, Ciraci S. A study of graphite surface with stm and electronic-structure calculations. Surf Sci 1987;181(1-2):126–38.
[54] Rabe JP, Buchholz S. Commensurability and mobility in 2-dimensional molecular-patterns on graphite. Science 1991;253(5018):424–7.
[55] Soler JM, Baro AM, Garcia N, Rohrer H. Interatomic forces in scanning tunneling microscopy—giant corrugations of the graphite surface. Phys Rev Lett 1986;57(4):444–7.
[56] Stolyarova E, Rim KT, Ryu SM, Maultzsch J, Kim P, Brus LE, et al. High-resolution scanning tunneling microscopy imaging of mesoscopic graphene sheets on an insulating surface. P Natl Acad Sci USA 2007;104(22):9209–12.
[57] Ferrari AC, Meyer JC, Scardaci V, Casiraghi C, Lazzeri M, Mauri F, et al. Raman spectrum of graphene and graphene layers. Phys Rev Lett 2006;97(18):187401.
[58] Ferrari AC, Robertson J. Interpretation of Raman spectra of disordered and amorphous carbon. Phys Rev B 2000;61(20):14095–107.
[59] Casiraghi C, Pisana S, Novoselov KS, Geim AK, Ferrari AC. Raman fingerprint of charged impurities in graphene. Appl Phys Lett 2007;91(23):233108.
[60] Ferrari AC. Raman spectroscopy of graphene and graphite: disorder, electron-phonon coupling, doping and nonadiabatic effects. Solid State Commun 2007;143(1-2):47–57.
[61] Calizo I, Balandin AA, Bao W, Miao F, Lau CN. Temperature dependence of the Raman spectra of graphene and graphene multilayers. Nano Lett 2007;7(9):2645–9.
[62] Durkop T, Getty SA, Cobas E, Fuhrer MS. Extraordinary mobility in semiconducting carbon nanotubes. Nano Lett 2004;4(1):35–9.
[63] Du X, Skachko I, Barker A, Andrei EY. Approaching ballistic transport in suspended graphene. Nat Nanotechnol 2008;3(8):491–5.
[64] Van Lier G, Van Alsenoy C, Van Doren V, Geerlings P. Ab initio study of the elastic properties of single-walled carbon nanotubes and graphene. Chem Phys Lett 2000;326(1-2):181–5.
[65] Reddy CD, Rajendran S, Liew KM. Equilibrium configuration and continuum elastic properties of finite sized graphene. Nanotechnology 2006;17(3):864–70.
[66] Kudin KN, Scuseria GE, Yakobson BI. C2F, BN, and C nanoshell elasticity from ab initio computations. Phys Rev B 2001;64(23):235406.
[67] Frank IW, Tanenbaum DM, Van der Zande AM, McEuen PL. Mechanical properties of suspended graphene sheets. J Vac Sci Technol B 2007;25(6):2558–61.
[68] Poot M, van der Zant HSJ. Nanomechanical properties of few-layer graphene membranes. Appl Phys Lett 2008;92(6):063111.
[69] Dikin DA, Stankovich S, Zimney EJ, Piner RD, Dommett GHB, Evmenenko G, et al. Preparation and characterization of graphene oxide paper. Nature 2007;448(7152):457–60.
[70] Peres NMR, Guinea F, Castro Neto AH. Electronic properties of disordered two-dimensional carbon. Phys Rev B 2006;73(12):125411.
[71] Gusynin VP, Sharapov SG, Carbotte JP. Unusual microwave response of Dirac quasiparticles in graphene. Phys Rev Lett 2006;96(25):256802.
[72] Peres NMR. The transport properties of graphene. J Phys-Condens Mat 2009;21:323201.
[73] Ni ZH, Wang HM, Kasim J, Fan HM, Yu T, Wu YH, et al. Graphene thickness determination using reflection and contrast spectroscopy. Nano Lett 2007;7(9):2758–63.
[74] Yu CH, Shi L, Yao Z, Li DY, Majumdar A. Thermal conductance and thermopower of an individual single-wall carbon nanotube. Nano Lett 2005;5(9):1842–6.
[75] Berber S, Kwon YK, Tomanek D. Unusually high thermal conductivity of carbon nanotubes. Phys Rev Lett 2000;84(20):4613–16.
[76] Nika DL, Pokatilov EP, Askerov AS, Balandin AA. Phonon thermal conduction in graphene: role of Umklapp and edge roughness scattering. Phys Rev B 2009;79(15):155413.

[77] Guo ZX, Zhang D, Gong XG. Thermal conductivity of graphene nanoribbons. Appl Phys Lett 2009;95 (16):163103.
[78] Ghosh S, Calizo I, Teweldebrhan D, Pokatilov EP, Nika DL, Balandin AA, et al. Extremely high thermal conductivity of graphene: prospects for thermal management applications in nanoelectronic circuits. Appl Phys Lett 2008;92(15).
[79] Seol JH, Jo I, Moore AL, Lindsay L, Aitken ZH, Pettes MT, et al. Two-dimensional phonon transport in supported graphene. Science 2010;328(5975):213–16.
[80] Hsu IK, Pettes MT, Bushmaker A, Aykol M, Shi L, Cronin SB. Optical absorption and thermal transport of individual suspended carbon nanotube bundles. Nano Lett 2009;9(2):590–4.
[81] Cai WW, Moore AL, Zhu YW, Li XS, Chen SS, Shi L, et al. Thermal transport in suspended and supported monolayer graphene grown by chemical vapor deposition. Nano Lett 2010;10(5):1645–51.
[82] Yavari F, Kritzinger C, Gaire C, Song L, Gullapalli H, Borca-Tasciuc T, et al. Tunable bandgap in graphene by the controlled adsorption of water molecules. Small 2010;6(22):2535–8.
[83] Szafranek BN, Schall D, Otto M, Neumaier D, Kurz H. High on/off ratios in bilayer graphene field effect transistors realized by surface dopants. Nano Lett 2011;11(7):2640–3.
[84] Morozov SV, Novoselov KS, Schedin F, Jiang D, Firsov AA, Geim AK. Two-dimensional electron and hole gases at the surface of graphite. Phys Rev B 2005;72(20):201401.
[85] Wang XR, Ouyang YJ, Li XL, Wang HL, Guo J, Dai HJ. Room-temperature all-semiconducting sub-10-nm graphene nanoribbon field-effect transistors. Phys Rev Lett 2008;100(20):206803.
[86] Li XL, Wang XR, Zhang L, Lee SW, Dai HJ. Chemically derived, ultrasmooth graphene nanoribbon semiconductors. Science 2008;319(5867):1229–32.
[87] Choi SM, Jhi SH, Son YW. Controlling energy gap of bilayer graphene by strain. Nano Lett. 2010;10 (9):3486–9.
[88] Min HK, Sahu B, Banerjee SK, MacDonald AH. Ab initio theory of gate induced gaps in graphene bilayers. Phys Rev B 2007;75(15):155115.
[89] Zhang YB, Tang TT, Girit C, Hao Z, Martin MC, Zettl A, et al. Direct observation of a widely tunable bandgap in bilayer graphene. Nature 2009;459(7248):820–3.
[90] Balog R, Jorgensen B, Nilsson L, Andersen M, Rienks E, Bianchi M, et al. Bandgap opening in graphene induced by patterned hydrogen adsorption. Nat Mater 2010;9(4):315–19.
[91] Boukhvalov DW, Katsnelson MI, Lichtenstein AI. Hydrogen on graphene: Electronic structure, total energy, structural distortions and magnetism from first-principles calculations. Phys Rev B 2008;77(3):035427.
[92] Denis PA. Band gap opening of monolayer and bilayer graphene doped with aluminium, silicon, phosphorus, and sulfur. Chem Phys Lett 2010;492(4-6):251–7.
[93] Geim AK, Grigorieva IV. Van der Waals heterostructures. Nature 2013;499(7459):419–25.
[94] Novoselov KS. Nobel lecture: graphene: materials in the flatland. Rev Mod Phys 2011;83(3):837–49.
[95] Ponomarenko LA, Geim AK, Zhukov AA, Jalil R, Morozov SV, Novoselov KS, et al. Tunable metal-insulator transition in double-layer graphene heterostructures. Nat Phys 2011;7(12):958–61.
[96] Gorbachev RV, Geim AK, Katsnelson MI, Novoselov KS, Tudorovskiy T, Grigorieva IV, et al. Strong Coulomb drag and broken symmetry in double-layer graphene. Nat Phys 2012;8(12):896–901.
[97] Ishigami M, Aloni S, Zettl A. Properties of boron nitride nanotubes. Aip Conf Proc 2003;696:94–9.
[98] Chen Y, Zou J, Campbell SJ, Le Caer G. Boron nitride nanotubes: pronounced resistance to oxidation. Appl Phys Lett 2004;84(13):2430–2.
[99] Lei WW, Portehault D, Liu D, Qin S, Chen Y. Porous boron nitride nanosheets for effective water cleaning. Nat Commun 2013;4 1777-1-7
[100] Rubio A, Miyamoto Y, Blase X, Cohen ML, Louie SG. Theoretical study of one-dimensional chains of metal atoms in nanotubes. Phys Rev B 1996;53(7):4023.
[101] Shi YM, Hamsen C, Jia XT, Kim KK, Reina A, Hofmann M, et al. Synthesis of few-layer hexagonal boron nitride thin film by chemical vapor deposition. Nano Lett 2010;10(10):4134–9.

[102] Britnell L, Gorbachev RV, Jalil R, Belle BD, Schedin F, Mishchenko A, et al. Field-effect tunneling transistor based on vertical graphene heterostructures. Science 2012;335(6071):947–50.

[103] Ponomarenko LA, Belle BD, Jalil R, Britnell L, Gorbachev RV, Geim AK, et al. Field-effect control of tunneling barrier height by exploiting graphene's low density of states. J Appl Phys 2013;113 (13):136502.

[104] Liu Z, Song L, Zhao SZ, Huang JQ, Ma LL, Zhang JN, et al. Direct growth of graphene/hexagonal boron nitride stacked layers. Nano Lett 2011;11(5):2032–7.

[105] Mishchenko A, Tu JS, Cao Y, Gorbachev RV, Wallbank JR, Greenaway MT, et al. Twist-controlled resonant tunnelling in graphene/boron nitride/graphene heterostructures. Nat Nanotechnol 2014;9 (10):808–13.

[106] Chen XF, Lian JS, Jiang Q. Band-gap modulation in hydrogenated graphene/boron nitride heterostructures: the role of heterogeneous interface. Phys Rev B 2012;86(12):125437-1-8.

[107] Lee GH, Yu YJ, Lee C, Dean C, Shepard KL, Kim P, et al. Electron tunneling through atomically flat and ultrathin hexagonal boron nitride. Appl Phys Lett 2011;99(24):243114.

[108] Britnell L, Gorbachev RV, Jalil R, Belle BD, Schedin F, Katsnelson MI, et al. Electron tunneling through ultrathin boron nitride crystalline barriers. Nano Lett 2012;12(3):1707–10.

[109] Georgiou T, Jalil R, Belle BD, Britnell L, Gorbachev RV, Morozov SV, et al. Vertical field-effect transistor based on graphene-WS2 heterostructures for flexible and transparent electronics. Nat Nanotechnol 2013;8(2):100–3.

[110] Britnell L, Ribeiro RM, Eckmann A, Jalil R, Belle BD, Mishchenko A, et al. Strong light-matter interactions in heterostructures of atomically thin films. Science 2013;340(6138):1311–14.

[111] Heersche HB, Jarillo-Herrero P, Oostinga JB, Vandersypen LMK, Morpurgo AF. Bipolar supercurrent in graphene. Nature 2007;446(7131):56–9.

[112] Levendorf MP, Ruiz-Vargas CS, Garg S, Park J. Transfer-free batch fabrication of single layer graphene transistors. Nano Lett 2009;9(12):4479–83.

[113] Lin YM, Jenkins KA, Valdes-Garcia A, Small JP, Farmer DB, Avouris P. Operation of graphene transistors at gigahertz frequencies. Nano Lett 2009;9(1):422–6.

[114] Farmer DB, Chiu HY, Lin YM, Jenkins KA, Xia FN, Avouris P. Utilization of a buffered dielectric to achieve high field-effect carrier mobility in graphene transistors. Nano Lett. 2009;9(12):4474–8.

[115] Lin YM, Dimitrakopoulos C, Jenkins KA, Farmer DB, Chiu HY, Grill A, et al. 100-GHz transistors from wafer-scale epitaxial graphene. Science 2010;327(5966):662.

[116] Ponomarenko LA, Schedin F, Katsnelson MI, Yang R, Hill EW, Novoselov KS, et al. Chaotic dirac billiard in graphene quantum dots. Science 2008;320(5874):356–8.

[117] Zhou SY, Gweon GH, Fedorov AV, First PN, De Heer WA, Lee DH, et al. Substrate-induced bandgap opening in epitaxial graphene. Nat Mater 2007;6(11):770–5.

[118] Oostinga JB, Heersche HB, Liu XL, Morpurgo AF, Vandersypen LMK. Gate-induced insulating state in bilayer graphene devices. Nat Mater 2008;7(2):151–7.

[119] Dan YP, Lu Y, Kybert NJ, Luo ZT, Johnson ATC. Intrinsic response of graphene vapor sensors. Nano Lett 2009;9(4):1472–5.

[120] Robinson JT, Perkins FK, Snow ES, Wei ZQ, Sheehan PE. Reduced graphene oxide molecular sensors. Nano Lett 2008;8(10):3137–40.

[121] Tang LH, Wang Y, Li YM, Feng HB, Lu J, Li JH. Preparation, structure, and electrochemical properties of reduced graphene sheet films. Adv Funct Mater 2009;19(17):2782–9.

[122] Ohno Y, Maehashi K, Yamashiro Y, Matsumoto K. Electrolyte-gated graphene field-effect transistors for detecting pH protein adsorption. Nano Lett 2009;9(9):3318–22.

[123] Huang B, Li ZY, Liu ZR, Zhou G, Hao SG, Wu J, et al. Adsorption of gas molecules on graphene nanoribbons and its implication for nanoscale molecule sensor. J Phys Chem C 2008;112(35):13442–6.

[124] Wehling TO, Novoselov KS, Morozov SV, Vdovin EE, Katsnelson MI, Geim AK, et al. Molecular doping of graphene. Nano Lett. 2008;8(1):173–7.
[125] Leenaerts O, Partoens B, Peeters FM. Adsorption of H(2)O, NH(3), CO, NO(2), and NO on graphene: a first-principles study. Phys Rev B 2008;77(12):125416.
[126] Zhu Y, Murali S, Cai W, Li X, Suk JW, Potts JR, et al. Graphene and graphene oxide: synthesis, properties, and applications. Adv Mater 2010;22(35):3906–24.
[127] Winter M, Besenhard JO, Spahr ME, Novak P. Insertion electrode materials for rechargeable lithium batteries. Adv Mater 1998;10(10):725–63.
[128] Yoo E, Kim J, Hosono E, Zhou H, Kudo T, Honma I. Large reversible Li storage of graphene nanosheet families for use in rechargeable lithium ion batteries. Nano Lett 2008;8(8):2277–82.
[129] Pan DY, Wang S, Zhao B, Wu MH, Zhang HJ, Wang Y, et al. Li Storage Properties of Disordered Graphene Nanosheets. Chem Mater 2009;21(14):3136–42.
[130] Wang CY, Li D, Too CO, Wallace GG. Electrochemical properties of graphene paper electrodes used in lithium batteries. Chem Mater 2009;21(13):2604–6.
[131] Lee JK, Smith KB, Hayner CM, Kung HH. Silicon nanoparticles-graphene paper composites for Li ion battery anodes. Chem Commun 2010;46(12):2025–7.
[132] Jang BZ, Zhamu A. Processing of nanographene platelets (NGPs) and NGP nanocomposites: a review. J Mater Sci 2008;43(15):5092–101.
[133] Park S, Ruoff RS. Chemical methods for the production of graphenes. Nat Nanotechnol 2009;4 (4):217–24.
[134] Hu HT, Wang XB, Wang JC, Wan L, Liu FM, Zheng H, et al. Preparation and properties of graphene nanosheets-polystyrene nanocomposites via in situ emulsion polymerization. Chem Phys Lett 2010;484 (4-6):247–53.
[135] Ramanathan T, Abdala AA, Stankovich S, Dikin DA, Herrera-Alonso M, Piner RD, et al. Functionalized graphene sheets for polymer nanocomposites. Nat Nanotechnol 2008;3(6):327–31.
[136] Salavagione HJ, Martinez G, Gomez MA. Synthesis of poly(vinyl alcohol)/reduced graphite oxide nanocomposites with improved thermal and electrical properties. J Mater Chem 2009;19 (28):5027–32.
[137] Liang JJ, Huang Y, Zhang L, Wang Y, Ma YF, Guo TY, et al. Molecular-level dispersion of graphene into poly(vinyl alcohol) and effective reinforcement of their nanocomposites. Adv Funct Mater 2009;19 (14):2297–302.
[138] Kalaitzidou K, Fukushima H, Drzal LT. A new compounding method for exfoliated graphite-polypropylene nanocomposites with enhanced flexural properties and lower percolation threshold. Compos Sci Technol 2007;67(10):2045–51.
[139] Ganguli S, Roy AK, Anderson DP. Improved thermal conductivity for chemically functionalized exfoliated graphite/epoxy composites. Carbon 2008;46(5):806–17.
[140] Kim H, Macosko CW. Morphology and properties of polyester/exfoliated graphite nanocomposites. Macromolecules 2008;41(9):3317–27.
[141] Verdejo R, Barroso-Bujans F, Rodriguez-Perez MA, de Saja JA, Lopez-Manchado MA. Functionalized graphene sheet filled silicone foam nanocomposites. J Mater Chem 2008;18(19):2221–6.
[142] Cai DY, Yusoh K, Song M. The mechanical properties and morphology of a graphite oxide nanoplatelet/polyurethane composite. Nanotechnology 2009;20(8).
[143] Ansari S, Giannelis EP. Functionalized graphene sheet-poly(vinylidene fluoride) conductive nanocomposites. J Polym Sci Pol Phys 2009;47(9):888–97.
[144] Kim H, Macosko CW. Processing-property relationships of polycarbonate/graphene composites. Polymer 2009;50(15):3797–809.
[145] Thostenson ET, Li CY, Chou TW. Nanocomposites in context. Compos Sci Technol 2005;65(3-4):491–516.

[146] Park S, Ruoff RS. Chemical methods for the production of graphenes. Nat Nanotechnol 2010;5 (4):217–24.
[147] McAllister MJ, Li JL, Adamson DH, Schniepp HC, Abdala AA, Liu J, et al. Single sheet functionalized graphene by oxidation and thermal expansion of graphite. Chem Mater 2007;19(18):4396–404.
[148] Yu QK, Lian J, Siriponglert S, Li H, Chen YP, Pei SS. Graphene segregated on Ni surfaces and transferred to insulators. Appl Phys Lett 2008;93(11):113103.
[149] Steurer P, Wissert R, Thomann R, Mulhaupt R. Functionalized graphenes and thermoplastic nanocomposites based upon expanded graphite oxide. Macromol Rapid Commun 2009;30(4-5):316–27.
[150] Wang SR, Tambraparni M, Qiu JJ, Tipton J, Dean D. Thermal expansion of graphene composites. Macromolecules 2009;42(14):5251–5.
[151] Rafiee MA, Rafiee J, Yu ZZ, Koratkar N. Buckling resistant graphene nanocomposites. Appl Phys Lett 2009;95(22):223103.

CHAPTER

ARRANGEMENTS OF CARBON-BASED STRUCTURES 10

10.1 INTRODUCTION

Carbon nanorings have been observed in single-walled carbon nanotubes (SWCNTs) grown by laser vaporization [1]; the diameters of these circular structures range between 300 and 500 nm and their widths between 5 and 15 nm. It appears that there are no topological pentagon–heptagon defects in these structures, as kinks that could be created by such defects are not observed. Therefore, they can be imaged as the bending of a straight SWCNT into a ring by connecting its two ends to form the carbon nanorings. These nanoring structures may be ideal nanodevices due to their fascinating mechanical and physical properties.

Nanorings have been extensively subjected to theoretical and modeling studies. Liu et al. [2] employed atomistic simulations using the second-generation reactive bond-order potential [3] and revealed equilibrium structures with highly regular ripples on the inside wall of large nanorings at smaller ring diameters. Subsequently, some of these ripples developed into buckles. Tight-binding and semiempirical quantum mechanics have been used to investigate a small CNT ring [4], in which the topological connections between CNTs were proposed theoretically by Dunlap [5]. In classical molecular dynamics (MD) simulations based on Brenner's interaction potential [6] carried out by Huhtala et al. [7,8], using a ring diameter from 22 to 700 nm, it was revealed that the critical tube radius is dependent on the buckling radius where the ring buckled.

In the simulations of Liu et al. [9], order-N nonorthogonal tight-binding MD simulations were employed to analyze the structural evolution and electrical conductance of a large carbon nanoring in terms of the changes in the angular position along the circumference. Minimum energy configurations based on the first generation reactive bond-order potential were used by Hod et al. [10] and Han [11], respectively, to analyze the structures of carbon nanorings. Continuum mechanics were also utilized to study the atomic and electronic structures of large and small carbon tori by Meunier et al. [12], who used the elasticity theory to examine the stability of a large ring.

10.2 CARBON NANORINGS

A study on the stability of zigzag nanorings in the previous work has not been reported. We employ the MD simulation to analyze the structures and energetic of both armchair and zigzag nanorings, in which a force field function is introduced to compute the intermolecular interactions of the carbon atoms. A new definition on judging the stability of carbon nanorings is proposed to calculate the critical ring diameter. Ring diameters ranging from 6 to 70 nm are used to study the armchair nanorings and diameters up to 101 nm to study the zigzag nanorings.

Mechanical Behaviors of Carbon Nanotubes. DOI: http://dx.doi.org/10.1016/B978-0-323-43137-8.00010-X

10.2.1 MD SIMULATION

COMPASS is an ab initio force field, that is, most of its parameters are derived based on ab initio data. In general, this parameterization procedure can be divided into two phases: ab initio parameterization and empirical optimization. In the first phase, partial charges and the valence parameters are derived by fitting them to the ab initio potential energy surfaces. At this point, the vdW parameters are fixed to a set of initial approximated parameters. In the second phase, the emphasis is on optimizing the force field to yield good agreement with the experimental data. A few of the critical valence parameters in the study reported herein are adjusted based on gas phase experimental data. More importantly, the vdW parameters are optimized to fit the condensed-phase properties.

In MD simulations, velocity Verlet integration [13] is used with a time-step size of 1.0 fs; the constant-temperature, constant-volume ensemble is obtained by controlling the thermodynamic temperature; the initial velocities of the particles are set randomly from the Maxwell–Boltzmann distribution; and the Nosé method [14] is utilized to control the temperature. The initial temperature of the system in this study is set at 298.0K.

10.2.2 CARBON NANORINGS

The initial atomic configurations of nanorings are obtained by bending a straight tube in space so that the tube ends can be connected, thus forming a closed ring. For a straight nanotube with a length of L, the diameter of its geometrically mapped ring is $D = L/\pi$, and the length of the carbon–carbon (C–C) single bond is 1.42 Å. It is more difficult to obtain a ring or a tube with a circular cross-section after the simulation, although the ring diameter is up to 100 nm and the length of the bending tube is more than 300 nm. Here, we define a nanoring is stable if there is no buckling, and the cross-section of the ring is similar to that of the tube with the largest length used in this paper. The largest lengths of different types of nanorings are different, e.g., the largest length of tube (5, 5) is 112.5 nm. Comparing the cross-section of the ring with this tube makes it much easier to obtain the length in the CNT experiment.

10.2.3 ARMCHAIR NANORINGS

Four armchair configurations, (5, 5), (6, 6), (8, 8), and (10, 10), of CNTs are considered in the present study, and the diameters, d, of these four tubes are 0.678, 0.814, 1.085, and 1.356 nm, respectively. Various initial ring diameters, D, that range from 6 to 70 nm are employed to estimate the critical diameter, D_c, for a stable nanoring.

The strain energy–diameter relationships of the four armchair configurations are obtained and shown in Fig. 10.1, in which strain energy is defined as the difference in potential energy between the ring and a corresponding straight nanotube with the same carbon atom numbers. The strain energy per atom of the four armchair rings, (5, 5), (6, 6), (8, 8), (10, 10), approaches to zero at ring diameters of 39, 46, 52, and 62 nm, respectively (see Fig. 10.1). Dash and arrow in Fig. 10.1 indicate the critical diameters' location of the four nanorings.

It is evident that strain energy decreases with ring diameter. Also, under the same ring diameter, specifically before 25 nm, the strain energy per atom clearly decreases with the diameter of the

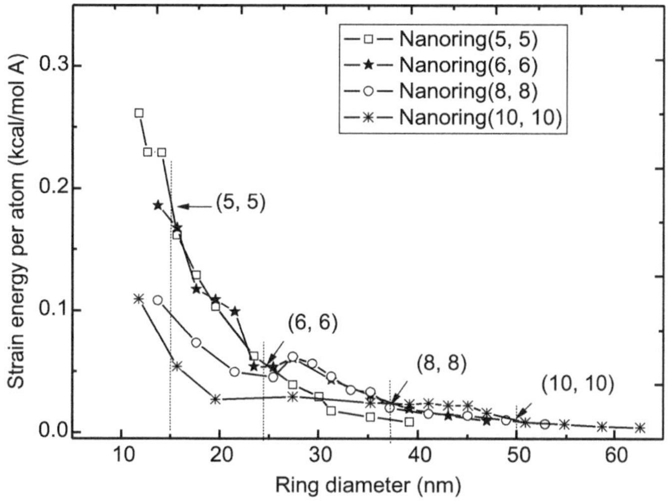

FIGURE 10.1

Relationship between the strain energy per atom relative to the straight tube and diameter for a ring [15].

straight tube. This shows that the average potential energy of a ring becomes close to that of its corresponding straight tube with an increased tube diameter.

For our analysis of stability, we take nanoring (10, 10) as an example for the armchair configurations. When 6 nm $< D <$ 40 nm, buckling can clearly be seen, and its cross-section is relatively flat (see Fig. 10.2A and C); when $D >$ 50 nm, no buckling occurs, and its cross-section becomes similar to that of the tube with the largest length used. Fig. 10.2 shows the features of a nanoring (10, 10) with and without buckling and the cross-sections of the ring and the tube. The selected cross-sections are away from the kinks. Fig. 10.2D–F describe the cross-section shapes of nanoring (10, 10) with three different diameter D, and Fig. 10.2G is the cross-section of nanotube (10, 10) with the largest tube length $L =$ 196.76 nm used in nanoring (10, 10) simulation, which is easier to obtain in CNT experiments. It can be seen from Fig. 10.2D–F, the shapes change from flat ellipse to square and gradually approach to that of the comparison tube (Fig. 10.2G). When the difference of cross-section shapes between ring and the comparison tube is less than one percent, they are considered as similar. The square-shaped cross-section should be the energy minimization configuration of nanoring (10, 10).

10.2.4 ZIGZAG NANORINGS

Zigzag configurations of (5, 0), (7, 0), (9, 0), and (10, 0) are used here to analyze the stability of zigzag nanorings with tube diameters of 0.391, 0.548, 0.705, and 0.783 nm, respectively. Initial ring diameters of from 8 to 101 nm are used. Fig. 10.3 shows the stain energy per atom as a function of the ring diameter for the four zigzag nanorings. It indicates that the strain energy per atom increases with the corresponding tube diameter under the same ring diameter.

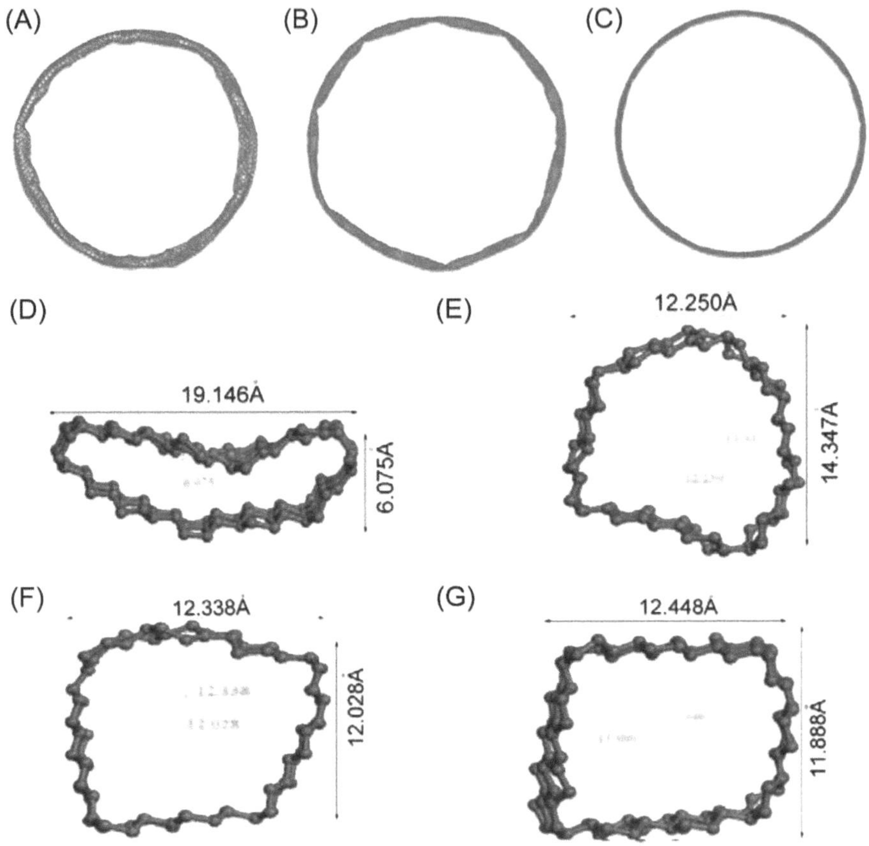

FIGURE 10.2

The shape of armchair nanorings (10, 10) with and without buckling and the cross-sections of the ring and the comparison tube: (A) ring with $D = 11.74$ nm and 6000 carbon atoms; (B) ring with $D = 27.4$ nm and 14,000 carbon atoms; (C) ring with $D = 54.80$ nm and 28,000 carbon atoms; (D) the cross-section of ring A; (E) the cross-section of ring B; and (F) the cross-section of ring C; (G) the cross-section of the comparison tube (10, 10) with length $L = 196.76$ nm and 32,000 carbon atoms [15].

The features of ring (7, 0) with and without buckling are shown in Fig. 10.4A–C, of which (A) and (B) are unstable, and (C) is a stable ring with $D = 50.85$ nm. Fig. 10.4D–F are the cross-section of (A), (B), and (C), respectively, and Fig. 10.4G is the cross-section shape of nanotube (7, 0) with tube length $L = 191.70$ nm, bending the tube will obtain a ring with $D = 61.05$ nm. From Fig. 10.4D–F we can see that their shapes gradually approach to (G). The cross-sections of zigzag nanoring are selected from a place close to the kink, while the shape of cross-section away from kink has few changes comparing with that of the corresponding nanotube.

As described above, when there is no buckling occurring and the shape difference between ring and the comparison tube (Fig. 10.4G) is less than one percent, the ring is considered as stable.

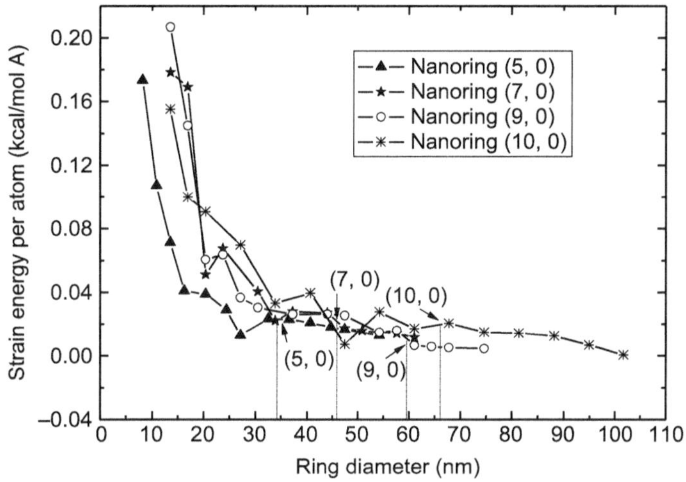

FIGURE 10.3

Relationship between the strain energy per atom of a ring relative to a straight tube, where D is the ring diameter [15].

The plum-shaped and symmetric cross-section is the characteristics of stable zigzag nanoring, which may be a minimum energy feature for zigzag nanotube.

According to the definition of a stable ring, i.e., there is no buckling and its cross-section is similar to that of the comparison tube, we obtain the critical diameters of nanorings. The critical diameters for armchair nanorings (5, 5), (6, 6), (8, 8), and (10, 10) are approximately 15, 24, 37, and 50 nm, respectively (shown in Fig. 10.1), and those for zigzag nanorings (5, 0), (7, 0), (9, 0), and (10, 0) are 34, 46, 59, and 66 nm, respectively (shown in Fig. 10.3).

The critical diameter of ring can be considered as the linear function of its corresponding tube diameter depicted in Fig. 10.5. In addition, the simulation results indicate that a zigzag nanoring has greater critical diameter than an armchair nanoring with the approximate tube diameters.

In order to clearly explain this point, we take zigzag nanoring (10, 0) with ring diameter 15.73 nm and armchair nanoring (6, 6) with ring diameter 15.66 nm as examples, where the diameters of the corresponding nanotube are 0.783 and 0.814 nm, respectively, and the ring critical diameters are 66 and 24 nm, respectively. Firstly, the radial cross-sections uniformly distribute in armchair nanoring, but not in zigzag nanoring, in which the largest interval between cross-sections is 1.16 times that of armchair nanoring. It can be seen in Fig. 10.6A and B that the armchair nanoring has the more compact structure with the approximately same ring diameter. The compact and uniform structure makes the armchair nanoring more stable.

Secondly, when two straight nanotubes under the approximate same diameter, such as armchair (6, 6) and zigzag (10, 0), are bended to nanorings, the axial bonds existing in a zigzag nanotube makes a larger discrepancy in bond length than that of an armchair. Fig. 10.7 shows the bond distribution of both nanorings, in which the vertical axis expresses the probability density of bond length, and the transverse represents the bond length. The larger average bond length of a zigzag

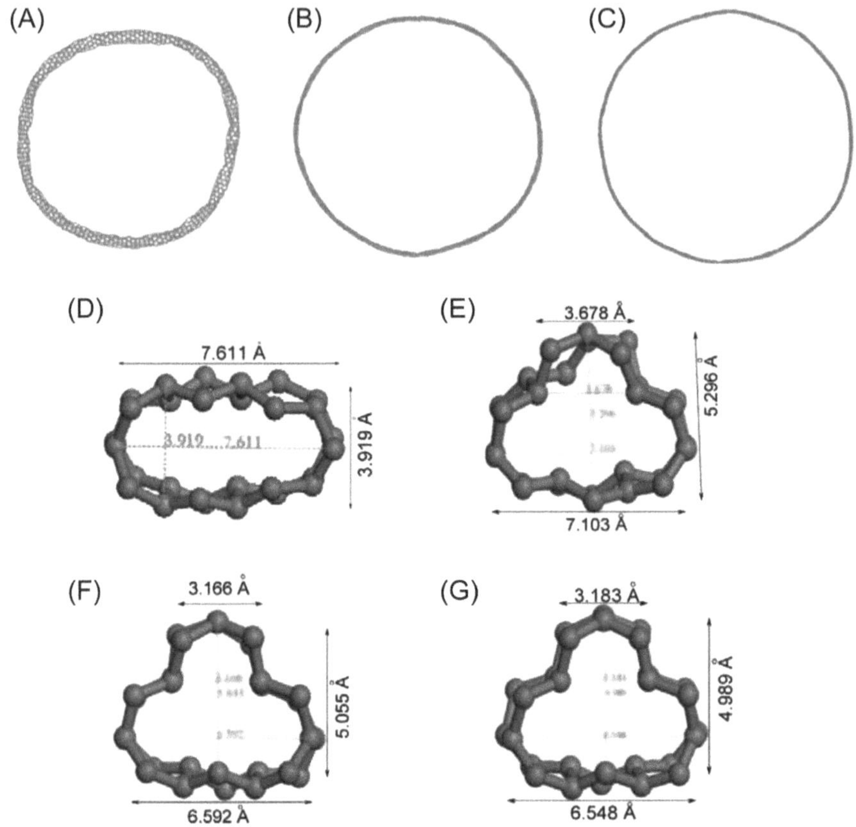

FIGURE 10.4

The shape of zigzag nanorings (7, 0) with and without buckling and their cross-sections: (A) ring with $D = 10.17$ nm and 2100 carbon atoms; (B) ring with $D = 30.51$ nm and 6300 carbon atoms; (C) ring with $D = 50.85$ nm and 10500 carbon atoms; (D) the cross-section of ring a; (E) the cross-section of ring b; (F) the cross-section of ring c; and (G) the cross-section of the comparison tube (7, 0) with length $L = 191.70$ nm and 12600 carbon atoms [15].

nanoring is another unstable factor. From the abovementioned two points, to some extent it explains that an armchair nanoring is more stable than a zigzag nanoring.

Buckling will severely affect the electrical and mechanical properties of nanorings, therefore it is crucial to study their buckling behaviors. Molecular mechanics (MM) simulations are employed to study the buckling behavior of carbon nanorings (5, 5) formed by bending an armchair nanotube (5, 5) and subjected to a displacement-controlled loading on the carbon atoms in its two symmetrical sides. The buckling shapes and critical tension displacements of nanorings with ring diameters ranging from 3 to 20 nm that are subjected to different ranges of tension are investigated.

MM simulations of nanorings are conducted to find the minimum energy configuration on the potential energy surface using a conjugated gradient algorithm. The simulations are carried out by a

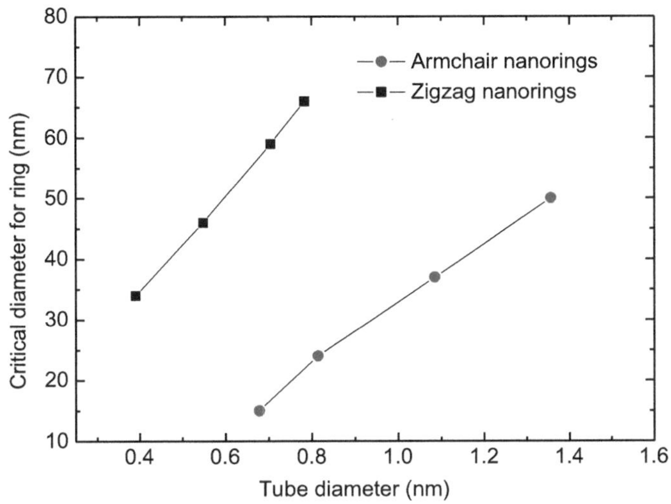

FIGURE 10.5

Critical nanoring diameters versus the nanotube diameters for armchair and zigzag nanorings [15].

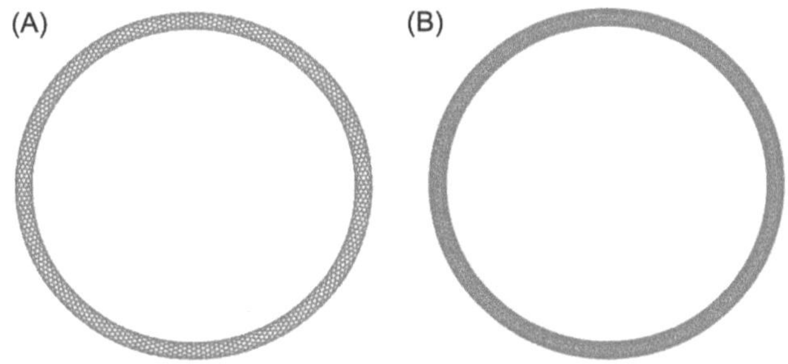

FIGURE 10.6

The shape of both nanorings (A) zigzag nanoring (10, 0) with 4640 carbon atoms and $D = 15.73$ nm; (B) armchair nanoring (6, 6) with 4800 carbon atoms and $D = 15.66$ nm [15].

quasistatic displacement update scheme, as in several other works [16,17]. Because the critical buckling strain of CNTs is highly sensitive to the displacement increment [18–20], we use four different displacement increments on each side of the ring, with 0.1, 0.2, 0.4, and 0.6 Å, to investigate their effects on the buckling behavior of nanorings (5, 5). It is found that the four displacement increments do not have a crucial impact on the buckling shapes and critical tension displacements, which are the largest accumulated displacements before buckling occurs, if a small increment with or less than 0.01 Å is applied when the accumulated displacement approaches to the critical tension displacement. The magnitude of CNTs [18–20] and nanorings may induce this discrepancy in the effect of a displacement increment on the buckling behavior. Mainly focusing on the relative buckling behavior

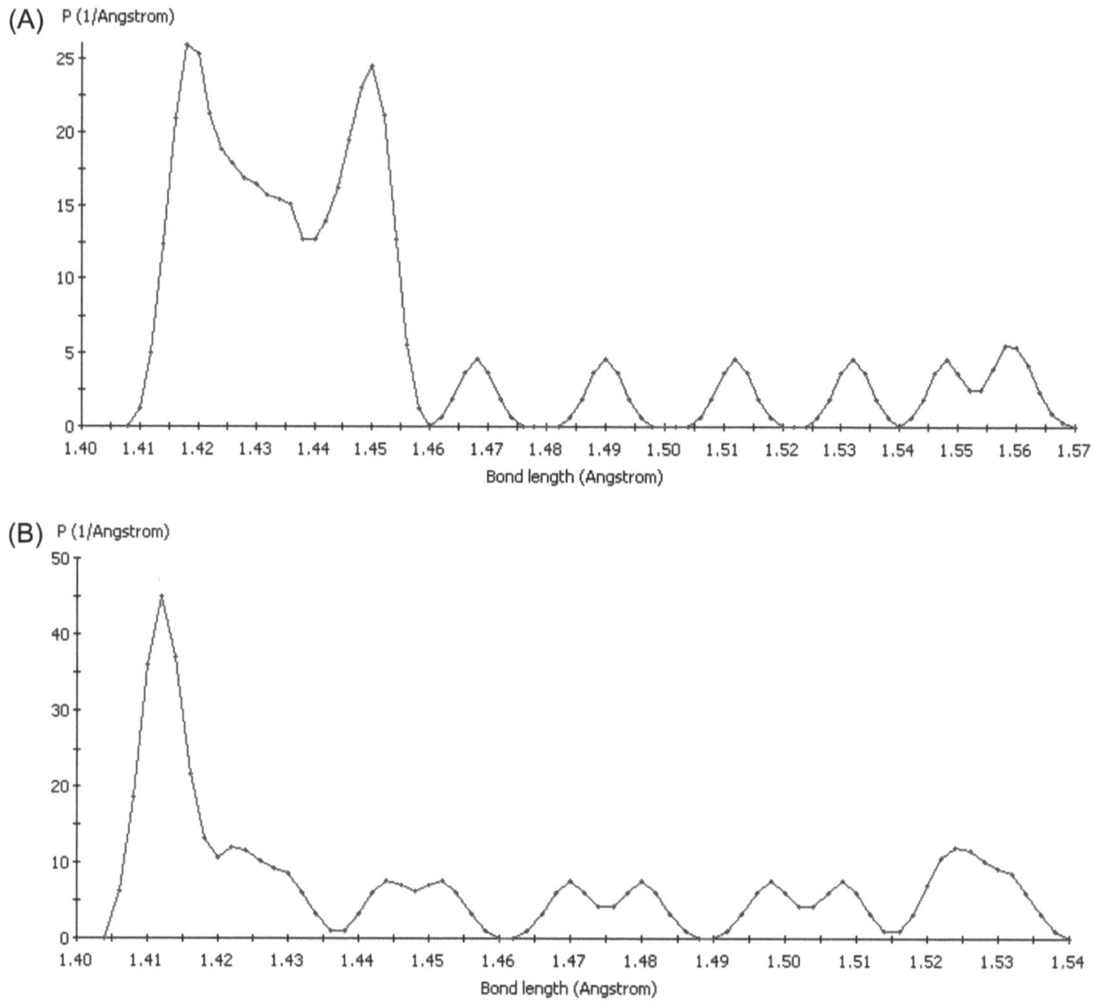

FIGURE 10.7

The Schematic of bond-length distribution (A) zigzag nanoring (10, 0) with 4640 carbon atoms and $D = 15.73$ nm; (B) armchair nanoring (6, 6) with 4800 carbon atoms and $D = 15.66$ nm [15].

among different type of nanorings (5, 5), we choose the displacement increment 0.2 Å to analyze the critical tension displacements and buckling shapes of nanorings (5, 5) with different ring diameters.

A carbon nanoring (5, 5) is formed by bending a straight nanotube (5, 5) with a carbon–carbon (C–C) single bond length of 1.42 Å and connecting its two ends together. After the nanoring has been established, it is subjected to an energy minimization process. The carbon atoms on the two symmetrical sides of the optimized nanoring are then clamped and the nanoring is optimized to obtain its initial potential energy. The incremental displacement step of 0.2 Å is applied to the clamped carbon atoms on each side, and the potential energy is collected through the minimization process at each incremental step until buckling occurs. Taking a nanoring (5, 5) with 50 units (1000 carbon atoms) as an

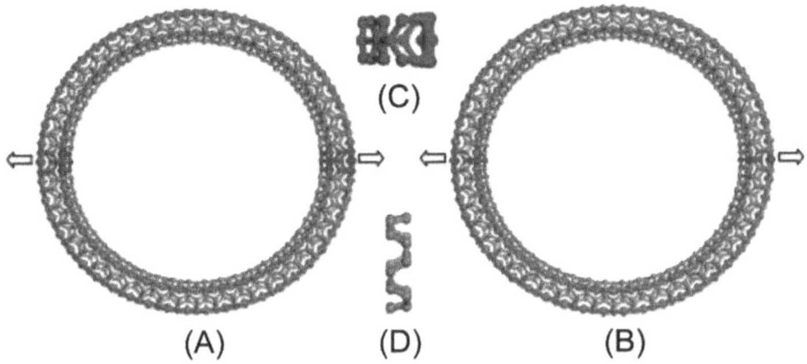

FIGURE 10.8

Sketch of a clamped nanoring (5, 5) with 50 units: (A) three circles of carbon atoms on the two sides are clamped and subjected to tension; (B) one circle of carbon atoms on the two symmetrical sides is clamped and subjected to tension; (C) the fixed three circles of carbon atoms on the right side of ring a; (D) a unit of a nanoring bended by nanotube (5, 5) [21].

example, Fig. 10.8 shows the directions of horizontal tension on the nanoring when three circles and one circle of carbon atoms are clamped on each side, respectively. The two models are optimized structures of clamped nanorings and have not been subjected to tension deformation. Each circle has 10 atoms, representing the number of atoms on the cross-section of the nanoring (5, 5), and each unit (shown in Fig. 10.8D) of the nanoring (5, 5) has two circles of carbon atoms, i.e., 20 atoms.

Nanoring buckling can be determined from the spontaneous drop in its strain energy, which is defined as the difference between the potential energy collected at each tension deformation and the initial potential energy of the clamped nanoring; buckling shapes are considered as the local collapse near the fixed boundary atoms. Bending a nanotube and connecting its two ends to form a nanoring lengthens the C–C bond, especially in the outer radius. Nanorings with longer bonds, where the interactional force of atoms is weak, are considered to be relatively unstable. Thus we analyze the bond length of the initial nanoring configuration before studying the buckling behavior. Further, only nanorings in which the longest bond is less than 1.8 Å in length and the corresponding ring diameter (D) is larger than 3 nm are investigated in this work.

10.2.5 CRITICAL TENSION DISPLACEMENTS

The tension displacement–strain energy curves of a 50-unit nanoring (5, 5) (see the Fig. 10.8 models) with a diameter of approximately 3.9 nm are shown in Fig. 10.9. The critical tension displacement is approximately 1.4 nm when three circles of atoms are subjected to tension and 1.62 nm when one circle of atoms is subjected to tension. It is similar to the tension behavior of the macroscopic materials with ring shapes in that the larger the stretched range of a nanoring is, the easier the nanoring will be to buckle, i.e., the critical tension displacement of three circles stretched is smaller than that of one circle. It is also observed from Fig. 10.9 that the strain energy of one circle and three circles stretched are close to each other when the tension displacement is less than 0.4 nm, then the two curves deviate when the tension displacement is larger than 0.4 nm until buckling occurs, and the strain energy of three circles is higher than that of one circle. The deformation of the three circles stretched is more obvious than that of one circle stretched under the same

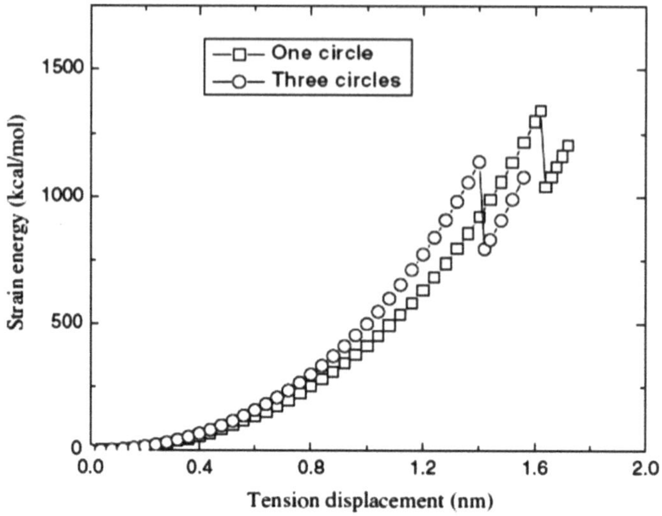

FIGURE 10.9

Strain energy of a 50-unit nanoring (5, 5) under uniform tension when one circle of atoms and three circles of atoms are stretched on each side of the ring, respectively [21].

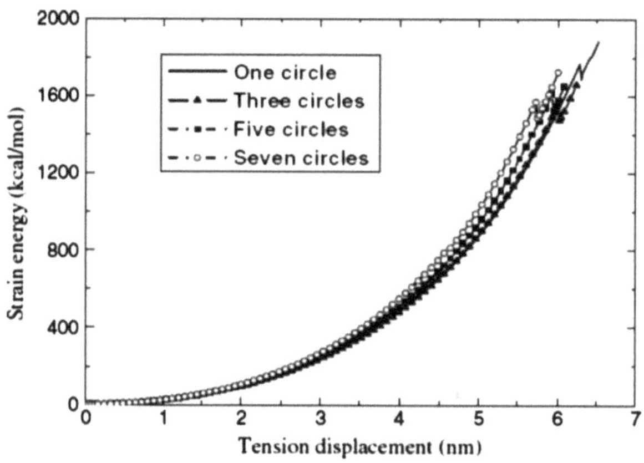

FIGURE 10.10

Strain energy of a nanoring (5, 5) with 180 units subjected to uniform tension, with one circle, three circles, five circles, and seven circles of atoms stretched on each side of the ring, respectively [21].

extension of displacement, thus the former strain energy is higher than the latter, especially in larger tension displacement.

One circle, three circles, five circles, and seven circles of carbon atoms stretched on each side of the nanoring (5, 5) with 180 units (3600 atoms) are respectively subjected to uniform displacement-controlled tension and their corresponding buckling behaviors are also investigated. Fig. 10.10 shows the relation between the strain energy and the tension displacement of the four cases, where the corresponding critical tension displacements of one circle, three circles, five circles, and seven circles of carbon atoms stretched are 6.28, 6, 5.92, and 5.72 nm, respectively. It is similar to Fig. 10.9 in that the four tension–strain energy curves are close to each other when the extension of displacement is small, and they will gradually deviate after the tension displacement is larger than 2.5 nm. It also elucidates that the strain energy with a larger number of circles stretched is higher than that of a smaller number of circles under the same accumulated displacement.

According to the simulations of nanorings (5, 5) with 50 units and 180 units, the critical tension displacement decreases with the increase of the number of stretched circles. The ratio r of the critical tension displacement to ring diameter is 0.415 and 0.359 for 50-unit nanorings with one circle and three circles stretched, respectively, and the difference between the two ratios is 0.056. Similarly, the ratios for 180-unit nanorings with one circle, three circles, five circles, and seven circles are 0.446, 0.426, 0.42, and 0.406, respectively; and the difference between the ratios of one circle and seven circles is 0.04. It suggests that a ring with 180 units is more flexible than a ring with 50 units due to its larger ratio *r*, and more circles are needed to be stretched for the former ring in order to arrive at the same ratio r as the latter, compared with their corresponding nanorings of one circle stretched.

According to the above analysis, it is clear that there is no essential difference in studying the buckling behavior of a nanoring when atoms of different circles are stretched for the same ring. Therefore, in the analysis of the other nanorings with ring diameters ranging from 3 to 20 nm, only one circle of atoms is stretched. The relationship between the critical tension displacement and the ring diameter is shown in Fig. 10.11A and the average potential energy of the initial clamped nanoring versus ring diameter is shown in Fig. 10.11B. The turning points in Fig. 10.11A and B are the same at a ring diameter of approximately 4.31 nm, which is the diameter of the nanoring with 55 units. As is well known, the lower the initial potential energy per atom, the more stable the structure, and the lower the critical tension displacement, the more unstable the structure, and therefore the similarity of the two turning points in Fig. 10.11A and B is not coincident. In addition, the critical tension displacement increases with ring diameter in nanorings with diameters ranging from 3.1 to 4.31 nm, for which the corresponding cross-sections are flat and the nanoring is like a belt, as in Fig. 10.8. When the ring diameter is larger than approximately 4.6 nm, the critical tension displacement also increases with the diameter. Fig. 10.11A shows that very small tension results in the buckling of the nanoring with 60 units, which has the smallest critical tension displacement at 0.24 nm. This indicates that the 60-unit nanoring is very unstable when two sides are stretched. The cross-section transformation of nanorings with one circle stretched is shown in Fig. 10.12; it can be seen that the shape of a nanoring with 60 units, i.e., 4.7 nm ring diameter, is a transition state from flat to triangular, and the shapes become gradually similar with the increase of the number of units when it is larger than 80 units. The transition of cross-section induces the changes of the average potential energy and critical tension displacement, and it is expressed as the two turn points in Fig. 10.11A and B. Moreover, according to the change of the average potential energy in Fig. 10.11B, it indicates that the ring with $D = 4.31$ nm is more stable than the ring with $D = 4.7$ nm, where the ring has the smallest critical tension displacement in Fig. 10.11A.

10.2.6 BUCKLING SHAPES

The buckling shapes of the nanorings are considered as the local collapse near the fixed boundary atoms, and there is a spontaneous drop in their corresponding strain energy. First, the nanoring with

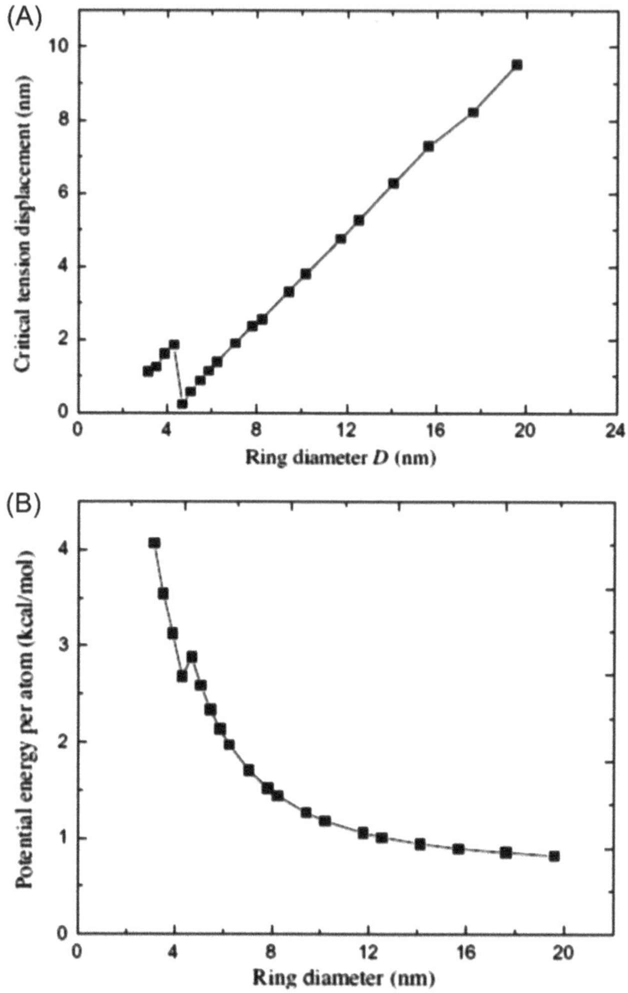

FIGURE 10.11

Sketch of (A) the critical tension versus ring diameter, (B) the initial potential energy per atom versus ring diameter [21].

180 units is investigated with one circle, three circles, five circles, and seven circles of carbon atoms subjected to tension, respectively. The corresponding buckling shapes are shown in Fig. 10.13. The buckling range gradually increases from Fig. 10.13A–D and the buckling structures are all symmetrical, i.e., independent of the number of circles stretched. The four buckling shapes suggest that the buckling range decreases with the increase of the number of atom circles stretched; it is caused by the displacement-controlled tension on the each side of the ring, where a larger number of atom circles stretched will bring out more influence on the deformation for the same nanoring, thus the affected buckling range will become bigger.

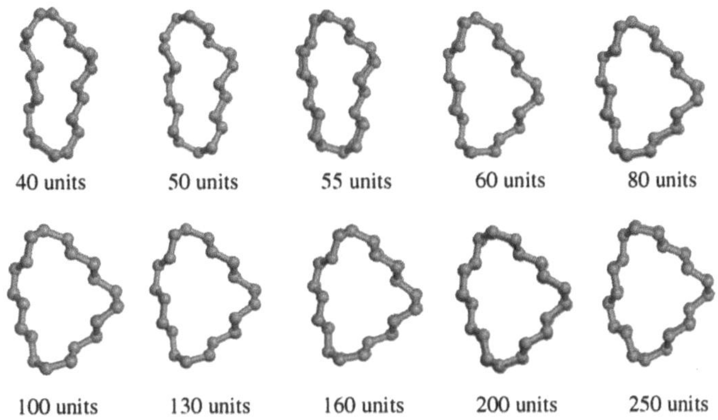

FIGURE 10.12

The cross-section shapes of nanorings (5, 5), where the number under each shape is its corresponding unit numbers [21].

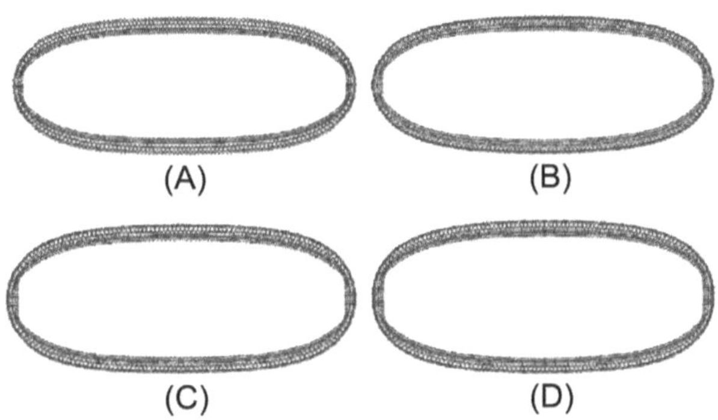

FIGURE 10.13

Buckling structures of a nanoring (5, 5) with 180 units: (A) one circle of atoms stretched; (B) three circles of atoms stretched; (C) five circles of atoms stretched; and (D) seven circles of atoms stretched [21].

Other buckling shapes of nanorings with even numbers of units and one circle stretched on each side are also studied. 60-, 70-, 80-, and 90-unit nanorings are analyzed (shown in Fig. 10.14), and their critical tension displacements are found to be 0.24, 0.88, 1.4, and 1.9 nm, respectively. The local collapse near fixed boundary atoms in Fig. 10.14A and B are much smoother and not very obvious, but they are considered as the buckling shapes due to the spontaneous drop in their strain energy. It is observed that the extent of the collapse is gradually obvious and severe from Fig. 10.14A–D, from which it can be deduced that the ratio of the area of the collapse near fixed atoms to ring diameter decreases with the increase of the ring diameter when one circle is stretched on its corresponding nanorings with the number of units larger than 60. Furthermore,

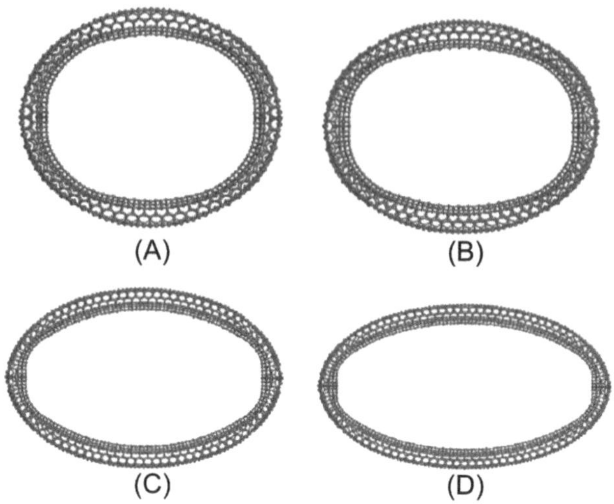

FIGURE 10.14

Buckling shapes of a nanoring (5, 5): (A) with 60 units and $D = 4.7$ nm; (B) with 70 units and $D = 5.48$ nm; (C) with 80 units and $D = 6.26$ nm; (D) with 90 units and $D = 7.05$ nm [21].

Fig. 10.14 also shows that the buckling shapes of nanorings with even units are symmetrical about the ring center.

Next, nanorings with odd numbers of units are examined. The buckling shapes and the relationship of strain energy versus tension displacement of 45-unit and 55-unit nanorings are presented in Fig. 10.15; this shows that they have unsymmetrical buckling structures about the ring center. When nanorings with 45 units and 55 units are subjected to displacement-controlled tension, they buckle first at one side, as in Fig. 10.15A and C where the critical tension displacements are 1.26 and 1.86 nm, respectively, which are respectively corresponding to the first turn point in Fig. 10.15E and F; and with continued extension of displacement buckle again at the other side, as in Fig. 10.15B and d where the critical tension displacements are 1.38 and 1.98 nm, respectively, which are respectively corresponding to the second turn point in Fig. 10.15E and F. The second critical tension displacement approaches the first. Unsymmetrical buckling structures in nanorings with the two odd numbers of units occur when a circle is stretched on each side of the nanoring because different sides of the same unit are being stretched, whereas in even-unit nanorings the same side is being stretched.

The unsymmetrical buckling shapes also occur on the 105-, 155-, and 225-unit nanorings, where the left side of the former two nanorings buckles first as in Fig. 10.15A and C, and the right side of the latter nanoring buckles first. However, the buckling shapes of nanorings with 65 and 75 units look symmetrical. Subsequently we investigate the buckling change process of the 65-unit nanoring, as shown in Fig. 10.16, which depicts the structures under different optimization steps when the extension of displacement is 0.6 nm, where the critical tension displacement of 65-unit nanoring is 0.58 nm. The left side buckles first at 700 steps, as shown in Fig. 10.16B, and next the other side

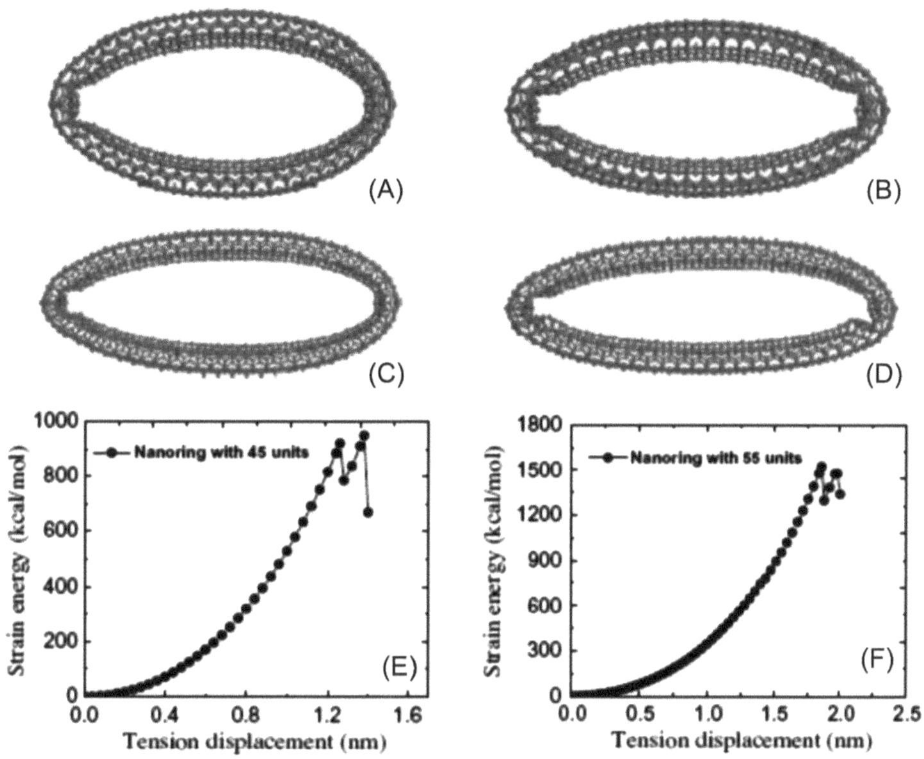

FIGURE 10.15

Buckling shapes of a nanoring (5, 5): (A) the initial buckling of a nanoring with 45 units at an accumulated displacement of 1.26 nm; (B) the secondary buckling of a nanoring with 45 units at an accumulated displacement of 1.38 nm; (C) the initial buckling of a nanoring with 55 units at an accumulated displacement of 1.86 nm; (D) the secondary buckling of a nanoring with 55 units at an accumulated displacement of 1.98 nm; € the relationship between strain energy and tension displacement of nanoring with 45 units; (F) the relationship between strain energy and tension displacement of nanoring with 55 units [21].

buckles at 1000 steps, as shown in Fig. 10.16D, to give a final buckling structure that is a symmetrical shape, as shown in Fig. 10.15F.

When the accumulated displacement approaches to its corresponding critical tension displacement, the larger displacement increment relative to its smaller critical tension displacement may induce the final symmetrical buckling structures for a 65-unit nanoring, similarly to a 75-unit nanoring. These symmetrical buckling structures for the above two nanorings still exist even if an increment with 0.00001 Å is stretched on each side, while a 0.01 Å increment will make the 105-unit nanoring buckle unsymmetrically, and 0.05 and 0.1 Å for nanorings with 155 and 225 units, respectively. The increment with 0.00001 Å is very small relative to ring diameter so that it barely has any effect on its critical tension displacement; hence we do not conduct further

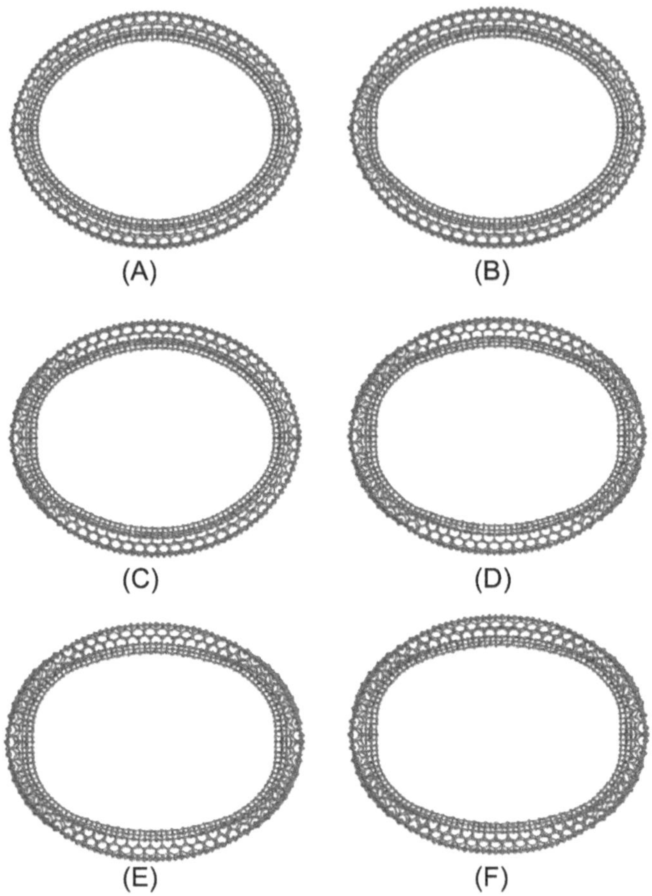

FIGURE 10.16

The buckling change process of a nanoring (5, 5) with 65 units at an accumulated displacement of 0.6 nm (A) the initial structure with incremental tension displacement of 0.1A° on each side of the last optimized nanoring; (B) the structure at 700 optimization steps; (C) the structure at 800 optimization steps; (D) the structure at 1000 optimization steps; (E) the structure at 2000 optimization steps; (F) the final optimized structure with a total of 4078 steps [21].

examinations on increments less than 0.00001 Å, but it is truly considered that the buckling structures of 65-unit and 75-unit nanorings are unsymmetrical if the appropriate displacement increment is applied.

According to the above simulations of nanorings with odd numbers of units, buckling structures stretching different sides of the same unit on each side of the ring lead to an unsymmetrical buckling shape, although it is not easily seen in nanorings with smaller critical tension displacement.

10.3 CARBON NANOCOILS

10.3.1 GEOMETRIC STRUCTURES

A carbon nanospring (*n*, *m*) with one pitch of length is formed by bending a SWCNT (*n*, *m*) into a closed circular shape that is cut into two pieces along the middle of two adjacent cross-sections, with one end fixed and the other stretched along the helical axis, which is perpendicular to the circular-centerline plane of the initial carbon nanoring [15].

The geometries of the general tubule nanosprings with right-handed and left-handed helices are shown in Fig. 10.17, in which the axis of the nanospring coincides with the global z-axis, and R, P, and r are the middle radii of the nanospring, pitch length, and tube radius, respectively. The spatial relationship of the nanospring centerline can be determined by Alfred [23]:

$$x = R\cos\theta, \quad y = R\sin\theta, \quad z = \frac{P}{2\pi}\theta \tag{10.1}$$

The vertical displacement shifts a pitch when θ rotates by an angle of 2π, and the spring rising angle α is defined as $\tan\alpha = P/2\pi R$.

Based on the structural characteristics of carbon tubules and the expression of the general spring (Eq. 10.1), we construct the geometric model of carbon nanosprings with various tube radii, spring radii, pitch lengths, and tube chiralities. Fig. 10.18A–C shows the three-turn carbon nanosprings

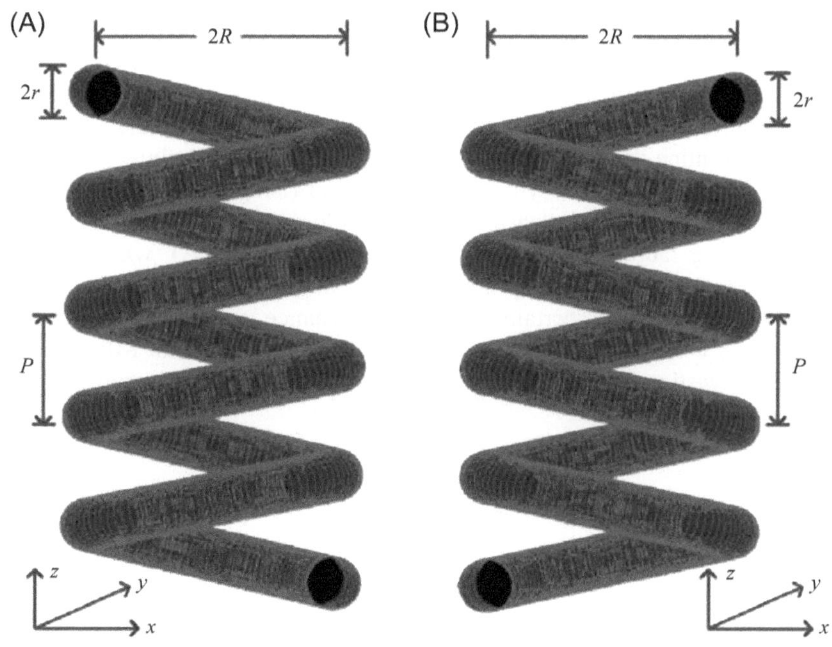

FIGURE 10.17

Geometric morphology of springs with: (A) a right-handed helix; and (B) a left-handed helix [22].

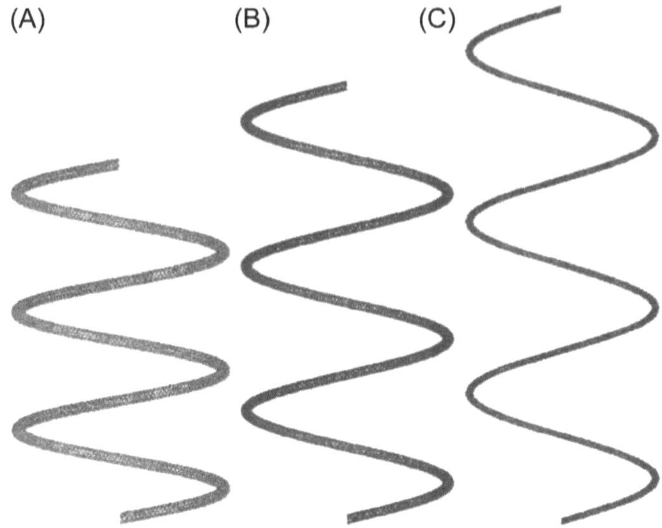

FIGURE 10.18

The geometric shape of carbon nanosprings: (A) spring (5, 5) with $R = 6.21$ nm and $a = 11.4$ degrees; (B) spring (6, 3) with $R = 6.59$ nm and $a = 14.3$ degrees; and (C) spring (7, 0) with $R = 8$ nm and $a = 19.5$ degrees [22].

formed from the nanorings (5, 5), (6, 3), and (7, 0), respectively. The three nanosprings are all right-handed helices that have different spring radii and rising angles $\alpha = \arctan(P/2\pi R)$. There is no difference between the right-handed and left-handed nanosprings in the context of structural stability analysis; thus only the right-handed nanosprings will be discussed in the following work.

In this study, we examine the structural behavior of armchair nanosprings (5, 5), (6, 6) and zigzag nanosprings (5, 0), (10, 0) with varying spring radii and rising angles. The nanosprings have a smaller tube-to-spring diameter ratio than those with topological defects. Carbon nanosprings with the same tube and spring radius have a maximum rising angle (MRA), which is decided by the structural characteristics of carbon nanosprings: i.e., carbon-atom ring needs to remain hexagonal and the C—C bond length has a certain range. For instance, when a nanospring (5, 5) with $R = 16$ nm has a C—C bond length in the range of 0.138—0.178 nm, its MRA is approximately 28.1 degrees. Before investigating the stable structures of all types of nanosprings, their corresponding MRAs need to be calculated.

10.3.2 THE MAXIMUM RISING ANGLES

Spring radii ranging from about 2 to 100 nm are considered for the four nanosprings. The relationships between the spring radii and their corresponding MRAs are shown in Fig. 10.19A (armchair nanosprings) and Fig. 10.19B (zigzag nanosprings), both of which indicate that the MRA increases approximately as a power function of the spring radius, and the increase in the MRA will slow down as the radius increases. The special functions of the MRA (α_m) in the model $\alpha_m = a \times R^b + c$ are shown in the caption of Fig. 10.19, in which R is the spring radius and

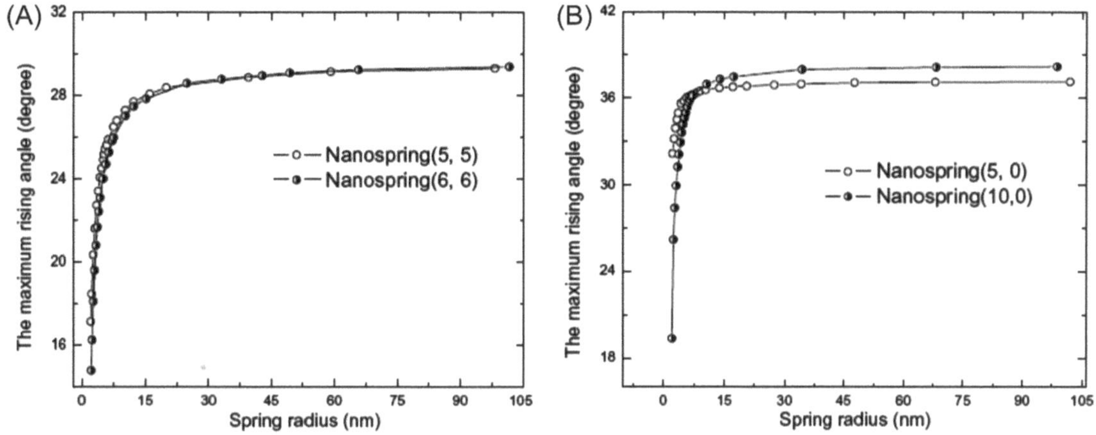

FIGURE 10.19

The relationship between the spring radius and the maximum rising angle: (A) $\alpha_m = -28.95R^{(-1.189)} + 29.22$ for the armchair nanospring (5, 5), and $\alpha_m = -36.67R^{(-1:196)} + 29.41$ for the armchair nanospring (6, 6); (B) $\alpha_m = -18:57R^{(-1.69)} + 37.01$ for the zigzag nanospring (5, 0), and $\alpha_m = -73.52R^{(-1:1956)} + 37.83$ for the zigzag nanospring (10, 0) [22].

a, *b*, and *c* are constants. For the four nanosprings, the regressed functions agree well with the simulated data despite some deviations in the large values of the spring radii. The deviations are all less than 1% within the radii range.

When each spring radius reaches about 100 nm, the MRAs of the four nanosprings (5, 5), (6, 6), (5, 0), and (10, 0) are around 29.3, 29.4, 37.15, and 38.19 degrees, respectively. Table 10.1 presents the values of the MRAs for various spring radii.

It can be observed from Fig. 10.19 and Table 10.1 that zigzag nanosprings have higher MRAs (α_m) than armchair nanosprings under the approximate spring radius, even if the tube radius of the armchair nanospring (6, 6), 0.407 nm, is much higher than that of the zigzag nanospring (5, 0), 0.195 nm. For each type of nanospring, when the spring radius becomes larger, the MRA increases as the tube radius increases; this trend is very clear for zigzag nanosprings when the radius is greater than 10.6 nm. However, when the spring radius is smaller, the MRA decreases as the tube radius increases. The C–C bond length of a nanospring with a small radius is greatly affected by the bended tube radius, especially in the outer part. Therefore, for small nanosprings with approximately the same radius, a nanospring with a large tube radius cannot have a higher rising angle than a nanospring with a small tube radius.

10.3.3 STABLE CHARACTERISTICS OF CARBON NANOSPRINGS

Before using the conjugate gradient algorithm to optimize a carbon nanospring, its two ends should be constrained; otherwise it will relax due to its high flexibility, as it will seek a state of minimum energy until it expands into a straight CNT. A nanospring is defined here as a stable structure if

Table 10.1 Parameters of the Spring Radius (R) and the Corresponding Maximum Rising Angle (α_m) for the Four Nanosprings [22]

(5, 5)		(6, 6)		(5, 0)		(10, 0)	
R (nm)	α_m (°)	R (nm)	α_m (°)	R (nm)	α_m (°)	R (nm)	α_m (°)
2.10	17.14	2.17	14.79	2.23	32.18	2.09	19.37
2.30	18.47	2.36	16.25	2.57	33.19	2.43	26.20
2.69	20.35	2.68	18.12	2.91	33.93	2.76	28.39
3.08	21.61	2.99	19.60	3.25	34.52	3.10	29.95
3.47	22.73	3.34	20.81	3.59	34.98	3.44	31.25
3.86	23.42	3.66	21.68	4.26	35.62	3.78	32.12
4.25	24.08	3.97	22.43	4.94	35.80	4.12	32.94
4.64	24.52	4.32	23.09	5.62	36.03	4.46	33.63
5.04	24.94	4.95	24.02	6.30	36.13	4.80	34.20
5.23	25.19	5.61	24.72	6.98	36.22	5.14	34.65
5.43	25.44	6.28	25.29	7.65	36.29	5.48	35.00
5.82	25.63	7.06	25.84	8.33	36.40	5.82	35.39
6.21	25.91	7.37	26.00	9.01	36.50	6.15	35.70
7.38	26.51	10.19	27.04	10.37	36.58	6.49	35.89
8.17	26.81	12.15	27.49	13.76	36.71	6.83	36.08
10.13	27.30	15.09	27.85	17.15	36.78	7.17	36.16
12.08	27.72	24.87	28.60	20.53	36.84	10.56	36.96
16.00	28.09	33.01	28.80	27.32	36.92	13.95	37.31
19.91	28.39	42.8	28.97	34.10	36.98	17.34	37.47
39.48	28.89	49.34	29.10	47.66	37.08	34.29	37.98
59.06	29.16	65.62	29.25	68.00	37.12	68.19	38.14
98.20	29.31	101.5	29.40	101.90	37.15	98.70	38.19

there is no turn-number change between the initial and the optimized configuration, ignoring the shape change. Flat and round carbon nanosprings are both considered to be stable.

10.3.4 EQUILIBRIUM STRUCTURES

Because a nanospring is constructed using a closed-tube ring, the equilibrium structure of the ring is similar to that of the constructed spring in terms of its cross-sectional shapes (viewed from the axial direction). The structural stability of nanorings has been fully discussed in [15], in which flat rings with kinks are obtained when the rings have small radii, and circular rings are produced when the rings have large radii.

The structure of a spring is determined by its coil shape, rising angle, and radius magnitude. Both flat and round carbon nanosprings are produced in CNT experiments [24,25]. Carbon nanosprings with different cross-sectional shapes each have their own mechanical and electric properties. For the structural stability of a carbon nanospring we mainly focus on the change in the number of turns; that is, if the spring's optimized structure has the same number of turns as the initial structure, we consider it to be stable. The equilibrium structure of a nanospring with a smaller spring radius is closely connected to its rising angle, whereas for a nanospring with a larger radius it is not.

First, the special equilibrium structures of small-radius nanosprings are studied. A nanospring with more turns is optimized into three main shapes: (1) a belt-like structure, with the same number of turns as its initial configuration; (2) a one-turn spring, in which the initial configuration is expanded into a spring with only one turn; and (3) an irregular structure, in which the spring cannot expand into a new spring due to the high number of turns. There is a critical rising angle (CRA) for a small-radius nanospring: when a spring's rising angle is less than the critical angle, the spring is optimized into shape 1, and when a spring's rising angle exceeds the critical value, the spring will have shape 2 or 3, depending on the number of turns and the tube radius. Taking a nanospring (5, 5) with a radius of 2.3 nm as an example (a pitch of 1000 carbon atoms), the initial configurations of the three-turn springs will be as shown in Fig. 10.20A and B, with rising angles of

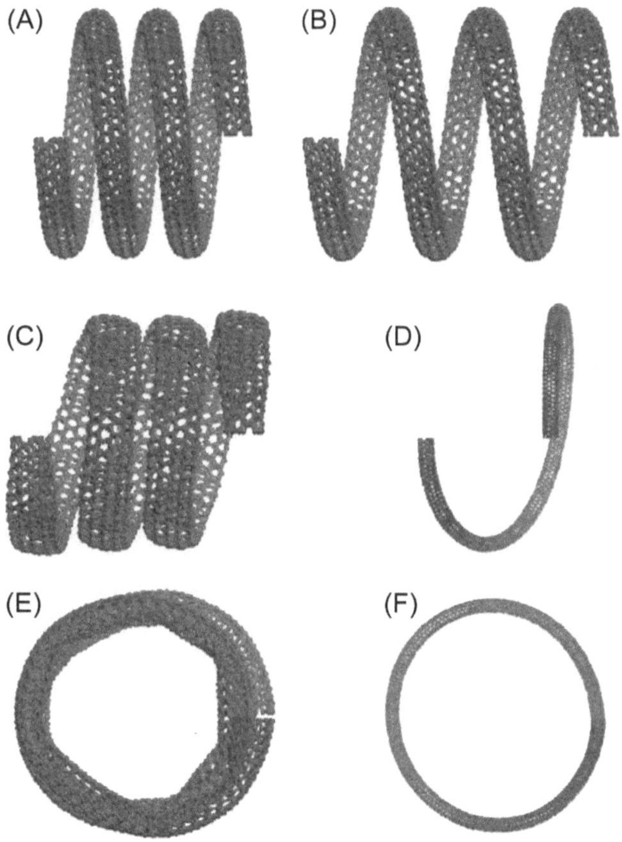

FIGURE 10.20

The initial configurations of (5, 5) nanosprings with the same radius of $R = 2.3$ nm and different rising angles: (A) $a = 5.1$ degrees, (B) $a = 7.7$ degrees; and their corresponding optimized structures: (C) and (D) present the side views of (A) and (B), respectively, and (E) and (F) present the front views of (A) and (B), respectively [22].

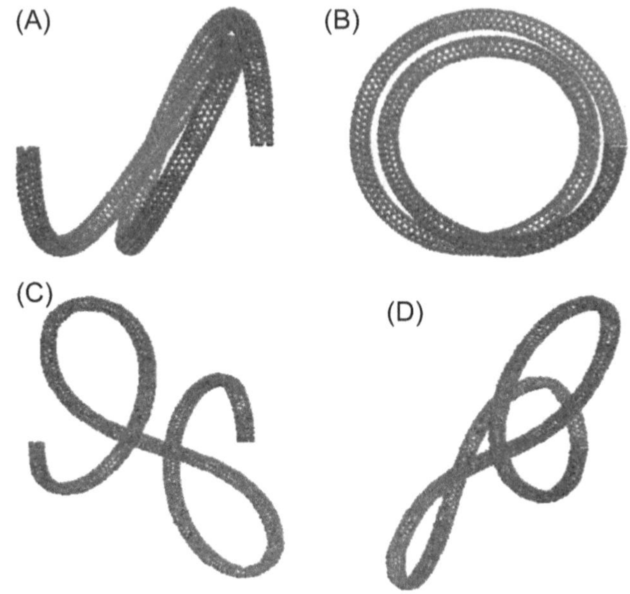

FIGURE 10.21

The optimized structure of (5, 5) carbon nanosprings with $R = 2.3$ nm: (A) and (B) present the side and front view of a four-turn spring, respectively; and (C) and (D) present the side and front view of a five-turn spring, respectively [22].

5.1 and 7.7 degrees, respectively. The constrained carbon atoms are depicted in red, and their optimized structures are presented in Fig. 10.20C and D (side views). It can be seen that the nanospring with a rising angle of 5.1 degrees has a belt-like shape (flat cross-sections), and the nanospring with a rising angle of 7.7 degrees has become a new spring with only one pitch (Fig. 10.20D). Fig. 10.20E and F presents front views of the two optimized nanosprings. Buckling occurs on the optimized nanospring (Fig. 10.20E), and the number of turns (three turns) has not changed before and after the MM simulation. For the spring with a rising angle of 7.7 degrees, its optimized shape (Fig. 10.20F) has no buckling because it becomes a large-radius spring.

When springs (5, 5) with $R = 2.3$ nm have four or five turns, their optimized structures are considered to be irregular coils (shape 3), as shown in Fig. 10.21. It can be observed from Fig. 10.21C (side view of the five-turn nanospring) that its initial configuration cannot be optimized into a new spring because of the large number of turns that can hardly be expanded into an arranged shape.

Second, carbon nanosprings with larger radii are examined. Similar to a large-radius nanoring, the equilibrium structure of a large-radius nanospring does not change much in terms of its shape, compared to the initial structure. Moreover, the larger the spring radius, the less change there is in its equilibrium structure, where the equilibrium shape includes the cross-section, the rising angle, and the number of turns. The initial and optimized structures of the zigzag nanospring (5, 0) are shown in Fig. 10.22. It can be seen that there are no real changes in the optimized nanospring except for a few deformations near the two ends, caused by the fixed carbon atoms of the two ends

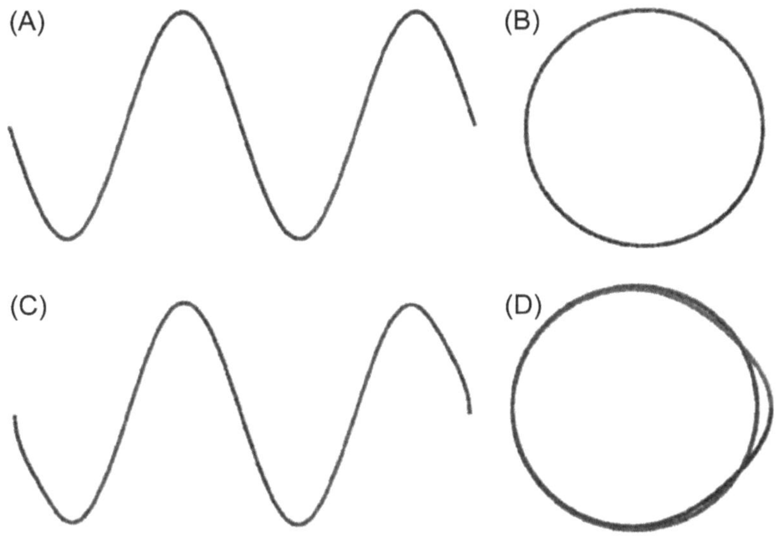

FIGURE 10.22

The initial configuration of a carbon nanospring (5, 0) with $R = 17.15$ nm and $a = 17.8$ degrees: (A) side view; (B) front view; and its equilibrium structure: (C) side view; (D) front view [22].

exerting force on their neighboring carbon atoms. As the number of turns increases, these end effects disappear in their distant locations in the nanospring. Thus, spring (5, 0) is viewed as a stable structure. When applying carbon nanosprings in the engineering and biomedicine fields, the end effect (constraint) can be resolved by other methods, e.g., by using biological molecules as gelators to fix the end of the nanosprings.

10.3.5 CRITICAL RISING ANGLES

From the foregoing analysis of equilibrium structures, it is apparent that a stable nanospring can have any spring radius as long as the rising angle does not exceed its corresponding MRA. However, for a small-radius spring, it will not be a stable structure unless its rising angle is less than the corresponding CRA. Table 10.2 lists the CRAs (α_c) for different nanosprings. It shows that nanosprings (5, 5) with $R = 2.3$ nm have a CRA of 7.52 degrees; thus, when the spring's rising angle is no greater than 7.52 degrees its optimized structure has the same number of turns as the initial one (such as Fig. 10.20C) and this spring is considered to be stable; when the spring's rising angle is greater than 7.52 degrees, its optimized structure generally does not have the same number of turns as the initial one (as in Figs. 10.20D and 10.21), and this spring is considered unstable.

We can also conclude from Table 10.2 that there is a critical radius for a given type of nanospring; when a spring radius is greater than the critical radius, the spring, called a large-radius spring, will be considered to be stable and independent of the rising angles. The critical spring radii for carbon nanosprings (5, 5), (6, 6), (5, 0), and (10, 0) are 4.25, 4.32, 3.59, and 4.46 nm, respectively, and their corresponding tube radii are 0.339, 0.407, 0.196, and 0.392 nm, respectively. It is

Table 10.2 Parameters of the Spring Radius (R) and the Corresponding Critical Rising Angle (α_c) [22]

(5, 5)		(6, 6)		(5, 0)		(10, 0)	
R (nm)	α_c (°)	R (nm)	α_c (°)	R (nm)	α_c (°)	R (nm)	α_c (°)
2.1	7.23	2.17	7	2.23	6.94	2.09	7
2.3	7.52	2.36	6.8	2.57	10.37	2.43	7.06
2.69	9.29	2.68	9.36	2.91	14.97	2.76	9.65
3.08	9.81	2.99	10.23	3.25	16.79	3.1	11.05
3.47	14.69	3.34	11.1	3.59	28.63	3.44	12.28
3.86	19.3	3.66	13.76			3.78	14.05
4.25	21.53	3.97	17.54			4.12	18.05
		4.32	19.73			4.46	21.92

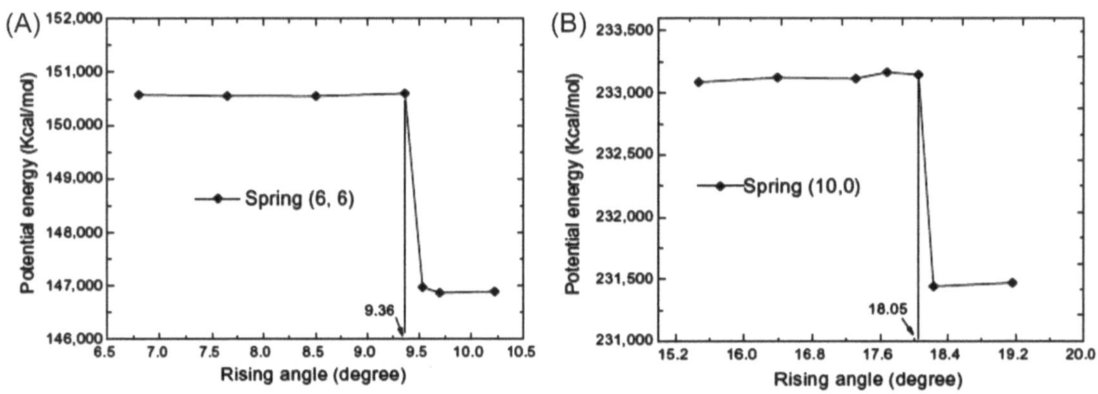

FIGURE 10.23

The relationship between the potential energy and rising angle for (A) an armchair nanospring (6, 6) with $R = 2.68$ nm; (B) a zigzag nanospring (10, 0) with $R = 4.12$ nm [22].

interesting that the critical spring radius of a nanospring increases with the increase in the tube radius and is rarely affected by the chiral type (armchair or zigzag).

Moreover, if a radius is smaller than the critical radius, the stability of the spring, called a small-radius nanospring, will be determined by the rising angle. The CRA can also be determined from the spontaneous drop in its potential energy that is expected from the number of turn changes in its geometrical structure. The relationship between potential energy and the corresponding rising angles of two-turn nanosprings (6, 6) and (10, 0) with a spring radius of 2.68 and 4.12 nm, respectively, are presented as examples in Fig. 10.23. The spontaneous decrease in the potential energy indicates the largely structural deformation, which mainly represents the change in the number of turns. The two turn points, 9.36 and 18.05 degrees (i.e., the CRAs), are indicated in Fig. 10.23A and B, respectively.

10.4 CARBON NANOCONES

The buckling of carbon nanocones (CNCs) is a key characteristic of their mechanical performance. Due to the special structure of CNCs, compressive buckling may be the typical buckling behavior which occurs easily in engineering applications. Here buckling behavior of SWCNCs under compressive loading is simulated. Initially, an unoptimized SWCNC with a side edge length of 7.48 nm or a height of 7.37 nm and an apex angle of 19.2 degrees is considered. Here the open CNCs are selected, obtained by cutting off the tips of its corresponding closed counterpart. The geometric parameters of an open CNC are depicted in Fig. 10.24. L_t is the side edge length of the cutting tip, L_s is the side edge length of the open CNC, α is the apex angle of the cone, and H is the height. The cutting tip's length L_t of this kind of CNC is set at 3.6 nm and in the following computational simulation of other kinds of CNCs, the same parameter is used, unless otherwise stated.

The arrangement of the mesh-free nodes on CNCs is a key factor that influences accuracy of numerical results. In the analysis of CNCs, the radius increases along the height and the top level suffers a larger stress than the bottom level, resulting in a much higher possibility of a buckling behavior appearing on the top surface. Therefore, a larger number of mesh-free nodes are needed to be arranged near the top end to capture the large deformation, while only a small number of mesh-free nodes are sufficient near the bottom end, where the deformation is sufficiently homogeneous. Due to the drastic deformation that occurred mainly along the axial direction in the compressive test, a sufficient number of axial mesh-free nodes should be arranged along the axial direction. In our simulation, the distance of the interfacing nodes increases along the side edge direction and this arrangement style is much more efficient than the uniform style [26]. This method has been confirmed to be efficient and effective compared with MD simulation [27]. Here

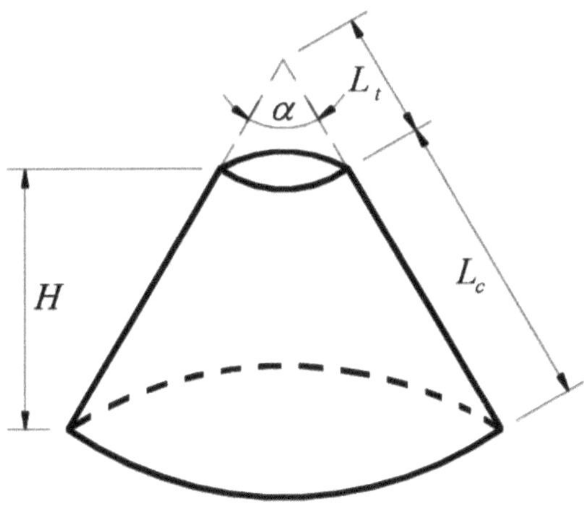

FIGURE 10.24

An open CNC is obtained by cutting off the tip of its closed counterpart. The geometrical parameters: L_t is the side edge length of the cutting tip, L_s is the side edge length of the CNC, α is the apex angle of the CNC and H is the height.

the effect of the apex angle and also the cutting tip's length on the critical strain and the buckling patterns is investigated.

10.4.1 THE EFFECT OF APEX ANGLE ON MECHANICAL BEHAVIORS OF CNCs

All five kinds of SWCNCs with different apex angles are selected as samples in the computational simulation for analysis of critical strain of CNCs. The side edge lengths are selected as constant $L = 7.48$ nm. It should be noted that geometrical parameters of CNCs are described in the unoptimized configuration, which is required to be treated with the minimum energy of the system to obtain the initial equilibrium configuration. As the cutting tip's length and the side edge length are the same for CNCs with different apex angles, a CNC with an increasing apex angle can be considered as being formed by reduplicating a 60-degree arcing horn fan-shaped graphite sheet. The height of a CNC with an apex angle of θ_i can be determined by:

$$H_i = \cos\,\theta_i \times L, \quad i = 1, 2, \ldots, 5 \tag{10.2}$$

where θ_i is the apex angle of ith kinds of the CNC.

Fig. 10.25 shows variation of average energy per atom versus compression ratios of all five kinds of CNCs with the same geometrical parameters. A distinct jump in energy appears at each energy curve, corresponding to the buckling of CNCs. An interesting phenomenon is that the critical axial compression displacements of all kinds of CNCs are 0.22 nm and thus the corresponding axial compression ratio increases as a function of $(1/\cos\,\theta_i)$. Moreover, it is expressly demonstrated that values of strain energy per atom at each compressive loading are very close to each other, which is slightly relevant to the apex angle but only to the compressive loading prior to buckling. At the prior to buckling stage, the strain energy–compression ratio curve follows a quadratic relationship and is subject to Hook's law:

$$W = \frac{1}{2} E_i {\varepsilon_i}^2 \tag{10.3}$$

Therefore, the relationship between axial Young's moduli of CNCs can be approximately expressed as:

$$\frac{E_i}{E_j} = \frac{\varepsilon_j^2}{\varepsilon_i^2} = \frac{\cos^2\theta_j}{\cos^2\theta_i} \tag{10.4}$$

where i and j denote ith and jth kinds of CNCs. The curves described in Fig. 10.25 clearly show that the apex angle has a reducing effect on axial Young's modulus of CNC and listed in Table 10.3 are axial Young's moduli of the five kinds of CNCs, ranging from 0.19 to 0.62 TPa. Fig. 10.26 depicts the corresponding buckling deformations of CNCs which demonstrate that an increasing number of fins on the surface, with the apex angle increasing, i.e., for the case of 19.2 degrees, there are two parallel fins on two flanks of the CNC symmetrically. For the case of 38.9 degrees, there are four fins allocated on corners of a quadrilateral with two vertical symmetrical axes, while there are eight fins on the surface for the case of 112.9 degrees CNC, allocated uniformly on the octagonal lattice. Each CNC has two levels of the similar buckling pattern along the height, except for the case of the CNC with the smallest apex angle.

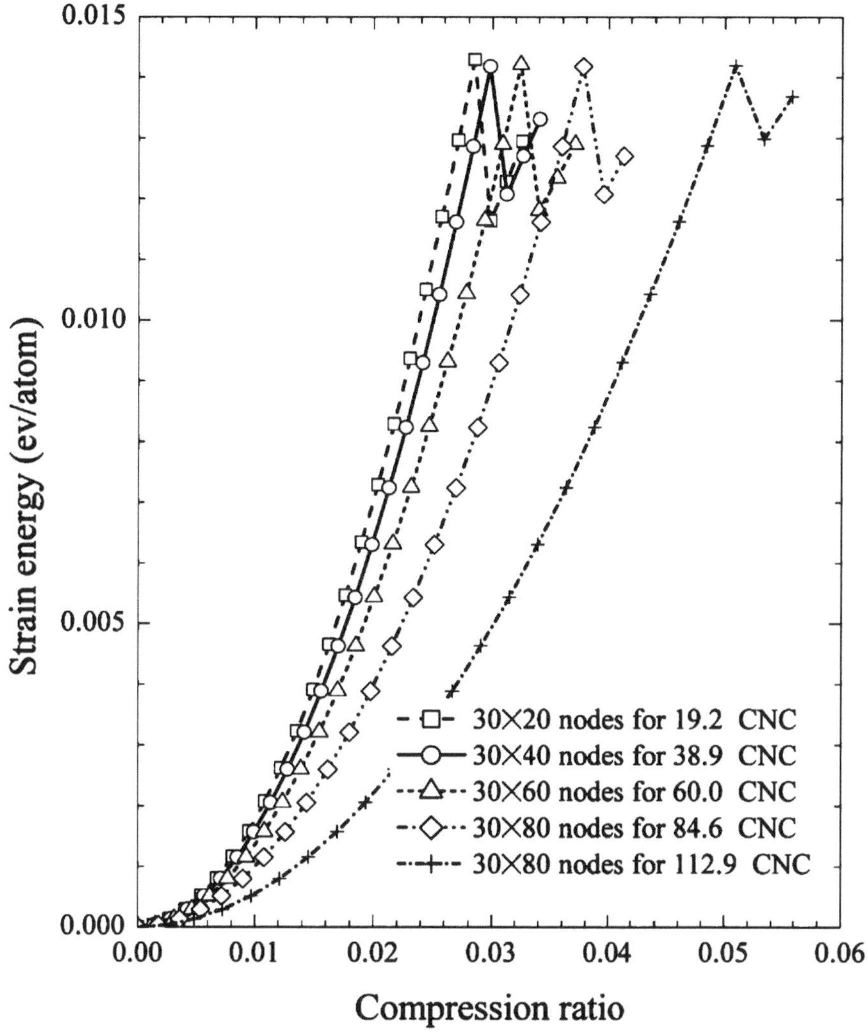

FIGURE 10.25

The average strain energy per atom versus the compressive ratio for CNCs with a same side edge length of 7.48 nm. The mesh-free nodes for the five kinds of CNCs are, respectively, 30 × 20, 30 × 40, 30 × 60, 30 × 80, and 30 × 80 [26].

Table 10.3 The axial Young's Moduli of CNCs With a Same Side Edge' Length of 7.47 nm [26]

Type of CNC	19.2°	38.9°	60.0°	84.6°	112.9°
Axial Young's Modulus (TPa)	0.61655	0.55884	0.47326	0.3489	0.19209

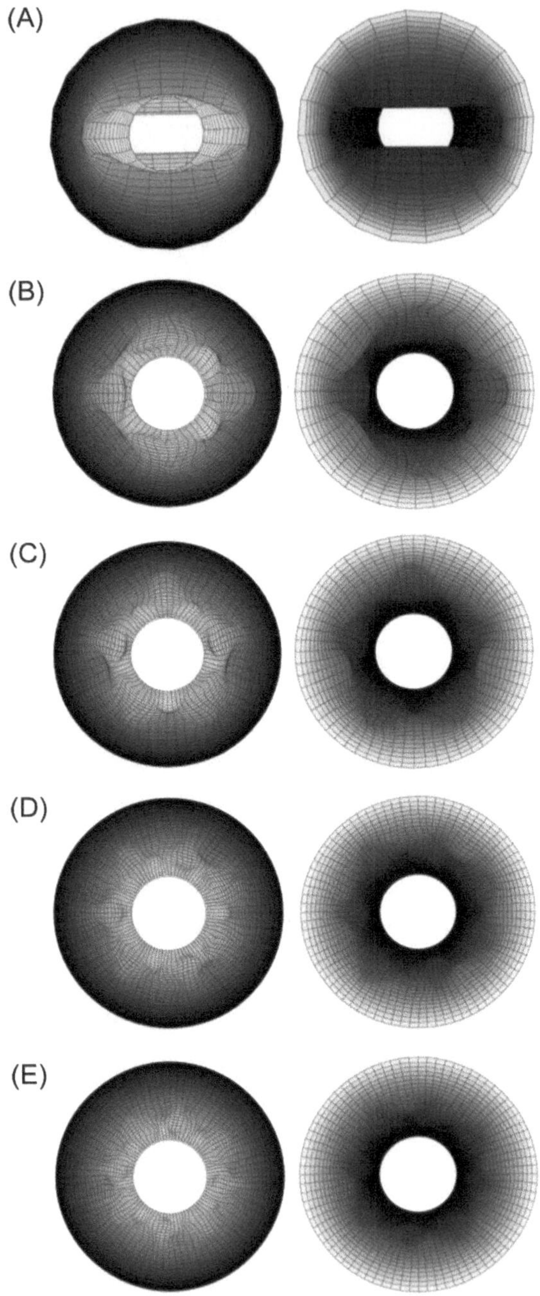

FIGURE 10.26

The top and bottom view of the buckling patterns of CNCs with a same side length. With an increasing apex angle, an increasing number of fins developed on the side surface of CNCs at the critical strain [26].

10.4.2 THE EFFECT OF CUTTING TIP'S LENGTH ON BUCKLING OF CNCs

To investigate the effect of the cutting tip's length L_t on buckling of CNCs, 19.2 degrees CNCs with a side edge length of 7.48 nm and various L_t ranging from 3.6 to 24 nm are selected as samples in our simulation. The same minimum potential energy of the system should be treated to obtain the initial equilibrium configuration. The loading on CNCs is exerted by reducing the height of the cone by 0.01 nm per step at the early stage and 0.001 nm per step when close to buckling. Due to the increasing L_t, corresponding to the increasing radii of CNCs at the same level, mesh-free nodes should increase along the circumferential direction.

Depicted in Fig. 10.27 are energy–strain curves of CNCs. The data demonstrate distinctly that the critical compression ratio decreases sharply with a sudden energy drop as the cutting tip's length L_t increases. Critical compression ratios of CNCs with L_t ranging from 3.6 to 24 nm are 0.0296, 0.0204, 0.0145, 0.0107, and 0.0090. The conclusion can also be applied to the energy drop. For the CNC with $L_t = 3.6$ nm, it has the largest energy drop of 24.89%, while others are 16.44%, 14.61%, 12.18%, and 11.52%. These results, i.e., the smaller critical strain and the larger energy drop as L_t increases, indicate that the CNC becomes much more unstable and easy to buckle as the cutting tip's length increases. Meyer et al. [28] experimentally found that suspended graphene

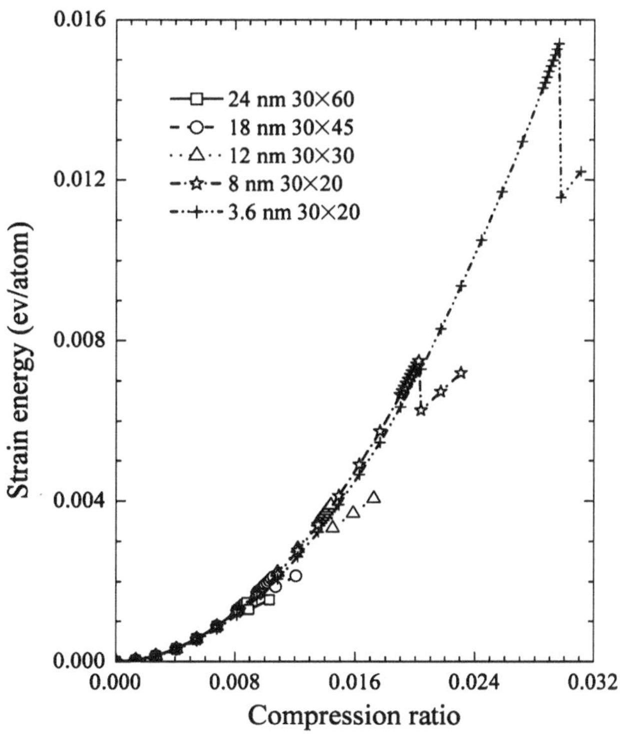

FIGURE 10.27

The average strain energy per atom versus the compressive ratio for CNCs with a same side edge length of 7.48 nm and an apex angle of 19.2 degrees. The cutting tip's lengths are 3.6, 8, 12, 18, and 24 nm, respectively [26].

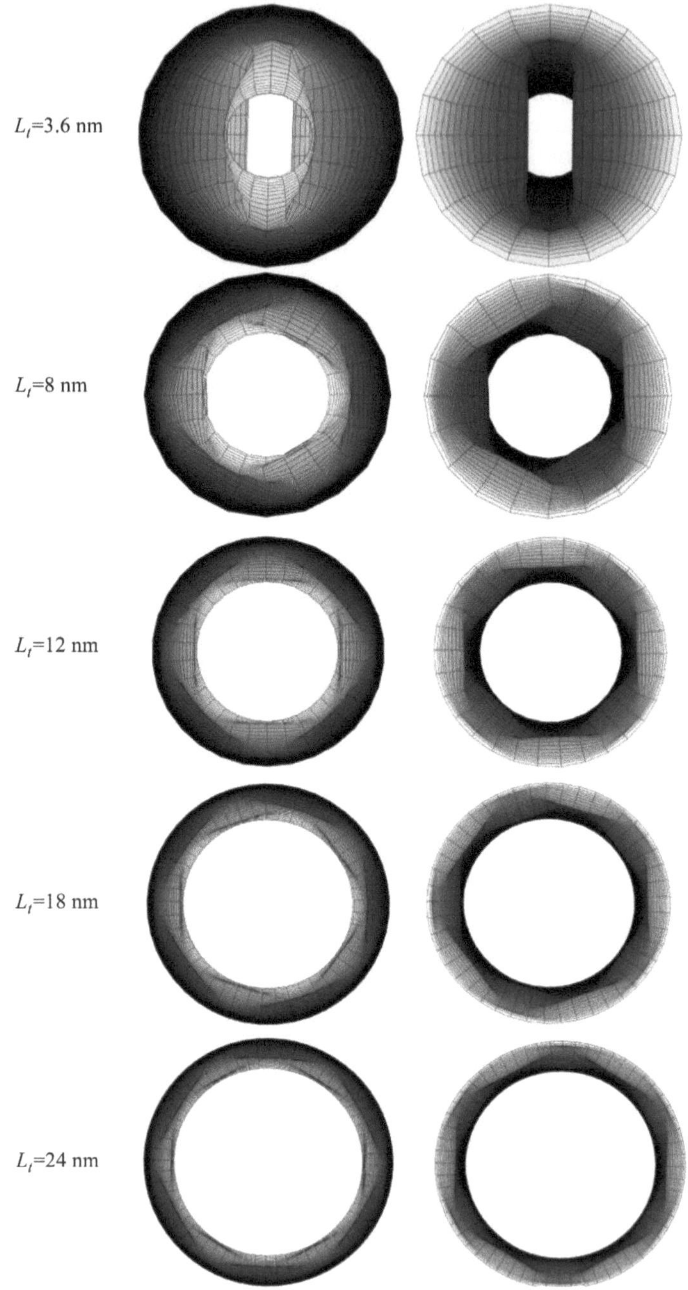

FIGURE 10.28

The buckling patterns of CNCs with the same apex angle of 19.2 degrees but various cutting tip's length ranging from 3.6 to 24 nm. As the cutting tip's length increases, the developed fins increase with no more drastic deformation but trend to ripples [26].

sheets are not perfectly flat by virtue of transmission electron microscopy (TEM). Fasolino et al. [29] interpret the experimental observation of ripples in graphene by a theoretical study. Graphite sheet can be viewed as a CNC or CNT with a very large radius. As L_t increases, radii of CNCs increase and thus the buckling tends to be close to that of graphene, which is a good explanation of the smaller critical strain and the smaller energy drop corresponding to smaller buckling deformation. It is concluded that as L_t increases, CNC becomes more and more unstable and buckles at a very small critical strain. An increasing L_t results in an increasing number of fins along the circumferential surface of the CNC, while the number of fins remains two, owing to the limitation of the side edge length. It can be predicted that with an increasing side edge length, CNCs will be divided into more levels to develop fins at the critical strain.

Prior to buckling, the curve of strain energy–compression ratio follows a quadratic relationship and ascends slightly quickly as L_t increases, which indicates that L_t has a slightly increasing effect on the elastic property of CNCs. The axial Young's modulus can be calculated from Eq. (10.3): 0.6165, 0.6420, 0.6595, 0.6622, and 0.6583 TPa. For the cases of CNCs with the largest L_t and the smallest one, the axial Young's modulus increases by 6.4%. Buckling patterns of the corresponding CNCs are described in Fig. 10.28. It is found that the fins that developed on CNCs increase and form into parallel, triangle, quadrilateral, pentagonal, and hexagonal shapes but tend to the ripples in the graphite sheet. This phenomenon occurs because SWCNCs and graphite sheets are essentially carbon-based hexagonal lattice structures, and the CNC approaches to the graphite sheet as the radius increases.

REFERENCES

[1] Liu J, Dai HJ, Hafner JH, Colbert DT, Smalley RE, Tans SJ, et al. Fullerene "crop circles". Nature 1997;385(6619):780–1.

[2] Liu P, Zhang YW, Lu C. Atomistic simulations of formation and stability of carbon nanorings. Phys Rev B 2005;72(11).

[3] Brenner DW, Shenderova OA, Harrison JA, Stuart SJ, Ni B, Sinnott SB. A second-generation reactive empirical bond order (REBO) potential energy expression for hydrocarbons. J Phys-Condens Mat 2002;14(4):783–802.

[4] Oh DH, Park JM, Kim KS. Structures and electronic properties of small carbon nanotube tori. Phys Rev B 2000;62(3):1600–3.

[5] Dunlap BI. Connecting carbon tubules. Phys Rev B Condensed Matter 1992;46(3):1933–6.

[6] Brenner DW. Empirical potential for hydrocarbons for use in simulating the chemical vapor-deposition of diamond films. Phys Rev B 1990;42(15):9458–71.

[7] Huhtala M, Kuronen A, Kaski K. Carbon nanotube structures: molecular dynamics simulation at realistic limit. Comput Phys Commun 2002;146(1):30–7.

[8] Huhtala M, Kuronen A, Kaski K. Computational studies of carbon nanotube structures. Comput Phys Commun 2002;147(1-2):91–6.

[9] Liu L, Jayanthi CS, Wu SY. Structural and electronic properties of a carbon nanotorus: effects of delocalized and localized deformations. Phys Rev B 2001;64(3):033412.

[10] Hod O, Rabani E, Baer R. Carbon nanotube closed-ring structures. Phys Rev B 2003;67(19):1–7.

[11] Han J. Energetics and structures of fullerene crop circles. Chem Phys Lett 1998;282(282):187–91.

[12] Meunier V, Lambin P, Lucas AA. Atomic and electronic structures of large and small carbon tori. Phys Rev B 1998;57(23):14886–90.

[13] Verlet L. Computer "experiments" on classical fluids. I. Thermodynamical properties of Lennard-Jones molecules. Phys Rev 1967;159(1):98–103.
[14] Nosé S. A molecular dynamics method for simulations in the canonical ensemble. Mol Phys 1984;52:255–68.
[15] Feng C, Liew KM. Energetics and structures of carbon nanorings. Carbon 2009;47(7):1664–9.
[16] Kuang YD, He XQ, Chen CY, Li GQ. Buckling of functionalized single-walled nanotubes under axial compression. Carbon 2009;47(1):279–85.
[17] Wang Q, Liew KM. Molecular mechanics modeling for properties of carbon nanotubes. J Appl Phys 2008;103(4).
[18] Cao GX, Chen X. The effect of the displacement increment on the axial compressive buckling behaviours of single-walled carbon nanotubes. Nanotechnology 2006;17(15):3844–55.
[19] Cao GX, Chen X. Buckling behavior of single-walled carbon nanotubes and a targeted molecular mechanics approach. Phys Rev B 2006;74(16).
[20] Zhang YY, Tan VBC, Wang CM. Effect of strain rate on the buckling behavior of single- and double-walled carbon nanotubes. Carbon 2007;45(3):514–23.
[21] Feng C, Liew KM. A molecular mechanics analysis of the buckling behavior of carbon nanorings under tension. Carbon 2009;47(15):3508–14.
[22] Feng C, Liew KM. Structural stability of carbon nanosprings. Carbon 2011;49(14):4688–94.
[23] Alfred G. Modern differential geometry of curves and surfaces with mathematica. Boca Raton, Boston, London, New York, Washington, DC: CRC Press LLC; 1998.
[24] Yang S, Chen X, Motojima S. Morphology of the growth tip of carbon microcoils/nanocoils. Diam Relat Mater 2004;13(11-12):2152–5.
[25] Wen YK, Shen ZM. Synthesis of regular coiled carbon nanotubes by Ni-catalyzed pyrolysis of acetylene and a growth mechanism analysis. Carbon 2001;39(15):2369–74.
[26] Yan JW, Liew KM, He LH. A mesh-free computational framework for predicting buckling behaviors of single-walled carbon nanocones under axial compression based on the moving Kriging interpolation. Comput Method Appl M 2012;247:103–12.
[27] Liew KM, Wei JX, He XQ. Carbon nanocones under compression: buckling and post-buckling behaviors. Phys Rev B 2007;75(19).
[28] Meyer JC, Geim AK, Katsnelson MI, Novoselov KS, Booth TJ, Roth S. The structure of suspended graphene sheets. Nature 2007;446(7131):60–3.
[29] Fasolino A, Los JH, Katsnelson MI. Intrinsic ripples in graphene. Nat Mater 2007;6(11):858–61.

Index

Note: Page numbers followed by "*f*" and "*t*" refer to figures and tables, respectively.

D

E

F

G

T

U

V

FSC
www.fsc.org
MIX
Papier | Fördert gute Waldnutzung
FSC® C083411

Kolibri 360 GmbH
Ferdinand-Jühlke-Straße 7
99095 Erfurt, Deutschland
produktsicherheit@kolibri360.de